MICROCONTROLLER TECHNOLOGY

The 68HC11 and 68HC12

FIFTH EDITION

Peter Spasov
Sir Sandford Fleming College

Upper Saddle River, New Jersey
Columbus, Ohio

Editor in Chief: Stephen Helba
Assistant Vice President and Publisher: Charles E. Stewart, Jr.
Production Editor: Alexandrina Benedicto Wolf
Production Coordination: Carlisle Publishers Services
Design Coordinator: Diane Ernsberger
Cover Designer: Kristina Holmes
Cover art: Digital Vision
Production Manager: Matt Ottenweller
Marketing Manager: Ben Leonard

This book was set in Times Roman by Carlisle Communications, Ltd. It was printed and bound by R.R. Donnelley & Sons Company. The cover was printed by The Lehigh Press, Inc.

Pearson Prentice Hall™ is a trademark of Pearson Education, Inc.
Pearson® is a registered trademark of Pearson plc
Prentice Hall® is a registered trademark of Pearson Education, Inc.

Pearson Education Ltd.
Pearson Education Singapore Pte. Ltd.
Pearson Education Canada, Ltd.
Pearson Education—Japan

Pearson Education Australia Pty. Limited
Pearson Education North Asia Ltd.
Pearson Educación de Mexico, S.A. de C.V.
Pearson Education Malaysia Pte. Ltd.

10 9 8 7 6 5 4 3 2 1
ISBN 0-13-112984-8

To my wife Renate

Preface

Microcontrollers are used in the industrial world to control many types of equipment, ranging from consumer to specialized devices. They have replaced older types of controllers, including microprocessors. Also, there is a growing need for off-line support of a computer's main processor. The demand will grow as more and more equipment uses more intelligence. Applications range from controlling engines in modern automobiles to controlling laser printers and other computer peripherals. One consumer application is a washing machine controller that adjusts the wash cycle based on load size, fabric type, and amount of dirt. Technology has evolved to the point where this same washing machine could be connected to the Internet. In principle it could shop for and order detergent when needed. We can envision a future with wearable computing where wristwatch-type devices could communicate with and control the washing machine using wireless networking.

This book illustrates how to use a popular microcontroller, the Motorola 68HC11, and to a great extent, the 68HC12 as well. The 68HC11 and 68HC12 microcontrollers are relatively easy to work with, yet they have most of the features essential for a complete control system. Thus the student of control automation can use them to work with control systems at the component level. The interested layperson can also use them as tools to understand and experiment with computer and data communications systems. Although the futuristic vision of the earlier paragraph would likely use newer processors, the 68HC11 is a great tool for learning about embedded control fundamentals because it is simple.

The goals of the fifth edition remain the same as those of earlier editions. The book provides the background knowledge needed to understand and use microcontrollers in general, specifically the 68HC11/12. It starts at an introductory level, explaining the applications and origins of microcontrollers. Next, a programmer's view of the machine is developed. Finally, machine hardware and how to connect it to the outside world for control applications are described.

You will find the book useful in learning how to use the 68HC11/12. Original manufacturers' data books do not provide the background explanations that a novice needs. Experienced users will find the book to be a useful accompaniment to the manufacturer's original documentation.

The book can be used as a text in introductory, interfacing, and industrial control courses. To use the book, you should have an understanding of digital logic and numbering systems. Some computer programming knowledge is useful but not absolutely essential. Although this book was written to be used in college courses, the interested hacker or hobbyist should find the material useful for self-study. Others with a formal education should also find useful information. The book can accompany hands-on exercises using a microcontroller training kit or simulation software. Where possible, concepts are presented generically, to help you understand other microcontrollers and related devices.

To permit flexible use of the book, we have organized it into five parts:

Part 1. Microcontroller technology is introduced in Chapter 1: what it is, how it evolved, and how to use it. The chapter continues with explanations of microcontroller terminology and parts. Finally, essential memory concepts are reviewed.

Part 2. Chapters 2 to 4 cover programming concepts, the language used to instruct the microcontroller, and how to use registers and memory. How to produce, use, and document programs are also discussed.

Part 3. Chapters 5 to 7 cover chip operation. Topics include the system bus, operating modes, clocked operation, and memory technology.

Part 4. Chapters 8 to 12 deal with the subsystems for parallel, serial, programmable timer, and analog interfacing. The basic software techniques to use these systems are presented. Also introduced are some common hardware designs used to connect the microcontroller to sensors and actuators. An overview of all subsystems without the processor-specific details is presented in Chapter 8.

Part 5. Control methods are illustrated in Chapter 13, where it is shown how programs use the subsystems for control applications. Chapters 14 and 15 cover the industry to date, presenting a survey of typical applications, choices in choosing microcontrollers, and characteristics of other microcontrollers.

Typically, an introductory course would include Parts 1 and 2. We also recommend including Chapter 8. Advanced or more intensive courses would include Parts 3 and 4. Either type of course can use Part 5, provided that Chapter 8 is also covered.

Although the book contains some data sheets, it is not a complete reference. The reader should refer to the literature listed in Appendix C for complete data.

■ FIFTH EDITION CHANGES

Because of the increasing frequency of embedded devices that can be connected to the Internet, Chapter 10 contains more information on networking communications. It describes Ethernet and Internet interfacing. Chapter 15 eliminates the section on the Intel 8051 core. Instead, features of the HCS12 family, the next generation in the evolution of the 68HC11 family tree, are discussed.

The accompanying CD-ROM omits some of the tools in the fourth edition because updates can be downloaded from the Internet.

■ HOW TO USE THIS BOOK

Most important to success when using this book is an enthusiasm for computer and related technology. If you are already familiar with assembly language programming, you may wish to skim Chapters 1 and 2 for new information instead of reading them in detail. If you are not familiar with digital logic and binary numbers, you should also read Appendix F.

We use generic principles whenever possible. After studying this book, you should be able to work with any microcontroller, but it is necessary to cover a specific microcontroller in depth to have a general understanding. We have chosen the 68HC11; however, there are many variations of the 68HC11. When explaining specific details, we generally refer to the 68HC11E9, unless we state otherwise.

To understand microcontrollers, you also have to know a lot of *jargon*. You cannot understand the concepts without knowing the jargon—but you cannot understand the jargon without knowing the concepts. We get around this dilemma by introducing details step by step.

The appendices provide references and further clarification of terms used in the book. Appendix A lists the details of the 68HC11 instructions used throughout the book. You will probably use it often. Appendix B is the quick reference section. Use it as a cross-reference to look up other 68HC11 details as needed. Appendices A and B are necessary if you do not have 68HC11 data sheets or a manual.

Most terms are explained in the text as required. New terms are indicated in italic type when they first occur in the text. You may wish to reread earlier sections of the book to review terms when used in another context. The glossary provides an additional reference.

A lot of abbreviations are used in the technical literature. Unfortunately, these frustrate many readers. In practice, most technical data books use abbreviations, often without explanation. In this book each abbreviation is defined the first time it is used, and many are redefined upon subsequent appearances. You can also find a list of abbreviations after the glossary.

Sources of further information on some topics are listed in Appendix C. Also listed are suppliers for some of the commercially available products mentioned in the book. The author's Web site, fleming0.flemingc.on.cd/~pspasov/MCU/mcu.htm, is another source of information.

Some conventions are defined in Appendix D. Since there are many ways to do the same thing, we felt it necessary to define a *standard way*. The conventions are those used by many in the industry. For example, we use Motorola's dollar sign ($) convention to refer to a method of coding numbers (in this case the sign does not indicate dollars). This ensures that we all understand the sign to have the same meaning.

The *header files* used by some program examples are listed in Appendix E. Note that a header file is explained in Appendix E.

The basics of digital logic and number conversions are explained in Appendix F. You should read this section early if you are not already familiar with these topics.

If you need basic information about waveforms, refer to Appendix G.

Finally, we included a reference to Internet resources and the enclosed CD-ROM in Appendix H.

68HC12 Compatibility. You may also use the book for the 68HC12. However, we will omit details of features that are not shared by the 68HC11. Chapter 15 provides an overview of features specific to the 68HC12. Also note that the machine code for both processors are different, although source code written for the 68HC11 will work with the 68HC12. In other words, you can assemble assembly source code and compile C source code for the 68HC11 to work in either the 68HC11 or 68HC12. Specific machine language details covered in Chapter 2 will not apply for the 68HC12 although the principles remain the same. The reader may also use the Internet to search for development boards and other tools targeted for the 68HC12.

■ SUPPLEMENT

Instructors can contact Prentice Hall for a copy of the *Instructor's Solutions Manual with PowerPoints* to accompany this book.

■ ACKNOWLEDGMENTS

I would like to thank Hugh K. McGugan of Motorola Training Organization, Canada, for technical literature, review of some chapters, and the assistance he always gave willingly and promptly. Without his help this book would not exist. I thank also Alan D. Niemi of Lake Superior State University, Sault Sainte Marie, Michigan, for his valuable comments and advice when reviewing sections of the preliminary manuscript. Helpful advice was also received from V. S. Anandu of Southwest Texas State University, San Antonio, Texas, and from Jeffrey Rankinen of Pennsylvania College of Technology during review of the manuscript. Also, I thank my students and teaching colleagues whose experiences led me to make some of the changes in this edition. Any deficiencies are mine because I did not always follow their advice.

I am grateful to Hitoshi Doi and Chuck Westfall of Canon Incorporated, Japan, for valuable information about their model EOS cameras. In addition, I thank Richard Man of ImageCraft, USA, for providing information as well as the demo version of the ICC11 C cross compiler. Similarly, Harry Broeders of Rijswijk Institute of Technology, Netherlands, provided a special version of the THRSim11 simulator. Peter Fabo of Comenius University, Bratislava, Slovakia, provided information about his VT11 module (covered in the fourth edition). For valuable comments, suggestions, and corrections, I express sincere appreciation to Charles Edward Nunally of Virginia Tech, Blacksburg, VA; Jack Levine of California State University, Los Angeles, CA; Kurt Schneider of EDV-Beratung und Schulung, Ober-Ramstadt, Germany; Patrick J. O'Connor of DeVry Institute of Technology, Chicago, IL; and Trung Do of University of Illinois at Chicago, Chicago, IL. In particular, I want to acknowledge the numerous suggestions offered by Kurt Schneider. Clive Maynard of Curtin University of Technology, Australia, and Cheah Soon Hooi of KDU College Penang, Malaysia, pointed out improvements and corrections for the fifth edition. Thanks to the editorial and production people who provided excellent and professional support. They are Charles Stewart, Mayda Bosco, Maria Rego, and Alex Wolf of Prentice Hall, and Kelli Jauron of Carlisle Publishers Services.

Finally, a very special thanks to my wife Renate, who provided invaluable emotional support. I also thank my eleven-year-old daughter Emilie, eight-year-old daughter Hannah, and five-year-old daughter Anika, who all provide much joy.

Peter Spasov
Sir Sandford Fleming College

Contents

9   Parallel Input/Output   291

10   The Serial Subsystems   355

1 Microcontroller Concepts

OBJECTIVES

After completing this chapter, you should be able to

- Describe in general terms what a microcontroller is and how it can be used.
- Use some terminology and conventions applicable to microcontroller development.
- Describe a block diagram in terms of a central processing unit (CPU), memory, registers, and a bus.
- Identify the basic types of semiconductor memory and their general applications.
- Determine addresses and addressing ranges.
- Interpret memory maps.

■ 1.1 WHAT IS A MICROCONTROLLER? AND WHAT IS IT USED FOR?

The Swiss Army Knife.   "Choose," the voice says. What do you choose? You may take only one tool for your journey. The journey will take you through forests, plains, mountains, and deserts. You need something small, light, and useful. You choose the Swiss Army knife and the voice "nods" respectfully. It is a good choice. Depending on the model, it has assorted blades, saw blade, scissors, can opener, corkscrew (most important!), tweezers, toothpick, and possibly more.

"What is a microcontroller?" the voice asks. You say: "It is a (computerized) Swiss Army knife." The voice is satisfied. Indeed, the microcontroller bears much resemblance to the Swiss Army knife. It is a single device, in this case a chip. The microcontroller follows instructions, reads information, stores information, communicates, measures time, and switches things on and off. It also does other things, depending on the model. Despite its power and versatility, you may not hear much about microcontrollers. Most computer literature describes

1

the processor chips used in computers. But most single-chip programmable devices are microcontrollers.

If you are the type of person who likes to take things apart, you will find microcontrollers in all kinds of places. The most common place is under the hood of almost any car produced since 1985. Consumer items include televisions, compact disc players, washing machines, telephones, and microwave ovens. Office computers use microcontrollers in addition to their main processor to control peripherals such as keyboards and printers. Automated manufacturing systems use microcontrollers in production equipment such as robots and conveyor lines.

Now, we will take three views of microcontrollers: general, technical, and specific. After we consider the general view, we will consider the technical and specific aspects of microcontrollers.

1.1.1 The General View

One could say that a microcontroller is a programmable single-chip integrated circuit (IC) that controls the operation of a system. Often, the system to be controlled is a machine. One of the most common applications is automotive control. This includes engine fuel injection control, transmission control, suspension and ride control, instrument displays, and braking systems. In fact, the 68HC11 microcontroller managed some engine control functions of first-place winners of Indianapolis 500 races for five years in a row.

Refer to Figure 1.1 for an application in controlling air/fuel mixture in an automobile engine. The amount of air and fuel and the timing of the ignition spark determine the fuel efficiency and amount of exhaust emission.

The microcontroller unit uses the sensor inputs to control the ignition timing for maximum fuel efficiency and minimum exhaust emissions. It also looks at how fast the inputs change. The exhaust gas sensor measures the oxygen level so that the system can correct for any error. The microcontroller can be programmed to handle conditions when the engine is cold or warm, and when it is accelerating or cruising. Also, the driver could instruct the unit to optimize speed and acceleration performance if necessary. With some extra inputs, the microcontroller unit could maintain constant speed, such as in a cruise control.

We went through this example in some detail to illustrate what could be involved in a control system using a microcontroller. For a more detailed understanding of air/fuel ratio control, refer to reference 1 in Appendix C.

Automatic cameras also use microcontrollers. They control exposure and focus. Another application is the computer mouse. It typically uses a microcontroller to read the mouseball movement, sense the push-button positions, and handle communications with the computer.

1.1.2 Terminology and Conventions

We will be using words, abbreviations, and acronyms specifically used in the computer and microcontroller industry. Although we have minimized the use of technical jargon, its use is sometimes unavoidable. When we first use a term, we italicize it; you will generally find a definition in the glossary. You will have to understand some terms in context when you first encounter them. You may also check the index to locate other sections where a term is used.

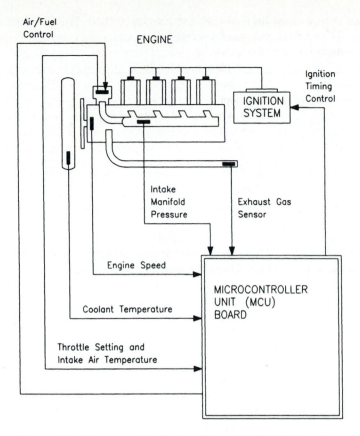

Figure 1.1 Microcontroller Used to Control Air/Fuel Mixture in an Automobile Engine

In the technical description that follows (Section 1.1.3), we use a number of technical terms. Since the computer and microcontroller field is changing rapidly, not all terms are used in the same way by everybody; but the differences are minor and can usually be understood.

We cannot predict the background of every reader, of course. For the benefit of some, we have included an introduction to digital logic in Appendix F. When and whether you consult Appendix F depends on your background. Those who are already familiar with this subject will probably want to skip it. Others, who have not encountered the subject previously, may consult it any time. If you encounter unfamiliar terms, you may wish to consult Appendix F and then return to where you were. Some of these terms are *binary, bit, byte, word,* and *hexadecimal* (or *hex*).

Microcontrollers use and store information in the form of binary numbers. We will be using the binary, hexadecimal, and decimal numbering systems. Any variations of these systems will be described as required. We mention one now, the *ASCII code.*[1] The ASCII code is a binary code. Each number represents a printable character such as 'A', 'a', '#', and

---

[1]ASCII stands for "American Standard Code for Information Interchange."

special control functions such as linefeed (LF) and backspace (BS). The Quick Reference (Appendix B) includes an ASCII chart that shows the hex codes for each (standard) character.

PRACTICE

Find the ASCII codes for ACK, FS, L, and j.

Solution

$06, $1C, $4C, $6A
Note that the dollar sign ($) means that the numbers are in hex. Refer to Appendix F if you need to find out more about hex numbers.

Using microcontrollers involves a lot of bit manipulation. When we say *set* a bit, we mean to make it a logic high (one). To *clear* or *reset* a bit means to make it a logic low (zero). Other conventions are introduced as required. Appendix D lists some of the conventions that we used in the book.

1.1.3 The Technical View

A *microcontroller* is a single-chip device that contains memory for program information and data. It has logic for programmed control reading inputs, manipulating data, and sending outputs. In other words, it has built-in *interfaces* for input/output (I/O) as well as a central processing unit (CPU). We often refer to the device (the chip) as a microcontroller unit (MCU). The built-in interface capability is used for sensors, actuators, and communications. There are several categories of interfaces, which are explained in Part 4.

In practice, a microcontroller has other parts. Also, many applications require other chips in addition to the microcontroller chip. These are called *support chips*. It is likely that more support chips will become part of the microcontroller chip in the future.

Sections 1.3 and 1.4 are technical descriptions of the internal sections of a microcontroller and how they are used. These descriptions are generic.

1.1.4 The Specific View

There are many manufacturers of microcontrollers. To understand and apply general concepts, it is necessary to cover one specific type in detail. You may use this specific knowledge to understand similar features of other microcontrollers. The specific type we cover is the Motorola MC68HC11. People often simply call it the 68HC11. The 68HC12 microcontroller is the next generation beyond the 68HC11. Many of the points covered in this book will also apply to the 68HC12.

■ 1.2 HISTORY

1.2.1 Early Computing Machines

Microcontrollers developed from *microprocessors,* which in turn developed from *computers.* Computers originated more than 5000 years ago. During prehistory, human beings learned to count and build tools. Stonehenge, an astronomical calculator to predict eclipses, was built in Europe. In the Orient, the abacus, an adding device with beads on a rod, was built.

In 1617, John Napier invented a system to perform multiplication and division based on logarithms. Others called these rods Napier's bones. Gottfried Wilhelm Leibniz completed a mechanical machine in 1694 that could add, subtract, multiply, divide, and perform square roots. It demonstrated the advantage of the binary over the decimal system for mechanical computers.

1.2.2 Digital Computers

In 1835, Charles Babbage proposed the world's first digital *computer*. The "Analytical Engine" was to use punched cards for data and instructions. It included program control by means of *looping* and *branching* with arithmetic and storage units. Lack of funding prevented it from being built. In Chapter 2 we discuss looping and branching. During the 1850s, George Boole developed symbolic logic, including the binary operators (AND, OR, etc.). This is a major feature of digital computers.

A major milestone in the history of computers was a research paper written by John von Neumann and others in 1946. It strongly influenced the design of future computers. Most computer and all microcontroller designs are based on the von Neumann computer. Von Neumann suggested that instructions be numerical codes instead of special wiring. In Chapter 2 we look at the numerical codes used by the 68HC11.

With *software* concepts developed, the following years saw the development of the necessary *hardware*. The transistor was invented in 1948. The first integrated circuit (IC) was built in 1959. We often refer to an IC as a chip. Chips are fabricated out of semiconductors, usually silicon. Since then, *semiconductor* chip manufacturers have fabricated more and more complex circuits on a single chip. For more historical details, refer to reference 2 in Appendix C.

1.2.3 Microprocessors

The early history is obscured by legal battles. On December 28, 1970, Gilbert Hyatt submitted a patent application for a computer on a chip, the microprocessor. In the following year, he and his partners went their separate ways. The partners founded a new company called Intel. During that year, 1971, the new company produced the first microprocessor, the Intel 4004. It contained 2250 transistors and handled binary data as 4-bit words. Later, Intel introduced the 8008, an 8-bit device that incorporated other advanced features. A larger bit size means that a microprocessor can process more data.

By 1974, several chip manufacturers offered microprocessors. Others offered products controlled by microprocessors. The most common microprocessors were the Intel 8080, 8085, Motorola 6800, Signetics 6502, Zilog Z80, and the Texas Instruments 9900 (a 16-bit device). By 1978 16-bit microprocessors had become common. They include the Intel 8086, Motorola 68000, and Zilog Z8000. Since then, microprocessor manufacturers have continually been developing microprocessors with advanced features and new architectures. These include 32-bit devices such as the Intel Pentium, Motorola DragonBall, and many microcontrollers using the ARM core.

Other companies using microprocessors in their products had to spend a lot of time designing hardware and software. To sell their newer products, microprocessor manufacturers tended to make them compatible with previous systems so that previously written software could be used. Finally, in 1990, Gilbert Hyatt was awarded a patent for the invention of the microprocessor, but the controversy continues.

1.2.4   Microcontrollers

Generally, microprocessors are evolving in two directions, performance and integration. The performance direction emphasizes increased and faster processing power with the ability to store more data. Computers use these microprocessors as CPUs. However, many control applications benefited by having a reduced chip count and did not need the increased computing power. There was a wish for more integration, so chip manufacturers developed CPU chips with built-in memory and interface circuits.

The general public is more aware of the former type of microprocessors because of the popularity of personal computers. However, more microcontrollers are sold than powerful microprocessors because they are used in many machines, instruments, and consumer products. A chip manufacturer will typically sell ten times as many microcontrollers as general-purpose microprocessors. Industry needs skilled people to design, test, build, and service microcontroller-based systems.

1.2.5   The Motorola MC68HC11

Development.   Because of the effort required to rewrite software, semiconductor manufacturers usually maintained compatibility with earlier microprocessors. Using their 6800 microprocessor as a base, Motorola developed the 6808 and 6802 microprocessors. Using these earlier designs, they built the 6801 microcontroller. It included some memory and interfacing circuits and became popular in automotive control systems.

Later, in 1985, Motorola developed the 68HC11, a microcontroller that is upwardly compatible with 6800-type microprocessors. The 68HC11 included more features. Then, in 1996, Motorola introduced the 68HC12 that included additional features such as more instructions. Yet, the 68HC12 is upwardly compatible, meaning one can use software created for the 68HC11. One primary distinction between the 68HC11 and earlier processors is in the method of fabrication. The 68HC11 uses high-density complementary metal-oxide semiconductor (HCMOS) technology. When a microcontroller is fabricated using this process, it is smaller and faster. It also uses less power and has a high tolerance for noisy signals. These are important factors in control applications. Further enhancements resulted in the HCS12 microcontroller.

Versions.   Several versions of the 68HC11 are available. Additional versions are being offered continuously. Also, other manufacturers supply the 68HC11. It is common practice to have other manufacturers build a chip designed by a competitor. This is called *second sourcing* and is done to ensure a reliable and stable supply of parts.

Each version has its own part number. It is not possible to list and describe all of them. An example is the MC68HC11E9. If you intend to design a 68HC11 system, consult your local Motorola applications engineer as to what is most suitable for your application. We will point out some differences in the specific types, but in this book we refer to the MC68HC11E9 most of the time.

Applications.   Some commercial products that are or were controlled by the 68HC11 are listed below.

Automotive:
Chrysler transmission and engine control modules

Ford digital instrument clusters

Jeep Cherokee drive and emissions control

Chevy/Delco GEN-111 Engine Control Module (used in the Indianapolis 500, winner five times in a row)

Others:
Canon EOS model automatic cameras

AIM portable gas detector for emergency personnel

Conners, Inc., hard disk drive controller

Laser Wand, a scanner for reading product codes

Motorola Sabre Radio, a walkie-talkie

Motorola multikeyset office phone system

StairMaster's Crossrobic 1650 LE, a physical exercise machine

AT&T Mobil 3040/3050 cellular telephones

Fiock ski binding designed to reduce knee injuries

You may be familiar with some of them; you may even use one. Other organizations or individuals use the 68HC11 for specialty products such as the following:

Transmitter for radio-controlled model airplanes produced by hobbyists

Automatic phone tester produced by a telephone company

Remote TV controller for the disabled persons produced by a research institution

Rug Warrior experimental home robot

Terrarium Control System simulating a tropical rain forest

Application Development.   Once a product is designed, consumers such as ourselves simply use the product. Companies and organizations producing the products have to design the control system. It is not simply a matter of plugging in a chip. A development team will have to determine what devices to connect to the microcontroller and how to connect them (the hardware). They will also have to write the instructions for the microcontroller to execute (the software). To do this, designers use prototyping tools.

One simple and inexpensive tool is the M68HC11EVBU evaluation board. It is an ideal system for students and hobbyists to experiment with in order to learn how to use the 68HC11. Figure 1.2 shows this board. The board allows one to test a control system. However, most users will use third-party systems, some of which are equivalent to the EVBU. Motorola no longer produces the EVBU. In Chapter 4, we discuss application development in more detail.

Wire-wrap area

Microcontroller chip

Input/output connectors

Communications connector

Figure 1.2   Microcontroller (Universal) Evaluation Board. The wire-wrap area can be used for a prototype of the device being controlled. For a microcontroller application, connect the center input/output connector to the device being controlled. For communications with a personal computer or terminal, use the connector on the left. (Courtesy Motorola, Inc.)

■ 1.3 TOP-DOWN VIEW OF MICROCONTROLLER SYSTEMS

1.3.1   The Microcontroller System

In this section we use the automotive application of Figure 1.1 to illustrate some new concepts. Figure 1.3 shows a microcontroller system block diagram. It includes a microcontroller with necessary support components. The *buffers* and converters condition I/O signal levels if necessary. For example, the engine control system of Figure 1.1 has an ignition system requiring high-voltage pulses. But a microcontroller uses logic signals that are nominally 0 volts (V) for logic 0 and 5 V for logic 1. A converter would have to change the microcontroller logic signals to a higher-voltage pulse.

Chapter 5 covers the flow of logic signals along a *bus*. For now, consider the bus to be a highway on which different types of signals travel. CPUs use three types of signals: *data, address,* and *control*. Data signals represent instructions and values of different vari-

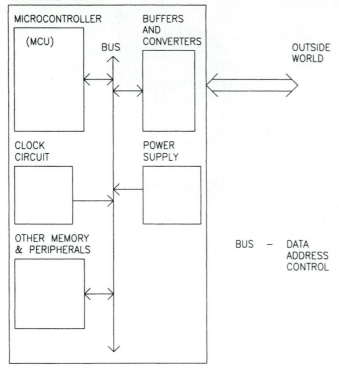

CONTROLLER BOARD

MICROCONTROLLER

(MCU)

BUS

BUFFERS
AND
CONVERTERS

OUTSIDE
WORLD

CLOCK
CIRCUIT

POWER
SUPPLY

OTHER MEMORY
& PERIPHERALS

BUS — DATA
ADDRESS
CONTROL

Figure 1.3   Block Diagram of a Typical Microcontroller System

ables, such as temperature. Addresses indicate where data is stored. Control signals coordinate microcontroller operation with associated chips. Do not confuse these with signals sent to control the application.

The CLOCK circuit generates a fixed-frequency signal that provides timing information for the entire system. This is necessary for microcontroller operation. Often, the CLOCK circuit is a crystal connected to two pins of the microcontroller. The power circuit converts incoming power such as 115 V alternating current (ac) or 12 V direct current (dc) to the nominal 5 V dc required to operate the microcontroller.

1.3.2   The Microcontroller Unit (MCU)

Figure 1.4 shows a generic block diagram of a microcontroller unit (MCU). Internally, it has three basic parts: the *central processing unit (CPU), memory,* and *registers.* They are connected by an internal bus. Externally, it has pins for power, input/output (I/O), and some special signals. In this section we consider only the I/O pins, which are grouped into units called I/O ports.

The CPU controls the operation of the microcontroller. We look at the CPU in more detail later. For the 68HC11, the CPU is essentially equivalent to the 6800 microprocessor.

Memory is where data and program code are stored. Physically, there are different types of memory: *read-only memory (ROM), random-access memory (RAM),* and

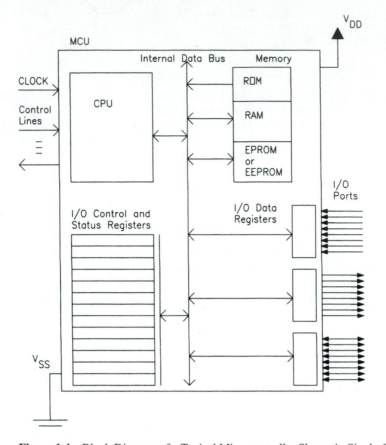

Figure 1.4   Block Diagram of a Typical Microcontroller Shown in Single-Chip Mode

electrically erasable programmable ROM (EEPROM). Some microcontrollers use *erasable programmable ROM (EPROM)* instead of ROM. In Section 1.4 we discuss the topic of memory.

Registers are used to handle specialized information. Essentially, registers are the equivalent of a workbench. It is a place where CPU works on (modifies) a binary number. There are I/O registers and CPU registers. CPU registers are covered later. There are three basic types of I/O registers: data, control, and status. As we shall see, these registers, together with the CPU and I/O *ports,* are used in I/O operations.

Each I/O data register holds I/O data associated with its corresponding I/O port. An I/O port is a collection of I/O pins on the chip that represents a unit of data. Usually, I/O ports have eight lines to transfer a byte of data. These ports can be input only, output only, or programmable to be either. The directions are always with respect to the chip. For example, "input" means that data enters the chip from the outside.

An example of input operation is the following. The microcontroller reads the coolant temperature. The temperature sensor is connected to an I/O port. The data from the sensor is transferred to that port's register. The microcontroller sends a copy of the register contents to the CPU to be processed. Another possibility is that it sends the data to memory for storage.

Output operations are similar. The CPU sends a data byte to an output register to be stored. This byte is transferred to the output port to be sent to a device connected to it. For example, the microcontroller sends the temperature data to the instrument panel for the driver to see.

The I/O control and status registers, respectively, control and monitor the microcontroller I/O process. For example, the microcontroller measures the engine speed by counting the pulses sent by the speed sensor over a set time period. A data register counts up by one every time an edge is detected. Pulses have rising and falling edges. One could program the microcontroller to count either, both, or neither edge. True, it does not make sense to count "neither" edge. However, you are perfectly free to do so if you wish! Counting both edges is fine if you divide the count by 2.

To specify what edges to count, you would write a program to set or reset the appropriate bits in the appropriate register. We use the manufacturer's data sheets to determine which bits in which register to modify and how to modify them. The Quick Reference (Appendix B) lists these bits and registers for the 68HC11 (Table 3.1).

Microcontrollers have internal timers. Each one sets or resets a special bit, called a status bit, in its respective status register when the programmed time period has expired. When the program detects a change in the bit, it reads the appropriate data register to find out how many pulses have arrived. From this it calculates the speed and does something with information. It may use the information to correct engine operation. It could also send it to the instrument panel.

The microcontroller uses other external pins for power and control and, if necessary, data and address lines. Since most microcontrollers are complementary metal-oxide semiconductor (CMOS) or variations thereof, the power pins are labelled V_{DD} for the positive supply voltage and V_{SS} for ground. CMOS is a type of integrated circuit that we describe in Chapter 6.

The CLOCK, which was already mentioned, is one of the control lines. Each control line has its own special function. One line, called $\overline{RESET}$, is used to bring the microcontroller to an initial condition. This is useful if the system locks up and does nothing more.

Sometimes, a microcontroller requires more memory or I/O ports than are available in the chip itself. In this case it will need external data and address line connections as shown in Figure 1.5. Typically, some pins can be used either as I/O ports or as external data and address lines. The microcontroller can be set up to work either way. How it is set up is called the microcontroller's mode of operation. Figure 1.4 shows single-chip mode and Figure 1.5 shows expanded multiplexed mode. If using the latter, extra chips need to be installed for memory and/or I/O ports. It is called *multiplexed* because data and address signals share some of the lines.

1.3.3   The Central Processing Unit (CPU)

Figure 1.6 shows a block diagram of the CPU. It executes program instructions. Note that the CPU has its own registers. The program counter (PC) is a special register that tells the CPU where to get an instruction or data byte. The other registers store specialized data or address information.

The instruction decoder tells the arithmetic and logic unit (ALU) what to do with the data. The control sequencer manages the transfer of instruction and data bytes along the internal data bus. The address register sets the condition of the address bus. The external

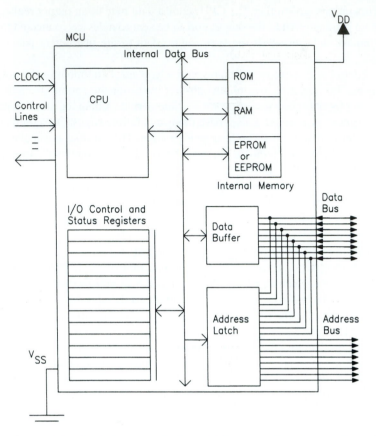

Figure 1.5 Block Diagram of a Typical Microcontroller Shown in Expanded Multiplexed Mode

address bus selects a specified location in memory. The data driver conditions data signals to be sent to or from memory or I/O registers.

We cover CPU operations in more detail in Section 2.4.

■ 1.4 MEMORY CONCEPTS

1.4.1 Organization

Figure 1.7 reviews how digital data is stored. Data is stored as a binary number (base 2), although we often represent it using hexadecimal (base 16). The *bit* (binary digit) is the smallest digital unit. It is either a 1 (high) or a 0 (low). A *byte* is defined to be 8 bits. The following is an example of 2 bytes shown as binary numbers:

```
00101010
01110011
```

Binary closely represents actual data storage since each 1 or 0, respectively, represents a semiconductor cell being on or off. However, it is more convenient to show data using the hexadecimal numbering system. Each hexadecimal digit represents 4 bits. Hence,

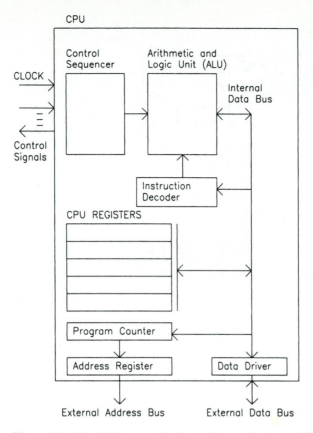

CPU

Control Sequencer

Arithmetic and Logic Unit (ALU)

CLOCK

Internal Data Bus

Control Signals

Instruction Decoder

CPU REGISTERS

Program Counter

Address Register

Data Driver

External Address Bus

External Data Bus

Figure 1.6   Block Diagram of a Central Processing Unit (CPU)

we can write a byte using 2 hexadecimal digits. The former 2 bytes are then represented in hexadecimal as follows. (The Number Systems section in Appendix G explains binary and hexadecimal numbers in detail.)

2A
73

Normally, the byte is the smallest addressable unit. However, as we shall see later, a program can address individual bits in I/O registers. The word size in bits varies depending on the processor. Its size is the number of bits the CPU can work on at a time. Figure 1.7 also shows the use of the binary and hexadecimal numbering system. Motorola's convention is to use the percentage sign (%) and dollar sign ($) prefixes, respectively. Refer to Appendix G for numbering system conventions. Unless otherwise specified, we follow these numbering conventions. For example, we can show the former 2 bytes in binary and hexadecimal as follows, %00101010, %01110011, $2A, $73.

Figure 1.7 also shows the logical organization of memory. An address indicates the storage location of a data byte. The address bus carries a signal consisting of 2 bytes. These bytes form a 16-bit address, which points to a data byte. In this case, the 2-byte address bus carries the signal $2A73. It is addressing (pointing to) a location whose

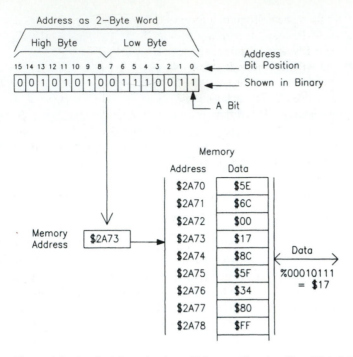

Figure 1.7 Logical Organization of Memory Showing How Digital Data Is Stored. Note the Motorola convention for binary and hex numbers.

contents are $17. The number $17 may represent an instruction or a numerical value used for computation.

PRACTICE

What are the contents of address $2A77 in Figure 1.7?

Solution

$80

1.4.2 Semiconductor Memory

The main types of memory are:

Read-only memory	ROM
Random-access memory	RAM
Erasable programmable ROM	EPROM
Electrically EPROM	EEPROM

ROM is used for permanent program information that can never be erased. The semi-conductor manufacturer fabricates the memory with the data. It cannot be altered later. Microcontroller ROM normally holds the final application program.[2]

RAM is read and write memory. It stores data that continually changes during microcontroller operation.[3] The memory is volatile, meaning that it loses all data when it loses power. "Random access" means that each location can be accessed in the same amount of time, independent of its address. This differs from sequential memory, such as tape drives.

EPROM is nonvolatile. Also, a program cannot write data to it as it can to RAM. However, it is possible to put data in it by using a special programming procedure. One erases an EPROM by exposing it to ultraviolet (UV) light for a period of time. Then it can be reprogrammed.

Microcontrollers use EPROM for two purposes. EPROM can hold customized calibration data. For example, a gas detection instrument may need different parameters because of sensor and converter tolerances. Although all the gas detectors have the same program, they may need different parameters. The other purpose is to develop the application in the first place. The developer uses a microcontroller that has EPROM but no ROM. He or she tests programs in EPROM, where they can be modified as explained in the preceding paragraph.

EEPROM differs from EPROM in that one can erase it electrically instead of using UV light. This is an advantage. The 68HC11 is the first microcontroller to have an internal EEPROM. This contributed to its popularity. In fact, it can program and erase the EEPROM using only a 5-V dc supply. Traditionally, EEPROMs required approximately 25 V dc for programming and erasing.

In Chapter 6 we explain semiconductor memory technology in more detail.

■ 1.5 MICROCONTROLLER MEMORY MAP

A memory map is a diagram that shows a computer's available addresses and how they are used. Figure 1.8 shows a memory map for the MC68HC11E9. It shows *default* conditions for single-chip and expanded multiplexed modes. The default condition is the condition of the chip, as supplied by the manufacturer, when powered up. By programming a special register, one can change the default settings. In single-chip mode, the external (EXT) addresses are not available. An application memory map would also show program and data information in addition to the memory type.

Note that the standard method to indicate the size of a block of memory is to state the number of bytes. A common unit is the kilobyte (KB), which is 1024 bytes or approximately 1000 bytes. "KB" is the short form (see "Measurement Quantities" in Appendix D). For example, an 8KB ROM block has 8 kilobytes. This is equivalent to 8192 (8 × 1024) bytes.

An addressing range is the total number of unique addresses that a processor can generate. It is the largest possible size of physical memory that a microprocessor or microcontroller can use. The total addressing range for the 68HC11 is 64KB since it uses a 2-byte address. Note that 2 bytes is 16 bits. This means that an address can have 1 out of 65,536 (2^{16})

---

[2]Users of personal computers (PCs) may be familiar with an operating system's BIOS (basic input/output system). One of the ROMs in the PC motherboard holds the BIOS.

[3]When a personal computer (PC) is powered up or reset, the BIOS loads the operating system from disk into RAM. Also, the operating system executes programs in RAM. Note that some PCs hold the operating system itself in ROM.

$0000		512-Byte RAM
$01FF		
	Not available / External	
$1000		64-Byte Register Block
$103F		
	Not available / External	
$B600		512-Byte EEPROM
$B7FF		
	Not available / External	
$D000		12KB ROM
$FFFF		

Figure 1.8 Memory Map of the MC68HC11E9 for the Normal Operating Modes. Note that other versions have different memory models.

possible values. Future versions of the 68HC11 may have a larger addressing range. For example, the 68HC711K4 uses something called a memory *management unit* to give a 1-megabyte (1024KB = 1MB) addressing range.

PRACTICE

What is the addressing range of a processor chip with a 20-bit address bus?

Solution

$$2^{20} = 1{,}048{,}576 \text{ (a 1MB range)}$$

2 Programming

OBJECTIVES

After completing this chapter, you should be able to

- Describe the fundamental characteristics of software.
- Explain the types of programming languages, source code, and types of translation to executable form (machine code).
- Describe the fetch/execute operation of the central processing unit (CPU).
- Identify and use the central processing unit (CPU) registers for programming.
- Write assembly and C source code to use the 68HC11 addressing modes.
- Document source code.
- Write a program using a look-up table.
- Use a memory dump to interpret the results of program execution.
- Write a program to perform basic data handling, arithmetic, and test operations.
- Write programs that perform arithmetic operations using unsigned, signed, and floating point numbers.
- Determine and account for data ranges and overflow.
- Interpret the results of instruction execution using the condition code register and memory dump.
- Write programs using looping, branching, and bit manipulation.

◼ 2.1 ASSEMBLY AND OTHER PROGRAMMING LANGUAGES

A Deck of Cards. On your journey, you meet a person who gives you a deck of cards. Each card is a coded instruction. As you pull each card off the deck, the card tells you what to do next. The deck is like the set of instructions that make up a programming language.

Hardware.   When a microcontroller chip is fabricated at the semiconductor factory, it is a silicon sheet. This sheet has semiconductor regions diffused with different impurities to form transistors and other components. Pins are connected to form the inputs and outputs. This is the hardware.

What About the Software?   The software is the invisible part of the microcontroller. *Software* is the term used for the instructions that tell the microcontroller what to do. When you write instructions for a specific task, you are writing a program. This is also called *programming*. A program is a set of instructions written to perform a specific task. We can use an analogy to differentiate between hardware and software. Take this book, for example. The cover, pages, and ink patterns for text and diagrams make up the hardware of the book. The software is the message; it is what the book is explaining. *Software design* is like figuring out what to write. *Programming* is like the activity of writing down the words. A key point is that programming (coding) is only one aspect of designing software.

A Program.   Unfortunately, microcontrollers do not understand human languages yet. Instead, they use machine language and we use a programming language. Then we translate the programming language into machine language to produce the instructions for the final application. Consider the situation of converting temperature from Fahrenheit to Celsius. You could simply ask a friend, "It is 62 degrees Fahrenheit. What is that in Celsius?" A microcontroller (or any computer) could give you the answer only if someone (a programmer) had already told (instructed) the microcontroller *how* to convert. The *program* is the set of instructions that describe the steps required to do a task.

For temperature conversion, a program could look something like the following:

```
Load Fahrenheit
Celsius = (5 x (Fahrenheit - 32))/9
Display Celsius
```

Note we said "something like." In the remainder of this chapter we describe how to *formally* program a microcontroller. To do this, we also have to use hardware features of the microcontroller and of the central processing unit (CPU) in particular.

What You Will Learn.   After studying this chapter, you should be able to write assembly language source code to control the microcontroller. To do this, you will use knowledge of:

> Source code
> Machine code
> Translation from source code to object code to machine code
> The CPU programming model
> Opcodes
> Operands
> Addressing modes
> The microcontroller instruction set

You will also be able to understand the new terms just listed!

Machine Language. Each type of CPU has its own set of instructions that it recognizes and executes. The word *execute* is often used to indicate that the CPU is carrying out the instruction. These instructions are binary numbers written in a special code. (Section 1.4 introduces binary numbers.) The CPU will interpret this sequence of binary numbers as a code telling it what to do. This is similar to playing cards, where a card may be like an instruction. In Euchre, for example, a trump card instructs other players to lay down their trump cards if possible. This is the machine language. Each type of CPU has its own machine language. For example, the 68HC11 cannot run programs written in the machine language for the Intel 8051 microcontroller even though both use binary numbers. Similarly, you use a deck of cards differently depending on the game you are playing. Note that it is possible for one microcontroller to be upwardly compatible with another. For example, a 68HC16 can run a program written for the 68HC11.

An example of machine instruction in binary is:

```
%10000110
%01011010
```

Do not worry about what it means right now except to note that the % prefix indicates a binary number. Just note that this is what the CPU understands. People do not write binary instructions. Instead, they write in another language that gets translated into machine language.

Assembly Language. One of these languages is assembly language. In assembly language, the equivalent of the previous machine instruction is:

```
LDAA #$5A
```

Again, do not worry about what it means except to note that the $ prefix means hexadecimal. Assembly language has some advantages to be discussed later. However, it also has some drawbacks. First, each different type of CPU has its own assembly language. Thus you would have to use a different assembly language for every different type of CPU. Also, you must have detailed knowledge of the CPU in order to use it. Finally, it is still cryptic to many people. But some people enjoy programming in assembly.

High-Level Languages. Assembly language is not portable; that is, it can run on only one type of CPU. On the other hand, *high-level languages* are portable. You write a program using the rules of the language, then translate it into machine code for each type of CPU. What is different is the translator, not the program. These languages are high level because they are more humanlike than assembly, but not humanlike enough for some. Many people would like to see computer languages that can understand normal human speech. High-level languages currently in use include Basic, Pascal, FORTRAN, Ada, Cobol, C, and Java. This is not a complete list. For microcontrollers, C is the most common high-level language for microcontroller development. For example, the following C code will correspond to several lines of machine code:

```
celsius = (fahrenheit - 32) * 5/9;
```

It converts temperature. Most high-level languages use the asterisk (*) to symbolize multiplication. We will say more about high-level languages in Section 2.3.

■ 2.2 SOURCE CODE, OBJECT CODE, AND THE ASSEMBLER

In this section we deal primarily with assembly language programming. With assembly language, the programmer can directly control the microcontroller's resources. Here we cover source, object, and machine codes. Using a sample program, we illustrate the use of the microcontroller's instruction set, formatting rules, and the translation process. We also preview how the CPU executes instructions.

Source Code. This is the original program before it is translated. It is what the programmer writes (types or keys in). Usually, a source-code program is stored as a file. Note the following examples of programming instructions in source-code form:

```
LDAA #$5A                                is an example for assembly language.
celsius = (fahrenheit - 32) * 5/9;is an example for C language.
```

Efficiency. Usually, an assembly program is more efficient than one written in a high-level language. However, the efficiency of high-level languages is improving, and C is already quite efficient. Efficiency refers to how many machine instructions are required in a program. More efficient programs take up less memory. Also, because they are shorter, they run faster. For example, the program to control the air/fuel mixture of an automobile engine (see Figure 1.1) may require 4KB if it was originally written in assembly language. However, if it was written in Basic, the machine language program may have required 20KB after the Basic program was translated. Besides being five times smaller (in this case), the assembly code may also execute several times faster than the Basic program.

This is only a sample comparison. Another important factor in efficiency is the programmer's skill. A well-written Basic program might require only 8KB of machine code, whereas a poor assembly program could require 16KB to perform the same task. Program efficiency is an important factor in 8-bit microcontrollers because they do not have as much memory as do 16- or 32-bit machines.

2.2.1 Machine Language

Fetch/Execute Cycle. A CPU executes each machine instruction by performing a fetch/execute cycle. Figure 2.1 illustrates the fetch/execute cycle. The CPU fetches an opcode byte from memory and decodes it to find out what it is supposed to do. Then it executes the instruction. Typically, these instructions examine or modify memory or input and output (I/O). After the instruction is completed, the cycle resumes for the next instruction. In Section 2.4 we explain this process in more detail.

Example. To give an example of a machine instruction, we must know more about the microcontroller. Recall that a microcontroller has registers. Registers are special places to store data or special status information for the microcontroller to operate on. We introduce the 68HC11 CPU registers in Section 2.5.

For now, we will use one of these registers. It is called accumulator A. We will call it ACCA. In Section 2.1 we introduced an example of a machine instruction, which was

Memory

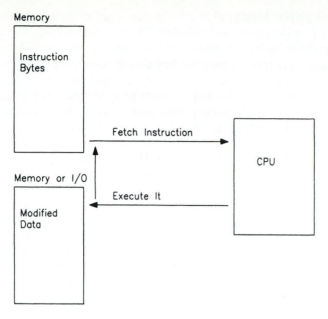

Figure 2.1 Fetch/Execute Cycle

```
%10000110
%01011010
```

This means "Put the number $5A in ACCA." The first binary number is the instruction, LOAD ACCA. The second number says what to load.

Normally, people do not write application programs using machine language. However, it may be necessary at times. The CPU only understands binary. When people use machine language, they use the hexadecimal equivalent, a number system based on 16 instead of 10. Hexadecimal is easier to read than are pages full of 1s and 0s. (The Number Systems section in Appendix F describes the hexadecimal system.) Thus the former machine instruction becomes:

```
$86
$5A
```

This code must be stored in memory. If the code starts in address $E000, we can show the machine code as follows:

```
$E000 $86
$E001 $5A
```

The left side shows the address and the right side shows the machine-code byte stored at that address. The CPU uses a special register called the program counter to keep track of these addresses during the fetch/execute cycle (see Figure 2.1).

Opcode and Operand (in Machine Code).   The instruction has two parts, the *opcode* and the *operand*. This is true for most machine language instructions. The opcode is the first part. It tells the CPU what to do. In other words, the opcode specifies what the operation is. In this case, the byte $86 is the opcode that tells the CPU to "Load ACCA." The operand tells the CPU what data to operate on. In more detail, the operand specifies where and/or how to get or put the data. The byte $5A is the operand. It is the data that the CPU will load into ACCA.

An 8-bit processor usually uses instructions consisting of bytes. The example instruction used 2 bytes, one for the opcode and one for the operand. With 8 bits, there are 256 possible combinations of opcodes. The 68HC11 uses 236 of these as 1-byte opcodes. We shall see later that it also uses some 2-byte opcodes. We shall also see that an instruction may have no operand bytes, or it may have 1, 2, or 3 operand bytes.

Two-Byte Opcodes.   Most 8-bit CPUs use 1-byte opcodes. However, the 68HC11 must use some 2-byte opcodes. There is a historical reason for this. The earlier 6800 microprocessor and 6801 microcontroller used only 1-byte opcodes. When Motorola designed the 68HC11, they wanted to include more instructions. However, they still wanted old 6800 programs to run in the 68HC11 (maintain compatibility). This increased the total number of opcodes to more than 256. Motorola solved this problem by using prebytes. These specially reserved bytes tell the CPU that the byte following it is an opcode, not an operand (or the following instruction). Using the prebyte system, the 68HC11 includes an additional 76 two-byte opcodes.[1]

2.2.2   Assembly Language

The many different possible hex opcodes make machine language awkward to use. *Assembly* language is the *mnemonic* form of machine language. Each opcode has a short-form word associated with it called a mnemonic. Thus the assembly language version of our example instruction is

```
LDAA #$5A
```

This translates to LOAD ACCA with the number (# means immediate) $5A. Assembly language has specified rules and formats. Each line of assembly language has a corresponding sequence of bytes in machine language. Since the machine code resides in memory, we must specify the address where the machine code starts. Typically, we use a special mnemonic to specify the start address of the program. This is the mnemonic "ORG," which is an abbreviation of "origin." The example instruction then becomes:

```
ORG    $E000
LDAA   #$5A
```

Translation/Assembly (Source Code to Object Code).   To create an assembly language program, we type it in using a text editor or word processor. The mnemonic form we type in is called *source code*. Then we run a special translation program called an *assembler*. It translates the *source* code into relocatable *object* format. Then we run another special program called a *linker,* which produces hex code that can be loaded into the microcontroller's

---

[1]You can refer to the *Programming Reference Guide* (Appendix H) to examine the opcode maps.

memory. Sometimes, other steps are required before machine code is produced. Note that the assembler and linker are also programs. We call this collection of programs required for translation an assembly language development system.

For microcontrollers you usually run the assembler and linker programs on a computer, not on the microcontroller itself. These programs produce the machine code that the microcontroller can execute. Also, the terminology can be confusing. Do not confuse *assembly, assembly language,* and *assembler.* An assembler translates (assembles) assembly language into object code that can be converted into machine code. Assembly is the process of translation.

Machine Code and Object Code. Machine code is the actual sequence of bytes shown in hex (stored in hardware in binary). They are the instructions that the CPU can fetch and execute. Why doesn't the assembler produce machine code directly? Why do we need to use a linker? The reason is that a typical application requires large programs. These take a long time to develop. One writes the source code in smaller sections that perform certain parts of the final application. Often, these sections need to be revised during development. A developer only has to reassemble the revised sections into their respective object codes. The linker then combines these object code sections with others to produce an executable machine language program. Object code itself is not executable. It has information that is necessary for linking. It usually does not specify the address of the machine code. In some situations, such as very simple programs, object code is the same thing as machine code. In Chapter 4 we cover assembly and linking in detail for a specific assembly language development system.

Line Assembly. Often, we want to enter short programs just to learn about various microcontroller features or to try out an idea. In these cases it may be more convenient to enter the code directly as hex numbers. Alternatively, we could use line assemblers. They translate source code directly into machine language. The line assemblers translate each line of source code after it is entered. Line assemblers are simple and do not have many features. However, these assemblers are good for quick experimentation. This type of assembly can also be known as an interactive assembler, command line assembler, or debugger.[2]

You may be using Motorola's EVB or EVBU board. It contains a built-in line assembler as part of its monitor program. You connect a terminal or PC to the board, invoke the assembler, and type in the source code. The result is the machine code, which is also the object code (for line assemblers).

Disassembler. A *disassembler* is a translating program that does the reverse. It reads machine language and displays the assembly source code. Most disassemblers display only the simple assembly source code used by line assemblers. The EVB board's assembler is also a disassembler.

2.2.3 Examples

To become more familiar with assembly language, we will look at an example (Listing 2.1). For now, concern yourself with the format, not with what the program is doing. We cover

---

[2]Users of MS-DOS may be familiar with the DEBUG utility. It includes a line assembler.

instruction set details later in the chapter. Note that instructions are indented. Also, we omit some details to keep it simple.

```
*Listing 2.1
*This is a demo program to introduce the format of an
*assembly language program for the 68HC11. It reads the
*input of the coolant temperature sensor of Figure 1.1 and
*subtracts an offset (a fixed number). It then stores
*this result for further processing. Another part of the
*program (not shown) will multiply this number to obtain the
*temperature in degrees Celsius. Then it sends the result to the
*dashboard display for the driver.

        ORG     $E000           ;specify start address

        LDAA    COOLANT_TEMP    ;get coolant temperature
        SUBA    #CT_OFFSET      ;subtract offset
        STAA    STORE_TEMP      ;store it for further processing
```

This three-instruction example illustrates the use of mnemonic opcodes, labels, and comments. Recall that ORG is a special code that specifies where the resulting machine code will start in memory. The opcodes are LDAA, SUBA, and STAA. We cover what they do later. Labels are the operands CT_OFFSET, COOLANT_TEMP, and STORE_TEMP. They refer to specific hex numbers, although the labels can be almost any word. You should try to use words that make the meaning clear. For example, the labels CAT and DOG, although perfectly legal, have no obvious meaning to anyone reading the program. Also, certain words are reserved and cannot be used as labels. For example, you cannot use LDAA as a label since the assembler would recognize this as an instruction. According to Motorola rules, CT_OFFSET refers to hex data because of the number sign (#) prefix. COOLANT_TEMP and STORE_TEMP refer to a hex address. Comments are the third feature of this example. If a line does not contain an opcode, it is necessary for an asterisk character to be the first character in the line. This indicates that the line is a comment. Otherwise, if a line contains an opcode, anything that is not an operand is a comment. It is a good convention to begin such comments with a semicolon. A comment is a section of the program that the assembler ignores. In other words, they are not translated into machine code. Although the CPU does not need comments, people need them to understand the program. Programmers who do not write comments or write them poorly get a reputation as poor programmers. Anyone can be a poor programmer. The challenge is to be a good one.

The same program can be rewritten as in Listing 2.2.

```
*Listing 2.2
*Illustrate Listing 2.1 without labels and with fewer
*comments

        ORG     $E000       ;specify start address
*                           ;of program code

        LDAA    $1031       ;load ACCA with contents
*                           ;of address $1031
        SUBA    #$20        ;and subtract 32, hex 20
        STAA    $D004       ;store result in
*                           ;address $D004
```

For Listing 2.1, how does the assembler know what hex numbers to use for the labels? Information about labels is also part of the object code. The linker converts labels to hex numbers. Since the CPU only understands machine language, this information about labels is not part of the machine code. To keep the discussion simple at this stage, Listing 2.1 does not show the details necessary to define labels. This is one of the details that we will leave until Chapter 4.

Remove the comments and the dollar sign from Listing 2.2 and you will have a form usable in a line assembler.

```
LDAA      1031
SUBA      #20
STAA      D004
```

Line assemblers assume that all numbers are hex. They also do not understand special instructions such as ORG. When using a line assembler, the programmer also states the start address. Therefore, the source code does not show an ORG instruction.

2.2.4   Manual Assembly

We can convert this program into machine code manually. To do this we will need to use data sheets since nobody memorizes all the opcodes. A good reference is Table 10.1 in the *Data Book.* Appendix A also shows this table as Table A.1. We call it manual assembly because you have to use a manual!

Table A.1 is an instruction set summary. It lists all the opcode mnemonics cross-referenced to their machine coding. It includes some other information as well. Now look up the machine code for LDAA. You will find that there are five possibilities, depending on the addressing mode for the operand. Each type of instruction has one or more address modes associated with it. In our example, LDAA and STAA use extended (EXT). SUBA uses immediate (IMM). In Section 2.5 we cover addressing modes.

Now, try to write the machine code. You should get the following:

```
B6      10      31
80      20
B7      D0      04
```

By convention, machine code is always hex. Therefore, we do not use the dollar sign. Also, each line is one instruction. Note that the second instruction requires 2 bytes and the others require 3 bytes each. The "Bytes" column in the instruction set summary shows how many bytes there are in the instruction.

Now, let's say you want to start the code at address $E000. This would be:

```
E000      B6 10 31
E003      80 20
E005      B7 D0 04
```

Note that each line shows one instruction. The address column (left side) shows the start address of the first byte of each instruction.

Normally, one does not assemble manually but uses an assembler instead. However, it is important to know how to do it since microcontroller application design requires a good knowledge of the microcontroller. To program in assembly, you need to know a lot about the CPU. However, another option is to program in a high-level language. This is the topic of Section 2.3.

2.2.5   The Simulator

A microcontroller simulator is a software tool that permits users to simulate the operation of a microcontroller. This book contains the demo version of the THRSim 11 simulator as described in Appendix H. As illustrated in Figure H.1 (of Appendix H), the simulator shows the machine code and the source code of the example program shown in Listing 2.1. It also shows other features such as memory, registers, and resulting execution of the program. These are things that will be covered later. For now note that there are minor differences in the addresses shown in the machine code. For instance, the machine code begins at $FF00 instead of $E000. This is due to the requirements of the demo version. With the full version, one can simulate the program exactly as shown. Another thing to note in Figure H.1 is the result of running the program. The coolant temperature of $47 is stored in address $1031. The result of subtracting $20 from $47 is $27, which is stored in address $1004 (instead of $D004).

■ 2.3   USING HIGH-LEVEL LANGUAGES

2.3.1   General Concepts

Software engineering includes *designing* database applications, operating systems, graphic applications, and artificial intelligence, among many others. Software designers usually write the programs for these applications in a high-level language. Note that we said *design*—not writing or coding. The design process involves feasibility studies, definition of details in the program or system of programs, and testing the system as well as programming. The term *programming* refers to the process of writing the program instructions.

Assembly language has different uses. Many developers tend to use assembly language for applications more directly related to hardware, such as I/O control, signal processing, and *CRT* video generation. However, more people are using high-level languages for these applications. In many cases, a developer will use both. He or she may write part of the program in a high-level language and write certain sections that have to run very fast in assembly.

As mentioned previously, C is a popular language used for microcontroller software *development*. Remember that the CPU always executes machine language. We use a language such as C or assembly to produce the machine language. Why is C so popular? Basically, it is because it combines the best of both worlds. C has the convenience of a high-level language. Also, C has features to allow direct control of I/O, which is very important for microcontroller applications.

Most high-level languages are compiled languages. A few are interpreted. For compiled languages, a compiler translates the source code into object code. It translates the entire code. The object code is then linked with other object modules to form machine code. This is similar to the operation of an assembler. For interpreted languages, an interpreter translates a line of source code and then executes it. It does not save the object code.

Compiled languages run faster because they can execute repeatedly after one translation. However, if you make a change to the program, you have to go through the entire compiling and linking procedure to test the change. You may already have used a compiled language "apparently" without going through all of these steps. This is because the development system performs these steps automatically.

Interpreted languages allow one to test changes quickly. However, such a program runs more slowly because it has to be translated each time you run the program. Traditionally, Basic has been an interpreted language, but compiled versions are now available.

We deal with C in the remainder of this section. However, most of the concepts also apply to other high-level languages. There are several steps to producing executable machine code using C. First, *design* the program, unless someone has already designed it for you. Then write the source code using the *syntax* rules of the C language. Compile the program to produce object code. Link the object code. Usually, you will have to link your object code with other modules stored in a C library. These library object codes typically implement standard functions such as printer output, keyboard input, and trigonometric functions. Your code need only use them. After linking, you can run and test the program using a debugger. The debugger can display program variables, and it allows you to control program execution.

2.3.2   The C Language for Microcontrollers

What you get out of this section will depend on your knowledge of C. If you have no prior knowledge of C, you may read this section to get an idea of what C is like, but it does not teach it. To study C programming, you should refer to any of the many books on the subject, such as reference 4 in Appendix C. Below we point out some features that are typical for microcontroller software.

An early standard for the language was defined by Kernighan and Ritchie in 1978 when they published *The C Programming Language,* which contained "The C Reference Manual." Most C compilers for microcontrollers follow this standard. *ANSI* defined a more comprehensive standard that included more features. This standard was published in 1989 as the American National Standard for Information Systems—Programming Language C, X3.159-1989. To give you an idea of what C looks like, we'll examine a section of a C program. This is shown in Listing 2.3.

```
/*Listing 2.3
  An excerpt from a C program. "Overhead" such as heading information
  and variable declarations not shown */

time()
    /* function time() increments sec, min, hour based
       on the 24-hour clock format */
{
sec += 1;
if (sec >= 60)
  {sec = 0;
  min += 1;
  if (min >= 60)
    {min = 0;
    hour += 1;
```

```
      if (hour >= 24)
         {hour = 0; }
      }
   }
}
```

Listing 2.3 is part of a program to run a 24-hour clock. You may not be familiar with the *syntax* of the C language. Basically, it goes like this: Every second, the instructions following "time()" are executed. The number of seconds increments by one. The seconds count increments until it reaches 60. Then it goes back to zero and the minutes count increments by one. This occurs every 60 seconds. When the minutes count reaches 60, it also goes back to zero (resets), and the hours count increments by one. This repeats until the hours count increments to 24. When it does, the hours count is reset to zero and the entire cycle repeats.

Even if the syntax is new to you, you should be able to tell that the program does this. It is normally easier to tell what a C program does than what an assembly program does. The program deals more directly with logical structure. An assembly program to do the same thing would have instructions to load and manipulate registers or the contents of memory locations.

This program has three variables: sec, min, and hour. We could have called them almost anything, such as A, B, and C. However, it would not be so obvious to see what the program does. Also, the algorithm shows a typical structure of computer programs. If a condition is true, then do this, or else if it is false, do that. A complete C program also requires some overhead for defining variables and telling the compiler what to do. We have shown only a section of a C source code, to give you an idea of what it is like. Since it can be written such that its purpose is obvious, some assembly language programmers will write comments in C. These comments explain what their program is doing.

After writing the C source code, you run a cross compiler to generate a machine language program that will run in the target processor. In this case the target is the 68HC11. It is not big enough to have its own resident C compiler. Typically, we use a cross compiler running in a personal computer (PC).

A free compiler is the ICC Demo for Windows from ImageCraft. The demo version has restrictions but it is still useful for learning and experimenting. For more comprehensive C development, you can order the commercial version from ImageCraft. In some of the following descriptions about programming, we will include some segments of C code. These will illustrate how to perform equivalent operations using C instead of assembly. It also serves to familiarize you with C programming, and in particular for the 68HC11 microcontroller. Note that advantages of programming with C may not be apparent with the short examples used in this book. Developers typically use C to write large full-blown programs particularly where intensive data processing and logic is required. To simplify the C examples, we may omit the overhead code necessary to make them complete for execution. Appendix H describes how to use the demo version of the C compiler.

Consider the following C program that is equivalent to the example program for processing coolant temperature (Listing 2.1).

```
/* C Listing 2.1 */

/* Declarations */
unsigned char StoreTemp, CoolantTemp;
```

```
#define CT_OFFSET 0x20
void main(void)
{
    /*Main body of code */
    StoreTemp = CoolantTemp - CT_OFFSET;
}
```

We will explain aspects of the program shortly. The main (or core) of the program is the line:

```
StoreTemp = CoolantTemp - CT_OFFSET;
```

This is the part that is essentially equivalent to Listing 2.1. The code is more readable for humans and its functionality is clearer. Next, we consider some of the overhead aspects of the C program. Consider the declarations section. Comments are enclosed in /* and */. The words unsigned char are used to inform the compiler that 8-bit memory locations will be used. Note that the source code does not indicate the address for these locations. The word #define is used to assign a constant value of 20 hex to the symbolic name CT_OFFSET. The preceding 0x in a number indicates that the number is hex.

The other section starts with the line void main(void), followed by opening braces {, and then the main code, which in turn is followed by closing braces }. The format with respect to separate lines and indentations are free form. The style shown above is a preference for the purposes of indicating the program's purpose. The following single line would also technically work.

```
void main(void) {StoreTemp = CoolantTemp - CT_OFFSET;}
```

We can compile the program of C Listing 2.1 with the ICC Demo. The compiler will create an equivalent assembly language source code, which we can then assemble into machine code. In this instance, ICC Demo compiled the C source into the following assembly source (with some assembly overhead omitted).

```
ldab _CoolantTemp
clra
subd #32
stab _StoreTemp
```

Equivalent Assembly Code. The point of this example is to show that a C compiler will produce assembly language code. This specific result is not identical to the assembly program of Listing 2.1 but it does have the same functionality. For now, you may prefer to ignore the details of the example. For instance, it uses lowercase instead of uppercase. The compiler chose to use accumulator B instead of accumulator A for loading and storing (ldab and stab). It also decided to use both accumulators for subtraction (clra and subd #32). Note that decimal number 32 is equivalent to hex 20. Also, it assigns the actual memory addresses for use by variables CoolantTemp and StoreTemp. The leading underscore for variables refers to the use of local scope, a topic not covered here.

Efficiency.   The assembly program created by the C compiler is less efficient than the original assembly program of Listing 2.1. The compiler has created machine code that is larger and would take longer to execute. In this case, the differences are not significant. Despite the lower efficiency, most developers prefer C because it is easier for coding the logical and data processing parts. In some cases, a developer will use both depending on the need for program efficiency in any section of the final product. For example, a developer may directly use assembly for parts that have to run fast such as real-time handling of I/O.

■ 2.4 FETCH/EXECUTE OPERATION OF THE CENTRAL PROCESSING UNIT (CPU)

Up to this point we have seen what a program is and some of the things it does. Now we examine how the central processing unit (CPU) executes an instruction. The CPU carries out an instruction by activating certain logic circuits and transfers data along internal bus lines. This sequence of activities is controlled by instructions stored internally in the CPU's control sequencer (see Figure 1.6). These instructions are called a *microprogram*. Microprogramming is a detail that we do not examine in this book. Normally, only chip designers write microprograms. Do not confuse microprogramming with machine language. A microprogram tells the CPU how to execute a machine language instruction.

Recall that in Section 2.2 we talked about the fetch/execute cycle (see Figure 2.1). Figures 2.2 to 2.5 show how a CPU executes a typical instruction. Most CPUs are synchronous, meaning that transfers along buses and register updates occur every cycle of a clock signal. The clock is a square-wave signal usually with a frequency above 1 megahertz (MHz). In general, a fetch/execute cycle may require many clock cycles, depending on the instruction.

To illustrate fetch/execute cycles, we show a very simple program with two instructions. The instuctions are `Clear Accumulator A` and `Load Accumulator A` with the number $5C. The first is an instruction to change the value of the byte in accumulator A to zero. The second changes the accumulator A value to the number $5C. In assembly language source code, these instructions (with comments) are:

```
CLRA              ;this clears ACCA to $00
LDAA      #$5C    ;this loads ACCA with $5C
```

Let's say that the machine code begins at address $E000. An assembler would translate this source code to the corresponding machine code.

```
E000    4F
E001    86 5C
```

Note that the first instruction uses 1 byte, which is an opcode byte. The second instruction uses 2 bytes, for the opcode and operand, respectively. Now examine Figure 2.2. Memory contains the program's machine code starting at address $E000. The program counter (PC) is an important register. At the beginning of the cycle, it always "points" to the address where the next instruction byte can be found. In this example it contains $E000, the address of the CLRA instruction. The CPU transfers the number $E000 to the address register (AR). The AR sends this number out to the address bus, which in turn selects ad-

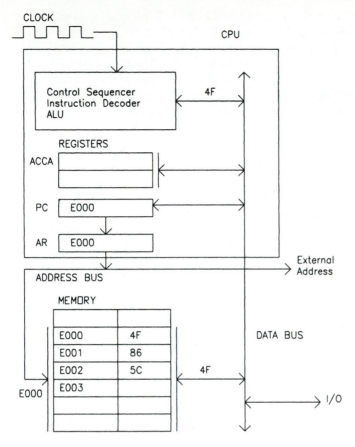

Figure 2.2 CPU Operation. Fetching the first instruction (CLRA); note that AR is the address register.

dress $E000 in memory. Because $E000 is selected, memory puts the byte $4F onto the data bus. The control sequencer reads what is on the data bus ($4F in this case) and decodes the data to be a CLRA instruction. The PC has also incremented by one. This completes the fetch portion of the fetch/execute cycle.

Now refer to Figure 2.3. Executing this instruction requires an additional clock cycle, but it does not require any data from memory. Therefore, the PC does not increment and memory is not selected. This is why the data bus has irrelevant data. We examine physical bus activity when we deal with hardware in Chapter 5. The control sequencer causes all bits in ACCA to be reset to zero. After this, the CPU is ready to fetch the next instruction.

Figure 2.4 shows the fetch of the next instruction. Since the PC has $E001, the CPU selects address $E001 in memory. Memory responds by putting $86 on the data bus. The control sequencer reads this byte and determines that it is an LDAA instruction. Also, it knows where to get the data. In this case the PC increments by one.

Figure 2.5 shows the execution phase of the LDAA instruction. Since the PC was incremented to $E002 during the preceding cycle, the CPU fetches the data $5C. Because the control sequencer has determined that it should execute an LDAA instruction, it places the data $5C into ACCA.

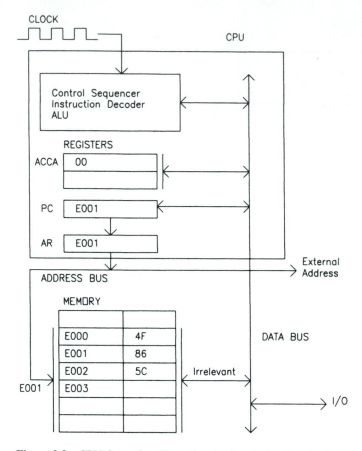

Figure 2.3 CPU Operation. Executing the first instruction (CLRA).

The description you have just read applies to most CPUs. Register names and mnemonics differ for different types of CPUs. We have used 68HC11 terminology in our description. In the next section we look at its CPU registers and examine the topic of addressing modes. We have also ignored some hardware details. These are covered where appropriate. To summarize, note that the PC increments after each clock cycle. Note that an instruction may require several clock cycles to complete its fetch/execute cycle, depending on the type of addressing mode used.

■ 2.5 THE INSTRUCTION SET AND ADDRESSING MODES

In a sense, a CPU uses registers like a craftsperson uses a workbench. A cabinetmaker can work on only one part at a time. He or she places a part (data) on the workbench (register). To work on another part (other data), he or she must first remove the former part. The cabinetmaker also uses specialized tools (instruction set) such as chisels, planer, drill bits, and sandpaper. Similarly, the CPU places data in specialized registers to process the data.

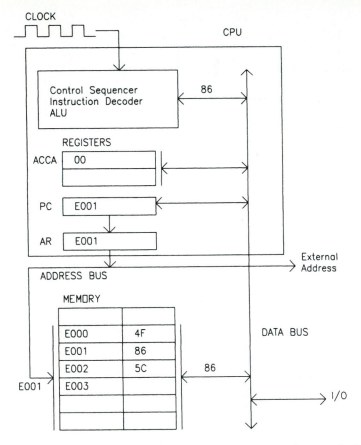

Figure 2.4 CPU Operation. Fetching the second instruction opcode (LDAA#).

Consider the automobile engine described in Section 1.1.1. The microcontroller will have to load a binary number representing oxygen level. Then it compares this number to another number representing the desired oxygen level. The outcome determines subsequent actions.

Figure 2.6 shows the programming model of the 68HC11. You can find this same drawing as Figure 1.2 in the *Data Book*. A programming model shows the registers in the CPU plus some of their details. To program the processor, you will have to become familiar with the model. We have already mentioned accumulator A (ACCA) and the program counter (PC). The 68HC11 also has an accumulator B (ACCB). One may use both of these 8-bit registers as a single 16-bit register called double accumulator D (ACCD). We cover other registers when appropriate. They have specialized functions.

2.5.1 Instruction Set References

A list or table showing all of the instructions is called the instruction set. Refer to Appendix A to see these instructions. We will look at some of the instructions. It is impractical to look

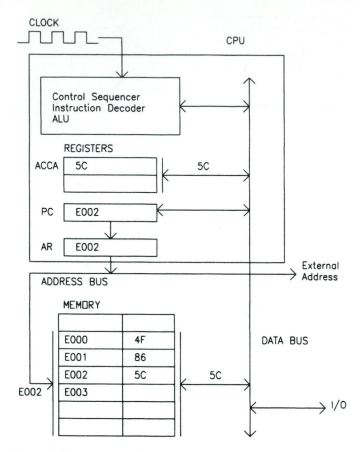

Figure 2.5 CPU Operation. Fetching the second instruction operand ($5C) and executing the instruction.

at each one. Once you understand how to use the instruction set, you can find out how to use any instruction when you need to. After becoming familiar with the 68HC11, you should use the *Data Book* and other literature listed in the "Sources" section of Appendix C. You will probably use Table 10.1 of the *Data Book* (same as Table A.1 in Appendix A) most often. We often refer to this table as the *instruction set summary.*

If you need more details than are available in the instruction set summary, refer to Appendix A of the *Reference Manual.* It gives all of the instruction set details. This may be the only section of the *Reference Manual* you will use. This book is very difficult reading. It tells you more than most people want to know.

PRACTICE

Use the instruction set summary to find the opcode for instructions CLRA, NEGB, and INX.

Solution

You should find the following:

Source code	Hex opcode
CLRA	4F
NEGB	50
INX	08

2.5.2 Types of Instructions

Appendix A lists assembly language instructions in alphabetical order. We could also categorize the instructions as follows:

Data handling

Arithmetic

Logic

Data test

Jump and branch

Condition code

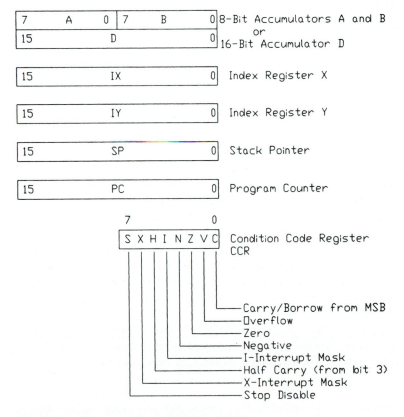

Figure 2.6 Programming Model of the 68HC11

In Section 2.2.3 the program of Listing 2.1 shows examples of data handling and arithmetic instructions. LDAA and STAA are data handling and SUBA is arithmetic. In the fetch/execute example of Section 2.4, the CLRA instruction is arithmetic and LDAA is data handling. Later we look at more examples.

2.5.3 Addressing Modes

When programming in assembly language, one must use the addressing modes of the CPU. Look at the instruction set summary. You will note that each instruction has associated with it one or more addressing modes. An addressing mode specifies how the instruction finds data. The addressing modes available for the 68HC11 are:

Inherent

Immediate

Extended

Direct

Indexed

Relative

We cover these in this section, except for the relative mode (see Section 2.8).

Recall that the operand specifies where and/or how to get or put the data. As we shall see in future source-code examples, the operand specifies which addressing mode is used by the opcode. However, in machine code, the opcode specifies the addressing mode.

Playing Cards Analogy. Consider a deck of cards to be like a machine code. Each card in the deck may be an opcode or an operand. Each suit (hearts, clubs, spades, or diamonds) indicates the addressing mode. The top of the deck is like the program counter. As you remove cards from the top of the deck, the program counter changes. We will occasionally refer back to this analogy when discussing the different addressing modes.

2.5.4 The Prebyte

Before we cover addressing modes, we will look at the prebyte. Most opcodes are a single byte. However, by examining the instruction set summary, you will note that some opcodes consist of 2 bytes. There is a historical reason for this. The earlier 6801 microcontroller (mentioned in Section 1.2.5) has fewer than 256 opcodes; hence all opcodes could be 1 byte long. When Motorola introduced the 68HC11, they added another index register (IY) and more instructions. Note that any instruction dealing with index register Y has 2-byte opcodes. In addition, Motorola wanted the 68HC11 to be downwardly compatible with the earlier processors.

To expand the instruction beyond 256 possibilities, some of the new instructions had to use 2 bytes. A few hex numbers were reserved to mean that any byte following them is also part of the opcode. These reserved numbers are called *prebytes*.[3]

---

[3]You can refer to the opcode maps of the *Programming Reference Guide*. Opcode map 1 has no prebytes. Opcode maps 2, 3, and 4 use prebytes $18, $1A, and $CD, respectively.

Find the opcode for the instruction INY.

Solution

Remember to use the instruction set summary. The opcode is 2 bytes, $18 followed by $08.

2.5.5 Inherent Addressing Mode

Coding and Operation. With inherent addressing, the opcode does not require an operand. In other words, all the information needed to carry out an instruction is contained (is inherent) within the opcode. Most inherent addressing instructions are 1 byte long. Others, with prebytes, are 2 bytes long. The name of the mnemonic opcode itself specifies where the operand data can be found. Inherent addressing always deals with registers instead of memory. Generally, it is used for internal operations such as clearing a register.

The fetch/execute example in Section 2.4 used the instruction `Clear Accumulator A`. It changes the value of the byte in accumulator A to zero. Since this instruction uses the inherent addressing mode, its source code is simply the mnemonic opcode CLRA. The corresponding machine code is 1 byte, which is $4F. Figure 2.7 shows this inherent addressing example.

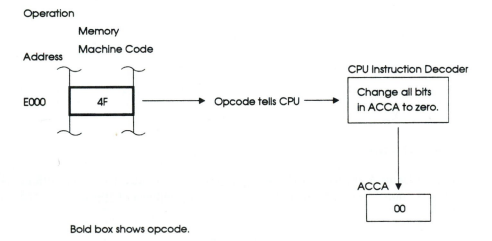

Figure 2.7   Inherent Addressing Example. Clear Accumulator A.

Listing 2.4 shows instructions using only inherent addressing. Like most listings in this chapter, it shows the machine code (on the left) and the source code (on the right).

```
*Listing 2.4
*Demonstrate inherent addressing mode

                    ORG      $E000 ;define start address
E000   4F           CLRA     $E000 ;clears ACCA
E001   5C           INCB     $E000 ;increment ACCB
E002   18 8F        XGDY     $E000 ;swap ACCD with IY
E004   18 09        DEY      $E000 ;decrement IY
```

To illustrate how the program executes, let's assign some initial values to the registers.

ACCA	$6A
ACCB	$80
IY	$1A47

Figure 2.8 shows how the microcontroller modifies the registers as explained by the comments in Listing 2.4. CLRA, INCB, and DEY are examples of arithmetic instructions. XGDY is an example of a data handling instruction. Note that ACCD refers to a combination of ACCA and ACCB.

Execution History.   Program developers may have to examine (trace) how a program executes. Typical development systems show the register values after each instruction is executed. This includes the program counter (PC). It contains the address of the *next* instruction to be executed. The developer can also examine applicable memory locations. We will show program execution in the form of an execution history. This is a modification of the program trace shown by a development system. The execution history of Listing 2.4 follows.

```
PC      ACCA    ACCB    IY       Operations

E000    6A      80      1A47     Initial condition
E001    00      80      1A47     0  → A
E002    00      81      1A47     B + 1  → B
E004    1A      47      0081     Y  → D, D → Y
E006    1A      47      0080     Y − 1  → Y
```

The Hearts Suit (♥).   When you pick a hearts card, simply keep the card. Don't pick up any others until it is your next turn.

2.5.6   Listing and Execution Conventions

In the execution history, note the arrow notation (→) for data transfer operations. This is the same notation as that used in the Instruction Set Summary (Appendix A). The general form is:

$$\text{source} \rightarrow \text{destination}$$

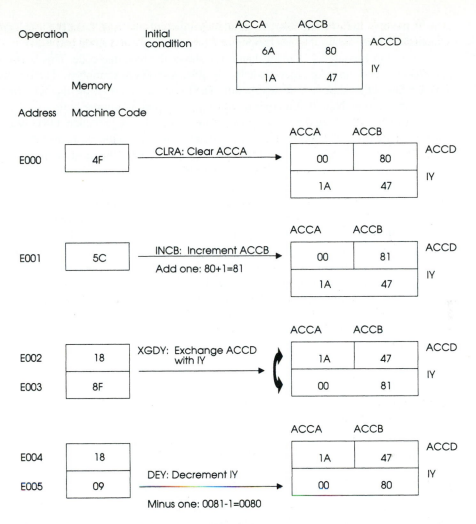

Figure 2.8 Execution of Some Inherent Mode Instructions

The transfer operation moves (or copies) the result in the source operation to the destination. For example, look up INCB in the instruction set summary. Note that it has the Boolean expression:

$$B + 1 \rightarrow B$$

This means that the contents of ACCB is incremented (add one to it), and the result is put back into ACCB. The execution history shows that ACCB originally contained $80. After executing INCB, it contains $81.

    Consider:

$$Y \rightarrow D, D \rightarrow Y$$

This is the operation for the instruction to exchange registers ACCD and IY (XGDY). The execution history shows this has happened. Compare lines (PC) E002 and E004.

Refer to Listing 2.4. The left-hand side shows the machine code. This is also called the *object code listing.* An object code listing always shows hex numbers. The first number on each line is the address of the instruction. The following numbers are the bytes that make up that instruction. Note that there is one line per instruction, not per byte.

Most listings show the source code to the right of the object code. ORG is a special mnemonic that specifies the start address. It means "origin." In this case, the program starts at $E000. We use $E000 for most of our examples because it is the most likely start address for a final application program.[4] However, you can start it at any available address. We prefix comments by the semicolon and hex numbers by the dollar sign.

An execution history also shows hex numbers. The first line shows register conditions before execution of the program. The initial register values shown are chosen arbitrarily. Following this, each line shows how the registers are affected by the execution of the corresponding instruction. Also, after the first line, the execution history shows the PC value *after* the instruction is executed. Recall that the PC increments after a clock cycle (Section 2.4). Also recall that an instruction fetch/execute cycle uses a few clock cycles. Hence once an instruction has completed execution, the PC points to the *next* instruction.

The column labeled "operation" shows the Boolean expression of what the instruction did. It is the same as the Boolean expression shown in the Instruction Set Summary (Appendix A). We repeat it in the execution history to help you understand what each instruction did.

Before execution of the program, the PC is pointing to the first instruction. When it executes CLRA, it clears ACCA to zero and increments the PC to $E001. The PC now points to the next instruction to execute, INCB. This instruction then executes to increment ACCB. Again, it increments PC to point to the following instruction. Execution proceeds similarly for the rest of the example.

PRACTICE

Write the execution history for the following:

```
*Practice
*We don't indicate what the instructions do so that
*you can practice looking up the instructions

            ORG    $E000 ;define start address
E000  5F    CLRB
E001  4A    DECA
E002  17    TBA
```

---

[4]Refer to the memory map of Figure 1.8. ROM starts at address $E000. It is available for both chip operating modes. If you are using a simulator, you can start your program anywhere. If you are using a trainer or development board, you will have to work within the restrictions of this hardware. For example, you would usually start test programs at address $0100 if you are developing an application using the Motorola EVBU board. However, the final application will probably still start at $E000.

The registers before execution are:

```
ACCA    $78
ACCB    $3B
```

Solution

```
PC      ACCA    ACCB    Operations

E000    78      3B      Initial condition
E001    78      00      0 → B
E002    77      00      A − 1 → A
E003    00      00      B → A
```

C Code Example. C Listing 2.4 shows how we can perform equivalent operations using C.

```c
/* C Listing 2.4 */

/* Declarations */
unsigned int Swap, VarD, VarY;
unsigned char VarA, VarB;

void main(void)
{
    /* Main body of code */
    VarA = 0;       /* Clears VarA */
    ++VarB;                 /* Increment VarB */
    /* Swap VarD with VarY */
    Swap = VarY;
    VarY = VarD;
    VarD = Swap;
    --VarY;     /*Decrement VarY */
}
```

A few explanations are in order. Using variables VarA, VarD, and so on for data manipulation corresponds to using the accumulators. Before using variables it is necessary to declare them. This makes the variables known to the compiler so it knows to assign them to memory locations. The declaration of unsigned int means that the variables are 16-bit values (type integer). We declare them as unsigned meaning that the corresponding binary numbers will be handled as unsigned. Section 2.7 will describe handling of unsigned and signed numbers (see also Tables F.5 and F.6 in Appendix F). The compiler will assign variables to memory locations instead of optimizing for direct manipulation by the accumulators.

Most C statements end in a semicolon (;). Every C program requires at least one *function,* named main. A function is a self-contained block of code. Generally, a C program will have several functions (the main function is mandatory). The line void main(void) identifies the start of the function (and is not ended with a semicolon). The word void refers to the situation where we do not require the main function to accept data from or transfer data to other functions. The braces, {and}, determine the beginning and end of the function body and enclose the statements making up the function. C also uses symbolic operators for increment and decrement, which are ++ and −−, respectively. To swap, it is necessary to use a variable, named Swap in this case, to temporarily hold a value during the swapping operation. There is no equivalent C statement to exchange two values.

2.5.7 Stopping a Program

In Listing 2.4 we have not shown which instruction starts at address $E006. The CPU will fetch this byte and execute it according to how it is decoded. It will continue fetching and executing until the CPU is told to stop. There are several ways to stop execution. There is an instruction called STOP, but it has a few complications that we would like to avoid for now. We will use the "magic" instruction BRA * (object code: 20 FE) to prevent the CPU from executing anything following it.[5] Technically, the program does not stop. The CPU continually executes the BRA * instruction. For now, do not worry about how it works. Later we will look at other techniques to stop a program. For C programming, it is only necessary to use the closing brace for the main function, as illustrated in C Listing 2.4 previously.

2.5.8 Immediate Addressing Mode

Coding and Operation. With immediate addressing, there is an opcode and an operand. Recall that the opcode specifies the action and the operand specifies the data that is to be acted upon (what to do and where to do it). The fetch/execute example of Section 2.4 also illustrates immediate addressing using the second instruction Load ACCA with the number $5C.

In source code, this is LDAA #$5C. The opcode LDAA is the instruction Load ACCA, but it does not say what to load. The operand #$5C specifies what to load. A prefix number sign (#) indicates immediate addressing—in source code. In immediate mode the operand *is* the actual data. In other addressing modes the operand tells the CPU *where* to find the data. In machine code the instruction has an opcode byte of $86 and an operand byte of $5C. Refer to the instruction set summary to see that source form LDAA has several different machine-code opcodes, depending on the addressing mode. The byte $86 specifies the opcode for LDAA for immediate (IMM) addressing mode—in machine code. Figure 2.9 shows this immediate addressing example. The term *load* actually means to "make a copy." The original operand byte still remains.

Technical Operation. The *Data Book* gives a brief technical explanation. In immediate addressing, the actual argument (another word for data) is contained in the byte(s) immediately following the opcode byte(s). Consider the instruction LDY #$B704, which is Load IY with the number $B704. Again, the prefix # indicates immediate mode in source code. Look up LDY in the Instruction Set Summary (Appendix A). The machine code for the opcode LDY in immediate mode are the two bytes $18 and $CE. The data is also 2 bytes, in this case $B7 and $04.

Program Example. Listing 2.5 shows instructions using immediate addressing.

---

[5]The earlier editions of this book used BRA $ instead of BRA * since they were based on using the Avocet™ assembler. Motorola uses the asterisk (*) instead of the dollar sign ($). We use '*' since we use the Motorola AS11 assembler throughout most of the book. See also the "Cross Assembly" section in Appendix D.

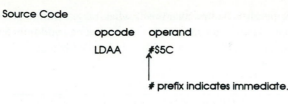

Source Code

opcode   operand

LDAA   #$5C

prefix indicates immediate.

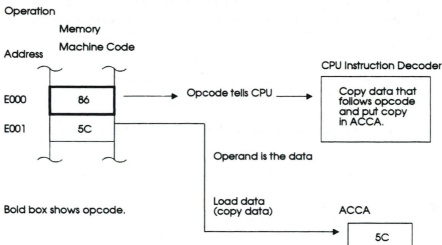

Figure 2.9   Immediate Addressing Example. Load Accumulator A (immediate).

```
*Listing 2.5
*Demonstrate immediate addressing mode
                  ORG      $E000         ;start address
E000    86 5C    LDAA     #$5C          ;load ACCA with $5C
E002    8B 02    ADDA     #$02          ;add $02 to it
E004    C6 17    LDAB     #$17          ;load ACCB with $17
E006    1B       ABA                    ;add it to ACCA,
*                                       ;note it is inherent
E007    CC 12 34 LDD      #$1234        ;$12 → ACCA
*                                       ;$34 → ACCB
E00A    20 FE    BRA      *             ;stop program
```

The execution history follows:

```
PC      ACCA     ACCB     Operation
E000    33       44       Initial condition
E002    5C       44       5C → A
E004    5E       44       A + 2 → A
E006    5E       17       17 → B
E007    75       17       A + B → A
E00A    12       34       1234 → D or 12 → A, 34 → B
```

The arithmetic instruction ADDA #$02 adds 2 to whatever was in ACCA previously and puts it back into accumulator A. Check the corresponding Boolean expression in the

Instruction Set Summary (Appendix A) to see that this is what it does. Similarly, ABA adds the contents of ACCB to A and puts it back into ACCA. Note that hex addition, $5E + $17 = $75, is equivalent to the decimal addition, 94 + 23 = 117. LDD is a double accumulator load instruction. It has a 2-byte operand.

PRACTICE

Write the object code and execution history for the following program. The initial conditions (hex) are:

```
PC          ACCA          ACCB        IX
E000        91            42          1234


*Practice
*No other comments shown, use Instruction Set
*summary.

    ORG       $E000
    LDX       #$3B25
    XGDX
    SUBB      #$12
```

Solution

```
*Object code solution

                          ORG        $E000
E000      CE 3B 25        LDX        #$3B25
E003      8F              XGDX
E004      C0 12           SUBB       #$12
```

The execution history follows:

```
PC          ACCA          ACCB        IX          Operations


E000        91            42          1234        Initial condition
E003        91            42          3B25        M:M + 1 → IX
E004        3B            25          9142        IX → D, D → IX
E006        3B            13          9142        B - M → B
```

Note that for line E003 (instruction LDX has executed), M:M + 1 refers to numbers $3B and $25, respectively. In general, M refers to an operand, which can be any addressing mode.

The Clubs Suit (♣). When you pick up any clubs card, pick up the next card immediately from the deck.

C Code Example. C Listing 2.5 shows how we can perform equivalent operations using C.

```
/* C Listing 2.5 */

/* Declarations */
```

```
#define IMMEDIATE3 0x17
unsigned char VarA,VarB;
unsigned int VarD;

void main(void)

{
        /*Main body of code */
        VarA = 0x5c + 2;
        VarB = IMMEDIATE3;
        VarA = VarA + VarB;
        VarD = 0x1234;
}
```

In this case, we have used the #define statement to assign a value (0x17) to a constant (IMMEDIATE3). A constant is a symbol with a fixed value. Hence VarB is assigned the value of 0x17. Note also the use of the arithmetic operator for addition (+). In particular note the following statement:

```
VarA = VarA + VarB;
```

In this statement, VarA is assigned its previous value added to the value of VarB. Hence VarA is assigned 75 which is 0x5e (its previous value) plus 0x17 (the value of VarB). The program also illustrates immediate addressing. Consider the following statement:

```
VarA = 0x5c + 2;
```

It turns out that the compiler generated the following assembly language statements as equivalents, with comments added by the author.

```
ldab #94      ;decimal 94 = 0x5e = 0x5c + 2
stab _VarA    ;store using addressing mode covered in following
              Section 2.5.9
```

An interesting point is that the compiler has already preprocessed the addition of constant values by adding 0x5c plus 2.

2.5.9 Direct and Extended Addressing Modes

Coding and Operation.   The direct and extended addressing modes are similar. In both cases the operand is an *address*, not the *data* itself. For example, LDAA $05 is a direct addressing mode instruction shown in source code. This is the instruction Load ACCA with the contents of address $05. Note the word *contents*. The operand tells the CPU *where* to find the data. If address $05 happened to contain the byte $FC, then ACCA contains $FC after execution. In 68HC11 assembly language, the term *load* means to "copy." In fact, the instruction LDAA $05 transfers a copy of data stored in address $05 onto the data bus. It loads this data into ACCA. The original data in address $05 remains. This is analogous to

Source Code

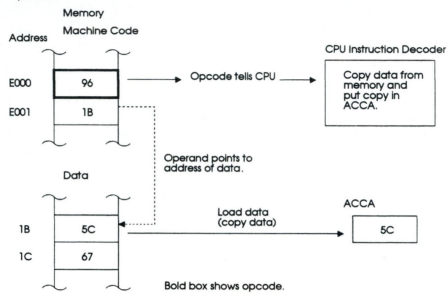

Figure 2.10 Direct Addressing Example. Load Accumulator A (direct).

faxing. The sender still has a copy of the original document after the receiver has a facsimile (copy) of the original.

In source code we indicate direct and extended by omitting the number sign (#) prefix in front of the operand. The difference between the two is that the operand is a 1-byte address (two-digit hex number) for direct mode and in extended mode the operand is a 2-byte address (four-digit hex number). For example, LDAA $0005 is extended mode, although it does the same thing as LDAA $05. Both operands refer to the same address of five. Refer to the Instruction Set Summary and see that the opcodes for LDAA for direct and extended addressing are $96 and $B6, respectively—in machine code. Figures 2.10 and 2.11 show examples of these addressing modes. Figure 2.11 shows the instruction Store ACCA. The term *store* also means to "copy." After STAA $6D00 is executed, address $6D00 has a copy and ACCA still contains the original data.

Addressing Ranges. It is not possible for an extended mode instruction such as LDAA $503E to use direct mode addressing. Because extended mode uses a 2-byte address, you can address anything within the 64-kilobyte addressing range (i.e., anything from $0000 to

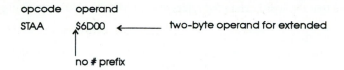

Source Code

opcode   operand
STAA $6D00 ⟵————— two-byte operand for extended

no # prefix

Operation

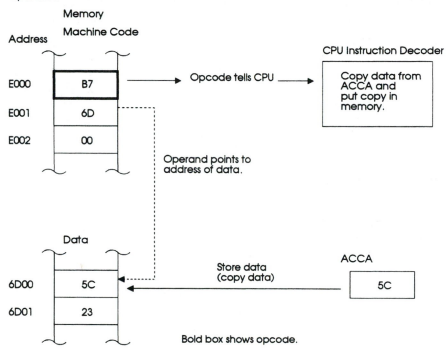

Figure 2.11 Extended Addressing Example. Store Accumulator A (extended).

$FFFF). Recall the memory map shown in Figure 1.8. With direct addressing you can only address 256 possible addresses ranging from $00 to $FF. Often the first 256 bytes are called page 0 because you can address them using a 1-byte address. The most significant byte is assumed to be $00. Some people refer to direct addressing as page 0 addressing. (A page of memory is 256 bytes.)

Why have a direct addressing mode when extended mode can access everything? It is more efficient to read data from page 0 or write data to page 0 using the direct addressing mode. You should note that direct addressing mode always uses fewer bytes than extended. Also, direct mode executes faster. It uses fewer clock cycles to execute (see Appendix A). If you have data that you need to access very often, it is better to store it in page 0 and use direct addressing. In fact, you can tell some C compilers to use direct addressing in some cases. We will also see that you could use direct mode for addressing input/output (I/O) registers for applications that require a lot of I/O.

Program Examples.   Listing 2.6 shows instructions using direct and extended addressing. Note that Figures 2.10 and 2.11 illustrate the first two instructions. Also note that now we can also write data to memory. This is not possible with inherent or immediate mode. Note that the word *expanded* refers to the 68HC11 chip's operating mode.

```
*Listing 2.6
*Demonstrate direct and extended addressing mode
*Program transfers data from addresses $1B,$1C to
*$6D00,$6D01
*Obviously, we must be using expanded multiplexed mode for
*the chip! (addresses $6D00,$6D01 unavailable in single-
*chip mode)

                           ORG    $E000      ;start address
E000  96   1B              LDAA   $1B        ;direct mode
E002  B7   6D   00         STAA   $6D00      ;must use extended
E005  F6   00   1C         LDAB   $001C      ;used extended though
*                                            ;possible to use direct
E008  F7   6D   01         STAB   $6D01      ;must use extended
E00B  20   FE              BRA    *          ;stop program
```

We must introduce another convention. An execution history will use parentheses around a number to mean the contents of an address. For example, "the contents of address $1B is $5C" is written as ($1B) = $5C. Following is the execution history. Recall that all numbers are hex.

```
PC      ACCA   ACCB   (1B)   (1C)   (6D00)   (6D01)   Operations

E000    AA     BB     5C     67     00       23       Initial condition
E002    5C     BB     5C     67     00       23       (1B) → A
E005    5C     BB     5C     67     5C       23       A → (6D00)
E008    5C     67     5C     67     5C       23       (001C) → B
E00B    5C     67     5C     67     5C       67       B → (6D01)
```

We could have performed the same data transfer using a more efficient program. Listing 2.7 shows a program using double accumulator instructions.

```
*Listing 2.7
*Demonstrate direct and extended addressing
*using double accumulator instructions

E000   DC   1B              LDD   $1B        ;direct mode
E002   FD   6D   00         STD   $6D00      ;extended mode
E005   20   FE              BRA   *          ;stop program
```

The execution history follows:

```
PC      ACCA   ACCB   (1B)   (1C)   (6D00)   (6D01)   Operations

E000    AA     BB     5C     67     00       23       Initial condition
E002    5C     67     5C     67     00       23       M → A, M + 1 → B
E005    5C     67     5C     67     5C       67       A → M, B → M + 1
```

Refer to the Instruction Set Summary (Appendix A). Note the columns labeled "Source Form(s)," "Bytes," and "Cycles." Now, count the number of bytes and clock cycles for each instruction in Listing 2.6 and 2.7. Total these numbers. You should find the following. Excluding the BRA * instruction, Listing 2.6 requires 11 bytes of program memory and 15 clock cycles to execute. Listing 2.7 requires 5 bytes of memory and 9 clock cycles to execute. This shows that programming style has an effect on efficiency.

PRACTICE

Write the object code and execution history for the following program. The initial conditions (hex) are:

PC	ACCA	ACCB	(4A)	(4B)	(50)	(51)	Operations
E01C	00	FF	3C	1D	42	B9	Initial condition

```
*Practice
*No other comments shown, use Instruction Set
*Summary.

    ORG     $E01C
    LDAA    $50
    LDAB    $0051
    STD     $4A
```

Solution

```
*Solution, show addressing mode and specific operation

                        ORG     $E01C     ;start address

*                                         ;direct mode

E01C  96  50            LDAA    $50       ;(50) → A
*                                         ;extended mode
E01E  F6  00  51        LDAB    $0051     ;(0051) → B
*                                         ;direct mode
E021  DD  4A            STD     $4A       ;A → (4A), B → (4B)
```

The execution history is:

PC	ACCA	ACCB	(4A)	(4B)	(50)	(51)	Operations
E01C	00	FF	3C	1D	42	B9	Initial condition
E01E	42	FF	3C	1D	42	B9	M → A
E021	42	B9	3C	1D	42	B9	M → B
E023	42	B9	42	B9	42	B9	A → M, B → M + 1

Forcing Extended Addressing. Some assemblers always use direct addressing for any addresses between $00 and $FF (Page 0), unless *forced* to use extended. This is true for the AS11 assembler that we will introduce in Chapter 4. In this case, we can use the greater than (>) symbol to force the use of extended addressing. Consider the following source code:

```
ORG        $E000
LDAB       $0005; assembler uses direct
LDAB       >$0005    ; force use of extended
```

Using the AS11 assembler will generate the following listing with machine and source code.

```
                   ORG     $E000
E000  D6 05        LDAB    $0005    ;assembler uses direct
E002  F6 00 05     LDAB    >$0005   ;force use of extended
```

Accessing Input/Output Ports. Now we can begin to handle data with the outside world. A microcontroller is not much use if it cannot interact with the outside world. In Section 1.3.2 we explained input/output (I/O) ports briefly. Each bit in an 8-bit port register corresponds to an external pin of the I/O port. Two of the ports in the 68HC11 are port B and port C. The addresses of their corresponding registers are $1004 and $1003, respectively. Port B is an output and port C is (normally) an input. The following instructions will send an output and read an input:

```
LDAA       #$A5 ;load data to output
STAA       $1004;send it to Port B
LDAA       $1003;read data from Port C
```

Chapter 8 introduces microcontroller I/O. Refer to Figure 8.2, which shows that pins PB0, PB2, PB5, and PB7 are driven high. They might be turning on lights or other devices. Port C might be connected to switches. The data read by the microcontroller indicates that switches PC0, PC1, PC5, and PC6 are turned on. This is a simple example. Chapters 8 to 12 will show the 68HC11's true power in handling I/O.

The Spades Suit (♠). The card analogy becomes more complicated. Spades can mean direct or extended. When you pick up a spades card, pick up the next card (address card) from the deck. The address card tells you where to pick up yet another card. For example, if the second card (address card) is a 4 of any suit, you pick up the fourth card down from the current top of the deck.

C Code Example. C Listing 2.6 shows equivalent operations using C.

```
/* C Listing 2.6 */

/* Declarations */
unsigned char LoadA, StoreA, LoadB, StoreB;
```

```
void main(void)
{
    /*Main body of code */
    StoreA = LoadA;
    StoreB = LoadB;
}
```

In this case, the compiler assigned the variables to consecutive locations in RAM as addresses $00 to $03 for StoreB, LoadB, StoreA, and LoadA, respectively. The following is part of the assembly code created by the compiler. It uses only accumulator B.

```
ldab_LoadA
stab_StoreA
ldab_LoadB
stab_StoreB
```

2.5.10 Indexed Addressing Mode

Complexity and Application. Each additional addressing mode we cover becomes more complex. We have seen that we can access data from registers (inherent mode), from the program code itself (the operand byte[s] in immediate mode), and from a specified address (operand byte[s] specifying the address in direct/extended mode). Indexed addressing uses one of the two index registers to specify an address. It is a powerful and useful addressing mode.

Why do we need to use an index register? We may wish to transfer or initialize a block of data. For example, we need to copy all the bytes from addresses $00 to $64 to corresponding bytes from addresses $2000 to $2064. In other words, we need to copy 100 bytes (Hexadecimal 64 is decimal 100). We could use something like LDD $00, STD $2000, LDD $02, STD $2002, and so on. However, this seems to be an awkward technique. We shall see later in Section 2.8 that we can repeat a series of instructions as often as we wish. This is called a *program loop*. Since an index register can change its value, we can use it to access different addresses.

Coding and Operation. Index register X or Y is used to *calculate* the *effective* address of the data. Refer to Figure 2.12. The index register *points* to the start address of a block of memory. The offset indicates the location (address) within the block. One source code example is LDAA $56,X, in Figure 2.13. The opcode LDAA is the standard mnemonic for Load ACCA. The operand shows a 1-byte offset, a comma (,), and the letter X. Refer to the Instruction Set Summary (Appendix A). The equivalent machine code is the opcode byte $A6, followed by the offset (the operand byte) $56. When executing, the CPU calculates the effective address by adding the offset to the contents of index register X. Let's say that the contents of index register X is $C500. The CPU will add the offset of $56 to the index register value of $C500 to calculate an effective address of $C556. Then the CPU will copy the data from address $C556 to ACCA. Figures 2.12 and 2.13 illustrate the operation when the contents of address $C556 is $AB.

Similarly, we can use the Y index register. Let's say that the next instruction is STAA $05, Y. The CPU calculates the effective address by adding the offset $05 to the contents of index register Y. If the contents of index register Y is $7B15, the CPU will add offset $05

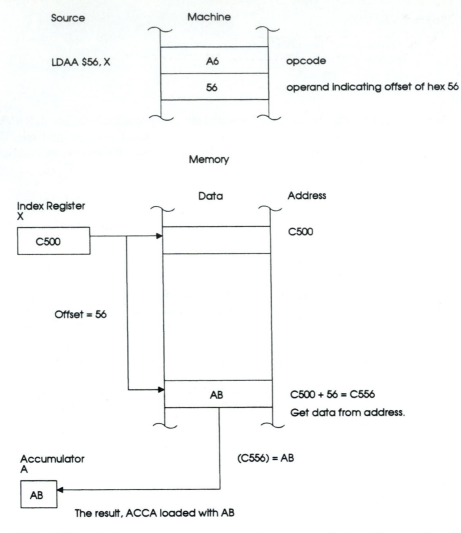

Source   Machine

LDAA $56, X

A6   opcode

56   operand indicating offset of hex 56

Memory

Data   Address

Index Register X

C500

C500

Offset = 56

AB   C500 + 56 = C556

Get data from address.

Accumulator A   (C556) = AB

AB

The result, ACCA loaded with AB

Figure 2.12   Indexed Addressing Mode Instruction. This example shows the execution of LDAA $56,X. Numbers are shown in hex.

to the index register value of $7B15 to calculate an effective address of $7B1A. Then the CPU will copy the contents of ACCA to the effective address of $7B1A. Figure 2.14 shows this operation. Note that ACCA was loaded with $AB from the previous LDAA instruction as shown in Figure 2.13.

Indirect Addressing.   The key point is that the CPU uses the index register to refer to the data indirectly. The index register refers to the address of the data. The CPU is not using the contents of the index register as the actual data. Some other microprocessors and micro-controllers use a similar method called indirect addressing. For example, the Intel 8051 mi-

Source Code

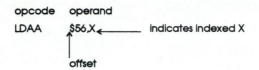

Operation

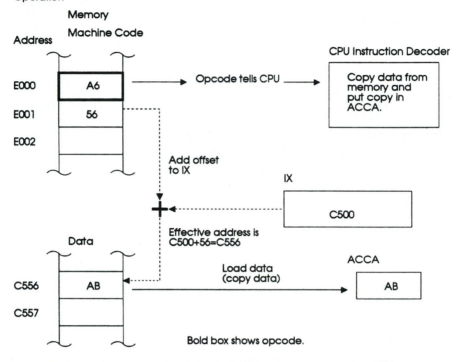

Figure 2.13 Indexed Addressing Example. Load Accumulator A (indexed X).

crocontroller can use the contents of any register to specify the address of an operand. The 8051 microcontroller is described in Section 15.3.

PRACTICE

Write the object code and execution history for the following program. The initial conditions (hex) are:

```
IX = 240E
IY = 2011

PC     ACCA  ACCB  (2020) (2030) (243A) (244A)  Operations
E000   AA    BB    01     02     03     04       Initial condition
```

```
*Practice
*No other comments shown, use Instruction Set
*Summary.

        ORG       $E000
        LDAB      $1F,Y
        STAB      $3C,X
```

Solution

```
*Solution


                              ORG       $E000
E000    18    E6    1F        LDAB      $1F,Y      ;(IY + $1F) → B
E003    E7    3C              STAB      $3C,X      ;B -> (IX + $3C)
```

The LDAB instruction is loading data from an effective address of $2030 ($1F + $2011 = $2030). The STAB instruction is storing data to an effective address of $244A ($3C + $240E = $244A). The execution history is:

```
PC      ACCA   ACCB   (2020)   (2030)   (243A)   (244A)   Operations

E000    AA     BB     01       02       03       04       Initial
                                                          condition
E003    AA     02     01       02       03       04       M → B
E005    AA     02     01       02       03       02       B → M
```

Program Example.   Listing 2.8 shows some instructions using indexed addressing. The operand of an indexed addressing mode instruction is a 1-byte offset. We can use either the X or Y index register (IX or IY). Either index register can be changed by incrementing, decrementing, or adding ACCB to it. In this way we can use the index registers to point to different areas in memory. There are other ways to modify the index registers that we will not mention now.[6]

```
*Listing 2.8
*Demonstrate indexed addressing mode
*for data transfer application

                    ORG     $E000    ;start address

E000    A6    00    LDAA    $00,X    ;load with indexed mode
E002    AB    01    ADDA    $01,X    ;add with indexed mode
E004    18    A7    20  STAA    $20,Y    ;store with indexed mode
```

---

[6]C programmers may recognize the principle of using index registers as being similar to using pointers (x = *ptr;), but indexed addressing is like computing a pointer value and fetching its contents (y = *(ptr + n);).

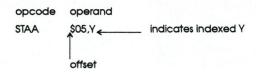

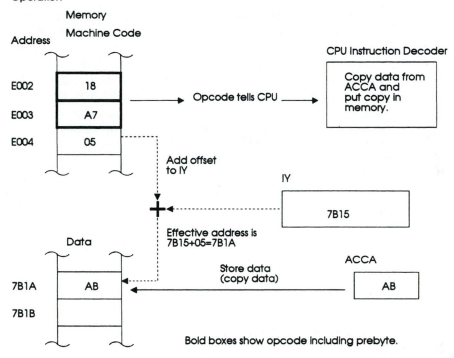

Figure 2.14 Indexed Addressing Example. Store Accumulator A (indexed Y).

```
E007    18   3A        ABY           ;an inherent mode
*                                    ;instruction to modify IY
E009    18   08        INY           ;another one which
*                                    ;increments IY
E00B    18   A7   30   STAA  $30,Y   ;again, store using
*                                    ;indexed mode
E00E    20   FE        BRA   $       ;stop program
```

Its execution history follows. The registers and memory locations have constant hex values
as shown.

```
ACCB = 80, IX = 2000, (2000) = 1A, (2001) = 45

PC        ACCA     IY        (B020)     (B0B1)     Operations
```

E000	32	B000	67	9C	Initial condition
E002	1A	B000	67	9C	$(X + 0) \rightarrow A$
E004	5F	B000	67	9C	$A + (X + 1) \rightarrow A$
E007	5F	B000	5F	9C	$A \rightarrow (Y + 20)$
E009	5F	B080	5F	9C	$B + Y \rightarrow Y$
E00B	5F	B081	5F	9C	$Y + 1 \rightarrow Y$
E00E	5F	B081	5F	5F	$A \rightarrow (Y + 30)$

Remember that the CPU adds the offset to the index register to form an address pointing to data. Consider the instruction ADDA $01,X. Index register X contained $2000 and ACCA contained $1A from the preceding instruction. The index register value of $2000 added to the offset of $01 forms the address $2001. The CPU fetches the contents of $2001 and adds it to what is already in ACCA. Thus it added $45 from address $2001 to the accumulator value of $1A to make up the sum of $5F, which it put back into ACCA.

You may find the indexed addressing mode difficult to understand. If so, review this section and consult the Instruction Set Summary (Appendix A). Also, it is very important to work through some examples using an evaluation board or simulator.

C Code Example.   C Listing 2.8 shows equivalent operations using C.

```
/* C Listing 2.8 */

/* Declare direct variables */
unsigned char VarA, VarB;
/* Declare pointer variables */
unsigned char *PtrX, *PtrY;

void main(void)
{
    /*Main body of code */
    VarA = *PtrX + *(PtrX+1);
    *(PtrY+0x20) = VarA;
    PtrY = PtrY + VarB;
    ++PtrY;
    *(PtrY+0x30) = VarA;
}
```

Section 13.6.7 will explain *pointers* formally. For now note that it is a special type of variable. It stores the address of another variable. To explain pointers we will explain some statements in detail.

```
unsigned char VarA, VarB;
```

This declares variables with `VarA` and `VarB` being assigned to fixed 8-bit addresses, such as $05 and $04 respectively.

```
unsigned char *PtrX, *PtrY;
```

This declares pointer variables `PtrX` and `PtrY`. The asterisk is known as a *dereference* operator and is used to declare the variable as a *pointer*. `PtrX` and `PtrY` are assigned to 16-bit addresses that will be used to contain 16-bit address numbers. The program will eventually access 8-bit data using these addresses, which `PtrX` and `PtrY` point to. For example, `PtrX` and `PtrY` are assigned $0000 and $0002, respectively.

```
VarA = *PtrX + *(PtrX+1);
```

Let's say `PtrX` contains 0x2000. Then `VarA` is assigned the contents of address 0x2000 added to the contents of address 0x2001, like the execution history shown previously. The unary operator * is known as the *dereference* or *indirection* operator. It means to access data by using the pointer to inform where the data can be found. In other words, the address of the data is the contents of the pointer.

```
*(PtrY+0x20) = VarA;
```

Let's say `PtrY` contains 0xb000. The value from `VarA` is stored to address 0xb020 (from 0xb000 + 0x20).

```
PtrY = PtrY + VarB;
```

Let's say `VarB` contains 0x80. Then `PtrY` is assigned the value of 0xb080 (from 0xb000 + 0x80).

```
++PtrY;
```

The value of `PtrY` is incremented, so it is now 0xb081.

```
*(PtrY+0x30) = VarA;
```

The value from `VarA` is stored to address 0xb0b1. The target address was calculated by adding the `PtrY` value of 0xb081 to the offset 0x30.

In this example, the resulting assembly language code generated by the compiler is complicated. This is because it uses memory locations for pointers instead of using the index registers directly. The benefits of this approach occur when larger programs with more data processing are developed.

Look-up Table Applications. Another common application for indexed addressing is a table look-up. Often, there is a need to perform a complex calculation or to look up a value from a nonlinear curve. For example, sensors measuring the flow rate of a fluid commonly provide a signal that is proportional to differential pressure. But the flow rate is proportional to the square root of the differential pressure as shown by the following equation:

$$F = K\sqrt{h} \tag{2.1}$$

where F is the flow rate, K a constant meter factor, and h the differential pressure.

There are two ways to determine square root. One way is to calculate it using one of several numerical techniques. The other way is to store numbers along with their square roots in a table. This table can be stored in memory. To find a square root, you look it up in the table. Let's say that the microcontroller reads a signal from a sensor and stores it in address $30. Also, it has a block of memory starting at address $B600, which contains the square root of all single-byte numbers. Then to find the square root, it looks up the value using the program shown in Listing 2.9.

```
*Listing 2.9
*Demonstrate using indexed addressing for a
*table look-up application

        ORG         $E000
        LDAB        $30             ;get stored differential
*                                   ;pressure signal
        LDX         #$B600          ;point to square root
*                                   ;table
        ABX                         ;look up its square root
        LDAA        $00,X           ;and load it to find flow rate
```

Consider a differential pressure sensor with a range of 0 to 4000 grams per square centimeter (g/cm^2) when the fluid flow in a pipe is between 0 and 200 liters per minute. Let's say that the differential pressure sensor output is 3440 g/cm^2. This is 86% of its full-scale value. Then the 8-bit signal is $DC. Note that $DC represents a decimal fraction of 220/256, or 86%. The LDAA instruction loads the contents of address $B6DC into ACCA. This address should have the value of its square root. For normalized numbers, this value is $ED (decimal fraction 237/256). Hence the flow rate is approximately 93% of full scale. If you are not familiar with scaling, binary fractions, or normalized numbers, refer to Appendix F about these topics. Figure 2.15a illustrates the use of a look-up table.

C Code Example. C Listing 2.9 shows equivalent operations using C.

```
/* C Listing 2.9 */

/* Declare a byte variable */
unsigned char Flow, DiffPress;
/* Declare an array of bytes */
unsigned char Table[0x100];
void main(void)
{
    /*Do look up */
    Flow = Table[DiffPress];
    /* Note that Table[0], Table[1], Table[2] up to
    Table [0xff] would normally be pre-assigned */
}
```

The chief feature of this program is the use of an *array*. Figure 2.15b illustrates an array. An array is a structured collection of elements that can be accessed individually by specifying the position of the element with an index value. For example, array Table has 256 elements. The element indexed (positioned) by 0xd0 has a value of 0xed.

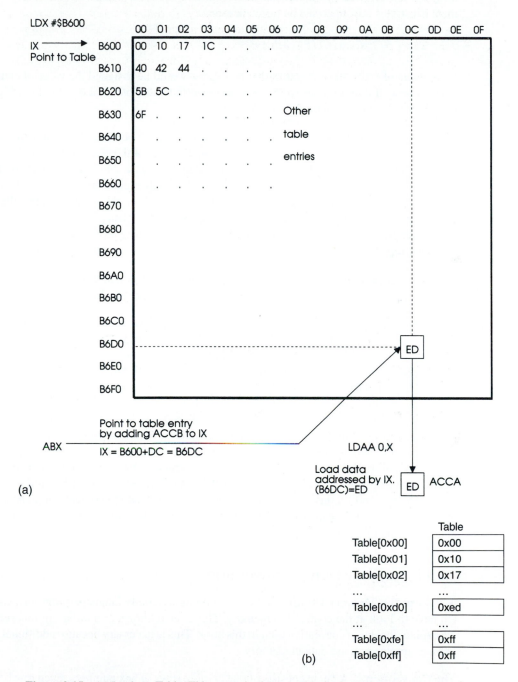

ACCB = DC   Loaded by previous instruction.

LDX #$B600

	00	01	02	03	04	05	06	07	08	09	0A	0B	0C	0D	0E	0F

IX ──→ B600 Point to Table

B600 00 10 17 1C .   .   .

B610 40 42 44 .   .   .   .

B620 5B 5C .   .   .   .   .

B630 6F .   .   .   .   .   .   Other

B640 .   .   .   .   .   .   table

B650 .   .   .   .   .   .   entries

B660 .   .   .   .   .   .

B670

B680

B690

B6A0

B6B0

B6C0

B6D0 ─────────────────────────────── ED

B6E0

B6F0

Point to table entry
by adding ACCB to IX

ABX ──────────
IX = B600+DC = B6DC

LDAA 0,X

Load data
addressed by IX.   ED   ACCA
(B6DC)=ED

(a)

	Table
Table[0x00]	0x00
Table[0x01]	0x10
Table[0x02]	0x17
...	...
Table[0xd0]	0xed
...	...
Table[0xfe]	0xff
Table[0xff]	0xff

(b)

Figure 2.15   (a) Look-up Table. This example shows a table for finding the square root of a fraction (8-bit normalized). (b) The Table Array

59

To declare an array, indicate its data type, unsigned char in this case, followed by the array name, followed by the number of elements in square brackets. The following statement illustrates how to access an array element:

```
Flow = Table[DiffPress];
```

If the value of `DiffPress` was 0×d0, then `Flow` would be assigned the value of element 0xd0 in array Table. In other words, it is assigned the value of `Table[0xd0]`, which is 0xed in this case (see Figure 2.15b).

The Diamonds Suit (♦). Now the playing card analogy becomes much more complicated. When you pick up a diamonds card, pick up the next card (offset card). Add the value of the offset card to the last hearts card you picked up before the diamonds card. If you have no hearts card, use the value zero. The sum is an address that tells you where to pick up yet another card. For example, if the second card (offset card) is a 4 of any suit and the last hearts card was the 3 of hearts, add them to make a sum of 7. Then pick up the seventh card down from the current top of the deck. In this case we are using the hearts suit to represent the index register.

Opcode or Operand? If you have difficulty visualizing how a microcontroller interprets a byte to be an opcode or an operand, read on about shuffling the deck.

Shuffling the Deck. How do you know whether a card is an opcode or an operand? It depends on how you shuffle the deck. If the 7 of clubs (immediate) is on top and the 4 of spades is next, the 7 of clubs is the opcode and the 4 of spades is an operand that is immediately picked up. If the position of the two cards is reversed, the 4 of spades is an opcode (direct or extended) and the 7 of clubs is an operand telling you where to pick up the next card. Machine code is similar to a deck of cards. Consider a sequence of 3 bytes: $86, $96, $86. The program counter points to the address of the first byte, which is $86. The microcontroller fetches it as the opcode for load accumulator A (LDAA) immediate. Hence the microcontroller will load $96 into accumulator A. Now consider the case when the program counter points to the address of the second byte, which is $96. The microcontroller will fetch it as the opcode for load accumulator A (LDAA) direct. Hence $86 (the third byte) becomes the operand. In this case, the microcontroller will load the contents of address $86 into accumulator A. The microcontroller will interpret a byte in memory as an opcode or an operand, depending on where it is located.

2.5.11 Memory Dump Convention

A Visit to the Memory Dump. When developing assembly language programs, we often have to look at the contents of memory. This is also known as a *dump*. In this section we explain the dump convention used in this book. This is necessary because additional program examples will use data in memory.

The Convention. A dump consists of one or more rows of hex numbers. Each row begins with a four-digit number that indicates the start address of the row. The contents (data) of memory are the 16 bytes (two-digit numbers) that follow the address. The address of the first byte is the start address of the row. The address of the second byte is the start address plus one. The address of the third byte is the start address plus two; and so on. The next row (if any) has a start address that is 16 added to the start address of the preceding row. We use this convention to indicate memory contents quickly. It is also similar to the way in which typical professional development systems show memory contents.[7] The EVB and EVBU boards have a command called Memory Display (MD). Its output is a row or block of rows as described. To illustrate this book's memory dump convention, consider the following dump:

```
0020    xx xx xx xx xx xx xx 54 — 68 65 20 44 75 6D 70 2E
0030    0D 0A xx xx xx xx xx xx — xx xx xx xx xx xx xx xx
```

Note that the start address of row 0030 has a least significant digit of zero. The dash (—) allows us easily to distinguish the first 8 data bytes from the last 8 bytes. For typical program examples, the specific value of all data bytes in a row is not important. We indicate the values of these as logic don't care (xx). In this example, the contents of address $0027 is $54, the contents of $0028 is $68, the contents of $0029 is $65, and so on. The contents of address $0031 is $0A. After that, it does not matter what the data is.

ASCII Data.   Typical development systems use a dump that shows memory contents in two portions: hex and ASCII. (Section 1.1.2 introduced the ASCII code.) The ASCII portion would appear beside the hex portion. In this book we show the hex portion only. If the don't care bytes in our previous example were $AA, the ASCII portion would be:

```
.......The Dump.
................
```

A dot (.) shows a nonprintable character or the printable ASCII character for period. Some systems may show equivalent extended characters instead of dots.

2.5.12 Addressing Review

Check your understanding of addressing modes by working on the following practice session.

PRACTICE

The initial memory dump is

```
00C0    AA 55 66 77 89 AD 08 45 — 12 4B D0 78 00 DE 33
```

Show the memory dump after the following program executes. Also state what addressing mode each instruction uses.

```
*Program that uses only data transfer instructions
*to review understanding of addressing modes.
*To check your understanding of assembly language,
*instructions have no comments.
```

---

[7]The DOS Debug utility has a command called Dump that displays the contents of a range of memory addresses as specified by the user.

```
        ORG     $E000

        LDAA    $C0
        STAA    $00C2
        CLRB
        STAB    $C1
        LDD     $C4
        STD     $00C6
        LDX     #$00C8
        LDAA    $00,X
        STAA    $03,X
        BRA     *
```

Solution

```
00C0    AA 00 AA 77 89 AD 89 AD — 12 4B D0 12 00 DE 33
```

*Solution program listing with comments

```
        ORG     $E000

        LDAA    $C0         ;Direct. ACCA <- $AA
        STAA    $00C2       ;Extended. ($00C2) <- $AA
        CLRB                ;Inherent. ACCB = $00
        STAB    $C1         ;Direct. ($00C1) <- $00
        LDD     $C4         ;Direct. ACCD <- $89AD
        STD     $00C6       ;Extended. ($00C6:C7) <- $89AD
        LDX     #$00C8      ;Immediate. IX <- $00C8
        LDAA    $00,X       ;Indexed X. ACCA <- $12, load from ad-
                            dress $00C8
        STAA    $03,X       ;Indexed X. ($00CB) <- $12, store to
                            address $00C8 + $03
        BRA     *           ;Relative. We will cover this later.
                            Good for you if you found
*                           ;this in the Instruction Set Summary.
```

■ 2.6 BASIC OPERATIONS

The instruction set can perform a variety of operations. In Section 2.5 we already covered basic data transfer operations. We survey other operations in the rest of this chapter to familiarize you with what is available.

2.6.1 Data Handling

We can divide data-handling instructions into three subcategories: data movement, alter data, and shift/rotate. Data-handling instructions involve loading, storing, and transfer operations. Examples of each type are LDY, STAA, and TSY. The alter data instructions include DEC, INY, and CLRA. The shift/rotate instructions move bits to the left or to the right. These can be further subdivided into rotate, logical shift, and arithmetic shift to the left or to the right. For example, Listing 2.8 illustrated the use of some common data-handling instructions. The data movement instructions are LDAA and STAA. INY is an instruction to alter data.

Now we will look at some examples of shift/rotate instructions. Before we do, though, we need to consider a bit in the condition code register called the carry bit. Motorola named it the carry bit, but it does not *always* mean an arithmetic carry. It may affect the operation of some instructions. Also, the carry bit (C bit) may be set or reset depending on the result of some instructions. Fortunately, you do not need to memorize which instructions. Once again, look at the Instruction Set Summary (Appendix A). The "Boolean Expression" column will tell you if the C bit affects an instruction. The "Condition Codes" column indicates whether an instruction affects the C bit. Sometimes, the Boolean expression column tells you how the instruction affects the C bit.

Figure 2.16 shows the shift/rotate operations operating on ACCA. They work similarly for ACCB or a byte in memory using the extended or indexed addressing modes. Some can also operate on the double accumulator D, ACCD. These instructions are ASLD, LSLD, and LSRD. When operating on ACCD, the CPU shifts 16 bits instead of 8. The left-hand column shows the condition before an instruction and what it does. The right side shows the result.

Figure 2.16(a) and (b) show the rotate instructions. They preserve all the bits. Hence, if you execute the same rotate instruction nine times in a row, you get the original number back. Rotate instructions are useful for examining bits. We will see later that a program can decide to do something depending on whether the C bit is 1 or 0. For example, the microcontroller reads in sensor information for a home security system. Each sensor sends an on or off signal to indicate whether an intruder has entered. We could have each sensor status represented by a bit. To examine each bit in a byte, the program executes a rotate instruction and examines the C bit. It does this for all bits. Depending on the C-bit status, the program shows that all is well or sends an alarm signal to a monitoring station. Hence we have maintained law and order.

Figure 2.16(c), (d), and (e) show arithmetic shift instructions. Shifting a byte left eight times fills it up with all 0s. Shifting right eight times fills it up with 1s or with 0s, depending on the status of the most significant bit (MSB). They are called arithmetic shifts because they multiply or divide by 2 for every shift. Recall that moving digits left or right in a decimal number causes multiplication or division by 10. Similarly, moving bits left or right in a binary number causes multiplication or division by 2. Look at the hex numbers before and after an operation. Figure 2.16(c) shows that two times $3B is $76, and Figure 2.16(e) shows that $40 divided by 2 is $20. Figure 2.16(d) may be confusing because the arithmetic shift-right instruction assumes signed arithmetic and preserves the MSB. In signed arithmetic the MSB is the sign bit. Figure 2.16(d) shows a negative number being divided by 2. We cover signed arithmetic in Section 2.7. For now, note that many programs use shift instructions when multiplying or dividing by powers of 2. The shift instructions execute faster than the multiply and divide instructions we will cover later.

Figure 2.16(c) and (f) show logical shifts. Each shift fills in 0 bits on the right or the left. Figure 2.16(c) shows an example of an instruction synonym. Look up the instructions ASLA and LSLA in the Instruction Set Summary (Appendix A). Note that they are identical. This means that you can use either in your source code. Logical shifts are useful in situations in which you wish to reset all bits to 0 after examining each one using the C bit. Alternatively, you could use logical shift right as an unsigned divide by 2.

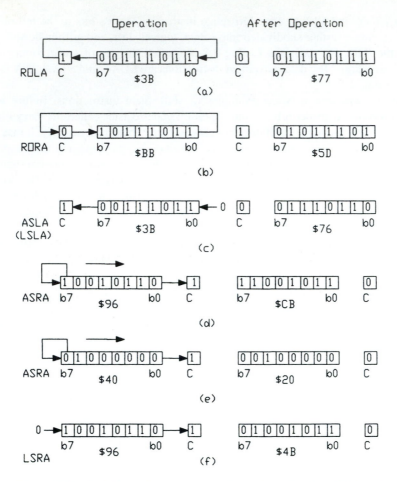

Figure 2.16 Shift/Rotate Instructions: (a) Rotate Left; (b) Rotate Right; (c) Arithmetic Shift Left; (d) Arithmetic Shift Right, −ve; (e) Arithmetic Shift Right, +ve; (f) Logic Shift Right. Note (c) is also a logic shift left.

PRACTICE

Indicate the contents of ACCA and ACCB and the C flag condition after each instruction is executed in the following sequence.

```
*Practice some data handling
    CLC                    ;this clears bit C
    LDD      #$AAC9        ;initial accumulator data
*                         ;from here on, predict the results
    ASRA
    ASRA
    LSLD
    RORB
*           ;whew! end of exercise.
```

Solution

Start from ASRA.

ACCA	ACCB	C
$D5	$C9	0
$EA	$C9	1
$D5	$92	1
$D5	$C9	0

2.6.2 Arithmetic

In this section we cover some of the arithmetic instructions. These will be the add and subtract instructions. In Section 2.7 we cover the rest after providing more information on numbering systems. In this section we consider only unsigned (positive) numbers.

For example, we want to add the following numbers:

$$
\begin{array}{ll}
\$00CC & 204 \\
+\ \$3276 & +12{,}918 \\
\hline
+\ \$3342 & +13{,}122
\end{array}
$$

We have shown the decimal equivalents to help you understand hex addition. The two 16-bit numbers are stored as 4 bytes starting from address $0000, as shown by the following dump:

```
0000    00 CC 32 76 xx xx xx xx - xx xx xx xx xx xx xx xx
```

We also want to store the result in addresses $04 and $05.

The 68HC11 can add or subtract 2 bytes or 2 double bytes. The following program section performs the addition using single-byte additions.

```
LDAA    $01     ;first add least significant bytes
ADDA    $03     ;without carry
STAA    $05     ;and store it
LDAA    $00     ;then add the next significant bytes
ADCA    $02     ;with carry
STAA    $04     ;and store it
```

Adding the first 2 bytes results in:

$$
\begin{array}{ll}
\$CC & +204 \\
+\ \$76 & +118 \qquad \text{(Decimal equivalent)} \\
\hline
\$142 & +322
\end{array}
$$

But when the second instruction executes, ACCA gets the value of $42 since it can only hold 1 byte. This result is stored. We need at least 9 bits to represent $142. The extra ninth bit is a carry. As a result, the ADDA instruction also sets the C bit to 1. Executing the STAA and LDAA instructions does not affect the C bit.

Using the ADCA instruction, the second addition becomes:

$$1$$
$$\$00$$
$$+\$32$$
$$\overline{+\$33}$$

Remember that the previous addition set the C bit, and the STAA and LDAA instructions did not change it. It is always important to consider how an instruction affects any condition code register (CCR) bit. It was important to initially add without carrying since it was possible that the C bit was set by an earlier operation. We wanted to ignore the carry for the least significant bytes.

After storing the second result, we have the number $3342 stored at addresses $04 and $05. You can extend this technique to add numbers represented by more than 2 bytes each or to add more than two numbers. We have shown the technique using single-byte additions to keep the example simpler. The following instructions also perform the same addition using double-byte additions:

```
LDD      $00      ;get first 2-byte number
ADDD     $02      ;add it to second 2-byte number
STD      $04      ;and store it
```

It uses ACCB to add the least significant bytes and ACCA to add the most significant. The ADDD instruction automatically adds ACCB first and adds any carry to the addition of ACCA.

Subtract instructions are similar except that the C bit is used for borrow operations. The following instructions will set the C bit to 1 because a borrow is required.

```
LDAA     #$02
SUBA     #$40
```

This results in the following operation:

$$-\$02$$
$$\overline{-\$40}$$

With a borrow, this becomes:

$$\$102$$
$$-\ \ \$40$$
$$\overline{\$C2}$$

If we had subtracted in reverse, we would have seen the following:

$$-\$40$$
$$-\$02$$
$$\overline{-\$3E}$$

When we look at signed number systems, we will see that $C2 represents negative $3E. We cover signed number systems in Section 2.7.

2.6.3 Logic

We can also perform the Boolean operations AND, OR, and EXCLUSIVE OR for each bit in a byte. Consider the following program section:

```
LDD  $06
ANDA #$A5    ;AND
ORAB #$A5    ;OR
EORB $08     ;EXCLUSIVE OR
```

Consider memory with the following hex data:

```
0000    xx xx xx xx xx xx 9D 9C - A5 xx xx xx xx xx xx xx
```

We show the results of executing the program in binary:

$$
\begin{array}{cc}
 & 10011101 \\
\text{AND} & \underline{10100101} \\
 & 10000101
\end{array}
\qquad
\begin{array}{cc}
 & 10011100 \\
\text{OR} & \underline{10100101} \\
 & 10111101 \;\rightarrow
\end{array}
\qquad
\begin{array}{cc}
 & 10111101 \\
\text{XOR} & \underline{10100101} \\
 & 00011000
\end{array}
$$

PRACTICE

Indicate the contents of ACCA and ACCB and the C-flag condition after each instruction is executed in the following sequence:

```
*Practice some logic
     CLC                 ;this clears bit C
     LDD     #$AAC9      ;initial accumulator data
*                        ;from here on, predict the results
     ANDA    #$C9
     ORAB    #$AA
     LDAA    #$AA
     EORA    #$C9
*                        ;whew! end of exercise.
```

Solution

Start from ANDA.

ACCA	ACCB	C
$88	$C9	0
$88	$EB	0
$AA	$EB	0
$63	$EB	0

Logic instructions are useful in many applications. Consider the security alarm system again. If a bit is 1, its corresponding sensor indicates that an intruder has passed by. To avoid false alarms, we may require that each one has two neighboring sensors in case one is faulty.[8] We AND the 2 bytes and then check if any of the resulting bits are 1 before calling in the cops! We can use EXCLUSIVE OR. If one of two neighboring sensors is faulty, the operation will set a bit to indicate that the two sensors send different signals.

PRACTICE

Write the object code and execution history for the following program. Note that C (by itself) represents the carry bit in the condition code register. The initial conditions (hex) are:

```
PC     C    ACCA    (1C)   Operations

E000   0    92      A3     Initial condition

*Practice
    ORG      $E000
    SEC
    ROR      $1C
    SBCA     $1C
    ANDA     #$5C
```

Solution

```
*Solution
                    ORG    $E000
E000   0D           SEC            ;set carry flag
E001   76 00 1C     ROR    $1C     ;rotate right, also must use
*                                  ;extended addressing mode
E004   92 1C        SBCA   $1C     ;subtract with carry and use
```

---

[8]Some local bylaws require that a security system use two different types of sensors for each alarm point. One may work on detecting a switch opening and the other may use infrared sensing. It helps to avoid problems such as the family cat snuggling up to a sensor.

```
*                           ;direct addressing mode
E006   84 5C      ANDA    #$5C   ;logical AND
```

The execution history is:

```
PC       C      ACCA     (1C)      Operations

E000     0      92       A3        Initial condition
E001     1      92       A3        1 → C
E004     1      92       D1        C → [b7 → b0] → C
E006     1      C0       D1        A − M − C → A
E008     1      40       D1        A · M → A
```

We will say more about how an instruction affects the condition code register bits later. For now, note that SBCA sets the C bit because a borrow was required and ANDA does not change the C bit.

Alarm Application.   Let's investigate a useful microcontroller application. In this section we describe an alarm system using only the instructions covered so far. The system uses port C as the input port (see Figure 8.2). Each pin is connected to an alarm switch. Listing 2.9a counts how many of the alarm sensors are activated.

```
*Listing 2.9a
*Count the number of alarm sensors that are on.

    ORG     $E000      ;start address of program
    CLC                ;make sure that bit C is initially off
    CLRB               ;make sure that ACCB is initially zero
    LDAA    $1003      ;get alarm inputs from port C
*                      ;and load them into ACCA

*Shift bits into ACCA (port C data) eight times and increment ACCB
*each time a high bit is found (by adding with carry).
    LSLA               ;C = PC7
    ADCB    #$00       ;add 1 if C set
    LSLA               ;C = PC6
    ADCB    #$00       ;add 1 if C set
    LSLA               ;C = PC5
    ADCB    #$00       ;add 1 if C set
    LSLA               ;C = PC4
    ADCB    #$00       ;add 1 if C set
    LSLA               ;C = PC3
    ADCB    #$00       ;add 1 if C set
    LSLA               ;C = PC2
    ADCB    #$00       ;add 1 if C set
    LSLA               ;C = PC1
    ADCB    #$00       ;add 1 if C set
```

```
        LSLA                    ;C = PC0
        ADCB        #$00        ;add 1 if C set

        BRA         *           ;stop program
```

Refer to Figure 8.2. Accumulator B will be $04 after the program has run.

PRACTICE

Indicate the contents of ACCA and ACCB and the C flag condition after each instruction in Listing 2.9a is executed using the port C data shown in Figure 8.2.

Solution

Start from LDAA $1003.

ACCA	ACCB	C
$63	$00	0
$C6	$00	0
$C6	$00	0
$8C	$00	1
$8C	$01	0
$18	$01	1
$18	$02	0
$30	$02	0
$30	$02	0
$60	$02	0
$60	$02	0
$C0	$02	0
$C0	$02	0
$80	$02	1
$80	$03	0
$00	$03	1
$00	$04	0

2.6.4   Other Operations

Data test operations include compare, test, and bit test instructions. They do the same as some arithmetic and logic instructions except that they do not change a register or memory location. They only change the appropriate condition code register bits. Examples of these are CMPA, TST, BITA, and BITB.

Branch instructions are another important category. We use them to cause a program to make a choice as to what instructions to execute next. We often use them with data test instructions because they use CCR bits to make the decision. Section 2.8 covers this topic. We include jump instructions in this category since they are similar to branch instructions. Examples of branch instructions are BMI, BEQ, and JMP.

The remaining category is the condition code instructions. Normally, we use the instructions mentioned previously. They change the bits depending on the outcome of its particular operation. Sometimes we may need to specifically set or reset a bit. Then we use one of the condition code instructions. Examples of these are CLC, SEV, and TAP.

2.6.5   Basic Arithmetic and Logic Operations in C

The following program shows basic arithmetic and logic operations using the C language.

```
/* C Listing Section 2.6.5 to illustrate
basic arithmetic and logic operations */

/* Declarations */
unsigned char VarA, VarB;
unsigned int VarD;

void main(void)
{
    /* Note: assignment operator is = */
    /* Arithmetic operators */
    VarA = 0x5a + VarB;        /* Addition */
    VarA = 0x5a - VarB;        /* Subtraction */
    VarD = 0x5a * VarB;        /* Multiplication */
    VarA = VarD / 0x5a;        /* Division */
    VarA = VarD % 0x5a;        /* Modulo Division */
    --VarA;                    /* Decrement */
    ++VarA;                    /* Increment */

    /* Bitwise Logic operators */
    VarA = 0x5a & VarB;        /* AND */
    VarA = 0x5a | VarB;        /* OR */
    VarA = 0x5a ^ VarB;        /* XOR */
    VarA = ~VarA;              /* One's complement */
    VarA = VarA >> 1;          /* Right shift (by 1) */
    VarA = VarA << 2;          /* Left shift (by 2) */ .
    /* Negate, which is same 2's complement to be
    covered in Section 2.7 to follow */
    VarA = -VarB;
}
```

With integers, division (/) results in a quotient. Modulo division (%) only applies for integers. It results in a remainder. For example, 7/2 results in 3 and 7%2 results in 1. With C, you can specify how many positions to shift. For example, 0x04 « 2 results in 0x10 (binary 0000 0100 shifted left by 2 is binary 0001 0000).

The compiler will generate assembly code that uses predetermined *subroutines* to implement operations such as multiply, division, and modulo division. Chapter 3 covers subroutines. Subroutines are reusable blocks of assembly language code. It is possible to use assembly language to implement operations that do not have direct opcode equivalents such as modulo division. Although, as we shall see in the next section, there are opcodes for multiplication and division, and they only apply for specific conditions. In order to create final machine code, it will be necessary to have a compiler *library* to define machine (or assembly) language equivalents of the more complex operations. The compiler handles the inclusion of the library code.

■ 2.7 MICROCONTROLLER ARITHMETIC AND THE CONDITION CODE REGISTER

Like most microcontrollers, the 68HC11 has arithmetic instructions such as add, subtract, multiply, and divide. With a few exceptions, these instructions operate on 8-bit data. By writing extra software, one can extend the bit size of the data. In Chapter 15 we briefly describe a variation of the 68HC11 that includes a math coprocessor for handling 16-bit data.

2.7.1   Two's Complement and the Sign Bit

We normally think of an 8-bit number as having a range of 0 to 255 and a 16-bit number as having the range 0 to 65,535 (see "Binary Ranges" in Appendix F). These are positive integers (although in *formal mathemetics,* 0 is neither positive nor negative). We can also call them unsigned integers. Sometimes, we need to use signed integers, negative and positive. However, computers use only binary digits, 0 or 1. To represent signed integers, we need another method of coding binary numbers. This system is called two's complement. We also use the Motorola convention of prefixing a binary number with the percent sign (%) and no prefix for a decimal number.

We can also represent a signed decimal system using only digits, with no plus or minus sign. This is the 10's-complement system. Figure 2.17 illustrates both systems. Imagine an odometer in your car that will run backward when driving in reverse! Assume that the odometer has a maximum reading of 999 kilometers. When you drive in reverse, it will start at zero, then will count down to 999, 998, 997, and so on. These numbers represent $-1, -2, -3$, and so on, respectively. The odometer has a total of 1000 possible integers. It is a three-digit system. You find the negative of a number by subtracting it from 1000. For example, $1000 - 5 = 995$. Thus 995 represents -5. A negative of a negative is positive. For example, $1000 - 990 = 10$. Note that 990 represents -10. Finding the negative is called finding the 10's complement of the number.

The two's-complement system for binary is similar to the 10's-complement system for decimal. Consider calculating $\$00 - \$01, \$00 - \$02$, and $\$00 - \$03$. You will get the results $\$FF, \$FE$, and $\$FD$, respectively. Figure 2.17(b) shows a circle representing this system for an 8-bit range. Moving counterclockwise represents going more negative. There are 256 possible 8-bit numbers. Thus you can calculate the two's complement (its negative) of a number by subtracting it from binary %100000000 (decimal 256). However, there is an easier way to find a two's complement. If you invert every bit in a binary number, you are finding its one's complement. To find the two's complement, add one to it. For example, find the two's complement of %00000010. The one's complement is %11111101. Add one to it to get %11111110. This number represents -2. The hex number for -2 is then $\$FE$. Alternatively, you can find a two's complement using a hex calculator. Subtract the num-

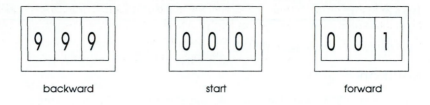

backward   start   forward

(a)   Odometer Readings

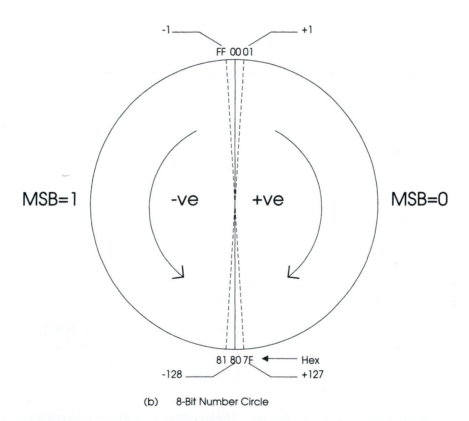

(b)   8-Bit Number Circle

Figure 2.17   10's and 2's Complement Systems: (a) Hypothetical Odometer Illustrates 10's Complement; (b) 8-bit Number Circle Illustrates 2's Complement

ber from $00 (or from $100 if including the borrow).[9] If using decimal, convert the hex number to unsigned decimal and subtract it from 256. Then convert this back to hex.

---

[9]Using a hex calculator may result in an answer of many Fs preceding the last digit, depending on how many digits it uses. For example, a six-digit calculator would display FFFFFF, FFFFFE, and FFFFFD, respectively. The calculator does not know that you are working with only 8 bits. Note that the calculator in Microsoft® Windows® does allow a selection of hex bytes (two hex digits), word (four hex digits), and double word (eight hex digits).

Find the two's complement of $D8.

Solution

The answer is $28. Note that $28 = 40 and $D8 = −40. The following are several possible methods.

  a. Using hex, $00 − $D8 = −$28. You can also convert $28 to 40 to prove that $D8 (or −$28) is −40.
  b. Using decimal, convert $D8 as an unsigned integer to 216. Then subtract, 256 − 216 = 40. Convert 40 to $28.
  c. Using binary, $D8 = %11011000. One's complement is %00100111. Add one to get two's complement of %00101000, which is $28.

Note that any negative number has its most significant bit (MSB) set to 1. The MSB of a binary number is also called the *sign bit.* You can subtract a number by adding its two's complement. For example, $76 − $7D is the same as $76 + $83. Both operations give an answer of $F9.[10] We can also have 16-bit signed numbers. In this case, bit 15, the MSB, is the sign bit. $FF would be a positive number and $FFFF represents −1.

Many 68HC11 instructions treat bytes as signed integers. These instructions will set the N bit (or N flag) of the condition code register (CCR) if the MSB is 1, meaning if the result is negative. The double accumulator instructions assume a 16-bit signed number. The instruction set details in Appendix A of the *Reference Manual* specify whether the sign bit is bit 7 or bit 15. Also, we may refer to CCR bits as flags. If an instruction produces a negative result, it sets the N flag to 1.

2.7.2   Carry, Overflow, Zero, and Half Carry

The carry, overflow, zero, and half carry bits are all in the CCR (see Figure 2.6). The flags H, N, Z, V, and C indicate the result of an arithmetic execution. The others, S, X, and I, are masking bits, which we discuss in later chapters. The Instruction Set Summary (Appendix A) indicates how each instruction uses and/or affects a flag, if any. We discuss next the common usage of these flags.

Unsigned Addition and Subtraction.   Unsigned bytes are those in the range 0 to 255. If the result of adding or subtracting exceeds the range, the CPU sets the carry flag to 1. If the result is within the range, the CPU clears the flag to 0 (Figure 2.18[a]). It is possible to exceed the range because the CPU is working with a fixed number of bits. Usually, the carry flag indicates a carry or borrow when adding or subtracting unsigned 8-bit integers. We can also consider the carry as the ninth bit of the sum when adding two 8-bit numbers. Thus the

---

[10]Again, using a hex calculator for $76 − $7D may result in an answer of many Fs, then a 9, depending on how many digits it uses. For example, a six-digit calculator would display FFFFF9.

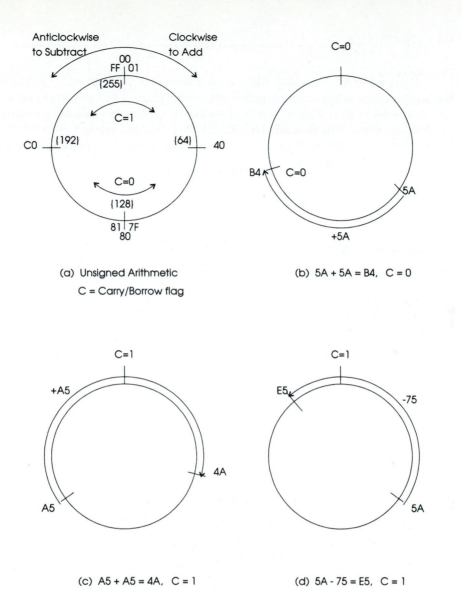

Figure 2.18 Unsigned Add and Subtract for 8-Bit Integers. Numbers in hex. Decimal numbers in braces {}.

result is acceptable for any addition. We have already seen in Section 2.6 how the carry flag (C) is used differently for some other instructions as well. Refer to the Instruction Set Summary (Appendix A) if you are in doubt as to how an operation will use or affect a flag. During addition, the half carry (H) sets if there is a carry from bit 3 to bit 4: in other words, when there is a carry from the low hex digit to the high hex digit. Only a few instructions affect the state of H. Only the DAA instruction is affected by the state of H. Let's look at some examples of addition.

```
        LDAA    #$5A
        ADDA    #$5A
```

The result in ACCA is $B4. Also, this would set H to 1 and clear C to 0. The operation set H because adding the digits in the right column generated a carryover to the left column ($A + $A = $14). Figure 2.18(b) shows that this operation generates a result within the unsigned range. With the carry bit it is possible to add and subtract using 9 bits. Consider $A5 + $A5 − $75 = $14A − $75 = $D5. The following program and RAM data illustrate instructions to do this.

```
        LDAA    $00         ;ACCA = $A5
        ADDA    $01         ;C:ACCA = $14A
        SUBA    $02         ;ACCA = $D5
```

The RAM data for this and the following examples is:

```
0000    A5 A5 75 5A 75 A6 77 77 - 73 65 5F AB D6 AE xx xx
```

In a sense, the program is using a 9-bit register to store the addition result. The 8-bit result in ACCA is $4A and the C flag is the ninth bit. The add instruction also cleared H to 0. Figure 2.18(c) shows that the range is exceeded. Consider subtraction, for example, $5A − $75 requires a borrow.

```
        LDAA    $03         ; ACCA = $5A
        SUBA    $04         ; ACCA = $5A - $75 = $E5 with C = 1
*                           ; because a borrow was required.
```

After borrowing, we have $15A − $75. The 8-bit result is $E5 and C is set to 1 because a borrow was required. The subtract instruction does not modify the H bit. Figure 2.18(d) illustrates the operation.

Zero Sum.   The zero (Z) flag bit sets to 1 whenever the result of an instruction is zero. This can occur if we add a number to its two's complement. For example, add $5A to $A6.

```
        LDAA    $03         ; ACCA = $5A
        ADDA    $05         ; ACCA = $00, C = 1, Z = 1
```

After execution, Z is 1 and C is 1.

Comparison.   We can use the compare instruction to check if two numbers are equal. It subtracts the two and sets Z to 1 if the result is zero. But unlike a subtract instruction, it does not put the difference in any register or memory location. Consider the following example to compare $77 with $77 and then subtract them.

```
        LDAA    $06         ;ACCA = $77
        CMPA    $07         ;Z = 1 because $77 - $77 = $00, but ACCA not modified
        SUBA    $07         ;Z = 1, but ACCA = $77 - $77 = $00
```

Both CMPA and SUBA will set zero (Z) flag to 1, but $77 remains in ACCA after CMPA executes. ACCA becomes $00 after SUBA executes. Look up CMPA and SUBA in the Instruction Set Summary (Appendix A). A comparison instruction checks if the difference between two numbers is zero. To do so, it subtracts the number but does not store the result. The numbers are equal if the Z flag is set.

Signed Addition and Subtraction.   Signed bytes are those in the range −128 ($80) to +127 ($7F).[11] If the result of adding or subtracting exceeds the range, the CPU sets the overflow (V) flag to 1 (Figure 2.19[a]). If the result is within the range, the CPU clears the flag to 0. As mentioned in Section 2.7.1, the CPU also sets the negative (N) flag if the most significant bit (MSB) is set.

An overflow sounds similar to a carry, but they are quite different. When C is 1, it means to add $100 to the result for an addition. For a subtraction, it means that $100 was borrowed. When V is set, the result does not make sense with respect to the sign. Overflow occurs in the following situations:

+ve plus +ve equals −ve

−ve plus −ve equals +ve

+ve minus −ve equals −ve

−ve minus +ve equals +ve

For example, adding the two positive numbers $70 and $60 gives a result of $D0. This result is negative according to the sign bit! Therefore, the overflow flag sets. However, the carry flag does not set. Consider $73 + $65 − $5F. The following program and RAM data illustrate instructions to do this.

```
    LDAA    $08
    ADDA    $09    ;ACCA = $D8, N = 1, V = 1, C = 0, note -ve result
*                  ;with +ve plus +ve
    SUBA    $0A    ;ACCA = $79, N = 0, V = 1, C = 0, note +ve result
*                  ;with -ve minus +ve
```

Again, the RAM data for this and the following examples is:

```
0000    A5 A5 75 5A 75 A6 77 77 − 73 65 5F AB D6 AE xx xx
```

With the first two instructions, adding two positive numbers produced a negative result. Then we subtracted a positive number from this negative result to get a positive result. The equivalent operations are:

8-bit hex operation	Unsigned decimal operation	Signed decimal operation	Signed result obtained
$73 + $65 = $D8	115 + 101 = 216	115 + 101 = 216	$D8 = −40!
$D8 − $5F = $79	216 − 95 = 121	−40 − 95 = −135	$79 = +121!

In both cases the result is outside the signed range −128 to +127. Therefore, the V flag set to indicate overflow. The C flag never set because the results are within the unsigned range of 0 to 255. Figure 2.19(b) illustrates the operations.

---

[11]The C language refers to unsigned 8-bit integers as the unsigned character data type. Signed 8-bit integers are the character data type. Other languages use similar terminology.

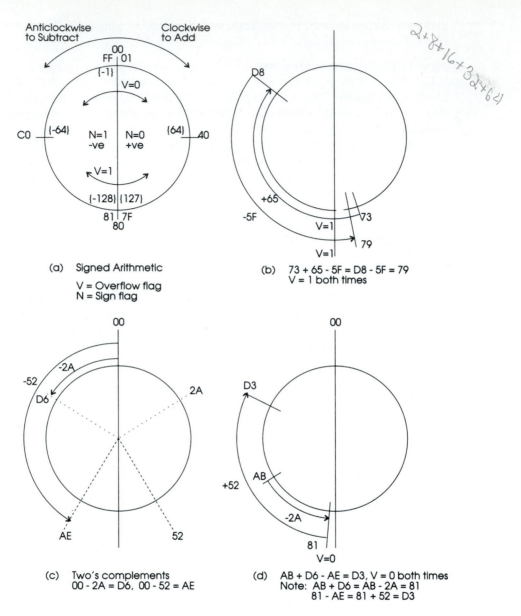

Figure 2.19 Signed Add and Subtract for 8-bit Integers. Numbers in hex. Decimal numbers in braces { }.

Another example shows a situation when V is cleared to 0 and C is set to 1.

```
LDAA    $0B    ;ACCA = $AB
ADDA    $0C    ;ACCA = $81, N = 1, V = 0,C = 1
SUBA    $0D    ;ACCA = $D3, N = 1, V = 0,C = 1
```

The equivalent operations are:

8-bit hex operation	Unsigned decimal operation	Signed decimal operation	Unsigned result obtained
$AB + $D6 = $81	171 + 214 = 385	−85 + (−42) = −127	$81 = 129!
$81 − $AE = $D3	129 − 174 = −45	−127 − (−82) = −45	$D3 = 211!

The unsigned results are outside their range of 0 to 255. For a view of the signed operations, refer to Figure 2.19. Part (c) shows the two's complement $D6 and $AE. Part (d) makes use of the complements to illustrate the arithmetic operations.

There are also cases when both set to 1 and when both set to 0. Appendix A of the *Reference Manual* gives the Boolean equations used by all instructions to set the CCR bits.

Boolean Formulas for CCR Flags. To the assembly language programmer, the CPU sets or clears the CCR flags based on the arithmetic criteria just described. Physically, the CPU executes Boolean logic in hardware. It just follows Boolean orders! Table 2.1 highlights the formulas.

Important! Note the context of the plus sign ($+$). It means plus or add for arithmetic and logic OR for Boolean operations (Table 2.1). For example, look up ADDA and ORAA in the Instruction Set Summary (Appendix A). The "Boolean Expression" column uses the sign for add in ADDA and logic OR in ORAA. For a reminder about Boolean (or digital) logic, see Section 2.6.3 and Appendix F

TABLE 2.1 BOOLEAN FORMULAS OF CCR FLAGS FOR ADDITION AND SUBTRACTION

Add (R = X plus M) — ADCA, ADCB, ADDA, ADDB

$H = X3 \cdot M3 + M3 \cdot \overline{R3} + \overline{R3} \cdot X3$ — Set if carry from bit 3.

$N = R7$ — Set if MSB of result is set.

$Z = \overline{R7} \cdot \overline{R6} \cdot \overline{R5} \cdot \overline{R4} \cdot \overline{R3} \cdot \overline{R2} \cdot \overline{R1} \cdot \overline{R0}$ — Set if result is $00.

$V = X7 \cdot M7 \cdot \overline{R7} + \overline{X7} \cdot \overline{M7} \cdot R7$ — Set if two's-complement overflow results.

$C = X7 \cdot M7 + M7 \cdot \overline{R7} + \overline{R7} \cdot X7$ — Set if carry from MSB.

Subtract (R = X minus M) — SBCA, SBCB, SUBA, SUBB

$N = R7$ — Set if MSB of result is set.

$Z = \overline{R7} \cdot \overline{R6} \cdot \overline{R5} \cdot \overline{R4} \cdot \overline{R3} \cdot \overline{R2} \cdot \overline{R1} \cdot \overline{R0}$ — Set if result is $00.

$V = X7 \cdot \overline{M7} \cdot \overline{R7} + \overline{X7} \cdot M7 \cdot R7$ — Set if two's-complement overflow results.

$C = \overline{X7} \cdot M7 + M7 \cdot R7 + R7 \cdot \overline{X7}$ — Set if |M| > |X| hence borrow required.

Definitions		Flags	
·	Boolean AND	H	Half carry
+	Boolean OR	N	Negative
−	Boolean NOT	Z	Zero
Xi	Bit i (0 to 7) of accumulator (A or B)	V	Overflow
Mi	Bit (0 to 7) of memory (operand) byte	C	Carry
Ri	Bit i (0 to 7) of result		
MSB	Most significant bit (bit 7 for byte operations)		
| |	Absolute value		

You can check the operation of the Boolean formulas for all program examples in this section. Consider X + M = $73 + $65 = $D8 = R. In binary, this is

$$
\begin{array}{rl}
\text{bit number} \rightarrow & 76543210 \\
X & 01110011 \\
+ \quad M & +01100101 \\
\hline
R & 11011000
\end{array}
$$

The Boolean equations are

$$H = X3 \cdot M3 + M3 \cdot \overline{R3} + \overline{R3} \cdot X3 = 0 \cdot 0 + 0 \cdot \overline{1} + \overline{1} \cdot 0 = 0$$
$$N = R7 = 1$$
$$Z = \overline{R7} \cdot \overline{R6} \cdot \overline{R5} \cdot \overline{R4} \cdot \overline{R3} \cdot \overline{R2} \cdot \overline{R1} \cdot \overline{R0} = \overline{1} \cdot \overline{1} \cdot \overline{0} \cdot \overline{1} \cdot \overline{1} \cdot \overline{0} \cdot \overline{0} \cdot \overline{0} = 0$$
$$V = X7 \cdot M7 \cdot \overline{R7} + \overline{X7} \cdot \overline{M7} \cdot R7 = 0 \cdot 0 \cdot \overline{1} + \overline{0} \cdot \overline{0} \cdot 1 = 1$$
$$C = X7 \cdot M7 + M7 \cdot \overline{R7} + \overline{R7} \cdot X7 = 0 \cdot 0 + 0 \cdot \overline{1} + \overline{1} \cdot 0 = 0$$

Appendix A in the *Reference Manual* lists all formulas for all instructions.

9-Bit Results.   The carry bit can act as the MSB of a 9-bit integer result when adding two unsigned bytes. However, the overflow bit cannot act as the MSB of a 9-bit integer result when adding two signed bytes. It is possible to obtain a signed 9-bit result using extra logic. Equation (2.2) shows the Boolean operation that can be used to set a ninth bit to make a valid 9-bit signed integer result for both addition and subtraction. Normally, one would not work with 9-bit numbers.

$$MSB = N \cdot \overline{V} + \overline{V} \cdot V \tag{2.2}$$

What's Your Sign?   Does the CPU care whether data is signed or unsigned? The CPU will modify N and V even if the numbers are unsigned. It will modify C if the numbers are signed. Remember that the CPU's internal logic generates the flags based on Boolean equations (see Table 2.1). It is up to the programmer to use the data as signed or unsigned. This is analogous to buying take-out chicken: You get the salt and pepper packages whether you use them or not. The programmer chooses to heed or ignore the flags.

PRACTICE

Write the execution history for the following program. The initial conditions (hex) are:

```
ACCA    (00)  (01)  (02)    HNZVC    Operations

AC      81    93    3B      11011    Initial condition

*Practice
    ORG       $E000
```

```
        CLRA
        LDAA    $00
        ADCA    $01
        ADCA    $02
```

Solution

ACCA	(00)	(01)	(02)	HNZVC	Operations	Notes
AC	81	93	3B	11011		Initial condition
00	81	93	3B	10100	$0 \rightarrow A$	H not affected
81	81	93	3B	11000	$M \rightarrow A$	H not affected
14	81	93	3B	00011	$A + M + C \rightarrow A$	+ve sum, V = 1
50	81	93	3B	10000	$A + M + C \rightarrow A$	+ve sum, V = 0

C Coding. To perform signed arithmetic in C, it is necessary to declare variables as signed. By default, variable declarations are signed. Hence the following will declare variables for signed arithmetic.

```
signed char VarA;      /* 8-bit signed */
char VarB;             /* 8-bit signed */
signed int VarD;       /* 16-bit signed */
int VarE;              /* 16-bit signed */
```

2.7.3   Binary-Coded-Decimal (BCD) Arithmetic

Binary-coded-decimal (BCD) is a binary number system that uses each group of 4 bits, starting from the left, to represent a decimal digit. For example, the BCD code for 256 is %001001010110. If we took this to be straight binary, we would also get $256 in hex. But $256 is 598 in decimal. Not all possible binary numbers can be used to mean BCD. For example, %00101010 is not a legal BCD number since %1010 represents 10, which is not a single decimal digit. BCD is useful in cases where we use BCD displays. If we send binary data such as %01011001 to such a display, it will show the number 59.

BCD is artificial in the sense that we cannot always do binary arithmetic and expect to get the right result. The 68HC11 has an instruction that will modify the result of an addition to make the result valid BCD. The CPU always adds in binary. It has no way of knowing that you really mean BCD, so you have to be careful. This instruction is decimal adjust ACCA, DAA. We must use ACCA for BCD addition. Also, to ensure proper results, you should use it immediately following one of the following instructions: ABA, ADD, or ADC. This is because it uses the C and H bits of the CCR as well as ACCA to determine a valid BCD result. If you are interested in more details, check DAA in Appendix A of the *Reference Manual.* The following instructions illustrate the use of DAA.

```
        LDAA    #$37
        ADDA    #$25      ;ACCA = $5C
```

```
DAA                 ;adjust, ACCA = $62
ADDA      #$15      ;ACCA = $77
DAA                 ;adjust, ACCA = $77, no change
```

Of course, the numbers to be added have to be valid BCD.

The 68HC11 has no instruction for BCD subtraction. What to do? Situations like these require that we write a sequence of instructions to do a trick. The following instructions subtract address $01 from address $00 in BCD. It uses two tricks.

```
LDAA      #$99      ;find nine's complement
SUBA      $01
SEC                 ;set up for 10's complement
ADCA      $00       ;add to 10's complement
DAA                 ;convert to BCD
```

Trick number 1 uses 10's complement. Subtract a number by adding its 10's complement. Trick number 2 shows how to calculate the 10's complement. Instead of subtracting a number from 100, subtract it from 99 (nine's complement) and add 1 afterward.

Let's say that addresses $00 and $01 contain $55 and $32. BCD subtraction should result in $55 - 32 = 23$. Alternatively, we can add the 10's complement to obtain a result of 55 + 68 = 23 plus carry. Let's check the execution of the instructions. After SUBA executes, ACCA is $67. Note that the 10's complement of 32 is 68, not 67. It is easier to subtract from $99 than $100 using 8-bit accumulators. To add the missing 1, the program sets the C bit to 1 for the next addition. After ADCA executes, ACCA is $BD. After DAA executes, ACCA is $23, the correct BCD result.

2.7.4 Multiplication

We have seen how each arithmetic shift to the left is equivalent to multiplying by 2. The MUL instruction multiplies the 8-bit unsigned binary value in ACCA with that in ACCB to obtain a 16-bit unsigned binary value in ACCD. However, if you are multiplying by powers of 2, you may use the ASL instructions since the MUL instruction requires 10 clock cycles to execute.

Note that the MUL instruction sets the C bit to 1 if bit 7 of ACCB is set to 1. This allows a program to use an ADC instruction to round the 16-bit result in ACCD to an 8-bit result in ACCA. To understand this, you have to realize that we can scale any binary system to represent decimal fractions between 0 and 1. The MSB represents 0.5, the next significant bit represents 0.25, and so on. If you are not familiar with binary fractions, refer to the section "Fractions and Normalizing" in Appendix F.

Consider the following example that is also illustrated in Appendix F:

```
*Multiply: $20 × $35

LDD       $00       ;bytes to be multiplied
MUL                 ;multiply them, ACCD = $06A0
ADCA      #$00      ;8-bit result, ACCA = $07
```

The memory dump is:

```
0000    20 35 7F 70 xx xx xx xx — xx xx xx xx xx xx xx xx
```

Let's analyze the operation using decimal fractions.

$$\$20 \times \$35 = \frac{32}{256} \times \frac{53}{256} = \frac{1696}{65{,}536} = \frac{6.625}{256} \approx \frac{07}{256}$$

Hence the 8-bit fraction of $07 is closer to the 16-bit fraction of $06A0 than the 8-bit fraction of $06. The program segment rounds the product to the nearest 8-bit fractional (or normalized) result. If desired, we can use MUL to multiply two 8-bit integers to get a 16-bit integer. Then the former example means $32 \times 53 = 1696$.

PRACTICE

Using the previous memory dump, find the result after the following instructions are executed.

```
LDD     $02
MUL
ADCA    #$00
```

Solution

After MUL executes $7F × $70, ACCD contains $3790. The carry bit set because the MSB of ACCB is set. After ADCA executes, ACCA contains $38.

2.7.5   Integer Division

The IDIV instruction divides the 16-bit numerator in ACCD by the 16-bit denominator in index register X (IX). After execution, it puts the quotient in IX and puts the remainder in ACCD. It considers all 16-bit numbers to be integers.

Consider the following contents of memory:

```
0000    00 06 00 04 xx xx xx xx — xx xx xx xx xx xx xx xx
```

Then the following instructions will divide 6 by 4.

```
LDD     $00     ;numerator = 6
LDX     $02     ;denominator = 4
IDIV            ;6/4 = 1 plus 2 remainder
```

After IDIV executes, ACCD has $0002 and IX has $0001. This instruction requires 41 clock cycles to execute. You may prefer to use ASR when dividing by powers of 2. The following instructions perform the same division:

```
LDAA    $01       ;load number 6
ASRA              ;divide by 2
ASRA              ;divide by 2 again
```

After execution, ACCA has $01 and the C bit is 1. It represents a fractional component of 1/2. However, IDIV is a more flexible instruction for dividing numbers. But note that IDIV assumes unsigned integers. ASRA assumes signed integers.

In the case of an attempted divide by zero, IDIV sets the C flag to 1 and the quotient (IX) to $FFFF. The numerator (ACCD) is indeterminate.

2.7.6   Fractional Division

The FDIV instruction divides a 16-bit fraction in ACCD by a 16-bit fraction in IX. Recall that *fraction* means that the 16-bit number ranges from 0 to 1. It places the quotient in IX and the remainder in ACCD. For example, $C00/$1000 is $C000. In decimal, this means that 0.046875/0.0625 is 0.75.[12] Often we use it to convert the remainder of an IDIV instruction into a fraction. Listing 2.10 shows the use of both division instructions.

```
*Listing 2.10
*Demonstrate division with fractional result
*Program divides a 2-byte number in address $00
*by another in address $02. It stores the integer result
*in $04 and the fractional result in $06

        ORG     $E000
        LDD     $00       ;load numerator
        LDX     $02       ;load denominator
        IDIV              ;integer division
        STX     $04       ;store integer result
        LDX     $02       ;reload denominator
        FDIV              ;divide previous remainder to get
*                         ;fraction
        STX     $06       ;store fraction
        BRA     *         ;stop program
```

Let's say that we have the following in memory to divide 6 by 4:

```
0000    00 06 00 04 xx xx xx xx - xx xx xx xx xx xx xx xx
```

After IDIV executes, ACCD has the remainder $0002 and IX has the quotient $0001. The program then divides this remainder by 4. After FDIV executes, IX has $8000 and ACCD

---

[12]You can convert a 16-bit hex number into a fraction by first converting it to a decimal integer. Then divide by 65,536. We used this technique in Section 2.7.4.

has $0000. Does this make sense? $8000 is halfway between 0 and $FFFF and 2/4 is 0.5. The FDIV instruction actually calculated $0.0002 divided by $0.0004 to get $0.8, which is 0.5. Thus memory will have the following after execution:

```
0000    00 06 00 04 00 01 80 00 - xx xx xx xx xx xx xx xx
```

This represents the answer of $1.8 (decimal 1.5). We could add other instructions to round the result to a 16-bit integer or to an 8-bit integer without fractions.

FDIV assumes that the denominator is greater than the numerator. Overflow occurs when the denominator is less than or equal to the numerator. This condition sets the V flag to 1. In the case of an attempted divide by zero, FDIV sets the C flag to 1 and the quotient (IX) to $FFFF. The numerator (ACCD) is indeterminate.

PRACTICE

A block of data has the following hex data (bytes):

```
0000    80 03 00 53 00 00 00 00 - xx xx xx xx xx xx xx xx
```

Note that "0000" indicates a start address of $0000. What is in the block of data after the following program is executed? (Divide $8003 by $0053 or 32,771 by 83.)

```
*Practice for division operation

        LDD     $00     ;load numerator
        LDX     $02     ;load denominator
        IDIV            ;integer division
        STX     $04     ;store integer result
        LDX     $02     ;reload denominator
        FDIV            ;divide previous remainder to get fraction
        STX     $06     ;store fraction
        BRA     *       ;stop for now
```

Solution

```
0000    80 03 00 53 01 8A D4 D1 - xx xx xx xx xx xx xx xx
```

The answer is $018A. D4D1, which is 394.831 in decimal (0.831 = 54,481/65,536).

2.7.7   Floating-Point Numbers

High-level languages often use scientific notation to represent decimal numbers and perform arithmetic operations. Assembly languages use floating point when it is necessary. An example of a floating-point number is:

$$\frac{5.345 \times 10^5}{-3.45 \times 10^7} = -1.5493 \times 10^{-2}$$

Floating point is a method of coding binary numbers to represent decimal numbers using scientific notation. Most 8-bit microcontroller applications do not use floating point except when it is necessary to increase the dynamic range. For example, if a microcontroller positions a cutting tool over a range of 1 meter (m) with a precision of 1 micrometer (10^{-6}m), it will be necessary to use 1 million points. This means using 20-bit integers to represent each position. An alternative method is to use floating point to represent data. Also, floating point is necessary in 16- or 32-bit computers performing many numerical calculations, such as trigonometric functions.[13]

Floating-point numbers have three basic parts: the mantissa or fraction, its sign, and the exponent. The exponent is already a signed number. For example, we represent 534,500 as $+0.5345 \times 10^6$. The mantissa is 0.5345. Its sign is positive and the exponent is 6.

There is a standard for representing floating-point numbers. It is called the IEEE 754, *Standard for Binary Floating-Point Arithmetic.* Motorola has an in-house standard for the 68HC11 because the IEEE 754 standard uses too much memory for an 8-bit machine. Figure 2.20 shows some formats of floating-point numbers.

Motorola 68HC11 Floating-Point Format.   (Skip this section if you are not interested in the format details.) Numbers are stored as a sign byte with three consecutive bytes as the mantissa and a byte as an exponent. A sign byte of $00 indicates a positive number and $FF indicates a negative number. The exponent indicates the position of the binary point as a power of 2. It uses an offset scale where $80 (halfway) is zero. Numbers $81 to $FF represent exponents of 1 to +127. Numbers $7F down to $00 represent exponents of −1 to −127. The mantissa always has a most significant bit of 1. It is a normalized 24-bit number (fraction). The "Fractions and Normalizing" section in Appendix F explains normalized 8- and 16-bit numbers. Simply extend the concept to 24 bits.

Consider the floating-point number (hex) 82 C90FDB 00. The exponent byte is $82, which represents 2^{+2}. The mantissa bytes are $C90FDB. It represents the number 13,176,795 divided by 2^{24}. The sign byte of $00 indicates a positive number. Hence the equivalent decimal number is:

$$\frac{13,176,795}{2^{24}} \times 2^2 = 0.785398175 \times 4 = +3.1415927$$

It happens to be the value of pi (π). Similarly, the hex number 85 A1C28F 00 represents:

$$\frac{10,601,103}{2^{24}} \times 2^5 = 0.631874985423 \times 32 \approx +20.22$$

Compare this with integer representation. The 24-bit mantissa means an accuracy of 1 part per 2^{24}. This is approximately 5.98×10^{-8}. The dynamic range is also larger. The range is $2^{\pm127}$, which is about $1.7 \times 10^{\pm38}$. A 5-byte integer (40 bits) would have an accu-

---

[13]PC users may be familiar with coprocessor chips. They are specially designed to perform floating-point calculations.

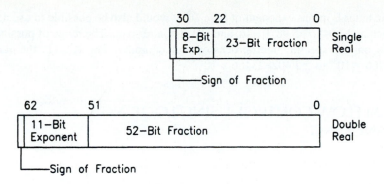

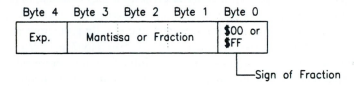

Format for the 68HC11 Floating–Point Package

Figure 2.20 Floating-Point Numbers

racy of 1 part per 2^{40}, which is approximately 10^{12}. However, the range is also approximately 0 to 10^{12}.

C Code Example. The following illustrates use of floating point in C.

```
/* C Listing Section 2.7.7 to illustrate
floating-point numbers */

/* Declarations */
float Numerator, Denominator, Result;

void main(void)
{
    Numerator = 5.345E+05;
    Denominator = -3.45E+07;
    Result = Numerator / Denominator; /* Result of -
    1.5493E-02 */
}
```

Most C compilers use the standard IEEE format for floating point and will use a library for floating-point operations. Note the format for coding a floating-point number in scientific notation. The letter E means exponent of base 10. It would also be possible to use decimal notations such as 17.045, −5.98, 0.346, and −0.405, and so on. The range of possible values depends on the specific microprocessor and compiler. For ICC11, the range is $\pm 3.402823466 \times 10^{38}$ to $\pm 3.402823466 \times 10^{-38}$.

■ 2.8 PROGRAM FLOW CONTROL USING LOOPING AND BRANCHING

2.8.1 Flow Control

Programs are logical. A microcontroller maintaining the temperature in your home would measure the temperature and then take an action. If the temperature is above the thermostat setting, turn off the furnace and turn on the air conditioner. If the temperature is below the setting, turn on the furnace and turn off the air conditioner. Another microcontroller in an assembly plant may command a spray gun to paint car parts. There are two possibilities. It may paint a predetermined number of parts; or it may continue to paint parts until there are no more parts to paint.

These are situations that require the program controlling the process to make decisions or repeat identical actions. Both are forms of program *flow control*. The temperature control system has to perform an *if-else* action. If the temperature is high, then do this; else do that. The paint application illustrates a *loop*. Repeat the action a certain number of times or repeat it until a certain condition is met, in this case, when there are no more parts.

Figure 2.21 shows flowcharts for both basic types. Flowcharts are a graphical technique to show program flow. A rectangle represents sequential processing operations, such as arithmetic, logic, or data movement. It has one entry and one exit. A diamond represents decision-making logic. There are two possible exit paths.

In high-level languages, there are variations of the structures shown in Figure 2.21. But the structures shown are easy to program using assembly language. For the temperature system, we can have

> If temperature greater than thermostat setting
>> then turn off heat (condition true)
>> else turn on heat (condition false)

For the paint application, we can have

> While parts are coming down the assembly line
>> paint them (condition true)
>> until there are no more parts (condition false)

If we wanted to paint only 100 parts, we could use the same structure:

> While the parts count is less than 101
>> paint them (condition true)
>> until 100 parts have passed (condition false)

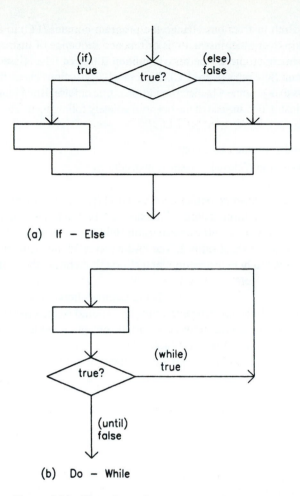

(a) If – Else

(b) Do – While

Figure 2.21 Flow Control

Actually, most programmers would use a different structure, called the *For.*

> For the first to the 100th part
>   paint them

Programmers using high-level languages will be familiar with all the flow control structures. In assembly language you can code all of them, but we concentrate on those shown in Figure 2.21. Both of these require using conditional branch instructions.

2.8.2 Conditional Branches

A program normally executes each instruction in sequential order, one after the other. It uses a branch or jump instruction to redirect the program flow (other assembly languages use different terminology). Both instructions change the program counter (PC) to a value that permits a program to skip over other instructions or repeat a sequence of instructions.

Conditional branch instructions check whether a condition is true or false. Based on its assessment, it redirects program flow (change the PC). The branch instructions check the condition code register (CCR) bits to determine whether a condition is true or false. Branching refers to the act of going to another instruction instead of the one immediately following it. The branch instruction does the following:

If true, branch there and continue execution
> else continue execution (following branch instruction)

To illustrate how a branch instruction works, look up BEQ (rel) in the Instruction Set Summary (Appendix A). The "Operation" column indicates the instruction's function. In this case, branch if a result is zero. The "Boolean Expression" tells us what the CCR bits are if the condition is true: in this case if the Z bit is 1. The "Addressing Mode" is relative and we will talk about it shortly. It is a 2-byte instruction, and although it checks the CCR, execution does not change it.

The relative address is a 1-byte signed number that determines how the PC should be updated if the condition is true. It indicates where the program should branch to. There are two views of the relative address, the source-code view and the object-code view. Since the source-code view is straightforward, we consider it first.

Following is program Listing 2.11. Ignore the object code for now.

```
*Listing 2.11
*Demonstrate branch instructions
*BEQ - Check if Z = 0
*If ($00) is zero then skip next two instructions
*      else load ($01) and store it in ($00)
*BRA - It's always true!
*If true then execute itself again.

                        ORG    $E000  ;start address
E000  96  00            LDAA   $00
E002  27  04            BEQ    THERE  ;if ACCA == 0 branch to THERE
E004  D6  01            LDAB   $01    ;($01) → →($00)
E006  D7  00            STAB   $00    ;when ACCA not 0
E008  97  01  THERE     STAA   $01    ;($00) → → ($01)
E00A  20  FE  HERE      BRA    HERE   ;always branch to HERE
```

Note that the source code uses two labels, THERE and HERE. When used on the left-hand side, they point to an instruction. When used with a branch instruction, they indicate where execution continues if the condition is true. Some assemblers require a colon after the label when it is used to indicate a location (on the left-hand side). If location $00 contains zero, execution of LDAA sets the Z bit to 1. Then when BEQ executes, it finds that Z is 1 and it redirects the program to continue from location THERE. Thus if the contents of address $00 is zero, the program puts zero in address $01. If not, it exchanges the contents of addresses $00 and $01. We have also shown another way to indicate BRA *. The use of the label HERE shows its true function, branch always to itself. The two equal signs ($==$) mean to *check* for equality (see *Relational and Bitwise Operators* in Appendix D).

When you use conditional branches, it is very important to ensure that you do not corrupt the CCR before executing the branch. Ensure that you actually test the condition. The test and compare instructions are the usual ways to do this. For example:

```
LDAA    $00     ;does not affect C bit
CMPA    #$00    ;does affect C bit
BHS     THERE   ;checks C bit
```

Some situations do not require a test or compare operation. But if in doubt, use it. So far we have seen how a branch instruction uses the CCR. However, we do not normally think in terms of CCR bits. The mnemonics tell us what arithmetic or logic conditions it checks. Motorola uses the words *higher* and *lower* to refer to unsigned numbers and the words *greater* and *less* to refer to signed numbers. Be careful.

Some instructions have synonyms (the same opcode). BCC and BHS have the opcode $24 and both check if C is clear (zero). You, as a programmer, may prefer to use one of these in particular situations because the mnemonic makes the meaning clearer. It so happens that to determine if a number is higher than or equal to another, you must compare them and check if C is clear. Any branch instruction that checks the V bit has meaning for checking signed arithmetic operations only. BEQ and BNE check only the Z bit and can be used for both signed and unsigned checks. Also, you can use BMI and BPL for both since they check only the MSB (flag N). However, the mnemonics will be misleading if the data is to be interpreted as unsigned.

If-Else.   Listing 2.12 illustrates the if-else structure. Again, ignore the object code for now. Also note that it does not show the instructions to perform the comparison, that is, what affects the C bit. These instructions should execute before those in Listing 2.12.

```
*Listing 2.12
*Demonstrate if-else
*if carry set then write 'T' to $00
*       else write 'F' to $00

*

*in this case, we define TRUE to be condition when carry
*is set and we define FALSE to be when carry is not set

                        ORG   $E000
E000   25  04           BCS   TRUE    ;true if C set
E002   C6  46   FALSE   LDAB  'F'     ;else false if C clear
E004   20  02           BRA   SKIP    ;and skip next
*                                     ;instruction
E006   C6  54   TRUE    LDAB  'T'     ;executed if true (C = 1)
E008   D7  00   SKIP    STAB  $00     ;store appropriate result
E00A   20  FE   HERE    BRA   HERE    ;stop program
```

Listing 2.12 illustrates another feature of assembly language. We can use ASCII characters as byte operands if we enclose the character in single quotes. The assembler translates the character to its ASCII code in hex. Thus 'T' (T for true) becomes $54 and 'F' (F for false) becomes $46.

C Code Example. C Listing 2.12 shows how we can perform equivalent operations using C.

```
/* C Listing 2.12 */

#define TRUE 1
#define FALSE 0

/* Declarations */
char Carry, VarB;

void main(void)
{
    if (Carry == 1)
            VarB = FALSE;
    else
            VarB = TRUE;
}
```

In this program, if the variable `Carry` has a value of 1, `VarB` is assigned with `FALSE`, otherwise when `Carry` has any value other than 1, `VarB` is assigned with `TRUE`. The statement if (Carry == 1) checks whether `Carry` is equal to 1. The equal operator == is used to check for an equality relationship. There are other relationship operators such as greater than (>) which are listed in Appendix D. The if statement can be of several forms. One of these is the if-then-else form. The following shows the general syntax:

```
if (relationship expression)
    Statement1
else
    Statement2
```

When the relationship expression is true, Statement1 is executed, otherwise Statement2 is executed.

PRACTICE

Write a program that sets the carry flag (C) if the N flag is clear, else it clears the C flag.

Solution

```
*Solution 1
        ORG     $E000   ;start address
        BPL     N_ZERO  ;if N clear, (if N not clear)
N_ONE   CLC             ;else clear C, (then clear C)
        BRA     HERE
```

```
N_ZERO    SEC                ;then set C, (else set C)
HERE      BRA    HERE        ;stop for now

*Solution 2
*ACCA = number to be tested

          ORG    $E000
          SEC                ;assume N clear and set C
          TSTA               ;check if assumption true
          BPL    SKIP        ;if N clear then skip
          CLC                ; else clear C, assumption false
SKIP                         ;program continues
HERE      BRA    HERE        ;stop for now
```

Solution 1 illustrates using an inverted definition for true and false as shown in parentheses ( ). It also uses labels that illustrate the condition causing an instruction (or sequence of instructions) to execute. Solution 2 illustrates an "assume first, then check assumption" technique.

Sometimes an if-else performs an action only if the condition is true. For example:

```
     BPL       N_ZERO    ;if N not clear
     CLC                 ;then clear C
N_ZERO                   ;else continue with rest of program
```

Note that we can define a true condition as the opposite of what causes a branch to occur.

Do-While.   Listing 2.13 shows a *do-while* loop. Program loops perform repeated actions. A loop may require a variable to keep track of the number of times it executes (loops). We also introduce the terms *source* and *destination*. These are self-explanatory. Often, the program requires pointers to access data. The program illustrates all of these concepts.

```
*Listing 2.13
*Demonstrate a do-while (DOWH) loop
*
*Move 16 bytes starting at address $C000 to a block
*at address $C500. Use the following strategy.
*    do a transfer of a byte while counter is not zero
*
*We also introduce some common entities, counters and
*pointers. In this case, ACCB is a counter
*IX is a source (src) pointer and IY is a destination (dst)
*pointer. When we say "move," it really means "copy"!
*----------------------------

                              ORG    $E000
E000   CE  C0  00            LDX    #$C000   ;initialize src pointer
E003   18  CE  C5  00        LDY    #$C500   ;initialize dst pointer
E007   C6  10                LDAB   #$10     ;initialize counter
```

```
E009   A6   00          DOWH   LDAA   $00,X    ;move byte from src
E00B   18   A7   00            STAA   $00,Y    ;to dst
E00E   08                      INX             ;increment src pointer
E00F   18   08                 INY             ;increment dst pointer
E011   5A                      DECB            ;decrement count, to
*                                              ;indicate another move
E012   26   F5                 BNE    DOWH     ;Do-while count not zero
E014   20   FE          HERE   BRA    HERE     ;stop program when done
```

Some programmers will point out that this is really a *for* loop.

> for count = 16 to 1, decrement count by one
>> and move byte
>> and increment pointers by one

For consistency, we shall stick to our terminology. By minimizing the possible number of structures, we minimize programming bugs when we design programs in the future.

C Code Examples.   C Listing 2.13a shows one way to perform equivalent operations in C.

```
/* C Listing 2.13a */

/* Declarations */
char Count, Mem[0×510];

void main(void)
{
     Count = 0×10;
     while (Count > 0)
{
          Mem[Count+0×500] = Mem[Count];
          Count−;
}                       /* end of while loop */
}
```

The program shows the use of the C while statement. It is a type of loop statement structure, which causes repeated execution of statements within the loop while the condition is true. The general form is:

> while (Conditional Expression)
>> Statement

While the conditional expression is true (while Count is greater than 0), the statement is executed. When the condition becomes false, program execution continues beyond the while loop. In this case the statement within the loop is a *compound* statement because it consists of several statements enclosed within braces.

C Listing 2.13.b illustrates another type of C loop known as the for statement.

```
/* C Listing 2.13b */

/* Declarations */
char Count, Mem[0x510];

void main(void)
{
    for (Count = 0×10; Count >= 0×01; Count−)
            Mem[Count+0×500] = Mem[Count];
    /* end of for loop */
}
```

The C for statement is a type of loop statement structure, which causes repeated execution of statements within the loop for a predetermined number of times in a specified way. The general form is:

for (Initialization Statement; Conditional Expression; Control Expression)
        Statement

The statement is repeatedly executed a number of times as specified by the expressions within the parentheses. The initialization statement assigns the initial value to the loop counter variable. The conditional expression specifies the condition for which the loop will repeat. The control expression specifies how the counter will be modified for each time the statement is repeated. In this case, Count is initialized to 0×10 and decrements during each iteration (repetition) of loop execution as 0×10, 0×0f, 0×0e, and so on until 0×01.

PRACTICE

Write a program such that ACCB specifies the number of times to shift ACCA left.

Solution

```
*Solution

        ORG     $E000
        CMPB    #$00    ;if ACCB is zero
        BEQ     DONE    ;then don't enter loop
*Loop begins here
DOWH    LSLA            ;do a shift
        DECB            ;decrement loop counter
        BNE     DOWH    ;do-while counter not zero
DONE    BRA     DONE    ;stop for now
```

2.8.3 Relative Addressing

When you go to Papa Joe's Variety Store, you turn left at Division Street. Your friend coming from the opposite direction has to turn right at Division Street. Left and right are relative directions. They indicate direction with respect to the person following them.

Branch instructions use the relative addressing mode. A relative address indicates how many addresses the branch should take with respect to itself. Since a branch can be forward or backward, the relative address is a signed number. Forward is positive and backward is negative. Also, the relative address is a 1-byte number. Hence its range is 127 addresses forward and 128 addresses backward (or +127 to −128; refer to Figure 2.17[b]).

Any assembler will calculate the relative address for you if the branch instruction and destination are within range of one another. If they are out of range, the assembler will give you an error message. For example, the instruction BRA $E200 at address $E010 is not possible since there are more than 127 addresses between $E200 and $E012. Why $E012? The relative address is actually *relative to the address after the branch instruction.* Although an assembler will calculate relative addresses for you, we will take a look at the object-code view of relative addressing in order to understand it better.

Sometimes the word *offset* is used for relative address. If the branch condition is true, this signed 8-bit number is added to the contents of the PC to form a new address. Otherwise, control proceeds to the following instruction. But remember that when an instruction executes, the PC points to the next instruction before executing it. Most branch instructions are 2 bytes long. Some special ones are 4 or 5 bytes long. The equation for calculating the destination address when the branch condition is true is:

$$PC_{new} = PC_{old} + T + rr \qquad (2.3)$$

where PC_{old} is the address of branch instruction, PC_{new} is the destination address, and rr is the relative address. T is 2 for most instructions, 4 for BRCLR,BRSET for Dir,IX, and 5 for BRCLR,BRSET for IY.

You can manually calculate the relative address byte by:

$$rr = PC_{new} - PC_{old} - T \qquad (2.4)$$

Now check the object code of Listings 2.11, 2.12, and 2.13. The instruction HERE BRA HERE has an object code of $20 $FE. The relative address $FE is −2 in decimal. Since the instruction itself is 2 bytes long, it has to branch back two addresses in order to repeat itself. In Listing 2.11, consider BEQ THERE. It has a relative address of $04. Since BEQ, a 2-byte instruction, is at address $E002, we use equation (2.3) as follows:

$$PC_{new} = \$E002 + 2 + \$04$$

Remember, we are calculating a 16-bit number, hence we convert all numbers to 16-bit and *maintain the sign bit.*

$$PC_{new} = \$E002 + \$0002 + \$0004 = \$E008$$

Indeed, this is the destination address. Now for the next challenge. Calculate a relative address given a desired destination and branch instruction. Consider BNE in Listing 2.13, a backward branch.

$$rr = \$E009 - \$E012 - 2 = \$FFF5$$

We must convert it to an 8-bit number. Hence the relative address byte is $F5.

Be aware that the relative address is signed and that you must extend the sign when adding or subtracting to 16-bit numbers. The absolute 16-bit addresses are unsigned. Consider BRA in Listing 2.12.

$$PC_{new} = \$E00A + 2 + \$FE$$
$$PC_{new} = \$E00A + \$0002 + \$FFFE = E00A$$

Manual assembly with branch instructions is a (tricky) two-step process. Write down the source code using labels for branch instructions and assign a start address. To translate to machine language, write down all the hex opcodes and operands for each instruction and keep track of the addresses of each instruction. When you encounter a branch instruction, leave a space for the relative address byte (or fill in a $00) since you may not know the address of the destination at this time. Once this is done, examine the destination addresses and calculate the corresponding relative addresses and fill them in. Fortunately, we use assemblers most of the time. The assembler performs a similar process to calculate the relative addresses.

PRACTICE

Pretend that you are an assembler. Find the relative addresses for the following section of code.

```
*Practice, this is a section of arbitrary code
          BEQ     THERE
BACK      LDAA    $5C
          INX
          DEY
          BNE     BACK
THERE     LDAA    $5B
```

Solution

The relative addresses are:

```
THERE = $07, BACK = $F9
```

Let's see how we obtain this solution. It does not matter what the absolute start address is since we only need to find relative addresses. We also need to check only how many bytes each instruction requires in machine code. Then we have the following:

```
*Solution, illustrate number of bytes per instruction
*and their start addresses.
```

```
*Assume start address of zero
        BEQ     THERE       ;address $0 then 2 bytes
BACK    LDAA    $5C         ;address $2 then 2 bytes
        INX                 ;address $4 then 1 byte
        DEY                 ;address $5 then 2 bytes
        BNE     BACK        ;address $7 then 2 bytes
THERE   LDAA    $5B         ;address $9 then 2 bytes
```

Now calculate the relative addresses.

```
THERE rr = $09 − $00 − $02 = $07
BACK  rr = $02 − $07 − $02 = $F9
```

2.8.4 Secondary Memory Reference Instructions

Many I/O operations rely on whether certain bits are set or cleared. The secondary memory reference instructions are branch instructions that make a decision to branch based on the bit values. These instructions are Branch if Bit(s) Set (BRSET) and Branch if Bit(s) Clear (BRCLR). A special byte in the instruction called a *mask* specifies which bits to check. These instructions are useful when we want to branch if certain input pins are high and/or certain pins are low. In Table 2.2 we list the format of these instructions along with the Boolean conditions they check. These instructions can use direct or indexed addressing. The machine code has an opcode, operand, mask, and relative address (or branch displacement).

As an example, consider the following instructions:

```
LDX     #$1000
BRSET   $03,X $07 TRUE_IF_SET
BRCLR   $03,X $01 TRUE_IF_CLEAR
```

Index register X (IX) is loaded with $1000. The effective address is $1003 (offset $03 added to IX). Hence the BRSET instruction is referencing port C (see Figure 8.2). As we shall see, the branch is true only if inputs PC0, PC1, and PC2 are on. The status of the other port C pins make no difference. It is the mask byte of $07 (%00000111) that specifies which bits to check. In this case, test whether bits 0, 1, 2 are set. If the data at port C is the byte $1F(%00011111), the branch condition is true. If the data is the byte $1A (%00011010), the branch condition is false. For $1A, bits 0 and 2 are clear, hence all tested bits are not set. However, when BRCLR is executed, the branch does occur because bit 0 (PC0) is clear.

If you wish, check the Boolean conditions as shown in Table 2.2.

$\overline{\$1F} \cdot \$07 = \$E0 \cdot \$07 = \$00$, therefore the branch in BRSET occurs.

$\overline{\$1A} \cdot \$07 = \$E5 \cdot \$07 = \$05$, therefore the branch in BRSET does not occur.

$\$1A \cdot \$01 = \$00$, therefore the branch in BRCLR does occur.

TABLE 2.2   BIT MANIPULATION INSTRUCTIONS

Instruction	Mnemonic and format	Boolean expression
Branch if bit(s) clear	BRCLR (opr) (msk)(rel)	$? (\text{opr}) \cdot (\text{msk}) == 0$
Branch if bit(s) set	BRSET (opr) (msk)(rel)	$? \overline{(\text{opr})} \cdot (\text{msk}) == 0$
Set bit(s)	BSET (opr) (msk)	$(\text{opr}) + (\text{msk}) \rightarrow (\text{opr})$
Clear bit(s)	BCLR (opr) (msk)	$(\text{opr}) \cdot \overline{(\text{msk})} \rightarrow (\text{opr})$

Two similar instructions are Bit Clear (BCLR) and Bit Set (BSET). They clear or set bits of the operand based on which bits are set in the mask. This is useful for independently driving port pins low or high using BCLR and BSET, respectively. Their formats are:

```
BCLR (opr) (msk)
BSET (opr) (msk)
```

For example,

```
BCLR $04,X $A1
```

If IX is $1000, the effective address is $1004 (port B). The mask is $A1 (%10100001). If port B contained $7F (%01111111), prior to the instruction, it would have $5E (%01011110) after the instruction. Bits 0, 5, and 7 were cleared. Note that the source-code form of these instructions does not prefix the mask with the number sign (#). Some assemblers require the number sign. Some require extra commas. Check the documentation for your specific assembler.

Why Indexed Addressing?   This is an odd situation. Extended addressing mode is not available (probably a design decision to maintain compatibility with the 6800 microprocessor and/or to reduce the size of machine code). A common practice is to use an index register to point to the I/O registers. Refer to the memory map in Figure 1.8. Note that the register block begins at address $1000 (see also Table B.1 in Appendix B).

Then what is the use of direct addressing? Well, there are complexities in the 68HC11 we haven't even begun to hint at. It is possible to *remap* the register block to begin at address $0000 using another special register called INIT. Okay, let's drop the subject and continue.

Brake Light Example.   New cars have a brake light located at the center of the rear window. Consider a microcontroller program that turns on the brake light when the driver presses the brake pedal. In this case PC0 (bit 0 of address $1003) sets whenever the brake pedal is pressed. Otherwise, the bit is clear. The microcontroller turns the brake light on or off by setting or clearing PB1, respectively. The following illustrates the program instructions:

```
        LDX      #$1000     ;point to register block
        BCLR     $04,X $02  ;initialize light to be off (PB1 = 0)
*                          ;if brake pressed (PC0 == 1?)
*                          ;meaning branch if brake not pressed (PC0 == 0?)
        BRCLR    $03,X $01 NOBRAKE
        BSET     $04,X $02  ;then turn on light (PB1 = 1)
NOBRAKE                     ;else do not turn on light
*                          ;which means that the light is off
*                          ;rest of program code follows
```

```
*Note that initializing the light by turning it off
*ensures that the light output is defined to be off if
*the driver doesn't press the brake pedal.
```

PRACTICE

A table saw will not run unless two safety guards are in place. Bits 4 and 5 of address $100A (input port E) are the safety guard inputs. A bit is high if its corresponding guard is in position. Bit 2 of address $1008 (output port D) is the saw motor output. It is high to turn on the motor. Write the program.

Solution

```
*Solution
        LDX      #$1000     ;point to register block
        BCLR     $08,X $04  ;turn off saw for safety
*                          ;if both guards in position, then turn on motor
        BRSET    $0A,X $30 GUARD
        BRA      CONTINUE  ;else continue
GUARD
        BSET     $08,X $04  ;turn on saw motor
CONTINUE                   ;rest of program code follows
```

```
*NOTE that outputs are defined.
*It is dangerous to leave outputs undefined.
```

2.8.5   Jump Instructions

Do you remember how to go to Papa Joe's Variety Store? Simple. Go south on Division Street. Your friend coming from the opposite direction can follow the same directions. Words such as *north* and *south* are absolute (or relative with respect to the North Pole). Jump instructions specify the exact and absolute location for the next instruction to execute. Also, jump instructions are not conditional. When executed, they always set the PC to the specified value. Since jump instructions specify a 16-bit absolute address, they have the full 64-kilobyte range that is available to the CPU.

The following jump instruction tells the program to continue execution at address $E67F, the destination of the jump.

```
JMP $E67F
```

We could also represent the destination as a label. The source code would have to specify the value of this label.

```
JMP WAY_OVER_THERE
```

We can also create conditional jumps by using JMP with a branch instruction as illustrated by the following. This is useful for extending the range of a branch beyond its 8-bit unsigned limit.

```
START     CBA
          BEQ     SKIP
          JMP     NOT_EQUAL
SKIP      INCA
```

These instructions causes program execution to continue at location NOT_EQUAL if the contents of ACCA and ACCB are not identical. NOT_EQUAL could be anywhere in memory, assuming that memory is available at that address. We cover special jump instructions called jump to subroutine (JSR) in Chapter 3.

2.8.6   Relocatable Programs

A relocatable program is one that can start anywhere in memory without changing the machine code. Branch instructions are relocatable because their address is relative to the instruction itself. Jump instructions are not relocatable because if you shift the location of the program, you have to change the destination address of the jump instruction as well. In a typical microcontroller environment, relocatability is not important because the final program resides in a read-only section of memory such as ROM. But programs executing in a computer RAM have to be relocatable because the computer's operating system could load it anywhere in memory every time someone executes the program.

■ 2.9 SUMMARY

We develop microcontroller application software using a programming language. Assembly language and C are two common choices. We write source code for assembly language and translate it into object code using an assembler. Then we use a linker to create executable (hex) machine code. The CPU executes each instruction in sequential order unless a branch or jump instruction redirects it. The CPU fetches from memory each instruction it encounters and then executes.

Each assembly language instruction has a mnemonic opcode and may have an operand. The 68HC11 assembly language instruction set uses six addressing modes to specify where the operands are found:

Inherent
Immediate

Direct

Extended

Indexed

Relative

The instructions can perform basic arithmetic, logic, data handling, and a few specialized operations. The arithmetic operations affect the CCR bits for signed or unsigned binary arithmetic. Most use 8-bit operands and some use 16-bit operands.

Use conditional branch instructions to implement program flow control. Each branch instruction checks the status of certain CCR bits to determine whether a branch occurs. For details concerning the instruction set, refer to the Instruction Set Summary (Appendix A).

■ EXERCISES

1. (a) What is a programming language?

   (b) What is source code?

   (c) What is machine code?

   (d) What is a high-level language?

   (e) Describe assembly language.

   (f) Describe "assemble."

   (g) Describe an assembler.

2. Write down the object code for the following program, assuming that it is not necessary to force extended addressing.

   ```
   ORG       $E000
   LDAA      #$40
   LDAA      $40
   LDAA      $0040
   ```

   Indicate the addressing mode for each instruction.

3. Write the machine code for the assembly language source code of Listing ex2a.

   ```
   *Listing ex2a
   *Assembly exercise
   *and execution history exercise
   *for inherent, immediate, and extended addressing modes.
       ORG     $E000     ;start address of program
       LDS     #$00FF
       LDD     $C010
       STD     $C012
   ```

```
        LDY       #$20      ;note IY is a 16-bit register
        ABY
        ANDA      #$16
        BRA       *         ;its machine code is 20 FE
```

4. Write the execution history for the program of Listing ex2a (Question 3). The initial conditions (hex) are

```
C010  A5 B6 C7 D8 11 22 00 00 - xx xx xx xx xx xx xx xx

PC     ACCA  ACCB  IX     IY     SP
E000   AA    BB    1234   5678   0000
```

5. The program of Listing ex2b finds the square root of a single hex digit and stores the result. The result is an 8-bit representation of the square root as a 4-bit integer with a 4-bit fraction. For example, the square root of 5 is 2.237. The program would return a value of $24 representing $2.4 = 2 + 4/16.

   (a) Complete the square root table for the program.

   (b) Write down the object code for the program.

   (c) Write down the execution history when ($70) = $45 and ($71) = $67.

```
*Listing ex2b

*********************************************************
*FILE: ex2b.asm
*Chapter 2 Question 5
*demo indexed addressing mode and
*find square root of a hex digit
*using a look-up table
*********************************************************

*get a hex digit at $71
*find its square root
*and store in $72 as a 1-byte number
*that is the hex representation of
*the square root in decimal multiplied by 16
*i.e., a 4-bit integer and 4-bit fraction
*********************************************************

*Program Segment

*User's main code

        ORG     $E000
BEGIN
        LDY     #$0060     ;find data in $0070
```

```
*                        ;as $0060+$10 to demo
*                        ;index offset

     CLR    $10,Y        ;truncate data to
     BCLR   $11,Y $F0    ;one hex digit

     BSET   $10,Y $E0    ;point to look-up table
     LDX    $10,Y

     LDAA   $A0,X        ;look up square root

     STAA   $12,Y        ;store result

     BRA    *            ;end program for now
*End of program code
*********************************************************
```

6. The program of Listing ex2c illustrates arithmetic operations. Given the following initial conditions, each time the program runs:

```
ACCA    IX      HNZVC
AA      0457    11111
```

Show the state of ACCA and CCR bits HNZVC after each instruction executes for the following data in memory:

(a) 0000 02 22 3B xx xx xx xx xx - xx xx xx xx xx xx xx xx
(b) 0000 64 7A 3D xx xx xx xx xx - xx xx xx xx xx xx xx xx
(c) 0000 E7 D8 C0 xx xx xx xx xx - xx xx xx xx xx xx xx xx
(d) 0000 E D8 41 xx xx xx xx xx - xx xx xx xx xx xx xx xx

```
*Listing ex2c

*********************************************************
*FILE: ex2c.asm
*Chapter 2 Question 6
*Add and subtract byte integers.
*Can interpret as signed or unsigned.
*
*Exercise to predict results
*
*********************************************************
*-----------------------------
*User's main code
     ORG    $E000    ;start address of program
```

```
        LDX       #$00

        CLRA                ;Add without carry
        LDAA      $00,X
        ADDA      $01,X
        ADDA      $02,X

        CLRA                ;Add with carry
        LDAA      $00,X
        ADCA      $01,X
        ADCA      $02,X

        CLRA                ;Subtract without borrow
        LDAA      $00,X
        SUBA      $01,X
        SUBA      $02,X

        CLRA                ;Subtract with borrow
        LDAA      $00,X
        SBCA      $01,X
        SBCA      $02,X

        BRA       *         ;stop program

*End Program Segment
*----------------------------------
```

7. Write the machine code for the assembly language source code of Listing ex2d.

```
*Listing ex2d
*Assembly exercise
*and execution history exercise
*for inherent, immediate, direct, and extended
*addressing modes.
        ORG       $C000     ;start address of program
        LDX       #$3000
        LDY       $3C
        LDD       $3D
        ASLA
        STD       $3002
        MUL
        IDIV
        STX       $3000
        NOP
        RORA
        XGDX
```

```
        CMPA    $3D
        SBA
        NEGB
        SUBD    $3002
        ORAB    #$5A
        BRA     *               ;its object code is 20 FE
```

8. Write the execution history of the program of Listing ex2d (Question 7). The initial conditions (hex) are

```
0030   xx xx xx xx xx xx xx xx - 12 A4 34 56 B2 F1 89 A2
0040   98 65 32 01 02 3C 34 15 - xx xx xx xx xx xx xx xx
3000   34 5C 55 7E 00 00 00 00 - xx xx xx xx xx xx xx xx

PC        CCR    ACCA   ACCB   IX      IY      SP
C000      D0     89     7A     0000    0000    00FF
```

9. Complete the following table and show the Boolean expression checked by each branch instruction. Indicate any "out of range" relative addresses as xx.

Branch instruction	Address of branch instruction	Branch destination address	Branch relative address
BLS THERE	007F		15
BEQ THERE	007F		E9
BLO THERE		C100	6C
BNE THERE	056C	0517	
BLT THERE	E315	E296	
BRSET $1F,X $7E THERE	56C1		E1
BRCLR $10,Y $5C THERE	1715	174C	

10. Study the program of Listing ex2e. The hex contents of RAM are as follows:

```
0000   xx xx xx xx xx xx 5C 71 - 3E 89 15 B2 00 B2 6C CD
0010   FF xx xx xx xx xx xx xx - xx xx xx xx xx xx xx xx
```

Show the results of program execution for the following COUNT data stored in address $05.

(a) $01

(b) $05

(c) $0A

(d) $00

```
*Listing ex2e

***********************************************************
*FILE: ex2e.asm
*MAXIMUM UNSIGNED BYTE
*Find the largest unsigned byte in a block of data.
***********************************************************

*DATA STRUCTURE:
*      MAXRESULT     Address where result is stored
*                    Address $04
*      COUNT         Address that contains number of
*                    bytes in data block
*                    Address $05
*      FIRST         Address of first block element
*                    Address $06
*

*ALGORITHM:
*      1. Get COUNT
*      2. Set MAX=0, the current maximum
*      3. Compare each element to MAX
*      If element > MAX then MAX = element
*      Else MAX unchanged
*      4. Repeat step 3 for all elements
*      5. Put result, MAX into MAXRESULT

*-----------------------------------
*Program Segment
        ORG       $E000        ;start address of program
        LDS       #$FF         ;Initialize SP
        LDAB      $05          ;Get Count
        CLRA                   ;Maximum=Zero
*                              ;(Min. possible)
        LDX       #$06         ;Point to first entry
*                              ;in block
MAXM    CMPA      0,X          ;Is current maximum >
                               ;than current entry?
        BHI       NOCHG        ;Yes, keep maximum
        LDAA      0,X          ;No,replace maximum with
*                              ;current entry
NOCHG   INX                    ;Point to next entry
```

```
          DECB                          ;All entries in block
    *                                   ;checked?
          BNE        MAXM
          STAA       $04                ;Save Maximum
    HERE  BRA        HERE               ;Stop program

    *End Program Segment
    *------------------------------------
```

11. Describe how you improve the program of Listing ex2e (Question 10) to handle cases when the COUNT is zero.

12. Disassemble the following machine code. Assume that the PC is initially at $E000.

    ```
    E000   7F 50 00 7C 50 00 CE 10 – 00 18 CE 20 00 08 18 09
    E010   A6 00 18 A7 00 78 50 00 – 24 F3 01 01 01 01 01 01
    ```

13. Write a program to clear the first 20 bytes of RAM (i.e., set them to zero).

14. Write a program to transfer (copy) 400 bytes from a block starting at address $C000 to another block starting at $D000.

15. Write a program that replaces all lowercase letters with uppercase letters. Initially, IX points to the start address of a block of ASCII character bytes. Assume that all bytes have bit 7 reset (no parity). A byte of $04 marks the end of the block. Note that this type of block is often known as a character string. The byte $04 is the ASCII code for control D. It is often used to mark the end of a string.

16. Write a program to find the smallest signed byte in a block of data. It should store the result in address $00. Address $01 holds the block length in bytes and $02 is the address of the start of the block.

17. Write a program to add two 32-bit unsigned integers. The program should set or clear the C and Z bits depending on the 32-bit result. Using the Motorola ordering convention, each 4-byte integer has the most significant byte first (in the lowest address) with the rest following. The two 32-bit input integers are stored at addresses $00 and $04, respectively. The program should store the sum at address $08.

18. Write a software time-delay program. It runs (delays) for 15 seconds. Assume that the 68HC11 is using a 2-MHz clock.

19. Write a program to control traffic lights at an intersection. Each traffic light is turned on or off by setting or clearing a bit in address $1004 (this is an I/O port address). The bit address of each light is:

Bit	Light	
0	N-S RED	N-S means north-south
1	N-S YEL	E-W means east-west
2	N-S GRN	
3	E-W RED	
4	E-W YEL	
5	E-W GRN	

During the first minute, a red light and the opposite green light turn on. During the next 15 seconds, the red light stays on while the opposite yellow light turns on. During the following minute, the green light turns on while the opposite red light turns on. Fifteen seconds later, the yellow light turns on while the opposite red light stays on. The entire cycle then repeats.

20. Write a program to control a sump pump. A sump well collects water runoff. The sump pump pumps out the well whenever it fills up. When the level rises to a *high limit,* the pump turns on. It stays on until the level drops to a *low limit.* Then the pump shuts off. It does not turn on again until the well fills up to the *high limit.* Data is stored in the following addresses:

Address	Data
$00	High limit
$08	Level reading
$0F	Pump control
$10	Low limit

The level has a range of $00 to $FF, representing empty to full. You can set the low and high limits to values within the range (e.g., $10 and $F0). Sending a value of $00 to address $0F turns on the pump. Sending a value of $FF shuts it off.

21. A microcontroller system uses a digital display to show a decimal number counting down. The display accepts binary data in BCD format and displays it as decimal. Write a program to display a countdown. ACCA holds the down counter preset. It can be any valid BCD integer less than 100. The program counts down every second. It starts with the preset and counts down by one until it reaches zero. The microcontroller sends the count to address $1004 each time it decrements the count.

22. A microcontroller system controls an extendable robot arm. Write a program to determine the arm projection on the horizontal plane. The system has sensors to measure the arm length (radial) and its angle with respect to the horizontal plane. The program has the following input information:

ACCA	arm length	zero extension = $00
		full extension = $FF
ACCB	angle	0 degrees = $00
		89.6 degrees = $FF
		90 degrees = $100, have to use $FF!

You may use a cosine look-up table that is scaled to contain only 64 entries for the cosine of 0 to 90 degrees. The program has the following output after execution:

ACCA	horizontal projection	zero = $00
		full = $FF
ACCB	cosine of angle	minimum = $00
		maximum = $FF

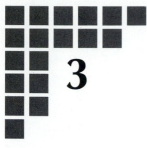

3 The Stack, Subroutines, Interrupts, and Resets

OBJECTIVES

After completing this chapter, you should be able to

- Describe and apply the structure of a stack for data storage.
- Write subroutines and code that call subroutines with parameters.
- Write and document source code in a modular fashion using subroutines.
- Describe how the microcontroller handles subroutine calls and use of local variables.
- Describe the concept of interrupts and resets.
- Write interrupt service routines.
- Write programs that use interrupt vectors, enable and disable interrupts, and use interrupt service routines.
- Analyze a memory dump to interpret interrupt operation and stacking the registers.
- Use software and CPU control interrupts.
- Avoid stack overflow.

It is crucial that you understand how the CPU uses the stack. Many software problems or bugs are caused by problems associated with the stack.

■ 3.1 INTRODUCING THE STACK

Think of a pile of playing cards. A player may draw from the top of the pile or add a card to the top of the pile. The central processing unit (CPU) also uses a similar procedure to use

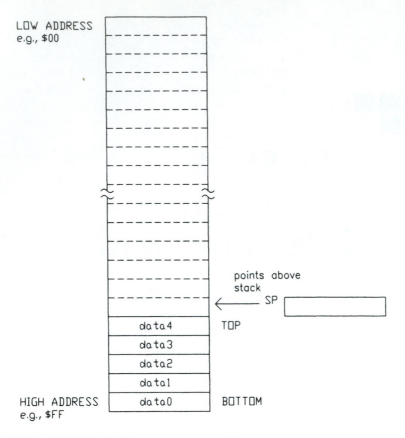

Figure 3.1 The Stack

the *stack*. The stack is a special area in memory. The CPU typically uses it to keep track of and store CPU register information during program execution. We can also use it for general data storage. Figure 3.1 shows its organization. The stack has a top and a bottom. Note that we consider the top part of stack memory to have a lower address. The bottom of the stack is fixed, but the top can change, moving up or down. During CPU operation, the stack changes size depending on operations that affect it.

Some people think of the stack as a pile of plates, each plate representing a byte of data. To put data into the stack, you *push* the plate onto the pile. To get data from the stack, you *pull* a plate from the pile. These operations are called push and pull to store and retrieve data, respectively. Also note that the term *pop* is sometimes used instead of *pull*. When playing cards, a player's hand is analogous to a register. The player draws a card from the top of the pile and puts it in his or her hand. This is like pulling data from the stack and storing it in a register. A card player may take a card from his or her hand and place it on the top of the pile. This is like pushing register data onto the stack.

This type of stack is known as a *last in, first out* (LIFO). In all LIFO stacks, the first pull operation will recover the last item that was pushed. Not all processors implement a LIFO stack the same way. The 68HC11 uses a 16-bit register called the stack pointer (SP) that indicates where to push or pull data. The 68HC11 also follows a postdecrement and preincrement convention. To store data, the 68HC11 pushes a byte onto the stack and then decrements the SP. To recover data from the stack, it increments the SP and then pulls the data. Note that the stack grows toward low memory addresses.

Let's look at this again for the 68HC11. The stack pointer (SP) always points to the address immediately preceding the top of the stack. When a CPU first begins operation, it initializes SP to some value. To change the stack size, you change the SP value. Decreasing the SP increases the stack upward (toward lower addresses) and increasing the SP contracts the stack downward (toward higher addresses).

Since we often write data to the stack, it makes sense to locate it in RAM. Figure 3.1 shows the bottom of the stack at $FF. This is typical for the single-chip mode (see the memory map of Figure 1.8).

■ 3.2 USING THE STACK TO STORE DATA

The 68HC11 has push and pull instructions that operate with the accumulators and index registers. The push instructions are PSHA, PSHB, PSHX, and PSHY. The pull instructions are PULA, PULB, PULX, and PULY. Refer to Figure 3.2 to see how these instructions operate. A push instruction copies data from a register and stores it into the address pointed at by the stack pointer (SP). Then it decrements the SP to increase the stack size. A push operation for 16-bit registers such as IX and IY performs this sequence twice. A pull instruction first increments the SP to decrease the stack size and then copies the data in the address pointed at by the SP. Then it stores the data in the applicable register. The pull instructions for the 16-bit registers perform this operation twice. When playing cards, a card may *physically* be removed from or placed on top of the pile. This differs from stack operations. After a pull, the byte still remains in memory. However, the stack pointer no longer points at it. This is much like burying treasure and throwing away the map.

Push Example.   Consider the following register values: ACCA = $AA, IX = $1234, and SP = $00F9. Also consider the following memory dump, where the bold type indicates the data currently pointed at by the SP.

```
00F0 00 01 02 03 04 05 06 07 - 08 09 0A 0B 0C 0D 0E 0F
```

Execution of Push ACCA (PSHA) causes $AA to be stored in address $00F9. The SP decrements to $00F8. Execution of Push IX (PSHX) causes $34 to be stored in address $00F8 and the SP decrements to $00F7. Then, execution continues with the high byte to store $12 in address $00F7 and decrement the SP to $00F6. The SP is pointing to the next available address in the stack. Memory now contains

```
00F0 00 01 02 03 04 05 06 12 - 34 AA 0A 0B 0C 0D 0E 0F
```

Pull Example.   Let's say that the CPU has executed several instructions. Some of these instructions have modified the contents of ACCA and IX. It can recover the earlier values using pull instructions. Execution of Pull IX (PULX) causes the SP to increment to $00F7.

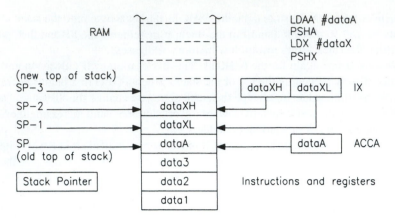

(a) Push data onto stack to store it

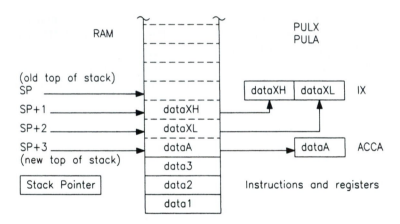

(b) Pull data from stack to retrieve it

Figure 3.2 Push and Pull Stack Operations

It copies the data in $00F7 and stores it in the high byte of IX. Then the CPU increments the SP to $00F8 and copies the contents of $00F8 into the low byte of IX. The top of the stack is now $00F8. Execution of Pull ACCA (PULA) causes the SP to increment to $00F9. The CPU copies the data in $00F9 and stores it in ACCA. The top of the stack is now $00F9. Memory now contains

```
00F0 00 01 02 03 04 05 06 12 - 34 AA 0A 0B 0C 0D 0E 0F
```

The contents of memory have not changed! Note the bold type. The top of the stack has changed. This means that future push instructions will overwrite the data used in the example.

Stack as a Storage Buffer. The CPU has only a few registers. Sometimes we push the register data onto the stack when we need to load the registers with other values to perform a certain operation. When the operation is finished, we pull the data from the stack to restore the accumulator to their previous values. It is important to pull data in the reverse order of the push operations, as demonstrated by PULX and PULA in Figure 3.2(b). We pushed ACCA first, then IX (Figure 3.2[a]) to store them. We assume that between the push and pull operations, both registers have changed. To recover them, we pull IX first, then ACCA (Figure 3.2[b]). The only time you do not pull in reverse order is when you deliberately want to change register values and not restore the original values. Be very cautious if you do this.

Refer to Listing 3.1. This program demonstrates using the stack as a scratchpad for an arithmetic calculation. It saves the value of x^2 on the stack because it uses both accumulators to calculate y^2. It pulls x^2 from the stack to add it to y^2. This is also an example of not pulling in reverse order.

```
*Listing 3.1
*SUM OF SQUARES
*Demonstrate push and pull operations
*Program calculates x squared plus y squared.
*It gets x and y as 8-bit numbers from addresses
*$1031 and $1032.
*It puts the 8-bit result in ACCA.


*    -------------------------------------------------
*
        ORG     $E000   ;start address of program
BEGIN   LDS     #$FF    ;You must define
*                       ;the stack first
        LDAA    $1031   ;Get first data, x
        TAB             ;and square it
        MUL
        ADCA    #$00    ;round it to 8-bit result
        PSHA            ;and save it
        LDAA    $1032   ;Get second data, y
        TAB             ;and square it
        MUL
        ADCA    #$00    ;round it to 8-bit result
        PULB            ;retrieve first result
        ABA             ;and add them
HERE    BRA     HERE    ;For now, stop program
```

As usual, the data is scaled from 0 to 1. For example, if x = $3B (0.2305) and y = $3C (0.2344), the result in ACCA will be $1C (0.1094). Performing the calculation in decimal with a calculator results in 0.1081. This example also serves to demonstrate the loss of precision when using 8-bit integer arithmetic.

There are other instructions that modify the stack. The instructions DES, INS, TSX, TSY, TXS, and TYS modify the SP. Other operations automatically modify the stack when they occur. Subroutines change the stack (Section 3.3). Resets and interrupts are special operations that modify the stack (Sections 3.6 to 3.11).

The following hex data is in memory:

```
0040     76 23 5C 77 00 00 B2 9E - 56 92 A3 9F 02 14 89 3B
```

Indicate what is in memory (includes the stack) after all instructions are executed. Also show the value of the stack pointer after executing each instruction.

```
*Practice stack instructions

      LDS     #$4A
      LDD     #$51C2
      LDX     #$3E06
      PSHB
      PSHX
      PULA
      PULY
*done for now
```

Solution

Contents of memory are

```
0040     76 23 5C 77 00 00 B2 9E - 3E 06 C2 9F 02 14 89 3B
```

```
*Solution includes additional comments

      LDS     #$4A       ;SP = $004A
      LDD     #$51C2
      LDX     #$3E06
      PSHB               ;($4A) <- $C2 , SP = $0049
      PSHX               ;($49) <- $06 , SP = $0048
*                        ;($48) <- $3E , SP = $0047
      PULA               ;SP = $0048 , A <- ($48)
      PULY               ;SP = $0049 , Yh <- ($49)
*                        ;SP = $004A , Yl <- ($4A)
```

■ 3.3 USING SUBROUTINES

You prepare a meal for guests. The main course is chicken cordon bleu and dessert is chocolate mousse. The program for meal preparation is

> Prepare main meal
>
> Prepare dessert

You can describe each step in the main program in more detail as subprograms (subroutines). Perhaps you would have to go to your library of recipe books (subroutine library) to prepare the courses. Instead of writing out all the details from a recipe book, you would refer to the appropriate sections. If you wanted to serve chicken teriyaki instead, you would simply refer to the chicken teriyaki recipe. Or if you prepare 10 instead of 4 meals, you modify the recipes. The recipes are *subroutines* that the meal preparation program is free to use when required.

A subroutine allows us to reuse code. It is a section of a program that may be used one or more times. The *main* program calls subroutines to perform certain steps. What if we modify the program of Listing 3.1 to add many squares? We can write the section of code to calculate the square of any number as a subroutine. The main program calls the subroutine whenever it needs to calculate the square of a number. This makes a program shorter and more understandable. Listing 3.2 shows such a program.

Listing 3.2 shows a main program using a subroutine called SQUARE to calculate the square of a number. The first loop in the main program increments a pointer to the data to be squared. It calls (executes) the subroutine after executing the instruction JSR SQUARE. JSR means "jump to subroutine." When the program executes this instruction, it continues at the first instruction of the subroutine. It executes the subroutine until the instruction RTS. RTS means return from subroutine. After RTS executes, the program continues at the instruction following JSR instruction. In this case it will resume at PSHA. Every time the program executes the first loop, it will also execute the instructions in subroutine SQUARE.

```
*Listing 3.2
*SUM OF SQUARES
*Demonstrate subroutines
*Program squares all the numbers
*in addresses $C000 to $C07F
*and puts the sum of the squares
*as a 16-bit result in IX.


        ORG     $E000       ;start address


*-------------------------------------------------
*Start MAIN

MAIN    LDS     #$CFFF      ;You must define
*                          ;the stack first


*                          ;squaring loop
        LDX     #$C000      ;init. data block
*                          ;pointer
LOOP1
        JSR     SQUARE      ;square it and
*                          ;point to next data
        PSHA               ;save square
        CPX     #$C080      ;squared all data?
        BNE     LOOP1       ;get more if not
*                          ;summing loop
        LDX     #$0000      ;sum = 0
        LDAA    #$80        ;i=$80
```

```
LOOP2    PULB                    ;get square(i)
         ABX                     ;sum = sum + square(i)
         DECA                    ;i=i-1
         BNE      LOOP2          ;repeat until i == 0

HERE     BRA      HERE           ;stop program

* End of MAIN
* -------------------------------------------------


*SUBROUTINE SQUARE
*calculates the square of an 8-bit number
*as a rounded 8-bit normalized result
*data pointer (IX) increments
*calling registers:
*        IX = address of data to be squared
*return registers:
*        ACCA = 8-bit square
*        IX=IX+1
*        ACCB is affected
*        others unaffected

SQUARE
         LDAA     $0,X
         TAB
         MUL
         ADCA     #$00           ;round it to 8-bit result
         INX
         RTS


*-------------------------------------------------
*End of program code
```

Calling registers are those that hold the input data for a subroutine. Return registers hold the output data of a subroutine. A data input/output process for subroutines is called *parameter passing*. If there are not enough registers for parameter passing, a calling program can pass more parameters using the stack. C compilers will produce an equivalent machine code that uses the stack for parameter passing. (To read about this topic, refer to "Stack Use" in Section 13.6.6.) When this program is assembled, it turns out that the first address of subroutine SQUARE is $E01B. Hence, the instruction JSR $E01B is equivalent to JSR SQUARE in this case.

Listing 3.3 and Figure 3.3 show another example to illustrate subroutines. Although it is an artificial example, it does illustrate more features.

```
*Listing 3.3
*Demonstrate subroutines
*Program does not do anything really useful
*other than to show use of subroutines
*The labels CALL0,1,2 are included only
```

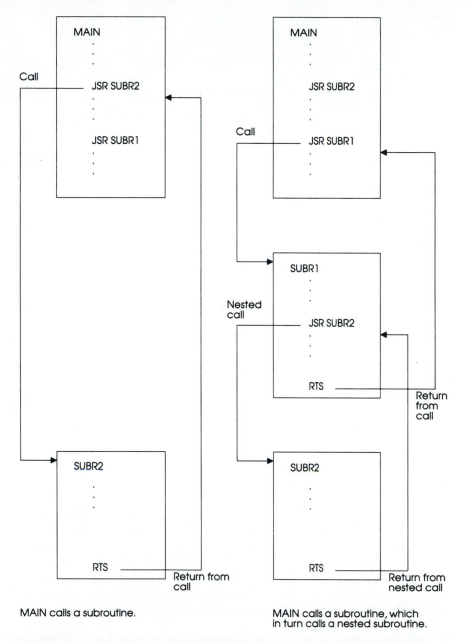

MAIN calls a subroutine.

MAIN calls a subroutine, which in turn calls a nested subroutine.

Figure 3.3 Process of Calling Subroutines. Note that MAIN calls the same subroutine (SUBR2) twice: directly and as a nested call (from SUBR1).

```
*to explain the program more easily

        ORG       $E000        ;start address
*-------------------------------------------------
*MAIN program
        LDS       #$FF         ;define stack
        LDAA      $05
CALL1   JSR       SUBR1        ;call SUBR1 first time, with call to
                               SUBR2 (nested)
CALL0   JSR       SUBR2        ;call SUBR2 second time (direct)
CALL2   JSR       SUBR1        ;call SUBR1 second time, call SUBR2
                               third time (nested)

HERE    BRA   HERE

*-------------------------------------------------
*DEFINE SUBROUTINES
*Define subroutine SUBR1
SUBR1   LDAB      $06
CALL    JSR       SUBR2        ;a subroutine can call another
*                              ;subroutine
RET     STAB      $06          ;RET marks the return for CALL
RTS                            ;miracle instruction
*                              ;always knows where to return


*Subroutine SUBR2
SUBR2   ABA
        MUL
        RTS                    ;again, the miracle instruction
```

Listing 3.3 shows that a program can use a subroutine more than once and that a subroutine can call another subroutine. In this case SUBR1 calls SUBR2. Also, when the program executes RTS in any of the subroutines, it knows where to return to. Initially, the CPU executes the first two instructions in the main program. Then it executes JSR SUBR1 at CALL1; it continues at SUBR1, itself executing LDAB $06 and then JSR SUBR2. After this, the CPU continues at SUBR2, executing ABA, then MUL. When it executes RTS, the CPU returns to SUBR1 to execute the instruction STAB $06. When it executes RTS in SUBR1, the CPU returns to the main program to continue at CALL0. It returns to execute SUBR2 and then returns to CALL2. From here it continues to execute SUBR1 (also SUBR2 a third time) and returns to HERE.

The question is: How does the CPU know where to continue when it executes RTS? You may have guessed: It is the stack. We examine how a subroutine uses the stack in Section 3.5. In the next section we examine how we as programmers use subroutines.

C Code Examples.   C Listing 3.3a introduces an example with user-defined functions. A *function* in C is a block of code that is a subprogram, similar to the use of a subroutine in assembly language. It turns out that main is a function, one which must be included in every C program. A programmer can create his or her own C function as illustrated in the listing.

```
/* C Listing 3.3a */

/* Function Prototype */
unsigned char Square(unsigned char);
```

```
/* Declare Variables */
unsigned int Sum, ResultA, ResultB;
unsigned char VarA, VarB;

void main(void)
{
      ResultA = Square(VarA);
      ResultB = Square(VarB);
      Sum = ResultA + ResultB;
} /* End Main */

/*Function Definition */
unsigned char Square(unsigned char Parameter)
/*Function returns unsigned normalized 8-bit square of Parameter
*/
{
      /* Declare Local Variable */
      unsigned int Result;
      Result = Parameter << 1; /* square it */
      return (Result);
} /* End Function Square */
```

The steps for programming user-defined functions are

> Create a function prototype
>
> Define the function
>
> Call the function from main or from within another function

Functions may use *parameters* for data input and may return a *value* for use by another function that called it, such as in the following example statement:

```
ResultA = Square(VarA);
```

Here the value in VarA is passed to the Square function and then is called. After executing, the function Square returns a value, which is then assigned to ResultA. Examine the function definition. To receive a parameter, it is necessary to define parameter(s) in the parentheses. In this case it is one parameter declared as unsigned char Parameter. A parameter is the placeholder for a value passed to the function when it is called. The function must have a name, which is Square in this case. If the function returns a value, it is also necessary to declare the data type being returned. In this case, it is unsigned char. Note that main is declared as void and its parameter is also void. The word *void* is necessary whenever a function does use a value. Most compilers require that functions be prototyped at the beginning of the program. The prototype only indicated the function name, the data type of its parameters, and the data type being returned, or void if none.

Functions in C are implemented in assembly language as subroutines. For example, the assembly output of the C program contains the statement jsr _Square, and there is a subroutine named Square. Equivalent assembly language operations for passing parameters and returning values can be complicated. Fortunately, the compiler figures this out for us. For those curious about these details, Chapter 13 will cover some of them.

C Listing 3.3b illustrates a C program that is functionally equivalent to the assembly language program of Listing 3.2. It uses an array and a loop.

```c
/* C Listing 3.3b */

/* Function Prototype */
unsigned char Square(unsigned char);

/* Declare Variables */
unsigned int Sum;
unsigned char Count, Data[0x80];

void main(void)
{
        /* Initialize variables */
        Sum = 0;
        Count = 0x7f;
        /* Add sum of normalized squares */
        while (Count >= 0)
        {
                Sum = Sum + Square(Data[Count]);
                Count- -;
        }
} /* End Main */


/* Function Definition */
unsigned char Square(unsigned char Parameter)
/* Function returns unsigned normalized 8-bit square of Parameter
*/
{
        /* Declare Local Variables */
        unsigned int Temporary; /* 16-bit holder for squaring */
        unsigned char Result;
        Temporary = Parameter * Parameter;
        /* Truncate to 8-bit normalized value */
        Result = Temporary >> 8;
        return (Result);
} /* End Function Square */
```

■ 3.4 MODULAR PROGRAMMING USING SUBROUTINES

3.4.1   A Top-Down Approach to Problem Solving

When you plan how to prepare dinner for your guests, you do not worry about how much salt to put in the sauce as a first step. Although everyone has their own style, most people probably do general planning first before worrying about the details. The first step of planning might be

> Decide what courses
> For each course
>> Look up recipe
>> Collect ingredients
>> Collect kitchen utensils
>> Follow recipe directions

Since this is not a cookbook, we will look at programming examples next. Before you can write down the final program instructions, you have to *design* the program first and do some other things. These steps are

> Understand the problem
> Design program
> Code program
> Test program

You may have to repeat earlier steps when encountering problems. The first step seems to be common sense, and it is, but many people have difficulty writing a working program because they do not fully understand the problem. It is even worse if you are writing a program for someone else and they do not understand the problem.

After understanding the problem, you design a program by developing its *algorithm.* An algorithm is a step-by-step description of how to solve a problem. You can consider it to be a recipe. To develop a program, you break it down into smaller programs. Each of these smaller programs is a module. You can break down each module into smaller programs or modules. Each of these can be broken down further, and so on. You do this until you can describe each module as a series of logical steps. You do not necessarily write down program code at this stage. In fact, many people will use an informal shorthand called *pseudocode.* The process of breaking down a program into smaller modules is called *stepwise refinement.* Some call this design process *structured programming.*

Pseudocode is a computerlike way of describing steps in an algorithm. You do not have to worry about the formal rules (syntax) of a programming language. Generally, when you have written a program in pseudocode, you can code it using any programming language. However, if you are planning to use assembly language, you may include specific register and memory operations as part of your pseudocode (after several steps of refinement).

Once you have decided on a target system and language, you can code your program. This involves translating pseudocode into programming language and deciding what memory and I/O registers to use. Probably, each module is a subroutine. Submodules would be other subroutines that a higher-level subroutine would call.

Testing the program is another big step. You have to choose test data carefully to check as many possibilities as possible to avoid bugs. It is usually impractical to test all possibilities. A *bug* is an unintentional mistake in the program.[1] A bug could also be a hardware design error. Sometimes, a company will sell a system with bugs. The effects of the bugs are discovered later when customers use the system in untested situations. Usually, it is very difficult to find the cause of a bug. Once found, it can also be difficult to correct.

---

[1]Computer users may be aware of intentional bugs. These include worms and viruses.

We will show an example of top-down design. The Backwater Bottling Company wants a process controller that prints out a report at the end of every shift. To be realistic, we would have to specify many details, such as what the report contains. Also, the process controller would do other things besides printing out an end-of-shift report. To understand the problem, you need to know the details. As part of this process, you break it down into modules. These may be:

Read all flow sensor data every second

Store flow sensor data every second

Total flow sensor data every second

Detect end of shift

At end of shift send out total flows to printer

We cannot devote this book to designing the entire program, but we will look at some modules. We will break down the last module:

For each flow sensor total

      Send out explanatory text

      Send out total flow

and break down its last module:

For each total

      Get binary (hex) bytes

      Modify bytes to represent liters

      Convert to ASCII

      Add appropriate ASCII control codes

      Send ASCII codes to printer

We will concentrate on the module "Convert to ASCII."

For each byte in total

      Split byte into two hex digits

      Convert each hex digit to ASCII

Now concentrate on how to convert a hex digit to ASCII. (See Section 1.1.2 to understand ASCII.) Also, we should determine how to code each hex digit. We want to design a subroutine to perform the conversion. Due to other factors in the entire program design, we may wish to specify it as follows:

Subroutine HEXTOASCII:

Calling register:	ACCA holds the single-digit hex number (i.e., a byte's most significant digit is zero).
Return register:	ACCA holds the ASCII code of the digit.

Although printing flow in liters requires only decimal, we shall convert any byte from $00 to $0F to ASCII. This will make the subroutine more adaptable to be used in other program sections. Often, there is more than one way to design an algorithm. One technique is to use a lookup table such as we did for square roots in Section 2.5.10. We shall use a different technique.

To begin, study the relationship between hex digits and their ASCII representations. You will need to refer to an ASCII chart (see Appendix B). You should note that the codes for '0' to '9' are \$30 to \$39, respectively. Also, the codes for 'A' to 'F' are \$41 to \$46, respectively. From this we can begin to develop an algorithm as follows:

If a digit is between \$00 and \$09, add \$30.

If a digit is between \$0A and \$0F, add \$37.

From this, you could get

Add \$30 to a digit

If result is greater than \$39

then add another \$07

Refine this further into pseudocode. Also note the subroutine specifications.

ACCA = ACCA + \$30

ACCA > \$39?

If true, then ACCA = ACCA + \$7

else ACCA = ACCA

From the pseudocode, we have the assembly language code of Listing 3.4.

```
*Listing 3.4
*Subroutine HEXTOASCII
*Converts a single-digit hex number (0 to F)
*to ASCII
*Calling registers:
*    ACCA = a single-digit hex number to be converted
*Return register
*    ACCA = ASCII code of hex digit
HEXTOASCII
      ADDA    #$30
      CMPA    #$39    ; check if digit > '9'
      BLS     SKIP    ; add another 7 if it is
      ADDA    #$07
SKIP          RTS     ; return to calling routine
```

We could "improve" the subroutine to check that valid input (between \$00 and \$0F inclusive) is passed to it. For example, if ACCA had \$10 when the subroutine is called, it would give an improper result. We could design the subroutine to indicate that an improper result occurred. For example, HEXTOASCII could clear the C bit to indicate that the input was valid and set the C bit and not modify ACCA if the input was invalid. The calling program should check the C bit after the subroutine returns.

The main thing is to develop a consistent style of writing subroutines, such as using certain registers for parameter passing. This makes it easier to develop a complete application program. It also makes it easier to modify later. Also, you may find it difficult to develop programs. It is almost an art as well as a discipline. When you know more about the software and hardware and have seen more examples, it will become easier.

PRACTICE

Now try to develop the subroutine to split a byte into two hex digits.

Solution

You may come up with something like the program shown in Listing 3.5.

```
*Listing 3.5
*Subroutine UNPACK
*Converts a byte into two single-hex-digit bytes (MSD=0)
*Stores the unpacked result in 2 bytes in RAM
*Calling Registers
*    ACCB = data byte to be unpacked
*    IX = address to store unpacked result (MSD first)
*Return Registers
*    All registers unaffected
UNPACK
     PSHA                ;store ACCA
     TPA                 ;store CCR
     PSHA

     TBA                 ;shift upper digit into lower nibble
     LSRA
     LSRA
     LSRA
     LSRA
     STAA   0,X          ;and store it
     TBA                 ;next,
     INX                 ;work on lower digit
     STAA   0,X          ;first store both digits
     BCLR   0,X $F0      ;then remove upper digit
     DEX                 ;restore IX
     PULA                ;restore CCR
     TAP
     PULA                ;restore ACCA
     RTS                 ;return to call
```

Note that the program calling subroutine UNPACK will have the same register contents before and after calling it. The subroutine changes the contents of two memory locations that were pointed at by index register X. The value of each of the two bytes falls within the range $00 to $0F, a hex digit.

3.4.2 Subroutine Libraries

A subroutine library is a collection of subroutine source codes. Originally, they may have been developed by you or anyone else. You may use them when developing or modifying a program. Note that when we mention a program, it could be a subroutine or the main program since either can call other subroutines.

Subroutine libraries are useful since you do not have to redevelop sections of code. For example, you need to read inputs from an I/O port and convert the data to ASCII several times. If the library has a subroutine to do this, you can use the subroutine in your program: you have to use its proper name and link the subroutine library object code with your object code to make the final executable machine code. In Chapter 4 we explain this process in more detail.

To use floating-point operations, you may use the subroutines in Motorola's floating-point package for the 68HC11 (see Section 2.7.7). Note that due to the nature of floating-point numbers, you pass parameters through "software" registers that are physically part of page 0 RAM. The Motorola EVBU board comes with a disk that contains source code files. You could modify and assemble these files to make up a subroutine library. Other libraries are available from their computer bulletin board (Appendix C).

Your place of work, training facility, or school may have its own library. Note that a library could contain interrupt and reset routines as well as subroutines. To use a library, it must be sufficiently documented.

3.4.3   Documentation and Using Subroutines

Comments and Naming.   Reading a long section of code without comments is virtually impossible to understand. As an exercise you copy a long listing without its comments. Do not look at it for a week. Then read it and see if you understand it. Since long programs use subroutines, it is very important to document them. Refer to Listing 3.2 in Section 3.3. When the program calls the subroutine with the instruction JSR SQUARE, it has a comment to summarize why it is calling the subroutine. The subroutine name should reflect its function. We could have called it PERIWINKLE and the assembler would accept it. Although this name might be amusing, it is frustrating to use strange or cryptic names.

Parameter Passing.   One of the essential things to do when using subroutines is to transfer data. Some subroutines require data. The calling program (or subroutine) *passes* data to the subroutine being called. Some subroutines pass data to the calling routine. A *parameter* is the data that is being passed.

The conventions we have used are calling registers for parameters the subroutine needs and return registers for parameters for the calling program to use.[2] Others will use different names for the same thing. The Motorola floating-point package uses the words *input* and *output*. It is also possible to use absolute memory locations and/or the stack to pass parameters. The floating-point package uses memory locations called *floating-point accumulators*. In other parts of this chapter we say more about these parameter-passing techniques where appropriate.

Comprehensive Documentation.   For each subroutine, the documentation indicates its name, summary of operation, code size in bytes, the stack space it uses, other subroutines it calls, parameter passing (input and output), and error conditions. All subroutines use the same convention for input, output, and error conditions. To indicate an error, a subroutine could use CCR flags or an error code in a register.

Preserving Registers.   Subroutine SQUARE in Listing 3.2 affects both accumulators. We could modify the subroutine such that it returns their original value as shown in Listing 3.6. A subroutine can use the stack to store register values. When it returns to the calling routine, it restores the values. Listing 3.5 also shows the same technique.

---

[2]DOS programmers may recognize the convention of using calling and return registers as similar to that used by the DOS interrupt functions.

```
*Listing 3.6
*Demonstrate a convention for using subroutines
*using a modification of Subroutine SQUARE
*(See Listing 3.2).
*Program squares all the numbers
*in addresses $C000 to $C07F

      ORG  $E000         ;start address
* --------------------------------------------------
*Start MAIN

MAIN  LDS  #$FF          ;You must define
*                        ;the stack first


*                        ;squaring loop
      LDX  #$C000        ;init. data block pointer
*                        ;IX is the calling register
LOOP1
      JSR  SQUARE2       ;square it and note that
*                        ;original IX value returned
      INX                ;point to next data

      CPX  #$C080        ;squared all data?
      BNE  LOOP1         ;get more if not

HERE  BRA  HERE          ;stop program

*End of MAIN
* --------------------------------------------------


*SUBROUTINE SQUARE2
*calculates the square of an 8-bit number
*as a rounded 8-bit result
*calling registers:
*    IX = address of data to be squared
*return registers:
*    IX = address of 8-bit square
*    CCR affected
*    others unaffected

SQUARE2
    PSHA             ;preserve registers
    PSHB
*                    ;note that following instructions modify
*                    ;ACCA and ACCB
    LDAA  $0,X       ;get data to square
    TAB              ;copy it to ACCB
    MUL              ;square it
    ADCA  #$00       ;round it to 8-bit result
    STAA  $0,X       ;store result
```

```
     PULB                ;restore registers
     PULA
     RTS                 ;return, note that ACCA and ACCB
*                        ;contain their original values


*-------------------------------------------------
*End of program code
```

We could extend this technique to preserve any registers that a subroutine has to use. Remember to pull in reverse order (Section 3.2). Preserving registers prevents a subroutine from modifying them unexpectedly. You might have written the calling program to use the affected register assuming that it was not changed. Although preserving registers lengthens the program, it is preferable to having bugs. At the very minimum, document any registers or memory locations that a subroutine modifies. Also, we do not normally preserve CCR. Section 3.5.1 with Figure 3.4 looks at the stack use in more detail.

What the Subroutine Does.   This subroutine multiples two hex numbers and throws out the lower byte of the product after rounding the upper byte up or down. One might want to do this when squaring a fraction. For example, suppose $95 actually represents the fraction $95/$100 = $.95. In decimal, this is 149/256 = 0.582. Squaring $.95 gives $.56B9 $\cong$ $.57 = 0.3389. The actual decimal value (found by converting $56B9/$10000 or by squaring 0.582) is 0.33876. The difference is only 0.3% so, in this example, the approximation is a good one. (For small values, the approximation is not so good. In fact, for anything less than $0C, the result is zero.) *This explanation provided by Jack Levine of California State University.*

Documentation Again.   You can use a well-documented subroutine without knowing its source code. In fact, if designed and documented correctly, it should not matter what specific instructions the subroutine uses. The subroutine should do what it says it does, nothing more and nothing less. If you give a drawing to a machinist, you expect a machined part. You do not expect to tell the machinist how to do the job. Of course, the drawing has to be complete.

PRACTICE

A subroutine called OUTA sends the contents of accumulator A as an ASCII character. Write documentation comments for the subroutine. Write instructions to output the message "good." *Hint:* Use the ASCII chart.

Solution

```
*Subroutine OUTA
*Outputs an ASCII character
*Calling registers
*    ACCA = ASCII code of character
*Return registers
*    None except that CCR affected
OUTA
*    ; subroutine code follows, last instruction is RTS
```

The Stack, Subroutines, Interrupts, and Resets   129

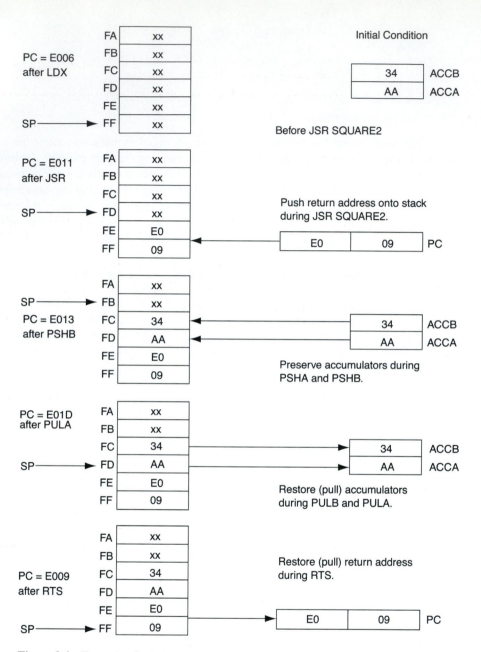

Figure 3.4 Example of a Subroutine Using the Stack. See Listing 3.6; note that all numbers are shown in hex; xx means any number.

```
*Send the message "good"

    LDAA   #$67    ;ASCII code for 'g'
    JSR    OUTA    ;send it
    LDAA   #$6F    ;ASCII code for 'o'
```

```
JSR     OUTA    ;send it
JSR     OUTA    ;send it again
LDAA    #$64    ;ASCII code for 'd'
JSR     OUTA    ;send it
```

3.4.4   Simple Parameter Passing

We have discussed calling and return registers. A calling register passes information to the subroutine to process, and a return register gives the subroutine results back to the calling program. We call this information-passing process *parameter passing* (some call it value or variable passing). Sometimes you need to pass more parameters than there are registers available.

A simple technique is to use fixed addresses in RAM. Each variable in the final application program has a specific address reserved for it. Naturally, these address reservations must be well documented. It is up to the programmer to research what these addresses are and to use them appropriately. We call this technique the absolute address technique.

This technique has a disadvantage if there is not enough memory to reserve places for every variable. Also, by using up more memory, there is less space for the stack. This increases the chance of stack overflow (see Section 3.11). For example, some addresses may be used rarely, yet you cannot use the address for anything else. General-purpose computer programs that execute in RAM have to use a modification of this technique since the operating system could load a program in any address. Subroutines using this technique are not *reentrant*. In Section 3.6 we describe reentrant subroutines.

Applications that have enough memory for all variables can use this technique. The EVBU board's monitor program uses reserved locations in page 0 RAM. For simpler applications, use the simpler techniques. In the following section we discuss using the stack for parameter passing. This is more flexible but requires more instructions. Hence a program executes more slowly than one using the absolute address technique.

PRACTICE

Write instructions to output the message "good" using subroutine OUTMSG. This subroutine outputs the ASCII codes stored in a block of memory starting at address $00. It stops when it encounters an address with the data byte of $0D (CONTROL M or carriage return).

Solution

```
*Send the message "good"
*First fill up data block for subroutine


    LDD     #$676F    ;ASCII codes for "go"
    STD     $00
    LDD     #$6F64    ;ASCII codes for "od"
    STD     $02
```

```
LDAA   #$0D      ;code to stop output
STAA   $04
```

*Then call subroutine to send message

```
JSR  OUTMSG
```

■ 3.5 SUBROUTINE OPERATION

Section 3.4 dealt with how a programmer uses subroutines. In this section we show how the CPU executes subroutines.

3.5.1 Subroutine Use of the Stack

The CPU uses the stack to preserve the return address. We, as programmers, may also use the stack to preserve some registers and to pass parameters. Whenever the CPU executes JSR or BSR, it pushes the address of the instruction following it onto the stack (2 bytes). When the CPU executes RTS, it pulls the address from the stack and puts it in the PC. Execution then continues. Appendix B shows instruction details as in Figure B.3, "Special Operations."

Refer to Listing 3.6 and Figure 3.4. The address of instruction JSR SQUARE2 is $E006, the address of INX is $E009, and the address of subroutine SQUARE2 is $E011. After the subroutine is called, it should return to the main program to execute INX. Let's see how this works. Calling the subroutine (execution of JSR) causes the CPU to push the return address (address of first instruction that follows a JSR or BSR instruction) onto the stack. Then it loads the PC with the subroutine address. To preserve registers, the subroutine code pushes the accumulators onto the stack. It uses (modifies) them and then pulls them off the stack to restore their entry values. The execution of RTS causes the CPU to pull the return address off the stack and to put it back into the PC. Execution now resumes at the instruction INX.

Can you see why a program should balance pushes with an equal number of pulls? If the subroutine did not include PULB and PULA, execution of RTS would pull $34AA and put it in PC. This would certainly cause program failure. Be careful. You don't want to change the SP without ensuring that execution will return to where you want it to be.

Parameter Passing via the Stack.   This is necessary if the subroutine must be reentrant. We say more about reentrancy after the topic of *interrupts* is introduced in Section 3.6. This stack use can become complicated if programming in assembly language. If you are programming in C, the compiler will probably translate the C functions into subroutines that use the stack for passing variables (parameters). Section 13.6.6 describes stack use in the C language. For those of you interested, you may also wish to use the technique described in Section 2.2.2 in reference 5. It uses instructions such as TSX, TXS, TSY, TYS, DES, and INS (see the *Instruction Set Summary* [Appendix A]) to implement the coding. People writing C compilers would have to know this or a similar technique.

Basically, the calling routine pushes parameters onto the stack. The called subroutine reads the parameters from the stack using the index register. Note that the TSX instruction transfers the contents of the SP plus one to IX. The subroutine returns with result parameters in the stack for the calling routine to pull off.

Local Variables.   One reason for using the stack instead of absolute memory locations is to save on RAM space. The memory space that a subroutine needs for its own temporary results is known as a *local variable*. A typical application program will call many subroutines, often with several layers of nesting. Hence, there will be many variables. Since only one subroutine can execute at a time, the program does not need space for each temporary variable all the time. The stack allows memory to be shared. For example, one subroutine needs a RAM byte for the result of an intermediate computation. After the subroutine returns, it does not need the space anymore. This allows another subroutine to use the same memory location for its own purpose.

For example, a subroutine that multiplies and adds several numbers will need space to store intermediate results. The pseudocode in Listing 3.7 illustrates this.

```
*Listing 3.7
*Subroutine pseudocode
*Calculate (A1 x A2) + (A3 x A4)
*Calling parameters: A1, A2, A3, A4
*Returns: RESULT


     temporary = A1 x A2
     RESULT = A3 x A4
     RESULT = RESULT + temporary
     return the RESULT
```

In this case, "temporary" is the local variable.

Another Stack Technique.   The following shows another technique. It breaks the rule for balancing pushes and pulls. The subroutine saves its results on the stack using more pushes than pulls. The calling program could get the results by pulling from the stack. Note that the subroutine could not easily use IY since it uses it to save the return PC value.

```
STARTSUBR
        PULY      ;to save return PC value in IY
        .         ;preserve other registers as required
        .         ;be careful, IY holds PC value
        .         ;main subroutine instructions follow
        .         ;and restore other registers
        PSHY      ;restore PC to new top of stack
        RTS       ;and return
```

An alternative to the last two instructions is JMP $00,Y. This subroutine returns to the calling program with the stack larger than before. We should avoid this alternative. There is a risk of stack overflow (see next section).

3.5.2   Nested Subroutine Details

We have already seen that a subroutine can call another subroutine, which in turn can call another one, and so on. We say that subroutine B is nested within subroutine A when subroutine A calls subroutine B. If subroutine C is nested within subroutine B, we can say that there are two levels of nesting. Although C is nested within B, it is still possible for the main program to call it directly without nesting. In Figure 3.3, SUBR2 is nested within SUBR1. In theory we can have infinite levels of nesting. In practice the level of nesting is determined by available RAM memory since each subroutine call pushes a return address on the stack. For each level, another return address is pushed on the stack. If the stack exceeds available memory, we have *stack overflow.*

Nesting refers to how a subroutine is called. When the main program calls subroutine B directly, it is not nested. When it calls subroutine A, which in turn calls subroutine B, then subroutine B is nested within A. Nested subroutines are a natural outcome of using the top-down approach to designing programs. In top-down design, you define modules. Each module is then made up of other modules, which in turn are made of more modules, and so on. The levels of subroutine nesting can reflect this structure.

Consider the program for the Backwater Bottling Company described in Section 3.4.1. The program may have a subroutine that generates an end-of-shift report. This subroutine calls a subroutine that prints out total flows. This subroutine calls a subroutine to get the total flow data from a specified sensor. Possibly, a calling register specifies which flow sensor. Then this subroutine could call a subroutine to convert flow data to ASCII. Finally (whew!), this subroutine calls the subroutine to convert a byte that falls within the range of $00 to $0F to ASCII.[3] If you count them all, there are four levels of nesting for the subroutines mentioned.

PRACTICE

Refer to Listing 3.3. Show the stack after each JSR instruction.

Solution

We need to find the addresses of labels first by assembling the source code:

```
CALL1  $E005
CALL0  $E008
CALL2  $E00B
HERE   $E00E
SUBR1  $E010
CALL   $E012
RET    $E015
SUBR2  $E018
```

---

[3]The subroutine to convert flow data to ASCII would have to call another subroutine first. The other called subroutine would "unpack" a byte into 2 bytes (see Listing 3.5).

The stack history follows. In this case, the memory dump has 8 bytes per row. Note that "xx" means unknown value.

```
JSR executed         SP     Hex contents of RAM

Initial              00FF   00F8   xx xx xx xx xx xx xx xx
CALL1                00FD   00F8   xx xx xx xx xx xx E0 08
CALL (in CALL1)      00FB   00F8   xx xx xx xx E0 15 E0 08
CALL0                00FD   00F8   xx xx xx xx E0 15 E0 0B
CALL2                00FD   00F8   xx xx xx xx E0 15 E0 0E
CALL (in CALL2)      00FB   00F8   xx xx xx xx E0 15 E0 0E
```

If you use many levels of subroutine nesting, you can see why it is very important to maintain the integrity of the stack. The stack is the only way for the CPU to know where to return to after returning from a subroutine call. Do not forget that the stack may also contain variables embedded between the return addresses for each level of nesting. Hence, the stack can grow significantly in size. If it grows too large, you will have stack overflow. We will remind you again in Section 3.11 since this is a very important point.

Note: Please ensure that you have a reasonable understanding of subroutines. The next subject, interrupts, is more complicated. If you are having difficulty, review the subject of subroutines.

■ 3.6 CONCEPT OF INTERRUPTS

Interruption in the Kitchen. You may be peeling potatoes. Suddenly the pot boils over—an interruption. You quickly remove the pot from the stove. Then you continue peeling potatoes.

Microcontroller Interrupts. An external event occurs to cause the CPU to suspend whatever it is doing—an interruption. The CPU handles the interrupt and then resumes what it was doing. To handle the interrupt, the CPU jumps to a special routine called an interrupt service routine. This differs from a subroutine call, since they are precoded in specified locations of the program. Interrupts can occur at any time since outside events trigger them.[4]

Real-time control software is software that must read input changes when they occur and change outputs quickly (within a time limit). For example, the MCU must stop the motion of a robot arm as soon as it detects that the arm is in the target position. The software to handle the inputs and outputs is often handled by using interrupts.

"Typical" Application Program. Without interrupts, the program executes continuously. Figure 3.5 shows a very, very simplified situation. It ignores subroutine calls and nested sub-

---

[4]To be more specific, we are talking about *asynchronous events. Asynchronous* means that something happens unpredictably. There is no regularity.

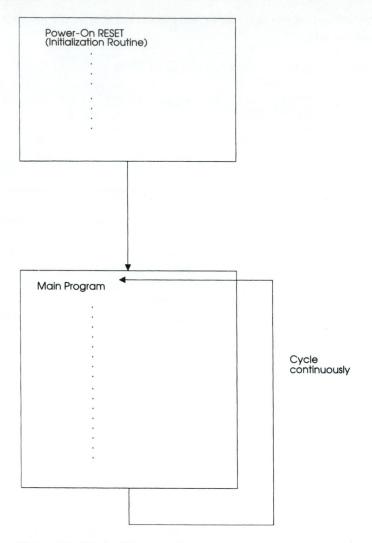

Figure 3.5 "Typical" Program Execution. (There really is no such thing as *typical*.)

routine calls. We will talk about power-on reset shortly. The diagram is also lying to some extent. Microcontroller systems often use interrupts to handle I/O in real applications.

3.6.1 Resets

A reset is a special type of interrupt. When a reset signal occurs, the CPU immediately jumps to begin execution of a routine to perform some initialization. Think of it as a "let's start over" routine. Unlike an interrupt, it does not return to the interrupted program. When an interrupt occurs, the CPU finishes executing the present instruction and then begins ex-

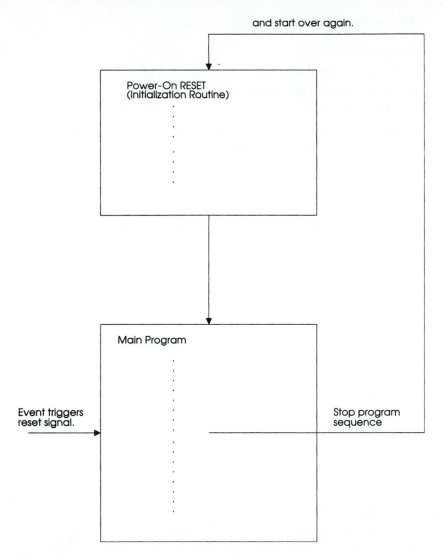

Figure 3.6 Reset. A reset causes transfer of control to restart the program.

ecution of the interrupt routine. The interrupt (not a reset) routine does return control back to the interrupted program.

A reset stops execution of the application program to do a special function. Usually, this function reinitializes the registers and memory. An example of a reset application is a situation when the program is waiting to read a sensor input. It executes a loop that it will exit when a reading is obtained. If the sensor is damaged, no signal will be received. The program then continually executes the loop. After a period of time, a timer generates a reset signal. The CPU responds by reinitializing itself and waits for further instructions. Refer to Figure 3.6.

The 68HC11 has four possible types of reset:

External $\overline{\text{RESET}}$ pin

Power-on reset (POR)

Computer operating properly (COP)

Clock monitor reset

Section 3.9.1 covers the first two; Section 3.9.2 covers the latter two. Note the convention as shown by $\overline{\text{RESET}}$. The line over the label means that a *low* signal at the reset pin causes a reset action. If a high signal causes the action, we would label the signal without the line above. You may recall that an overbar means inverse or complement in Boolean logic. Note that Figure 3.6 shows POR. It occurs when power is first applied to the microcontroller chip. The computer operating properly (COP) reset occurs if a special timer times out. The clock monitor reset occurs when the clock fails.

3.6.2   Interrupts

An interrupt suspends the execution of an application to do something else. But the application program can continue from where it left off after the interrupt routine has completed its task. An *interrupt request* is the occurrence of an interrupt signal. When it occurs the CPU must respond immediately to execute a routine specified by the interrupt. This specific routine is called an *interrupt service routine.* The microcontroller uses interrupts for critical tasks that cannot wait. Usually, these are input/output (I/O) operations. Refer to Figure 3.7.

An example of an interrupt application is a patient life support system in a hospital. The loss of a heartbeat signal generates an interrupt signal. Obviously, this is a critical situation. The microcontroller should respond immediately to warn us about the situation.

Typically, I/O operations use interrupts since the hardware cannot wait. For example, a keyboard may generate an interrupt whenever a key is pressed. Computers often execute a main program and read a keyboard input only when an interrupt occurs. It suspends the main program, reads which key was pressed, processes this information, and then returns to the interrupted program. Similarly, a printer generates an interrupt telling the CPU that it is available to receive data. Since a printer, a mechanical device, is slower than the CPU, it makes sense to let the CPU do other things instead of waiting for the printer.

Reentrant Subroutines.   Since real-time control software uses interrupts, subroutines should be reentrant. This refers to subroutines that can be called again before they have returned from a previous call. For example, a routine calls a subroutine to unpack a byte. During execution of this subroutine, an interrupt occurs. The interrupt service routine also calls the same subroutine to unpack a different byte. Hence, this subroutine has been entered again.

Note that a subroutine may need RAM space to store temporary (local) variables and parameters. If assigned absolute addresses were used for simple parameter passing (Section 3.4.4), the subroutine would not be reentrant. Calling it the second time would wipe out results for the first time it was called. Therefore, it is necessary for a subroutine to use the stack for local variables and parameters in order to be reentrant. Consider Listing 3.7 again. For the subroutine to be reentrant, the address for the variable "temporary" cannot be a fixed location.

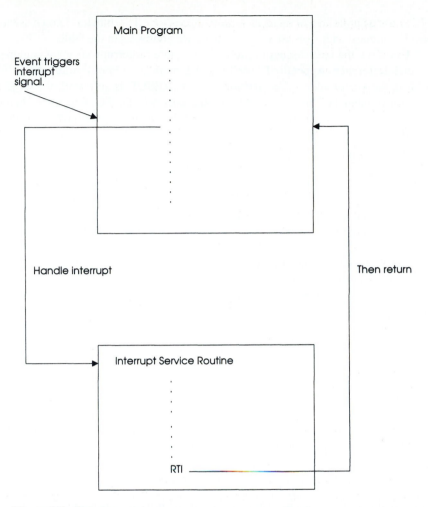

Figure 3.7 Interrupt

Different Interrupts. The 68HC11 supports many different types of interrupts. Some of these are generated by a signal at an external pin of the microcontroller chip. Others are generated when special bits set. A combination of hardware and software causes these bits to set. One type of interrupt is generated by software only. We examine all of these interrupts in the following sections. But first we should talk about *vectors*.

■ 3.7 INTERRUPT VECTORS

All resets and interrupts use *vectors*. A vector indicates the start address of reset or interrupt routines. In this section we sometimes omit the word *reset* since what applies to interrupts also applies to resets.

A *vector address* is a 2-byte memory location that stores a vector. The vector is the start address of an interrupt (or reset) routine. Typical Motorola processors such as the

68HC11 use a single area of memory to store the vectors. This block is known as the *vector table*. The highest addresses are reserved as vector addresses in the table.

To follow the next discussion, refer to the table "Interrupt Vector Assignments" in the Quick Reference (Appendix B) (or Table 9.2 in the *Data Book*). Addresses $FFFF and $FFFE is the vector address for $\overline{\text{RESET}}$. When $\overline{\text{RESET}}$ is activated, the CPU fetches the vector (contents of the vector address) and puts it in the PC. Execution then continues at the service routine address put in the PC. For example, we have the following in memory.

Address (hex)	Data (hex)	Interrupt source
FFFC	EA	COP Clock Monitor Fail (reset)
FFFD	87	
FFFE	EC	RESET
FFFF	04	

When $\overline{\text{RESET}}$ occurs, the CPU will execute a routine beginning at address $EC04. If a COP reset occurred, the CPU will execute a routine beginning at $EA87.

Figure 3.8 shows the operation of an interrupt called $\overline{\text{IRQ}}$ (interrupt request). For now, ignore the pseudovector. Note that $\overline{\text{IRQ}}$ has a vector address of $FFF2 and $FFF3 ($FFF2,F3). When the signal at pin $\overline{\text{IRQ}}$ goes low, the CPU responds by putting the vector contents, $00EE, in the PC. The CPU executes the instruction at $00EE, in this case, JMP $C15A.

PRACTICE

Indicate the start address for the $\overline{\text{RESET}}$, SWI, and $\overline{\text{XIRQ}}$ routines. The internal ROM contains the following hex data:

```
FFF0   00 EB 00 EE 00 F1 00 F4 - 00 F7 00 FA 00 FD E0 00
```

The internal RAM contains the following hex data (JMP opcode is $7E):

```
00E0   xx xx xx xx xx xx xx xx - 7E E0 00 7E C5 00 7E C1
00F0   5A 7E C1 9B 7E C3 67 7E - C2 56 7E E0 00 7E E0 00
```

Solution

RESET starts at $E000.
SWI starts at $00F4 and then jumps to $C367.
XIRQ starts at $00F1 and then jumps to $C19B.

The vector addresses are fixed. They are a permanent part of the chip hardware. But we can choose the start address of any interrupt routine by putting it into the vector addresses. These vector addresses usually end up being in ROM. Hence, you have to tell the chip manufacturer what data to put into these memory locations. If you are developing an application, you probably will not know this information until after you have finished de-

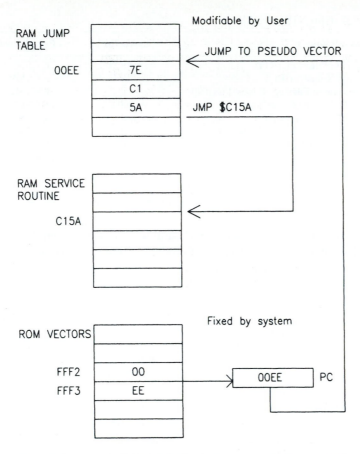

Figure 3.8 Vectors and Pseudovectors. This example shows an IRQ vector and pseudovector for the Motorola EVB board.

velopment. Yet you need the chip to develop the application. There are several solutions to this dilemma.

One of these solutions is that used by the EVB and EVBU boards. It uses something called a *vector jump table*. The contents of the vector jump table are called *pseudovectors* or jump vectors. These are in RAM. Each pseudovector reserves 3 bytes in RAM. At these locations, you can insert a jump instruction. You can have it jump to the address of your choice. Figure 3.8 shows how this works for the IRQ vector. The IRQ vector is $FFF2,F3. The pseudovector is $00EE. In this case its' 3 bytes contain the machine code for JMP $C15A. Hence, the actual IRQ service routine starts at $C15A. (An EVBU board would require extra RAM.) Your main program can set up the RAM jump table. For example:

```
*Initialize IRQ Jump Table

    LDAA  #$7E     ;Jump (extended) opcode
    STAA  $EE      ;IRQ pseudovector
    LDX   #$C15A   ;IRQ service routine start address
    STX   $EF
```

TABLE 3.1   VECTOR ASSIGNMENTS (IN HEX)

Vector address	Interrupt source	Vector	Pseudovector
FFF0, F1	Real-Time Interrupt (RTI)	00EB	7E C500
FFF2, F3	Interrupt Request (IRQ)	00EE	7E C15A
FFF4, F5	Nonmaskable Interrupt Request (XIRQ)	00F1	7E C19B
FFF6, F7	Software Interrupt (SWI)	00F4	7E C367
FFF8, F9	Illegal Opcode Trap (IOT)	00F7	7E C256
FFFA, FB	Computer Operating Properly (COP)	00FA	7E E000
FFFC, FD	Clock Monitor Fail (CMF)	00FD	7E E000
FFFE, FF	RESET	E000	xx xxxx

Table 3.1 illustrates this method in tabular form using the RAM data shown in the "Practice" section.

Other microcontroller trainer or evaluation boards use a similar system. If you are using microcontroller simulation, you can change the contents of the real vectors in the "simulated" ROM.

■ 3.8 INTERRUPT OPERATION

3.8.1 Interrupt Masks and Enables

Consider opening a door to be like an interrupt. Unlocking a door is like *enabling* the interrupt. Someone can open it anytime. Locking the door is like *masking* or disabling the interrupt. It prevents someone from opening the door.

For some interrupts you can *program* the microcontroller to ignore its interrupt request signal if you did not set a special bit called an *enable* bit. This is called *masking* an interrupt and these interrupts are called *maskable*. Each maskable interrupt has its own corresponding enable or *mask* bit. When an enable bit is set to 1, the CPU will respond to the interrupt request. If the enable bit is 0 and an interrupt signal occurs, the CPU will ignore it. The data sheets do not always use the same terminology. Depending on the interrupt source, they will use the words *interrupt enable* or *interrupt mask*. Essentially, they are opposites. To allow the CPU to respond to the interrupt request, you have to enable it. To make the CPU ignore the request, you mask the interrupt. Also, you can enable some interrupts by resetting the enable bit to 0 instead of setting it to 1. Similarly, you can mask an interrupt by setting the mask bit to 1 instead of resetting it to 0. You will have to refer to the data sheets to know how to enable/mask any specific interrupt.

Most of the time, you set or reset the enable/mask bits by writing to a specific I/O control or status register. In Section 1.3.2 we mentioned that a microcontroller has special registers to control I/O operation. The *Programming Reference Guide* is a useful guide to find the registers and bits.

Refer to the "Interrupt Vector Assignments" (Table B.2) in the Quick Reference (Appendix B). You will see that the 68HC11 uses many interrupts not covered so far but will be covered later. For now, concentrate on the vectors and masks. Most interrupts are maskable by the I bit in the CCR. That is, if the I bit is set, all these interrupts are masked. To mask them you can set the I bit using the instruction SEI. If you want to mask only some of these interrupts, you have to clear the I bit using the instruction CLI and then set the local mask bits for the interrupts desired.

To illustrate one example of using local mask bits, consider the interrupt Timer Overflow. For now, do not worry about what it means. It has a vector address of $FFDE,DF. Its local mask is TOI. There are two questions. Where is bit TOI? Do we set or reset TOI to mask the interrupt? To answer the first, look at the "Register and Control Bit Assignments" (Table B.1) in the Quick Reference (Appendix B). Search for the bit. You will find it as bit 7 of $1024, register TMSK2. To answer the second, you will have to look up the description of register TMSK2 in the *Data Book* or the *Programming Reference Guide*. You will find that TOI means "Timer Overflow Interrupt" enable. Hence, to mask the interrupt, you have to clear TOI. You can mask the interrupt by the following instructions:

```
LDX     #$1000        ;point to I/O registers
BCLR    $24,X $80     ;clear TOI bit
```

Note that the instruction mask (not the interrupt mask) for BCLR specifies that bit 7 of address $1000+$24 is to be reset.

PRACTICE

Enable the interrupt for SPI Serial Transfer Complete by setting bit SPIE.

Solution

```
LDX     #$1000        ;point to I/O registers
BSET    $28,X$80      ;set SPIE bit for local enable
CLI                   ;clear I to enable interrupts
```

3.8.2 Stacking the Registers

An interrupt service routine will probably change the register contents. Also, it needs to know where to return to when it has completed servicing the request. Unlike a subroutine, the return location for an interrupt is unpredictable. It should restore the interrupted program back to its original condition before the interrupt occurred. This means that all CPU registers must be restored. Congratulations if you thought that an interrupt uses the stack for this operation.

When an interrupt request occurs and it is enabled, the CPU will push all the CPU register values onto the stack before it fetches the vector. This process of pushing register values onto the stack is called *stacking the registers*. Refer to the "Interrupt Stacking Order" in the Quick Reference (Appendix B; Figure B.2). The CPU pushes PC, IY, IX, ACCA, ACCB, and CCR in this order. Hence, the top of the stack stores the CCR. Note that the CPU pushes the low byte of a 16-bit register before the high byte.

An interrupt service routine must use the instruction RTI to return control back to the program interrupted. When RTI is executed, it pulls the data from the stack to restore them to the CPU registers. Like subroutines, an interrupt service routine should balance the pushes and pulls; otherwise, it will not restore the CPU to its original condition.

When a reset signal occurs, the CPU does not stack the registers since a reset routine does not normally return control back to the program interrupted. A reset routine typically initializes the SP and sets up interrupt enable/mask bits to desired values.

PRACTICE

What is pushed into RAM after an interrupt request occurred? The CPU registers are (after current instruction executed)

ACCD	$5C76
IX	$1020
IY	$B67F
CCR	$81
PC	$E915
SP	$00D6

Solution

```
00C0   xx xx xx xx xx xx xx xx  –  xx xx xx xx xx xx 81 76
00D0   5C 10 20 B6 7F E9 15 xx  –  xx xx xx xx xx xx xx xx
```

Now, SP = $00CD and CCR = $91.

3.8.3 Interrupt Priority and Nested Interrupts

What happens if two interrupt signals occur at the same time? Or if a reset and an interrupt signal occur together? Interrupts and resets have assigned priority levels. If two or more unmasked interrupts or resets request service at the same time, the CPU will respond first to the one with the highest priority. Note that resets have higher priorities than interrupts (Quick Reference [Appendix B]). After stacking the registers, the CPU sets the I bit in the CCR to prevent any other interrupts from occurring. If the interrupt service routine does not reset the I bit, only nonmaskable interrupts or resets can interrupt this service routine. The interrupt service routine could clear the I bit to allow itself to be interrupted. Execution of RTI restores the original condition of the I bit. Then the lower-priority interrupt can be serviced.

It is possible to have nested interrupts. Many systems try to avoid this by not enabling other interrupts. If an interrupt service routine enables other interrupts (clears bit I), the software is more complicated and there is more danger of stack overflow. But nonmaskable resets and interrupts will always be able to interrupt a low-priority maskable interrupt.

It is possible to change the priority level of any one of the maskable (I-bit-related) interrupts using the highest-priority I interrupt (HPRIO at $103C) register. The *Data Book* explains in detail how to use this register.

■ 3.9 HARDWARE INTERRUPTS AND RESETS

3.9.1   Power-on and External Reset

Have you ever wondered how the microcontroller chip knows where to start executing a program? How does it initialize its registers? The RESET routine does this. Whenever the chip is powered up, it executes the RESET routine. It also executes the RESET routine whenever the external $\overline{\text{RESET}}$ pin is activated. Both of these use the RESET vector address $FFFE,FF. They have a higher priority than any other reset or interrupt.

There are some differences between the two. Power-on reset begins after power is supplied to the chip for 4064 clock cycles. A clock delay is built in to allow the clock oscillator to stabilize. This is important since some applications have critical timing requirements.

External hardware reset requires that the $\overline{\text{RESET}}$ pin be at logic low for at least six clock cycles. If it is, reset begins when the $\overline{\text{RESET}}$ pin is at logic high again. The $\overline{\text{RESET}}$ pin can be an output as well as an input. The clock watchdog (COP) reset and the clock monitor reset will drive the $\overline{\text{RESET}}$ pin low for four clock cycles when they occur. Hence, the six-clock-cycle delay enables the CPU to distinguish an external reset from the clock-related resets.

3.9.2   Other Processor Resets

There are also Computer Operating Properly (COP) failure and COP Clock Monitor Fail resets. We call them processor resets since both occur when something goes wrong with processor operation.

COP Failure Reset.   COP failure occurs when the CPU gets "hung up" for a period of time. This means that the CPU is not executing certain sections of code within an allotted time period. This could compromise the integrity of your control system. For example, your system is controlling a robot. It is supposed to control the robot's hand and monitor collision detectors for safety. Due to hardware failure or a software bug, the CPU is spending too much time executing the hand control part of the program and neglects the collision detector part. Thus, the 68HC11 can sense within limitations when it is not executing certain sections of code. It then triggers a COP failure reset.

Another situation may occur in an environment where there is a lot of electrical noise. This occurs in an automobile engine control system because of electromagnetic interference (EMI) generated by spark plugs and electric motors. This noise can change data in the microcontroller system such that the software can *run away*. When software is out of control, it can overwrite vital data in the system's memory, which in turn can cause further faults. One way to recover from this kind of fault is to turn power off and then on again to perform a power-on reset (POR). However, this is not practical in an automobile. Another way is for the system to detect software runaway and automatically trigger a reset.

This works as follows: The 68HC11 has an internal *watchdog* timer. Many other control systems also use some kind of watchdog. Basically, a watchdog is a timer that the system must continually reset before it times out. If it does not reset the timer, an error condition occurs. This is somewhat like timed chess. Each move is allowed a maximum time limit. Players must make their move before the time limit expires. If a player does not make a move in time, he or she loses the move. When a move is made within the time limit, the player resets the clock for the opposing player.

In the 68HC11, if the watchdog times out, a COP failure reset occurs. To prevent this, the application program must "reset" the watchdog timer on a regular basis. Then the watchdog timer starts timing all over again. To "reset" the watchdog timer, the program must write a $55 to the COP reset register (COPRST at $103A) followed by writing a $AA to COPRST. To enable the COP failure reset, that is, to determine whether or not a watchdog timeout causes reset, the power-up RESET routine should turn off the COP system disable (NOCOP as bit 2) bit in the configuration control register (CONFIG at $103F) to enable the COP system. The Motorola manuals use the word *reset* for two different actions when describing the COP failure reset. One refers to the software action of *clearing* the watchdog timer; the other refers to the watchdog timing out (COP failure).

CONFIG is a critical register. You cannot change its contents using normal store instructions because the register is actually an EEPROM cell that can be changed only by using a "programming" procedure. We describe this procedure in Chapter 6.

Assuming that the RESET routine has enabled the COP system, the application program can "reset" the internal watchdog timer as follows.

```
AGAIN   LDAA    #$55        ;first step to reset watchdog timer
        STAA    $103A       ;writing $55 to COPRST
*                           ;arms the clearing mechanism

*               .           ;many instructions
                .
                .
        LDAA    #$AA        ;second step to reset watchdog timer
        STAA    $103A       ;writing $AA to COPRST
*                           ;clears the COP timer
                .
*               .           ;many instructions
                .
        BRA     AGAIN       ;repeat application
*                           ;continuously
```

Most microcontroller programs execute continuously, repeating themselves (see Figure 3.5). It is best to space the two watchdog timer reset steps as far apart as possible. If the program hangs up in a section between the two steps, the watchdog timer times out. This causes the COP failure reset routine to execute. What should this routine do? That depends on many factors in your application.

Clock Monitor Reset.   This reset triggers if the CPU clock drops below a frequency of 10 kilohertz (kHz). To enable the clock monitor reset, you set the clock monitor enable (CME as bit 3) bit in the system's configuration options (OPTION at $1039) register. You may ask the obvious question: How can the system cause the reset if the clock fails completely? You need the clock to execute instructions in the first place! This is not much of a problem. The clock is controlled by a crystal. It takes time for a crystal to fail. When it fails, it can still generate approximately 1000 more cycles due to its internal capacitance. If lower power consumption is more important than speed, you would not enable this reset. The 68HC11 consumes less power when running at a lower clock frequency.

3.9.3   Illegal Opcode Trap

First, what is a trap? It is the part of the program that aborts program execution when an error occurs. Usually, it gives a message to the user to indicate (or hint) what the error is. Sometimes these are called error traps. For example, a divide by zero may be a trap. You have to write your own code to create most traps.

However, the 68HC11 comes equipped with one built-in trap, the illegal opcode trap. After the CPU fetches the first byte in an instruction, it decodes it to determine what the instruction is. Due to data corruption or a software bug, there may be the rare occasion when it cannot decode the byte. For example, the CPU has just finished executing an instruction and it fetches the byte $4E. This does not correspond to any instruction. In this case, it executes the illegal opcode trap (IOP) interrupt. Alternatively, the CPU may have fetched $18, which it decoded as a prebyte, then fetched the second byte as $11. Again, this does not correspond to any opcode.

What you write as an IOP routine depends on the application. If you are developing an application, you can use it as a development tool. The routine could halt the program and send you an error message.

3.9.4   Nonmaskable Interrupt (XIRQ)

You would use the nonmaskable (XIRQ) to handle the highest-priority interrupts. By default, when you power up the 68HC11 or RESET it, XIRQ is masked. That is, the X bit in the CCR is set. To unmask XIRQ, you use the TAP instruction to clear bit X. Once the program has cleared it, executing another TAP will not set X again. The only way to set X is to RESET the 68HC11. The following shows some instructions to reset X.

```
TPA             ;get original CCR
ANDA   #$BF     ;reset bit 6, (X)
TAP             ;put it back in CCR
```

Figure 3.9 illustrates the operation of XIRQ. One of the external pins is for the XIRQ input. For example, a printer control line is connected to it. To request service from the controller, the printer forces this line low. The controller responds by executing the XIRQ routine, which services the printer. Note that bits X and I also set temporarily while the service routine is executing. The processor restores these bits to their original condition after returning from the interrupt routine.

The sequence of operation is as follows: While the controller is executing the instruction LDAA #1C, the signal at pin $\overline{\text{XIRQ}}$ goes low. First, it completes executing the instruction. Then, because the X bit is zero, the microcontroller stacks the registers and sets the X and I bits in the CCR. This prevents other interrupts from interfering with the rest of this operation. The controller fetches the contents of vector address $FFF4,F5 and loads it into the PC. This happened to be $E71F for this example. It then executes the routine starting at address $E71F. If the XIRQ service routine does not reset the I bit, no other interrupt can interfere with its operation. When the service routine executes RTI, it pulls the stored register values from the stack. This includes the return value of $E125, the value of $1C in ACCA, and the original CCR contents (in which X = 0). Sometimes people refer to XIRQ as pseudo nonmaskable because you really can mask it by not clearing bit X in the CCR after RESET.

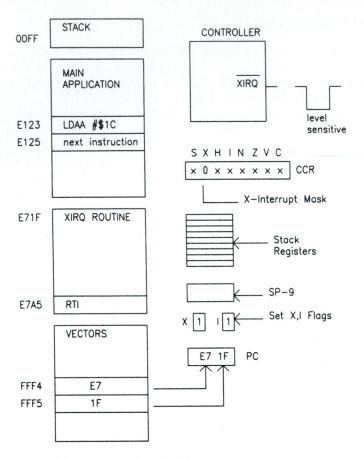

Figure 3.9 Example of an XIRQ Interrupt

Simulator Demonstration.   It would be particularly instructive to try an XIRQ simulation. Listing 3.8 shows a program that can be used to demonstrate XIRQ operation using the demo version of simulator. It more or less mimics the situation shown in the previous example of an XIRQ interrupt as shown in Figure 3.9. Due to the memory configuration of the demo simulator, it is necessary to use some different addresses from those used in the previous example. In this section, we will offer basic guidance for simulating the example.

```
*Listing 3.8
*Demonstrate xirq interrupt for use
*with demo version of simulator

      ORG  $FF00    ;start address for demo sim


*---------------------------------
*Start MAIN

MAIN
```

```
        LDS    #$FF     ;You must define
*                       ;the stack first
        LDX    #$1122   ;modify some registers for demo
        LDY    #$3344
        LDAB   #$55


        LDAA   #$1C     ;not possible to interrupt yet
*Enable XIRQ interrupt
        TPA             ;get original CCR
        ANDA   #$BF     ;reset bit 6, (X)
        TAP             ;put it back in CCR
        LDAA   #$1C     ;now possible to interrupt
        LDAB   #$C1     ;next instruction
HERE    BRA    HERE     ;end of main
*End of MAIN
*--------------------------------
RXIRQ
*RXIRQ        is the label to start XIRQ service routine
*rest of routine code follows it

        LDX    #$6677   ;modify some registers for demo
        LDY    #$8899
        LDAB   #$AA
        LDAA   #$BB
        RTI             ;return from RXIRQ
*--------------------------------


*Define vectors, assembler pseudo operations (directives)
*to be discussed in Chapter 4!

        ORG    $FFF4
VXIRQ   FDB    RXIRQ    ;Start address of RXIRQ
VSWI    FDB    HERE     ;point to HERE for now
VILLOP  FDB
VCOP    FDB    HERE
VCMF    FDB    HERE
VRES    FDB    MAIN     ;Start address for Reset


*--------------------------------
*End of Program Code
```

Run the simulator and open the file L3-8.ASM. Assemble it and examine the generated listing. View the following.

CPU Registers

Memory List at $FFF0. This will show you the vectors.

Memory List at $F0. This will show you the stack.

Other Pins. This will show you pin XIRQ.

Differences from Figure 3.9 are that the XIRQ vector is $FF18 instead of $E71F, and hence, the corresponding start address of the service routine RXIRQ is $FF18. The instruction to be interrupted will be at address $FF12 instead of $E123. The best way to monitor interrupt handling is to single step.

Single step to the first LDAA instruction (address $FF0C). Toggle pin XIRQ low. Single step the LDAA instruction. Note that the low XIRQ signal is ignored and execution continues with the next instruction. Toggle pin XIRQ high again.

Single step to the next LDAA instruction (address $FF12). Note now that bit X in the CCR is low. Single step the LDAA instruction and then toggle pin XIRQ low. Single step. Now program execution is at the beginning of the RXIRQ service routine. The PC should be $FF18. Observe the stacked registers by comparing the stack contents to the CPU registers. Toggle pin RXIRQ high (important because we want the signal high when returning back to the main program). Single step to RTI instruction. Note that the CPU registers were modified. Single step the RTI instruction. Note that the CPU registers were restored to their original values. Program execution will resume at the LDAB instruction in the main program (address $FF14). Feel free to repeat and modify the example to experiment with interrupts.

PRACTICE

Assume that bit X is clear and register SP is initialized to $004A. Write a program that outputs the ASCII character 'H' using subroutine OUTA (see "Practice" in Section 3.4.3) every time the signal at pin $\overline{\text{XIRQ}}$ goes from a logic high to a logic low. Some contents of memory are

```
FFF0   00 EB 00 EE 00 F1 00 F4 - 00 F7 00 FA 00 FD E0 00
```

Solution

```
*Solution
    ORG    $E100      ;arbitrary start address
*                     ;set up XIRQ pseudovector
    LDAA   #$7E       ;store opcode for JMP
    STAA
    LDD    #$E10B     ;then store RXIRQ start address
    STD    $F2
HERE                  ;just waiting  for interrupts
    BRA    HERE

*Service routine RXIRQ (interrupt XIRQ)
*Outputs 'H' when a high-to-low transition
*occurs at pin XIRQ.
RXIRQ                 ;note routine start address is $E10B
    LDAA   #$48       ;get ASCII code for 'H'
    JSR    OUTA       ;then send it
    RTI               ;and return from interrupt

*Note that the stack should start at an address lower than
*the vector jump table if both are using the internal RAM.
```

C Code Example. The following C Listing illustrates an interrupt example. In this case, the program increments a memory location whenever the XIRQ signal is low. One thing to note is the use of `asm` statement to embed assembly language statements in the C source code. In this case, we are setting up a jump vector for use in Buffalo and enabling the XIRQ interrupt. The final asm statement is a fudge of sorts. The compiler translates functions into subroutines. We are forcing an inclusion of the RTI instruction. The compiler will add an RTS instruction after it. The RTS instruction becomes redundant because it will never be executed when the program is run. In practice it may be appropriate to code interrupt service routines directly in assembly to make them more efficient.

```
/* C Interrupt Example */

/* Declarations */
unsigned int Data;   /* Data to be incremented */

void main(void)
{
        Data = 0;       /* initialize data */
        /* set up jump vector for buffalo */
        asm(" ldaa #$$7e");
        asm(" staa $$f1");
        asm(" ldd #_rxirq");
        asm(" std $$f2");
        /* enable XIRQ */
        asm(" tpa");
        asm(" anda #$$bf");
        asm(" tap");
        while (1);      /* Loop forever */
}

void rxirq(void)
/* Interrupt Service Routine for XIRQ*/
{
        Data++;
        asm(" rti");  /* This is an interrupt service routine */
}
```

3.9.5 Interrupt Request (IRQ)

Interrupt request (IRQ) is maskable by setting bit I in the CCR. Two different interrupt sources will use the same vector address of $FFF2,F3. These are a low signal at external pin, $\overline{IRQ}$, and a parallel I/O operation. We shall discuss only the former in this section. In Chapter 9 we discuss parallel I/O. Listing 3.9 illustrates both masking and unmasking.

You can mask IRQ using the instruction SEI. Similarly, you can unmask it using CLI. It is also possible to program pin $\overline{IRQ}$ to be edge sensitive or level sensitive. If it is edge sensitive, a high-to-low transition causes interrupt. When the IRQ service routine returns back to the interrupted program, a low signal at $\overline{IRQ}$ does not cause another interrupt. For another interrupt to occur the signal at $\overline{IRQ}$ has to go high and then low again.

To make IRQ edge sensitive, you have to set the IRQ Select Edge Sensitive Only (IRQE at 5) bit to 1 in the OPTION (at $1039) register. Some bits in OPTION are time protected. That means that you can change the bit only within the first 64 clock cycles after a RESET. After 64 cycles, it is a read-only bit. This means that only the RESET routine can make IRQ edge sensitive. This minimizes the possibility of accidental changes to IRQ *configuration. Configuration* is a term meaning to set up an operation a certain way.

3.9.6   Interrupt Configuration

To set up interrupt operation, it is necessary to configure. During the configuration steps it is prudent to disable interrupts. Typically after configuration is complete, interrupts are enabled as required. Listing 3.9 demonstrates a simple IRQ configuration. The program increments memory location $0000 every time there is high-to-low transition of pin IRQ.

```
*Listing 3.9
*Demonstrate IRQ Configuration


        ORG    $E000       ;modify start address if necessary
*---------------------------------
*Start MAIN

MAIN
        LDS    #$FF         ;You must define
*                          ;the stack first
        SEI                 ;mask (disable) IRQ interrupt
        LDX    #$1000       ;point to register block
        BSET   $39,X $20    ;turn on bit OPTION
        CLI                 ;unmask (enable) IRQ interrupt
        NOP                 ;let interrupt drive program
HERE    BRA    HERE         ;end of main

*End of MAIN
*---------------------------------
RIRQ
*RIRQ is the label to start IRQ service routine
*rest of routine code follows it
        INC    $00
        RTI                 ;return from RIRQ


*  -----------------------------------
*Define vectors, assembler pseudo operations (directives)
*to be discussed in Chapter 4!


        ORG    $FFF2
VIRQ    FDB    RIRQ         ;Start address of RIRQ
VXIRQ   FDB    HERE         ;point to HERE for now
VSWI    FDB    HERE
VILLOP  FDB    HERE
VCOP    FDB    HERE
VCMF    FDB    HERE
VRES    FDB    MAIN         ;Start address for Reset
        *---------------------------------
*End of program code
```

3.9.7 Interrupt Polling

A microcontroller has a lot of internal I/O systems built in. Each one can generate a unique interrupt. Hence, you would not normally need to use interrupt polling. However, we discuss it here in case you need to use it.

Let's say you have three devices that can generate an external interrupt request. Each one will drive an output line low when it wants service. You can connect these outputs to an AND gate and then connect the AND gate output to $\overline{IRQ}$ (or $\overline{XIRQ}$). Note that normally all inputs to the AND gate are high. Hence, its output is also normally high. When one of the devices requests service, the signal at $\overline{IRQ}$ goes low and sets (or resets) an interrupt request bit in its status register. Now the microcontroller knows that at least one of these devices has requested service. It has to determine which one. To do this, it uses a technique called *polling*. It reads the status register of each device to determine which one has set the interrupt request bit. When it finds the set bit, the service routine will jump to the appropriate code to service the device.

■ 3.10 SOFTWARE AND CPU CONTROL INTERRUPTS

3.10.1 Software Interrupt (SWI)

Software interrupt (SWI) is a special instruction that has several applications. When it is executed, it behaves like an interrupt by stacking the registers, setting the I bit, and fetching the contents of its vector address $FFF6,F7. You can use it to simulate hardware interrupts during system development. You can also use the SWI routine for debugging software. For example, a development system may allow you to set *breakpoints* in your program. A breakpoint marks the location of an instruction. When the program executes the program, it will halt execution of the program when it reaches the breakpoint. This allows a developer to see what effect the program has up to this point. For example, the EVB board implements breakpoints by placing an SWI instruction at each address specified as a breakpoint. The SWI service routine handles the breakpoint functions.

Another use for SWI is as a subroutine that preserves all the registers automatically. If you want to use SWI for more than one subroutine function, you will have to set up a consistent system. For example:

ACCA = function code number
IX = parameter

The SWI service routine would initially check ACCA to determine which function it should execute. It would then jump to the section of code for that function. Each function would have an RTI instruction at the end.[5]

---

[5]DOS assembly language programmers may recognize this use of SWI as being similar to using the DOS INT function calls.

3.10.2 CPU Control and Monitoring

We mentioned in Section 3.9.2 that the microcontroller itself has to be controlled. Some internal operations affect the operation of the CPU itself. We have already discussed the COP failure reset and the clock monitor fail reset. The real-time interrupt (RTII) generates an interrupt whenever a specified period of time expires. The *Data Book* describes how to enable and set up this interrupt.[6] Some applications do not need to be executing a program all the time but need to be "woken up" when required. By using either the WAI or STOP instruction, the 68HC11 can reduce its power consumption while waiting to be woken up. Motorola calls these low-power modes.

WAI Instruction.   When WAI is executed, it stacks the registers and waits for an unmasked interrupt, suspending execution. This reduces power consumption although the clock is still ticking. Hence, only an unmasked interrupt will wake up the controller. Some of the input/output (I/O) subsystems in the controller may still be active and cause the interrupt that terminates the WAI instruction. Note that I/O subsystems are introduced in Chapter 8. For now, consider them as specialized modules to handle I/O. The *Data Book* explains how to reduce power consumption further by disabling these systems.

STOP Instruction.   The effect of executing STOP depends on the S bit in the CCR. If S is set, the CPU treats STOP like a No Operation (NOP) instruction and continues on to execute any instruction following STOP. If the S bit is reset, all internal clocks in the CPU halt, thus halting execution. This puts the 68HC11 in the state of lowest power consumption short of turning off power completely. To wake up the controller, use a $\overline{\text{RESET}}$, $\overline{\text{XIRQ}}$, or $\overline{\text{IRQ}}$ signal. $\overline{\text{IRQ}}$ must be unmasked (I bit clear). The effect of an $\overline{\text{XIRQ}}$ signal depends on the X bit. If the X bit is set, processing will continue with the instruction immediately following the STOP instruction. If the X bit is clear, the processor performs an XIRQ operation. $\overline{\text{RESET}}$ causes the power-up reset operation.

■ 3.11 THE KISS OF DEATH: STACK OVERFLOW

We have already mentioned the importance of the stack several times. Many software problems occur when something corrupts the stack or the stack pointer (SP). We shall mention stack overflow because if the stack does overflow, the game is over. The only way to recover is to start all over again using power-on reset (POR) or by asserting the $\overline{\text{RESET}}$ line.

Stack overflow may occur because memory has a finite size. Hence, the stack can only have a finite size. Since the stack decrements the stack pointer to grow, the RESET routine should initialize the stack pointer to the highest available RAM address. A typical microcontroller application program is in nonvolatile memory such as ROM. In this case it is safe from writing to itself accidentally. Any modifiable variables (i.e., RAM variables) that are not saved in the stack could be saved in the lowest possible addresses. Stack overflow occurs when the stack pointer decrements enough such that push operations are overwriting these other variables.

---

[6]We will use "RTII" to indicate real-time interrupt to distinguish it from the return from interrupt instruction (RTI). The manuals will show "RTI" used for both.

If the stack continues to grow (e.g., many levels of nesting) the stack pointer will eventually decrement past $0000 to wrap around to $FFFF. At this point, pushing data does no good because this area has read-only memory.

If you have more pulls than pushes, there is a possibility of stack underflow. If you need to have program code in RAM, keep it as far away from the stack as possible. For example, the EVB board already initializes the user stack area to start from $004A to $0000. This is sufficient for most purposes. If you are using the EVB board, you should start your RAM code at address $C000 (user RAM space). If you find the page 0 RAM insufficient, initialize SP to $DFFF, the highest address in user RAM. This will not guarantee that you won't have stack problems, but it helps.

■ 3.12 SUMMARY

The stack is a reserved area in RAM for storing data. Some instructions and operations use the stack to store the CPU register states. The stack pointer (SP) register contains the address immediately above the stack. The CPU pushes data onto the stack to store it and pulls it from the stack to retrieve it. Pushing decrements the SP and pulling increments the SP. The RESET routine should initialize the SP value.

Subroutines are a miniprogram within a program used to perform a sequence of operations. This same sequence of operations usually has to be repeated several times. A "main" program calls a subroutine to perform a specific sequence of operations. A subroutine may call another subroutine. The latter subroutine is nested within the former subroutine. The program calls a subroutine using the JSR or BSR instruction. These instructions push the return address onto the stack. Each subroutine returns back to where it was called from using the RTS instruction. The RTS instruction pulls the return address from the stack and puts it back into the PC.

You should use a consistent convention when using subroutines. One of these is to use calling and return registers to pass information (parameters) to the subroutine and pass the results back to the calling routine. Other ways are to use reserved absolute RAM addresses or the stack. The latter is more flexible, yet it may be more complicated than necessary for your application. It is important to document subroutines. One should be able to use a subroutine in a code without knowing the subroutine's program code.

Some signals interrupt the execution of a program to execute a special service routine and then return back to the interrupted program. These are called interrupts. Resets are similar but do not return control back to the interrupted program. When an interrupt occurs, the CPU stacks the registers, sets the X (and sometimes I) bit in the CCR, and gets the service routine start address from an assigned vector. The service routine has an RTI instruction to return to the interrupted program. It pulls the stored register values from the stack to restore them to the registers. A reset does not stack the registers and "usually" does not have an RTI instruction.

An interrupt may or may not be masked by setting or clearing a mask/enable bit. Interrupts and resets have assigned priority levels. When two or more occur simultaneously, the CPU responds to the one with the highest priority first. Refer to the Quick Reference (Appendix B) or Motorola data sheets for a list of interrupts/resets, vectors, and mask bits.

Stack overflow or underflow is a common cause of software failure. You should avoid it.

■ EXERCISES

1. Draw the stack after execution of the following instructions in the program of Listing ex3a. Include the stack contents and the stack pointer.
   (a) JSR MWORD
   (b) JSR DCHEK
   (c) JSR WCHEK
   (d) RTS   ;in DCHEK
   (e) PULY   ;in MWORD
   (f) BRA DONE

   Assume the following initial hex numbers are in page0 RAM.

   ```
   0000  00 10 00 20 00 00 00 00 - xx xx xx xx xx xx xx xx
   ```

   ```
   *Listing ex3A

   ***************************************************
   *FILE: ex3a.asm
   *Demo program that moves a word from one block of
   *memory to another.  It moves only the first eight
   *characters of any word longer than eight characters.
   *A delimiter marks the end of a word.  Possible
   *delimiters are a comma or a whitespace character
   *such as a space, tab or carriage return.
   ***************************************************

   *-------------------------------------------------
   *Program Segment
           ORG     $E000
   *User's main code
   *User must enter src and dst addresses and
   *src bytes (the word with delimiter)

   MAIN
           LDS     #$4A
           LDX     >$0000     ;note extended addressing
           LDY     >$0002     ;mode
           LDAB    #$08
           JSR     MWORD
   DONE    BRA     DONE


   MWORD
           PSHA
           PSHB
           PSHY
   ```

```
AGAIN
        CMPB    #$0
        BEQ     EXIT
        LDAA    0,X
        JSR     DCHEK
        BEQ     EXIT
        STAA    0,Y
        DECB
        INX
        INY
        BRA     AGAIN

EXIT
        PULY
        PULB
        PULA
        RTS

*following subroutine sources taken from
*EVB buffalo monitor

DCHEK   JSR     WCHEK
        BEQ     DCHEK1      ;jump if whitespace
        CMPA    #$0D
DCHEK1  RTS

WCHEK   CMPA    #$2C        ;comma
        BEQ     WCHEK1
        CMPA    #$20        ;space
        BEQ     WCHEK1
        CMPA    #$09        ;tab
WCHEK1  RTS

*End Program Segment
*-----------------------------------------------------
```

2. Write down the pseudocode for the program of Listing ex3a. See also Question 3.

3. Complete the documentation of Listing ex3a. See also Question 2.

4. Write a subroutine that implements the documentation shown in Listing ex3b. Also write a main program that calls it and branches to PR if the string is printable. Otherwise, the main program branches to NOPR if a nonprintable character (other than CR) is found in the string.

```
*Listing ex3b
```

```
*Subroutine stlen(x,b,v)
*Determines the length of a string of characters
*up to a maximum length of 40 characters.  End of
*string is marked by a carriage return (CR).
*Calling Registers
*   IX = start address of string
*Return Registers
* ACCB = length of string
*      = $29 if string length > 40
*IX = address of first nonprint char, including CR
*    = start addr + 40 if CR not found
*Error Code
*V = 1 if nonprintable char found other than CR
*V = 0 if string is printable
```

5. Write a subroutine to save the first consecutive even bytes pointed to by IX onto the stack. It is not necessary to preserve any registers. *Hint:* Refer to Section 3.5.1.

6. Sometimes a program inputs a code to perform a certain task depending on what the code is. It probably uses a "command table" to look up the code and the start address of a subroutine that each code has associated with it. Often, the code can be several bytes long. In our case it will be only 1 byte. You will develop such a program. Address $00 contains an ASCII character that represents a code. Address $01 contains the operand that the command code operates on. These commands are as follows:

> D ⟨opr⟩ : double, 2×⟨opr⟩
> H ⟨opr⟩ : half, ⟨opr⟩/2
> N ⟨opr⟩ : negate, −1×⟨opr⟩

This program could have a look-up command table in memory with the following format.

```
START_TABLE
      ASCII code for 'D'
      high byte of D routine address
      low byte of D routine address
      ASCII code for 'H'
      high byte for H routine address
      low byte for H routine address
      ASCII code for 'N'
      high byte for N routine address
      low byte for N routine address
      $FF to mark end of table
```

If the user enters an unknown code such as 'S', the program should return an appropriate error code. Write the program using subroutines. You may wish to use the following instructions.

```
*                     ;IX contains command table
*                     ;address of code
    LDX   $01,X  ;get command start address
    JSR   $00,X  ;execute command
```

Although the tasks in this question are simple, you can extend this technique to perform complex tasks.

7. The initial value of PC is $E000. Memory contains the following hex data (machine code):

```
E000   4F A6 52 08 26 FC 20 FE - xx xx xx xx xx xx xx xx
```

What happens when this program executes?

8. Assume that the microcontroller has the program in Listing ex3a (Question 1) with initial conditions as follows:

```
CCR bits: S = 1, X = 0, I = 0
```

It also has initial hex data as follows:

```
0000     00 08 00 20 00 00 00 00 - 74 65 73 74 20 69 74 2E
0010     0D 00 00 00 00 00 00 00 - xx xx xx xx xx xx xx xx
FFF0     F0 00 F1 10 F2 22 F3 04 - F4 56 F5 30 F6 8A F7 54
```

An XIRQ signal goes low during execution of STAA $00,Y when ACCB is $06. Draw the stack before the interrupted instruction and before the XIRQ service routine executes its first instruction. Also show the CCR and the PC before the XIRQ routine executes its first instruction. Is there a danger of stack overflow?

9. Repeat Question 8 using an IRQ signal instead of an XIRQ signal. The IRQ signal goes low during execution of PULY in subroutine MWORD.

10. Write a RESET service routine starting at $E000 that enables the following:

> XIRQ
> External IRQ as edge sensitive
> COP Clock Monitor Fail
> Timer Output Compare 1
> Timer Input Capture 1
> Pulse Accumulator Overflow

It should disable all other interrupts and resets if possible. It should initialize the SP to $004A and jump to $C000 when it is done.

11. Rewrite the program of Listing ex3a (Question 1) such that subroutines DCHEK and WCHEK are SWI service routines. To execute DCHEK, load ACCB with $02 before executing SWI. To execute WCHEK, load ACCB with $01 before executing SWI. This may be a more complicated technique than necessary but it is a good brainteaser exercise in stack manipulation.

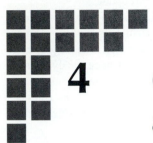

4 Cross Assembly and Program Development

OBJECTIVES

After completing this chapter, you should be able to

- Use the AS11 cross assemblers and assembler directives for source code development.
- Organize code into segments.
- Describe the assembly two-pass process.
- Describe and use the various assembly output files for object code and listings.
- Identify and determine development tools such as simulators, evaluation boards, and emulators.

■ 4.1 INTRODUCTION TO PROGRAM DEVELOPMENT

In this chapter we explain how we can use a computer to create assembly language programs for the microcontroller. The collection of computer programs used to do this is called a *development system*. Various development systems are available. Although they differ in details, they do the same thing using similar methods. Here we explain the use of a development system in general terms. We use a single specific system in the illustrative examples.

In this book we use terminology and describe procedures for a person using a personal computer (PC). However, many program developers also use computers running the UNIX or Mac operating system. This is especially true for a development system for Motorola processors. Essentially, both operating systems produce the same results.

First, you must create the *source code*. This must be a *text file*. A text file is one that contains only standard ASCII characters. Most word processor files also contain special nonstandard characters for such special features as letter size, font style, and italics. Hence, the normal word processor file is not a text file.

You can create the source code using a text editor program. Examples of these are Edit in DOS and Notepad in MS Windows.[1] Alternatively, you can use a word processor as long as you save the result as an ASCII text. Most word processors provide this option.

The next step is to use an *assembler* to translate the source code into *object code*. (At this point you may wish to review Section 2.2 which describes assembly.) The assembler reads the source-code file and produces an object-code file. The object code contains the machine code plus other information. But it does not necessarily contain the final address assignments for each instruction. It may reserve operand bytes instead of assigning the final operand addresses. The assembler also produces a *listing* file. The listing file is a text file that cross-references source-code instructions to their translation into hex.

Some development systems require a *linker*. The *linker* combines one or more object-code files to produce a *hex file*. The hex file contains information about the final machine code, including final address assignments. A *loader* converts the hex file into an executable form called a *binary file*. In Section 4.6 we describe these files in more detail.

If you are creating a 68HC11 assembly language program using a computer, you must use a cross assembler. A cross-assembler program runs on a system using one type of computer. But unlike a resident assembler, it assembles the source code for a different processor. For example, we use the Motorola AS11 Cross Assembler for illustration. It runs in an Intel 80x86-based operating system to produce object code for the Motorola 68HC11. A resident assembler for the 68HC11, on the other hand, is a 68HC11 program used to translate other source code to object (or machine) code.

In this chapter we use the Motorola AS11 Cross Assembler as the primary means to illustrate examples. It is a typical cross assembler. We will not cover every instruction or variation of instructions. For complete information about this assembler, refer to the *Cross Assembler Manual* (see "Manuals" and "Sources" in Appendix C). Motorola's Freeware Cross Assembler (AS11) is useful because it is a "free" assembler. The reader can obtain AS11 from the Web or the disk accompanying this book. Note that AS11 is usually supplied with evaluation boards (we'll say more about them later) such as the EVB and EVBU. The GNUAS6811 assembler is another free assembler. Other assemblers also exist and most have some different syntax requirements. For example, AS11 and GNUAS6811 have different rules for handling comments. The ICC11 C compiler produces assembly code that is compatible with the GNUAS6811. The THRSim11 simulator handles source code compatible with the AS11.

■ 4.2 FORMAT OF THE SOURCE CODE

The source code must follow very definite rules in order for an assembler to assemble it properly. If these rules are not followed, the assembler produces error messages depending on the type of error. It can catch errors in syntax (grammar) but not errors in program logic. If you are careless, the assembler could give you faulty results without warning.

---

[1]UNIX uses the vi editor.

Consider the following source code instruction:

```
LDS   FF   ;want to load SP with the value $00FF
```

The assembler will warn you of an error because it does not recognize "FF." The dollar sign ($) is missing. However, the assembler cannot correct the following error:

```
LDS   $FF   ;want to load SP with the value $00FF
```

In this case the program writer has accidentally used direct mode instead of immediate mode. The assembler cannot read your mind. Both addressing modes are legal. Listing 4.1 illustrates the format of a sample source code.

```
*Listing 4.1

*FILE list4a.asm
*demo source file
*-------------------------------------------
*MAXIMUM UNSIGNED BYTE
*Find the largest unsigned byte in a block of data
*
*DATA STRUCTURE
*     MAXRESULT    Address where result is stored
*     COUNT        Address that contains number of
*                  bytes in data block
*     FIRST        Address of first element in block

*ALGORITHM
*     1. Get COUNT
*     2. Set MAX = 0, the current maximum
*     3. Compare each element to MAX
*                  If element > MAX then MAX = element
*                  Else MAX unchanged
*     4. Repeat step 3 for all elements
*     5. Put result, MAX into MAXRESULT
*===========================================
* Define data and segment addresses

PAGE0       EQU    0
STACK       EQU    $FF
PROGSEG     EQU    $0100


*-------------------------------------------
*start data segment
*want page0 RAM to force direct addressing mode
*whenever possible

            ORG    PAGE0

RAM         RMB    4
MAXRESULT   RMB    1
COUNT       RMB    1
```

```
        *for convenience of testing, predefine some RAM data

FIRST   FCB    $5C,$67,$00,$31,$B3,$A5,$20,$80
        FCB    $88,$3C,$91,$91,$43,$CE,$34,$01


*end data segment
*---------------------------------------------
* Program Segment
*---------------------------------------------
        ORG    PROGSEG


        LDS    #STACK         ;initialize start of stack
        LDAB   COUNT          ;get count
        CLRA                  ;maximum=zero(min. possible)
        LDX    #FIRST         ;point to first entry in block
MAXM    CMPA   0,X            ;is current maximum greater
*                             ;than current entry?
        BCC    NOCHG          ;if yes, keep maximum
        LDAA   0,X            ;else replace with current entry
NOCHG   INX                  ;point to next entry
        DECB                 ;all entries checked?
        BNE    MAXM
        STAA   MAXRESULT      ;save maximum
        LDX    #MAXRESULT-2
        TAB                   ;set up calling registers
        JSR    UNPACK         ;unpack result

HERE    BRA    HERE           ;stop program


*---------------------------------------------
*Subroutine UNPACK (Same as Listing 3.5)
*Converts a byte into two single-hex-digit bytes (MSD=0)
*Stores the unpacked result in 2 bytes in RAM
*Calling Registers
*      ACCB = data byte to be unpacked
*      IX = address to store unpacked result (MSD first)
*Return Registers
*      All registers unaffected
UNPACK
        PSHA                  ;store ACCA
        TPA                   ;store CCR
        PSHA

        TBA                   ;shift upper digit into lower nibble
        LSRA
        LSRA
        LSRA
        LSRA
        STAA   0,X            ;and store it
        TBA                   ;next,
```

```
        INX                    ;work on lower digit
        STAA    0,X            ;first store both digits
        BCLR    0,X $F0        ;then remove upper digit
        DEX                    ;restore IX
        PULA                   ;restore CCR
        TAP
        PULA                   ;restore ACCA
        RTS                    ;return to calling routine

*end of program segment
*-------------------------------------------------
        END     ;(optional) end of source code,
*               ;assembler will not assemble
*               ;anything AFTER this point.
*-------------------------------------------------
```

The source code illustrates several features. Do not worry if these features seem to be confusing at first glance. We explain some of them in this section and others in the following sections. Each line of source code is organized into *fields* (or columns). These are

> label   operation   operand   comments

With one exception, any field is optional. You must always use an operand with an operation. Operands were discussed in Chapter 2.

Label Field. A label is an identifier to mark a line in a program. Another program line can use the label as an operand. For example, MAXM is a label. It marks the address of the instruction CMPA 0,X. The BNE instruction uses it as an operand. The assembler can calculate the relative address of the BNE instruction using this information. The label field must begin in the first column of any line. Other fields may start anywhere else.

Operation Field. There are two types of terms that may appear in the instruction field. These are the opcode and the pseudo-operation (pseudo-op). The opcode is the mnemonic form of a processor instruction such as those covered in previous chapters. Opcodes tell the target processor (68HC11 in this case) what to do. CMPA and BNE are examples.

A pseudo-op tells the assembler what to do. The assembler does not translate them into machine code for the target processor. We cover pseudo-ops in Section 4.4. An example of a pseudo-op is EQU. In this case it assigns a hex value of $FF to the symbol STACK. Note that STACK itself is put into the label field. The assembler will interpret the instruction LDS #STACK as equivalent to LDS #$FF.

The *operation field* is sometimes known as the *instruction field*.

Operand Field. An instruction does not necessarily need an operand. DECB and END are examples. An operand field identifies the data or an address for its corresponding instruction. Both opcodes and pseudo-ops can have operands. MAXM is the operand for the instruction BNE. $FF is the operand for the pseudo-op EQU.

Comment Field. Some assemblers will not permit a comment to begin in column 1 unless the first character is an asterisk (*). Other assemblers may use a different character to mark the beginning of a comment. For example, Motorola's Freeware Cross Assembler (AS11) also uses one or more spaces (whitespace characters) after the operand field (or after the instruction field if there is no operand field) to indicate the comment field. A preceding semicolon is optional. The following (without the semicolon) is a legal source code line in AS11.

```
LDS     #$FF     load SP with the value $00FF
```

However, by convention, we would use the following.

```
LDS     #$FF     ;load SP with the value $00FF
```

A comment *line* in AS11 must begin with the asterisk (*) character in column 1. For example,

```
*This is a comment line in AS11
```

Note that the assembler ignores comments. Comments are very important for program documentation. Imagine the program of Listing 4.1 without comments. Would it be more difficult to understand?

PRACTICE

Write a different instruction to stop the program. Use all fields.

Solution

```
DONE    JMP     DONE     ;this is one way to stop a program
```

End. The source code of some assemblers must have an END statement to tell the assembler when to stop assembling. Without it, the assembler would give an error message. AS11 does not require an END statement.

■ 4.3 CODE AND DATA SEGMENTS

We can organize the source code into blocks called *segments*. Segments specify sections of code, data, and reserved areas of memory. It is possible to mix everything into one segment. However, it is a good idea to specify separate segments for code, data, and special areas such as vectors, look-up tables, subroutine codes, and the stack. In some development systems the linker will assign the final segment start addresses. With AS11, there are no segments since the assembler itself assigns the final addresses without requiring a linker.

Figure 4.1 shows a typical organization of memory using segments for the 68HC11 and similar microcontrollers. Each segment has a purpose. Refer to Listing 4.1 to see how the segment "PROGSEG" is defined and started.

Segments are useful for modularized program development. For example, the subroutine UNPACK could be defined in another segment. This segment could be a subroutine library. The program developer only needs to consider the application program, not the subroutines that are already developed.

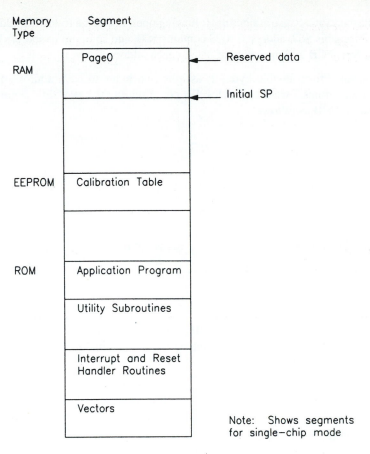

Memory
Type   Segment

RAM
        Page0 ◄─────── Reserved data

                ◄─────── Initial SP

EEPROM Calibration Table

ROM   Application Program

        Utility Subroutines

        Interrupt and Reset
        Handler Routines

        Vectors

Note: Shows segments
for single-chip mode

Figure 4.1   Organization of Memory Using Segments (Example)

■ 4.4 PSEUDO-OPERATIONS

Recall that a pseudo-op tells the assembler what to do. We have encountered some already. Note that some assembler writers refer to these as assembler directives. Pseudo-op ORG defines a start address in a segment. END marks the end of the source code for the assembler. EQU assigns a value to a symbol. Listing 4.1 shows examples for all of them.

Some pseudo-ops define data. Refer to the "PAGE0" segment in Listing 4.1 to see how they are used.

RMB.   RMB means "reserve memory bytes." Note that the PAGE0 segment begins at address $0000 as defined by the ORG directive. The assembler assigns address $0000 to the label RAM. It reserves 4 bytes, so the assembler assigns address $0004 to the label MAXRESULT, which has 1 byte reserved for it. Label COUNT is assigned to address $0005. Similarly, label FIRST is assigned to address $0006. RMB does not assign data to the addresses.

FCB.   FCB means "form constant byte." Due to the previous RMB pseudo-ops, the assembler assigned the label FIRST to the address $0006. The FCB pseudo-op assigns this address with the data $5C. It also assigns the following operands to the following addresses. Address $0007 contains $67, address $0008 contains $00, and so on up to address $0015, which contains $01. FCB assigns 1 byte of data to each address.

FDB. FDB means "form double byte." It assigns two bytes to each address. For the 68HC11 Motorola defines a word as 2 bytes. We could rewrite the "PAGE0" segment using FDB instead of FCB as follows.

```
PAGE0       EQU     0

            ORG     PAGE0
RAM         RMB     4
MAXRESULT   RMB     1
COUNT       RMB     1
```

* for convenience of simulation, predefine some RAM data

```
FIRST       FDB     $5C67, $0031, $B3A5, $2080
            FDB     $883C, $9191, $43CE, $3401
```

FCC. FCC means "form constant characters." It assigns a byte in memory for every character in a string. A string is a sequence of ASCII codes in which each code represents an alphanumeric character. We must define the string by enclosing it in double quotation marks. For example, we can store the phrase "Alarm 5A high!" at address $150 as follows.

```
ORG     $0150
FCC     "Alarm 5A high!"
```

The assembler would translate this as:

```
0150  41  6C  61  72  6D  20  35  41-  20   68  69  67  68  21  xx  xx
```

Fill, BSZ, and ZMB. These are directives to define a block of bytes to a constant value. The syntax for FILL is FILL val, nbytes. It sets a specified number of bytes (nbytes) to a specified value (val). ZMB (Zero Memory Bytes) and BSZ (Block Store Zeroes) are equivalent. They will initialize a specified number of bytes to zero. An example as a listing output follows.

```
0001                              *Demonstrate FILL, BSZ, and ZMB
0002
0003 0000                                     ORG     $00
0004
0005                              *-------------------------
0006
0007 0000 5a 5a 5a 5a 5a 5a                   FILL    $5A, 16
     5a 5a 5a 5a 5a 5a
     5a 5a 5a 5a
0008 0010 00 00 00 00 00 00                   BSZ     16
     00 00 00 00 00 00
     00 00 00 00
```

```
0009 0020 00 00 00 00 00 00                      ZMB    16
     00 00 00 00 00 00
     00 00 00 00
0010
0011                              *--------------------------
0012                              *End of program code
0013
0014
```

PRACTICE

A subroutine called OUTA (at address $FFB8) sends the contents of accumulator A as an ASCII character. Write instructions to output the message "good."

Solution

```
*Solution
*Practice in Section 4.4

*define data
OUTA       EQU       $FFB8

SOLN       EQU       $0100

ASCg       EQU       $67        ;ASCII code for 'g'
ASCo       EQU       $6F        ;ASCII code for 'o'
ASCd       EQU       $64        ;ASCII code for 'd'

*Send the message "good"

           ORG       SOLN

           LDAA      #ASCg      ;send 'g'
           JSR       OUTA
           LDAA      #ASCo      ;send 'o'
           JSR       OUTA
           JSR       OUTA       ;send it again
           LDAA      #ASCd      ;send 'd'
           JSR       OUTA
HERE       BRA       HERE       ;stop for now
```

The following alternative is also possible:

```
*Alternative Solution
*Practice in Section 4.4

OUTA       EQU    $FFB8
SOLN       EQU    $0100

*define data stored in RAM
```

```
          ORG     0
COUNT
          FCB     $04        ;number of characters in block
*                            ;then, the message block
MESSG
          FCB     $67, $6F, $6F, $64

          ORG     SOLN
*code to send message

          LDX     #MESSG     ;point to message block
          LDAB    COUNT      ;number of characters
NEXT
          LDAA    0,X        ;send a character
          JSR     OUTA
          INX                ;get next character
          DECB               ;until all characters sent
          BNE     NEXT
DONE
          BRA     DONE       ;stop for now

          END                ;stop assembly
```

Numbering System Conventions. The Quick Reference (Appendix B) lists the numbering system conventions. Listing 4.2 illustrates an example of their use. To aid your understanding of the results of the pseudo-ops, we have included explanations and hex results as comments. In normal assembly language programming, we would not predict results of assembly and include them as comments. This defeats the purpose of using an assembler. Study Listing 4.2 carefully.

```
*Listing 4.2
*Sample data segment to demonstrate data definition
*pseudo-ops
*
*start data segment
**-------------------------------------------
DATASEG    EQU     $C000

           ORG     DATASEG

*assigning values to symbols, can use decimal, hex, or ASCII

STACK      EQU     $FF
NUM1       EQU     55
NUM2       EQU     $55
NUM3       EQU     'U'

*RESULT STACK=$FF, NUM1=$37, NUM2=$55,
*NUM3=$55, ASCII code for 'U'
```

```
*RMB reserves uninitialized bytes of space

BEGINS      RMB      4         ;assigned to address $C000
RESULT      RMB      1         ;assigned to address $C004
COUNT       RMB      1         ;assigned to address $C005


*FCB defines 8-bit quantities
*can use decimal, hex, or ASCII
BEGIN       FCB      50,$50,'5'
            FCB      76,$76


*RESULT     BEGIN    assigned to address $C006
*hex code   C006     32 50 35 4C 76


*FCC (Form Constant Characters) can define a string
*a string is a sequence of bytes representing ASCII
*characters

STRING      FCC      "Hello my friend"
CR          EQU      $0D
LF          EQU      $0A
            FCB      CR, LF
*add carriage return (CR) and line feed (LF) to string


*RESULT     STRING   assigned to address $C00B
*hex code   C00B     48 65 6C 6C
*           C00F     6F 20 6D 79
*           C013     20 66 72 69
*           C017     65 6E 64 0D
*           C01B     0A


*FDB defines 16-bit quantities

            FDB      $6,6,$54C3,'x'
            FDB      'y'
            FDB      'z'


*RESULT
*hex code   CO1C     00 06 00 06
*           C020     54 C3 00 78
*           C024     00 79 00 7A


*demonstrate prefix convention

HEXP        FCB      $FB            ;hex
DECP        FCB      251            ;decimal
OCTP        FCB      @373           ;octal
BINP        FCB      %11111011      ;binary


*RESULT
*hex code   C028     FB FB FB FB
```

```
*demonstrate signed numbers and arithmetic

            FCB       $FB
UNSGN       FCB       251
SGN         FCB       -5
ADD1        FCB       248+3
SUB1        FCB       254-3

*RESULT
*hex code   C02C      FB FB FB FB FB

*use labels or symbols as operands
            FCB       NUM2          ;a symbol
            FDB       STRING        ;a label

*RESULT     note that label STRING was assigned address $C00B
*hex code   C031      55 C0 0B

*end of data segment
*-------------------------
```

There is a subtle distinction between labels and symbols. A label is determined by the assembler when using the FCB, FDB, or RMB pseudo-ops. In some development systems a label's final target address is defined by the linker. A symbol is defined by the EQU pseudo-op. Its value is fixed during assembly. It cannot be changed by the linker in those systems that support linking. One can think of labels as special symbols that can appear in a label field and an operand field.

PRACTICE

A subroutine called OUTA (at address $FFB8) sends the contents of accumulator A as an ASCII character. Write instructions to output the message "good."

Solution

```
*Solution
*Practice in Section 4.4

*define data
OUTA    EQU     $FFB8

SOLN    EQU     $0100

*Send the message "good"

        ORG     SOLN

        LDAA    #'g'      ;send 'g'
        JSR     OUTA
        LDAA    #'o'      ;send 'o'
```

```
        JSR       OUTA
        JSR       OUTA        ;send it again
        LDAA      #'d'        ;send 'd'
        JSR       OUTA
HERE    BRA       HERE        ;stop for now

        END                   ;stop assembly
```

As an alternative we could use the following:

```
*Alternative Solution
*Practice in Section 4.4

OUTA    EQU       $FFB8
SOLN    EQU       $0100

*define data stored in RAM

        ORG       0
COUNT
        FCB       4           ;number of characters in block
*                             ;note 4 = $4

MESSG
        FCC       "good"      ;ASCII string

        ORG       SOLN

*code to send message

        LDX       #MESSG      ;point to message block
        LDAB      COUNT       ;number of characters
NEXT
        LDAA      0,X         ;send a character
        JSR       OUTA
        INX                   ;get next character
        DECB                  ;until all characters sent
        BNE       NEXT
DONE
        BRA       DONE        ;stop for now

        END                   ;stop assembly
```

PRACTICE

Assume that bit X is clear and register SP is initialized to $004A. Write a program that outputs the ASCII character 'H' using subroutine OUTA at address $FFB8 (see "Practice" in Section 3.4.3) every time the signal at pin XIRQ goes from a logic high to a logic low. Start the code at address $E100 (see also Section 3.9.4).

Solution

```
*Solution
*Practice in Section 4.4

OUTA       EQU     $FFB8

JMPCODE    EQU     $7E        ;opcode for JMP

*next define vector jump table
*Chose EVB/EVBU board convention,
*it starts vector jump table at $C4

VECJUMP    EQU     $C4
           ORG     VECJUMP

           RMB     45         ;reserve space for 15 pseudovectors
JXIRQ
           RMB     3          ;reserve space for XIRQ pseudovector
*in general, should also reserve space for remaining
pseudovectors

*Now main code segment
SOLN       EQU     $E100
           ORG     SOLN

           LDAA    #JMPCODE ;set up vector jump table
           STAA    JXIRQ
           LDD     #RXIRQ
           STD     JXIRQ+1
HERE
           BRA     HERE     ;just waiting for interrupts

*Service routine RXIRQ (interrupt XIRQ)
*Outputs 'H' when a high-to-low transition
*occurs at pin XIRQ.
RXIRQ
           LDAA    #'H'       ;get ASCII code for 'H'
           JSR     OUTA       ;then send it
           RTI                ;and return from interrupt

*Define vector for XIRQ, though in general
*should define all vectors

VECTORS    EQU     $FFF4
           ORG     VECTORS

VXIRQ      FDB     JXIRQ      ;vector points to jump table entry

           END                ;end assembly
```

Some Other Cross-Assembler Pseudo-Ops. Some cross assemblers use different pseudo-ops. For example, the Avocet cross assembler includes pseudo-ops that are typically used for Intel microprocessors. The following shows a comparison.

ORG	Set location counter	Basically the same as the AS11 version
DB	Define byte	Similar to FCB
DW	Define word	Similar to FDB
EQU	Equal	Same as the AS11 version
DS	Define space	Similar to RMB

Motorola Syntax Differences. Syntax refers to the rules for writing source code. There are some differences between most of the code examples based on Motorola's AS11 and some other cross assemblers, such as Avocet. It uses commas to separate operand expressions for bit set or clear and bit test and branch instructions. Recall that AS11 uses a space character (see Section 2.8.4). The syntax for index addressing remains the same.

```
Avocet                          AS11


BCLR    $04,$02                 BCLR    $04 $02
BRCLR   $03,X,$02,THERE         BRCLR   $03,X $02 THERE
```

AS11 uses the asterisk to indicate the location counter. Consider the instruction to branch to itself forever.

```
    BRA     *       ;AS11 syntax for branching to itself
```

The Avocet version is the following.

```
    BRA     $       ;Avocet syntax for branching to itself
```

Both cross assemblers will assemble the following:

```
HERE
    BRA     HERE    ;Legal syntax for both Avocet and AS11
```

Both cross assemblers interpret nonprefixed numbers as decimal.

■ 4.5 THE ASSEMBLY TWO-PASS PROCESS

For an assembler to understand labels and symbols, it must work through the source code twice. This is known as a *two-pass process*. The first pass fills in all known address and hex code information line by line and leaves spaces for any unknown information. Also during the first pass, the assembler collects all possible information when it encounters labels or symbols defined in the label field. The second pass starts again from the first source code line and fills in the missing information left over from the first pass. During the assembly process, the assembler uses a *location counter* to point to the line on which it is working. This is like a machine-code program using a program counter register. Line assemblers need only one pass because they do not support the use of labels and symbols in the source code.

Listing 4.3 helps you look into this process in more detail. When the source code of Listing 4.1 is assembled, the assembler will produce a listing file for documentation as well as an object file for linking. Listing 4.3 is a modification of the *listing file.* To fit the listing on these pages, we split lines 0043 and 0044. Look at Listing 4.3 now and read the explanations that follow.

```
0001            *Listing 4.3
0002
0003            *Listing file produced when source code of
0004            *Listing 4.1 is assembled
0005            *-----------------------------------------
0006            *MAXIMUM UNSIGNED BYTE
0007            *Find the largest unsigned byte in a block of data
0008            *
0009            *DATA STRUCTURE
0010            *    MAXRESULT    Address where result is stored
0011            *    COUNT        Address that contains number of
0012            *                 bytes in data block
0013            *    FIRST        Address of first element in block
0014            *
0015            *ALGORITHM
0016            *    1. Get Count
0017            *    2. Set MAX = 0, the current maximum
0018            *    3. Compare each element to MAX
0019            *          If element > MAX then MAX = element
0020            *          Else MAX unchanged
0021            *    4. Repeat step 3 for all elements
0022            *    5. Put result, MAX into MAXRESULT
0023            *=========================================
0024            *Define data and segment addresses
0025
0026 0000       PAGE0      EQU    0
0027 00ff       STACK      EQU    $FF
0028 0100       PROGSEG    EQU    $0100
0029
0030            -----------------------------------------
0031            *start data segment
0032            *want page0 RAM to force direct addressing mode
0033            *whenever possible
0034
0035 0000                  \ORG   PAGE0
0036
0037 0000       RAM        RMB    4
0038 0004       MAXRESULT        RMB   1
0039 0005       COUNT      RMB    1
0040
```

```
0041          *for convenience of testing, predefine some RAM data
0042
0043 0006 5c 67 00 31 b3 a5 20 80
              FIRST     FCB    $5C,$67,$00,$31,$B3,$A5,$20,$80
0044 000e 88 3c 91 91 43 ce 34 01
                        FCB    $88,$3C,$91,$91,$43,$CE,$34,$01
0045
0046          *end data segment
0047          *----------------------------------------
0048          *Program Segment
0049          *----------------------------------------
0050 0100              ORG    PROGSEG
0051
0052 0100 8e 00 ff     LDS    #STACK  ;initialize start of stack
0053 0103 d6 05        LDAB   COUNT   ;get count
0054 0105 4f           CLRA           ;maximum=zero(min. possible)
0055 0106 ce 00 06     LDX    #FIRST  ;point to first entry in block
0056 0109 a1 00  MAXM  CMPA   0,X     ;is current maximum greater
0057          *                       ;than current entry?
0058 010b 24 02        BCC    NOCHG   ;if yes, keep maximum
0059 010d a6 00        LDAA   0,X     ;else replace with current entry
0060 010f 08    NOCHG  INX            ;point to next entry
0061 0110 5a           DECB           ;all entries checked?
0062 0111 26 f6        BNE    MAXM
0063 0113 97 04        STAA   MAXRESULT        ;save maximum
0064 0115 ce 00 02     LDX    #MAXRESULT-2
0065 0118 16           TAB            ;set up calling registers
0066 0119 bd 01 1e     JSR    UNPACK  ;unpack result
0067
0068 011c 20 fe   HERE BRA   HERE     ;stop program
0069
0070          *----------------------------------------
0071          *Subroutine UNPACK (Same as Listing 3.5)
0072          *Converts a byte into two single-hex-digit bytes (MSD=0)
0073          *Stores the unpacked result in 2 bytes in RAM
0074          *Calling Registers
0075          *   ACCB = data byte to be unpacked
0076          *   IX = address to store unpacked result (MSD first)
0077          *Return Registers
0078          *   All registers unaffected
0079          UNPACK
0080 011e 36           PSHA           ;store ACCA
0081 011f 07           TPA            ;store CCR
0082 0120 36           PSHA
0083
0084 0121 17           TBA            ;shift upper digit into lower nibble
0085 0122 44           LSRA
0086 0123 44           LSRA
0087 0124 44           LSRA
0088 0125 44           LSRA
0089 0126 a7 00        STAA   0,X     ;and store it
```

```
0090 0128 17          TBA            ;next,
0091 0129 08          INX            ;work on lower digit
0092 012a a7 00       STAA   0,X     ;first store both digits
0093 012c 1d 00 f0    BCLR   0,X $F0 ;then remove upper digit
0094 012f 09          DEX            ;restore IX
0095 0130 32          PULA           ;restore CCR
0096 0131 06          TAP
0097 0132 32          PULA           ;restore ACCA
0098 0133 39          RTS            ;return to calling routine
0099
0100          *end of program segment
0101          *----------------------------------------
0102            END                  ;(optional) end of source code,
0103          *                      ;assembler will not assemble
0104          *                      ;anything AFTER this point.
0105          *----------------------------------------
```

Each line begins with a line number. The next two columns to the left are the object code. It begins with the address followed by the address contents. The right columns show the corresponding source-code lines.

First Pass.   At segment "PAGE0" the assembler initializes the location counter. Then it increments the location counter to reserve bytes for RAM, MAXRESULT, and COUNT. It also records their addresses at $0000, $0004, and $0005, respectively. When it encounters FIRST, it records its address at $0006. Then it counts the number of FCB operand bytes and fills them in. The assembler records the end of the "Page0" segment at address $0015.

For segment "PROGSEG" the assembler sets the location counter (address count) to $0100. This is the beginning address of the segment. It translates the first opcode and operand as an LDS immediate. It assigns the opcode byte of $8E and reserves two bytes for the operand. Then it translates the second opcode and operand as LDAB direct. It assigns the value of $D6 to segment address $103 and reserves an operand byte at $104. The assembler translates each source code line similarly. Whenever it finds a label in the label field, it records the location. For example, it will determine that MAXM is a reference to segment address $0109. When the assembler encounters the instruction BNE MAXM, it puts in the opcode of $26 and reserves a byte for the relative address. When the assembler encounters the END pseudo-op, or the end of the source code (if END is missing), it stops the first pass.

Second Pass.   The assembler starts from the beginning. When it encounters LDS #STACK, it fills in the value of STACK at segment addresses $0101 and $0102 as $00FF. Note that we could have put the EQU statement near the end of source code. It does not matter because the assembler does not use the information until the second pass. The assembler fills in other missing operand information similarly. When it encounters a branch instruction (e.g., BNE MAXM) it calculates the relative address and puts it in. In this case since the label MAXM was recorded as segment address $0109 and the BNE instruction is at segment address $0111, the assembler calculates a relative address of $F6.

Final Target Addresses.   In this case the AS11 assembler calculated the final target addresses. Note that some other assemblers support linking. It would not be necessary to explicitly set a segment start address. For example, with the Avocet assembler we can begin a "PAGE0" segment by using the following statement:

```
SEG PAGE0
```

The operand addresses for FIRST and COUNT would not be finalized. When using a different development system that supports linking, we could link this file with another containing a "PAGE0" segment. The linker may assign the other "PAGE0" segment to start at zero and then the "PAGE0" segment in this file would follow it. The linker would also fill in the appropriate operands with the final target address values. Of course, relative addresses do not change. Otherwise, they would not be relative!

■ 4.6 ASSEMBLER OPTIONS AND PREPROCESSOR DIRECTIVES

Since the AS11 assembler does not have many options or directives, we will use the Avocet assembler. In Section 4.4 we noted its pseudo-ops. The Avocet assembler options and preprocessor directives may be considered to be a special kind of pseudo-op. We will not cover all options and directives; instead, we will show some of them to illustrate what they do.

The options must begin in column 1 with a dollar sign. They must occur at the beginning of the source code. One example of an option, $ALLPUBLIC, tells the assembler to include labels and symbols in the object code. This allows labels and symbols to be used in other documentation files. Some simulators and debuggers can use these documentation files to display these labels and symbols. Other options specify features to add or delete when assembling. These include titles for every page on a listing file, its page width, whether or not to produce an object file, and other things.

The preprocessor directives follow the assembler options. They begin with a percent sign (%). They tell the assembler to do something before beginning the assembly process. Listing 4.4 shows some preprocessor directives. It includes comments that explain what the directives do.

```
*Listing 4.4
;File list4d.asm

;Demonstrate Preprocessor Directives

$ALLPUBLIC ;an assembler option

;Preprocessor Directives begin here

;the INCLUDE directive

        %INCLUDE "d:\include\iolib.h"

;This inserts the contents of the header file "iolib.h"
;into this source code during assembly. The header file is
;in drive d:, subdirectory include.
```

```
;define a macro called 'NMUL'
;NMUL multiplies two 8-bit numbers in addresses
;indat and indat+1 and puts a scaled 8-bit result
;in outdat. indat and outdat are macro parameters.
;Macros are similar to subroutines

NMUL      %MACRO      indat, outdat
          LDD         indat
          MUL
          ADCA        #0
          STAA        outdat
          %ENDM       ;ends the macro definition


;will use the defined macro later in source code

;start Page0 segment
;----------------------------------------
          SEG         Page0

;example of conditional assembly
          %IF         sample eq 1

COUNT     DB          15              ;assembled if 'sample' = 1

          %ELSEIF     sample gt 1

COUNT     DB          20              ;assembled if 'sample' > 1

          %ELSE

COUNT     DB          7               ;assembled if 'sample' < 1

          %ENDIF                      ;ends conditional assembly

INDATA    DS          20
OUTDATA   DS          10

;start program code segment
;----------------------------------------

          DEFSEG      program,     START=$E000
          SEG         program

          LDS         #$FF
          NMUL        INDATA, OUTDATA      ;here we use the macro
          NMUL        INDATA+2, OUTDATA+1 ;use the macro again
          BRA         $                   ;stop for now

;also note the use of parameters with NMUL
```

```
;end of program code segment

        END
```

Now we need to explain what the preprocessor directives do.

Include. This inserts the contents of the specified file into the file of Listing 4.4 during assembly. These files are often known as header files. Normally, they contain a section of source code that would appear in many other source codes. For example, a company producing software may define standard labels in standard files. Anybody producing source code in the company would use the INCLUDE directive for the code. Generally, these are long files, which make their use worthwhile.

To conserve time and space, we will use a trivial example. Let's say that file "iolib.h" contains the following:

```
sample    EQU    5
```

Then, when we assemble the file "list4d.asm," this EQU statement is inserted into the file.

Macros. A *macro* is a label that replaces a block of instructions. Macros are similar to subroutines. Both execute a section of code when called or referred to. Unlike a subroutine, the assembler executes the macro sequence. The assembler substitutes the macro reference with a sequence of instructions. Like subroutines, macros can make the source code more readable. Both refer to a sequence of instructions. A subroutine saves on final machine-code space because its code is reused. A macro is faster but uses more memory since it is an inline substitution of an instruction sequence. When writing software, you will have to decide which to use. This decision is a trade-off between a macro's faster execution speed (no overhead in calling and returning) versus the code space saved when using subroutines.

To see how a macro works for the program of Listing 4.4, look ahead at the program of Listing 4.5. It shows an equivalent source code (assuming the former example for iolib.h). A macro defines a sequence of instructions. To define the macro, use the directive %MACRO. Then follow it with the desired sequence of instructions. To complete the definition, use the directive %ENDM. A macro can use parameters to reserve space for operands within the macro itself. When using the macro, you have to specify the parameters (see Listing 4.4).

Conditional Assembly. The purpose of using conditional assembly directives is to reuse sections of source code for flexibility. For example, a control systems supplier builds different systems. All application programs are similar, yet not entirely identical. The supplier's software developer uses the same source code files for all systems. These files contain all the code variations that are required. When it is necessary to produce executable code for one particular system, conditional assembly can be used. This allows the software developer to specify easily which sections of code to assemble and which sections to ignore.

Listing 4.4 shows only one example of conditional assembly directives. This is the if-then-else structure, which is similar to the program flow control structure described in Section 2.8.1. The assembler will include only one of the three possibilities of the label

COUNT, depending on the value of the symbol "sample." In this case, "sample" is defined in the header file. Since this defines "sample" to be 5, the assembler will define address COUNT to contain a value of 20 ($14). Normally, we would define the value of a conditional parameter such as "sample" outside the source code. We can do this by specifying it as an option to the assembler.

Equivalent Source Code. Listing 4.5 shows an equivalent source code to Listing 4.4 for the specific header file "iolib.h" described earlier.

```
*Listing 4.5

sample      EQU       5                 ;inserted by INCLUDE directive

            SEG       Page0

COUNT       DB        20                ;result of conditional assembly

INDATA      DS        20
OUTDATA     DS        10

            DEFSEG    prognam,   START=$E000
            SEG       prognam
            LDS       #$FF

            LDD       INDATA            ;result of using macro first time
            MUL
            ADCA      #0
            STAA      OUTDATA

            LDD       INDATA+2          ;result of using macro second time
            MUL
            ADCA      #0
            STAA      OUTDATA+1

            BRA       $

            END
```

In this section we have surveyed only the topic of assembler options and preprocessor directives. The examples are simpler than would be used in normal practical applications. For further information, refer to the *Cross Assembler Manual* (Appendix C).

PRACTICE

Write a section of code to set up a pseudovector for IRQ. The IRQ routine only executes RTI. Assume that segment "Solution" and the stack have already been set up. Use header file "hc11vec.h" (Appendix E), even though the header file is written in AS11 format.

Solution

```
*Solution
    INCLUDE  <hc11vec.h>  ;include vector file

;other code if applicable

    SEG      Solution       ;reenter solution segment if necessary

    LDAA     #$7E           ;set up jump table for IRQ
    STAA     JIRQ
    LDD      #RIRQ
    STD      JIRQ+1
    CLI                     ;enable interrupt
AGAIN                       ;just waiting for interrupt
    WAI
    BRA      AGAIN
RIRQ
    RTI                     ;not much of an interrupt routine is it?

    END                    ;end assembly
```

■ 4.7 HEX AND BINARY FILES

Recall that the assembler produces an object-code file and a listing file. Also the linker combines one or more object-code files to produce a *hex file*. A hex file is not machine code. It represents the machine code using ASCII characters. It is a text file that you can send to any printer and get text output if you should wish to do so. There are several standard formats of hex files. The most common are the Intel hex and Motorola S-record (also known as Motorola hex) formats. With Avocet's development system, you must run an assembler first and then run a separate linker. AS11 assembles and links. Hence, it produces a hex file directly.

Other software recognizes a hex file and converts it into executable machine code. If this executable machine code is put in a file, the file is known as a *binary file*. The binary file is a sequence of binary numbers that is the machine code. It is up to the user to download this file into the proper start address in the target processor (the 68HC11, in our case).

Uses. Let's say that you are developing your 68HC11 application program and wish to run it in order to test it. You could use an evaluation board such as Motorola's M68HC11EVBU board (EVBU board). This board contains the 68HC11 processor with RAM. You can connect the board to a personal computer (or a dumb terminal). With this setup, you can use a computer communications program (e.g., Procomm, Kermit, Windows Terminal) to send a Motorola hex file to the EVBU board. The EVBU board has a built-in program in ROM to convert the hex file into binary and load it into RAM.

Later you have finally developed and tested an application program and wish to put it into ROM. You would only do this for controlling equipment that will be manufactured in large quantities. Examples are automobiles and soft-drink vending machines. When ordering the microcontroller chip from the semiconductor manufacturer, you have to specify what to put in ROM. Only the manufacturer can put in the ROM code. One way to specify ROM data is to send a disk with a hex file. The *Data Book* provides other details about ordering custom chips.

It may cost several tens of thousands of dollars to pay for the overhead of creating the ROM pattern for your application. Obviously, you would make sure that the ROM code you specify will work for your application.

An alternative is to order a standard 68HC11 chip (not custom) and run it in expanded multiplexed mode. You can put the application program into external memory such as EPROM. To put data into EPROM, you need an EPROM programmer. An EPROM programmer usually has software to convert your hex file into binary format so that it can "burn" the pattern into the EPROM chip. This alternative could be used for equipment manufactured in small quantities, such as test instruments.

Simulators also use hex files. We will talk about simulators in a later section.

Format.　　We have mentioned the uses for hex files. Next we will talk about their format in detail. A hex file contains lines of ASCII characters representing hex digits. The lines begin with a special code, then a four-character start address, and then a sequence of characters (representing hex digits). Each line ends with a two-digit checksum, a carriage return character, and a linefeed character. The start address specifies where the data is to be loaded into memory. The checksum is used to check whether any bits have changed in other parts of the line. (We will explain more about checksums later.) There are two types of lines. One type, the *data record,* contains information about the machine code. The other, the end record, marks the end of the file.

Motorola S-Record Format.　　Following is an example of Motorola S-record format:

```
S11300065C670031B3A52080883C919143CE3401CE
S12301008E00FFD6054FCE0006A1002402A600085A26F69704CE000216BD011E20FE3607AD
S11701203617444444A7001708A7001D00F0093206323944
S9030000FC
```

This shows the hex file for the program of Listing 4.1. The data record contains

　　　　'S1'

Two-character data byte count

(Number of data bytes plus three)

Four-character load address

Data byte characters

Two-character checksum

Carriage return

Linefeed

Examine the second data record in the preceding example. The byte count is $23. This means that there are 32 data bytes. The load address is $0100. The first three data bytes are $8E, $00, and $FF. This is the machine code for LDS #$FF. The checksum is $AD. The carriage return and linefeed are nonprintable control characters. Their effect is evident; look at the third record beginning in column 1 on the next line. The end record contains

'S9'

Count, '03'

Four-character start address

Two-character checksum

Carriage return

Linefeed

Intel Hex Format. Following is an example of Intel hex format. It is also a hex file for a program *similar* to Listing 4.1. Using Avocet, the source for data and the main program were assembled into an object file. A separate file for subroutine UNPACK was also assembled. The two object files were linked to create the Intel hex format file. Also, the load address is $E000 instead of $0100.

```
:100006005C670031B3A52080883C919143CE3401D2
:20E000008E00FFD6054FCE0006A1002402A600085A26F69704CE000216BDE01E20FE3607F3
:14E02000361744444444A7001708A7001D00F0093206323969
:00000001FF
```

The data record contains

':'

Two-character data byte count

Four-character load address

Record type, '00'

Data byte characters

Two-character checksum

Carriage return

Linefeed

Consider the second data record. There are $20 data bytes and the load address is $E000. Following the record type, '00', are the data bytes. The first instruction is $8E, $00, $FF, the machine code for the instruction LDS #$FF. The checksum is $F3. The end record contains

':'

Two-character count

Four-character start address

Record type, '01'

Two-character checksum

Carriage return

Linefeed

Even though you are developing programs for a Motorola processor, your development system may use Intel hex format. This is the case for some simulators and EPROM programmers.

Checksum. You transmit a hex file to a target such as an evaluation board, programmer, or emulator (see Sections 4.9 and 4.10). There is a possibility that data will be corrupted. A bit may change state due to electrical noise. Also, bit errors could occur due to synchronization problems. There are several schemes to detect and correct errors. They vary in their degree of sophistication. These include parity check and cyclic redundancy check (CRC). In Chapter 10, which covers serial communications, we discuss some of these methods. Another method is checksum.

The checksum technique used by hex files can detect errors but cannot correct them. Basically, you keep on trying to send a hex file until you do not detect any errors. Data integrity is important since any bit error will result in a different program. This becomes a very expensive error if the hex file defines a ROM pattern in a custom chip.

For Motorola hex format, the checksum is the one's complement of the 8-bit sum of the data bytes, byte count, and address bytes. Consider the end record in the Motorola hex file example:

Add $03 + $00 + $00 = $03. The one's complement is $FC.

For Intel hex format, the checksum is the two's complement of the 8-bit sum of all the data bytes, byte count, record type, and address bytes. Consider the end record in the Intel hex file example:

Add $00 + $00 + $00 + $01 = $01. The two's complement is $FF.

The assembler (or linker) calculates the checksum for each record and puts it at the end. When the target system receives the hex file, it also computes the checksum for each record. If no bits have changed, the two checksums will be equal and the target will accept the hex file. If any bits have changed, the target's checksum will differ from the original checksum. The probability that bit errors will occur in a way that will cause equal checksums is very low.

■ 4.8 DOCUMENTATION FILES

Listing File. The assembler produces a listing file. In Section 4.5 we showed this as Listing 4.3. It showed both the source and object code and their relationship. For a large programming project, it is necessary to keep track of segments, addresses, and labels. With the AS11 cross assembler, we can also generate additional information in the listing file. To do this, we use the assembler with additional options.

Consider the following DOS commands to assemble the file for Listing 4.1. These examples assume that the AS11 executable file (AS11.EXE) and source file LIST4A.ASM are in the same directory. By convention, a source file uses an .ASM file extension and a listing file uses an .LST file extension.

The following command produces only the S-record hex file, LIST4A.S19:

```
AS11 LIST4A.ASM
```

The next command uses the L option to specify a listing. It displays the listing to the screen as well as creating hex file LIST4A.S19:

```
AS11 LIST4A.ASM -L
```

The next command produces both the hex file and listing file as shown in Listing 4.3 (see Section 4.5). These files are LIST4A.S19 and LIST4A.LST, respectively. It uses the greater than sign (>) to redirect the command output to the listing file:

```
AS11 LIST4A.ASM -L > LIST4A.LST
```

Cycle Count.   It is sometimes useful to know how many clock cycles each instruction uses. This is particularly true for control applications where fast response times are necessary. We can also show clock cycle counts in a listing file. The next command produces the hex file and a listing file that includes the clock cycle counts. It used the C option in addition to the L option for producing the listing:

```
AS11 LIST4A.ASM =L C > LIST4A.LST
```

Listing 4.6 shows part of the main program as part of this listing file. Note that it indicates the number of clock cycles in square brackets for each instruction.

```
*Listing 4.6, part of listing produced with cycle count option
0050 0100                      ORG    PROGSEG
0051
0052 0100 8e 00 ff   [3]       LDS    #STACK     ;initialize start of stack
0053 0103 d6 05       [3]       LDAB  COUNT      ;get count
0054 0105 4f           [2]       CLRA              ;maximum=zero(min. possible)
0055 0106 ce 00 06   [3]       LDX    #FIRST     ;point to first entry in block
0056 0109 a1 00       [4]  MAXM  CMPA 0,X        ;is current maximum greater
0057                             *                  ;than current entry?
0058 010b 24 02       [3]  BCC   NOCHG          ;if yes, keep maximum
0059 010d a6 00       [4]  LDAA  0,X            ;else replace with current entry
0060 010f 08           [3]  NOCHG INX            ;point to next entry
0061 0110 5a           [2]       DECB              ;all entries checked?
0062 0111 26 f6       [3]       BNE   MAXM
0063 0113 97 04       [3]       STAA MAXRESULT ;save maximum
0064 0115 ce 00 02   [3]       LDX    #MAXRESULT-2
0065 0118 16           [2]       TAB               ;set up calling registers
0066 0119 bd 01 1e   [6]       JSR   UNPACK    ;unpack result
0067
0068 011c 20 fe       [3]  HERE BRA   HERE     ;stop program
0069
```

Symbol Listing.   Sometimes it is useful to quickly find the addresses that correspond to symbols used in a program. We can use the S option to produce a symbol listing, as follows:

```
AS11 LIST4A.ASM -S > LIST4A.SYM
```

Listing 4.7 shows the resulting symbol listing for LIST4A.SYM, except that the comment line was typed in manually using a text editor.

```
*Listing 4.7

♀COUNT         0005
```

```
FIRST          0006
HERE           011c
MAXM           0109
MAXRESULT      0004
NOCHG          010f
PAGE0          0000
PROGSEG        0100
RAM            0000
STACK          00ff
UNPACK         011e
```

These symbols correspond to the source code for Listing 4.1 (Section 4.2) and the listing file of Listing 4.3 (Section 4.5). Actually, the symbol listing begins with a form feed (new page). The first character (♀) is how the Form Feed ASCII character shows up when the table is displayed using the DOS TYPE command.

Cross-Reference Table. The cross-reference table shows more information about the program symbols. It shows where in the listing file the symbol is defined and identifies where the symbol is used. We can use the CRE option to produce a cross-reference as follows:

```
AS11 LIST4A.ASM -CRE > LIST4A.CRE
```

Listing 4.8 shows the resulting cross-reference table for LIST4A.SYM, except that the comment line was typed in manually using a text editor. Again, the first character is a form feed (♀).

```
*Listing 4.8

♀COUNT        0005    *0039   0053
FIRST         0006    *0043   0055
HERE          011c    *0068   0068
MAXM          0109    *0056   0062
MAXRESULT     0004    *0038   0063  0064
NOCHG         010f    *0060   0058
PAGE0         0000    *0026   0035
PROGSEG       0100    *0028   0050
RAM           0000    *0037
STACK         00ff    *0027   0052
UNPACK        011e    *0079   .0066
```

To understand how to interpret the table, refer to both Listing 4.3 (in Section 4.5) and to Listing 4.8. The first number is the symbol address. The asterisk indicates the line on which the symbol is defined. The other numbers indicate where the symbol is used. Consider the following cross-reference for MAXRESULT:

```
MAXRESULT 0004 *0038 0063 0064
```

If you look at Listing 4.3, you should find that symbol MAXRESULT is defined on line 0038 as follows:

```
0038 0004 MAXRESULT RMB 1
```

Note that its address is $0004. By examining the cross-reference table of Listing 4.8, we can find that lines 0063 and 0064 use this symbol as follows:

```
0063 0113 97 04    STAA    MAXRESULT    ;save maximum
0064 0115 ce 00 02 LDX     #MAXRESULT-2
```

Putting It Together.   It is possible to generate a listing that contains all the information in one file. We can do this by using the assembler with all options, as follows:

```
AS11 LIST4A.ASM -L CRE S C > LIST4A.LST
```

Final Word.   We have not shown all the commands to use the assembler. The manuals that come with your development system will provide this information. We have shown sample outputs using AS11. Other development systems such as that supplied by Avocet Systems, Inc., generate similar outputs. The hex file formats conform to an industry standard; hence, you can mix hex files generated from different development systems.

■ 4.9 SIMULATION

Instead of running a program in the actual 68HC11 CPU, you can run it using a computer program that simulates the CPU. With simulation you can freely modify your program, modify registers, modify data, and set breakpoints. Breakpoints are points in the program where execution will stop. They are useful because you may want to examine the state of registers and memory after some instructions are executed.

An evaluation board or trainer also provides many of the features mentioned. However, a simulator usually shows the information more conveniently and allows easier access to registers and memory for modification. Appendix H provides information about the THRSim11 Simulator, including instructions on using it. A special demo version accompanies this book and the full version can be purchased if desired.

A simulator is very useful in testing and learning about software. However, a simulator does not run in real time. It generally tests only software. Some simulators also offer simulated hardware components to examine basic logic of hardware interfacing. In situations where there is a critical time relationship between the program and the external hardware, you need to test the program with the hardware. Sometimes, additional hardware simulation is possible, but most designers will test their systems using physical hardware.

■ 4.10 EVALUATION BOARDS AND EMULATION

Development.   Since microcontrollers control physical devices, developers need another tool to test the microcontroller (with its program code) connected to the I/O. You may have to correct mistakes in the hardware design or the program. Very likely, you will have to do both. When you just have a 68HC11 chip by itself, you need a way to enter, modify, and test programs. Also, you must connect to it external hardware such as sensors and actuators. Then you need to modify and test the I/O operations with the external hardware. This is very difficult to do if all you have is the circuit and some basic electronic instruments.

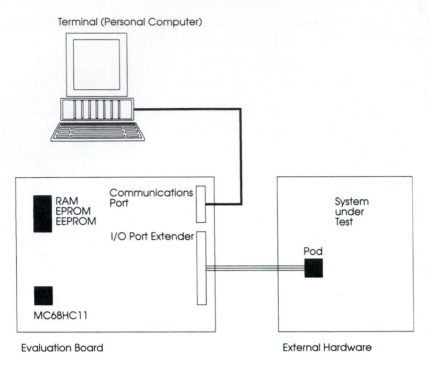

Terminal (Personal Computer)

RAM
EPROM
EEPROM

Communications
Port

I/O Port Extender

MC68HC11

Evaluation Board

System
under
Test

Pod

External Hardware

Figure 4.2   Testing the Design Using an Evaluation Board

Evaluation Board.   A relatively inexpensive way to do this is to use an evaluation board. Such a board contains the microcontroller chip with external pins for connecting to the external hardware. Refer to Figure 4.2, which shows a generic evaluation board used to develop a microcontroller system design.

Monitor Program.   The board has a built-in program called a monitor program, which permits the user to control its operation. You connect a terminal or computer to communicate with the board. The board allows you to monitor and modify registers and memory using the monitor program commands. Unlike using a simulator, you can only modify RAM or program EPROM or EEPROM data. An evaluation board provides the means to single-step programs and execute using breakpoints.

You can load a hex or binary file into the evaluation board's memory. The monitor program will translate the hex file into binary and load it into the addresses specified. The monitor program may also disassemble the machine code to show the source code produced. However, it will show only hex operands; it will not show labels or symbols. We will not describe the monitor programs in further detail since the manuals do that already. Two common monitor programs are Buffalo and PCbug11. Both can be obtained from the Internet.

Buffalo itself is a 68HC11 assembly language program. Its machine code resides in ROM or EPROM. PCbug11 is an MS-DOS program that runs in a personal computer. To use it, you must connect the MCU in bootstrap mode and connect its serial communications pins to the computer's serial port. The PCbug's monitor program resides in the personal computer. It communicates with a *talker* program that resides in the MCU internal RAM, external memory, or EEPROM. For further information, refer to the PCbug11 User's Manual. Section 5.8 describes the bootstrap mode.

Pod and I/O Port Connector.   The circuit you are developing will eventually have the microcontroller chip in it. During development, you insert a *pod* where the microcontroller chip will eventually go. The pod acts as a surrogate chip whose pins do the same thing as the microcontroller pins. Alternatively, you can connect wires from the I/O port connector to the circuit board you are developing.

Available Evaluation Boards.   Typically, you load your program code into memory in the evaluation board. With an evaluation board you are running a real microcontroller. Motorola used to supply an EVBU and EVBU2 (Universal Evaluation Board) to develop single-chip applications. Now, third parties supply equivalents. The EVBU restricted the user memory space to the built-in RAM and used a single 5-V supply. A *charge pump* circuit internally supplies the voltage required for computer communications. (In Chapter 10 we will say more about computer communications and charge pump circuits.) The EVBU board uses the Buffalo monitor program. The source code for Buffalo is also available for free from the Internet.

Typically, developers now use third-party boards, some of which run the Buffalo monitor. There are several sources of EVBU replacement or other types of boards, for both academic and commercial use. Appendix C lists some of them. For example, Wytec supplies the EVBPlus2 development board in both student and instructor versions. It also includes programmer and emulation features. More information is on the Internet at the URL http://www.evbplus.com. Wytec also supplies other kits including one for 68HC12 development, such as the DRAGON12 and miniDRAGON. HVW Technologies supplies both kits and fully assembled EVBU replacements. Axiom's CME-11E9-EVBU includes a power supply. It operates in expanded or single-chip mode as the SC68HC11EVBU2. Technological Arts supplies a MicroStamp11, which also can be inserted into a protoboard. The Miniboard, originally developed at the Massachusetts Institute of Technology (MIT), is very popular for robot control. Users have to build their own board and supply a C cross compiler. It requires a 6- to 36-V supply because of built-in stepper motor drive circuits. Instructions for building the board and obtaining both hardware and software are available from the Internet (e.g., URL http://www.bol.ucla.edu/~frank/68HC11.htm). Another MIT design is the Handy Board. Other board suppliers and additional updates will be posted at fleming0.flemingc.on.ca/~pspasov/mcu/mcu.htm.

Emulation.   Emulation is similar to using an evaluation board. Instead of using a circuit board with a real microcontroller, it uses a special pod connected to a computer or emulator control unit. It allows more control. The computer program can run a program, send signals to the pod, and read signals from the pod. You can choose whether to use real signals or to simulate them. An emulator may also show the source code with symbols.

With an emulator, you can test stages of the design. For example, if the 68HC11 uses external memory, you can simulate the memory before the external memory is actually built. You can test if the memory circuit works by using the emulator to check if it can read or write data. Another example is a situation in which the microcontroller responds to an interrupt to update a control signal to an actuator. You can simulate a signal at the interrupt pin before building the circuit that generates the interrupt. An example of an emulator is Nohau's EMUL-PC. It can provide source-level debugging in C.

More Information. There are more suppliers which we will not list here. For more information on Motorola and third-party tools, contact Motorola for their *Microcontroller Development Tools Directory* (MCUDEVTLDIR/D).

■ 4.11 SUMMARY

In this chapter we described the steps to produce an executable program. If you are experimenting with short programs for learning purposes, you will probably not go through all of these steps. For short programs, it may be more convenient to use a line assembler or enter hex code directly into the 68HC11 system with which you are working. Otherwise, do the following: Begin by entering the source code using a text editor. The source code may include a header file. The header file usually contains standardized symbol definitions. These may be vectors and I/O registers.

For an 8-bit microcontroller, you would typically use a cross assembler instead of a resident (or native) assembler to translate the source-code file into an object-code file. It also produces a listing file for documentation. The listing file shows the object code cross-referenced to source code. A software developer can also use a cross compiler to produce other object codes from a high-level language source code: for example, C.

The linker combines several object-code files to produce a hex file. Some of the object codes may contain standard subroutines or interrupt handler routines. Others may have been produced by a cross compiler. A software developer can use a combination of assembly and high-level language to produce executable code. The hex file is in a standard format. Typically, these are Intel hex or Motorola hex. The linker also produces a map file and symbol file for documentation purposes. The symbol file is useful for symbolic debuggers, simulators, and emulators. After the hex file is produced, you load it into a target system for testing and debugging. This may be a simulator, evaluation board, or emulator. Eventually, you convert a hex file into binary to load in the target processor's memory (internal or external).

Appendix C lists suppliers of cross assemblers, simulators, evaluation boards, and emulators. The Motorola AS11 cross assembler for the 68HC11 was used in this chapter for illustration purposes. Check out other alternatives to see which suits your purpose or budget. They will use similar principles. If you are developing programs in a high-level language such as C, there are development systems for these as well. Appendix C lists some of them.

The AS11 cross assembler we mentioned earlier is relatively simple to use. The instructions are in a text file called ASSEMBLER.DOC. This cross assembler produces Motorola S-record files directly from source code files by combining the assembly and linking process.

■ EXERCISES

1. Describe the following terms.
   (a) Assembly
   (b) Assembler
   (c) Assemble

2. Describe the differences between the following types of files.
   (a) Text file
   (b) Word processor file
   (c) Binary file

3. Write down the machine code for the following data segment (Listing ex4a) without using an assembler. Afterward, you may check your predicted machine code with the listing file produced by an assembler. Assume that uninitialized data is zero.

```
*Listing ex4a
*data segment definition
*-----------------------------
PAGE0      EQU     0
TAB        EQU     9
CR         EQU     13
LF         EQU     10

           ORG     PAGE0
BEGIN      RMB     10
           FDB     %01110110
           FCC     "Good Luck"
           FCB     TAB,CR,LF

           ORG     $40

           FCB     $40,64,'4'
*-----------------------------
    END
*-----------------------------
```

4. Write the source code for the program of Listing ex4b. Use labels or symbols for instruction operands. Start the program at address $E200. Set up a data segment reserving space for the program data.

```
*Listing ex4b
*Assembly exercise
*and execution history exercise
*for inherent, immediate, and extended addressing modes.

        LDS     #$00FF
        LDD     $C010
        STD     $C012
        LDY     #$20       ;note that IY is a 16-bit register
        ABY
        ANDA    #$16
        BRA     *          ;its machine code is 20 FE
```

5. Write the source code for the program of Listing ex4c. Use labels or symbols for the operands. The data segment contains the string "MC68HC11A8," which is to be copied by the program to the destination.

```
*Listing ex4c
*Demonstrate a do-while loop
*
*Move 16 bytes starting at address $C000 to a block
*at address $C500.
*-----------------------------
                          ORG   $E000
E000  CE  C0  00          LDX   #$C000    ;initialize src pointer
E003  18  CE  C5  00      LDY   #$C500    ;initialize dst pointer
E007  C6  10              LDAB  #$10      ;initialize counter
E009  A6  00          DO  LDAA  $00,X     ;move byte from src
E00B  18  A7  00          STAA  $00,Y     ;to dst
E00E  5A                  DECB            ;decrement count, to
*                                         ;indicate another move
E00F  20  F8              BNE   DO        ;do-while count not zero
E011  20  FE      HERE    BRA   HERE       ;stop program when done
```

6. Repeat Question 5 such that the data is in internal RAM. This will allow the 68HC11 to run in single-chip mode.

7. Describe the differences between the following file types.
   (a) Object file
   (b) Hex file
   (c) Binary file

8. Write down the ORG statements for the following (the source file is called ragtop.asm):

Segment name	Type
maincode	Starts at $D000
engine	Follows maincode
display	Follows engine
sensor	Absolute, code starts at $CE00

Hint: Assume each segment labels its last line for use by a following ORG statement.

9. Is it possible to rewrite the source for Question 8 so that
   (a) Segment display follows maincode?
   (b) Segment engine follows display?
   If so, how?

10. For the source code of Question 8, what could happen if the segment sensor has more than 512 bytes?

PART 3
Hardware

5 Bus Concepts and Modes of Operation

OBJECTIVES

After completing this chapter, you should be able to

- Explain the fundamental operations of a bus and tristate logic.
- Determine some control signal sequences for memory interfacing.
- Design and analyze a memory interface to satisfy a memory map using address decoding.
- Explain the modes of operation for the 68HC11 in terms of memory mapping and pin functions.
- Determine the mode of operation for some applications.
- Design a basic microcontroller circuit.

■ 5.1 INTRODUCTION

So far, we have looked primarily at programming the microcontroller. In this chapter we begin a description of the microcontroller's hardware. They include the basics of the system bus, tristate logic, address decoding, and the operating modes of the 68HC11. These are fundamental when interfacing a microcontroller to peripheral circuits.

We introduce *interfacing* in this section. *Interfacing* refers to building hardware and writing software to allow one device to exchange data with another device. For example, the microcontroller reads the coolant temperature in the automobile engine control system described in Chapter 1. In this case the interfacing hardware refers to the wiring, connections, and conditioning circuits that connect the temperature sensor to an input pin of the microcontroller. The software may be a subroutine that returns a binary number representing the temperature.

Note that interfacing requires both hardware and software. So far, you should be familiar with the CPU registers and instruction set of the 68HC11. Also, you should know how to write, load, and run a 68HC11 assembly language program. For interfacing, we need to have more software skills. But before acquiring these, it is necessary to understand its hardware. In this and the next chapter we concentrate on the hardware of the microcontroller chip itself, but, occasionally, we may have to sidetrack along the way.

■ 5.2 THE BUS

The first thing we will cover is the *bus*. Refer back to Figure 1.3. In Section 1.3.1 we described briefly the purpose and operation of a bus. A bus is a set of signal lines. The bus connects the microcontroller unit (MCU) to the other devices with which it must communicate. Sometimes, an external bus is simply the extension of a microcontroller's pins. In other systems, the bus lines carry signals that are modifications of the MCU pin signals. The microcontroller receives its power and clock signals through the bus. It sends and receives data using the bus. To select devices with which to exchange data, it sends address signals on the bus. A controller board such as that shown in Figure 1.3 shows the bus as a communications link for all board components. The bus is typically a series of etched copper lines on a printed circuit board connecting plug-in sockets for other printed circuit boards.[1] A bus can be divided into three separate buses that transfer data, address, and control signals. Note that we consider the power supply lines to be part of the control bus.

Many controller manuals include a system block diagram to show the system bus. The block diagram does not show detailed connections, nor does it show how the devices use the address and data lines. However, by looking at the block diagram first, it makes it easier to understand a complete circuit schematic later. A schematic shows each individual line and pin connection. Each line is dedicated for a single data, address, or control bit. By using a block diagram first, we will not get swamped with details when trying to understand how the system works. Figure 5.1 shows a block diagram of an expanded mode system. In this chapter we examine how to connect the MCU. In Chapter 7 we will look at details of using the various types of memory. In Chapter 8 we examine the PRU. In later chapters we examine various I/O ports and their subsystems, including COM (Chapter 10).

Actually, we have simplified the address and data bus considerably. Recall from Section 1.3.2 (and Figure 1.5) that many microcontrollers use a *multiplexed* address and data bus. This means that it shares the same physical lines for data bits and some address bits. The MCU in this case includes the extra circuits to split (demultiplex) the address and data signals into their respective buses.

Note that the bus lines are unidirectional or bidirectional, depending on the type and where it is connected. The microcontroller unit (MCU) sends address signals to other (peripheral) devices. A device recognizes its address and responds as specified by signals sent on the control bus. Data transfer takes place on the data bus for both read and write operations. When we talk about read and write or input and output, we describe the direction of data travel relative to the current bus master, typically the MCU. For example, a write or output operation means that the MCU sends data to a peripheral device. If this convention

---

[1]Users of personal computers (PCs) may recognize a PC bus as the expansion slots mounted on the motherboard.

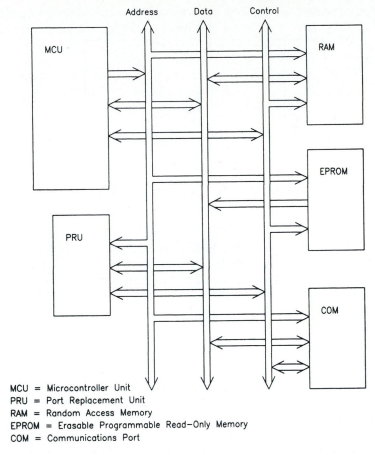

MCU = Microcontroller Unit
PRU = Port Replacement Unit
RAM = Random Access Memory
EPROM = Erasable Programmable Read—Only Memory
COM = Communications Port

Figure 5.1 An Expanded Mode System Block Diagram

is violated, it should be specifically stated as such. This may occur if we describe the operation of a peripheral chip by itself.

Also note how Figure 5.1 schematically shows a bus: as an outline of an arrow. Other diagrams may show a multiline bus as a single line with a slash. The slash may have a number beside it to indicate the number of lines in the bus. For an example of this format (without the number), browse ahead to Chapter 8 and look at Figure 8.1. The line labeled "I/O data" is a multiline bus. By convention, single lines without a slash refer to single connections.

Generally, there are two types of peripheral devices, memory or input/output (I/O). We shall examine these types of devices in more detail later. The EVB board includes random access memory (RAM) and erasable programmable read-only memory (EPROM) chips for various memory options. The port replacement unit (PRU) provides extra I/O connections for control purposes. The communications port (COM) provides the interface to monitor and control the EVB board using a terminal or personal computer.

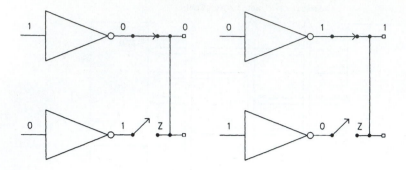

Figure 5.2   Tristate Analogy Using Mechanical Switches (Z = High Impedance)

Since the bus is not static and many devices share the same bus, it is necessary to control when and how a device can use the bus. For example, if the MCU needs to fetch an instruction from the EPROM, it addresses the EPROM, tells it to put data on the data bus, and then reads the data bus. The PRU should not put data on the bus at the same time. In the next few sections we discuss how this is done.

■ 5.3 TRISTATE

For an MCU to use only one device at a time, it *enables* the device, meaning that it gives permission to the device to use the bus. The reverse is also true. The MCU *disables* other devices, meaning that it refuses permission. To disable other devices, the MCU must effectively "disconnect" them from the bus. To do this, it puts them in a state called *high impedance*. This is like opening a switch. An open switch has infinite resistance, so it is high impedance. But there are no mechanical switches used to disconnect other devices. Instead, they are put into a high-impedance state using a logic signal.

A digital signal can have two states, high or low. If an output can also be high impedance, a third state, we say that the device is a *tristate* device. A device with a high-impedance output is also known to be tristated.

Figure 5.2 shows how tristate devices can share a common output line using a mechanical switch analogy. In both cases the upper inverter sends its output to a shared line while the lower is disabled by the open switch. If the switch is open, the output is disconnected (high impedance) independent of the input. If the switch is closed, the output is high or low depending on the input(s). Similarly, logic gates with multiple inputs could use the same technique. Figure 5.3 shows tristate inverters. A tristate device has extra input(s) used as enable signal(s). Typically, an enable signal is active low; that is:

A logic 0 enables.

A logic 1 disables.

Again, only one device should be enabled at a time.

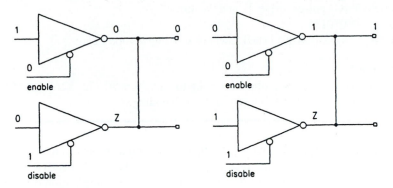

Note: Z = High Impedance

Figure 5.3 Tristate Logic

PRACTICE

An inverting tristate buffer has an active-low enable pin. Its input is presently high. Its output is connected to a 5-kilohm (KΩ) (pull-up) resistor, which in turn is connected to +5 V dc. What is the output when the enable signal is low? When it is high? *Note:* Figures 9.10 and 9.22 show examples of pull-up resistors.

Solution

Enable low, output is low. Enable high, output is high (high-impedance output pulled high by resistor).

Similarly, an enable signal can establish whether a device reads in input as well as sending output. Figure 5.4 shows a generic memory chip. Many memory chips have similar connections. If it is a writable memory chip, it will input the data signals on the data lines only when the $\overline{E}$ signal is low. Similarly, it will output data to the data lines only when $\overline{E}$ is low. The write signal, $\overline{W}$, determines the direction of data travel. If it is low, data is written to the chip. If it is high, data is read from the chip. Some chips have separate signals for reading and writing. The clock signal, $\overline{G}$, synchronizes the data transfer operation. We examine the memory interface more in the next section and again in Chapter 6.

PRACTICE

In Figure 5.4, what signals should the MCU send to write $17 to address $2A73?

Solution

Drive clock ($\overline{G}$) pulses; drive $\overline{E}$ and $\overline{W}$ low
Data = %00010111
Address = %xxx0101001110011, where x = logical don't care

(See also Figure 1.7.)

Note: The same solution applies if writing to addresses \$0A73, \$2A73, \$4A73, and so on. This is a preview of the next topic, address decoding.

Not all chips use the same labeling. The enable signal ($\overline{E}$) is often known as a select ($\overline{S}$) or chip select ($\overline{CS}$). Some enable or select signals are also active high. Many memory chips have more than one select line. Thus a unique combination of enable signals select the chip. Note that the clock signal is often labeled as *E*. Unfortunately, this can lead to confusion with an enable signal. You will have to study circuit schematics carefully to understand the context for a specific circuit.

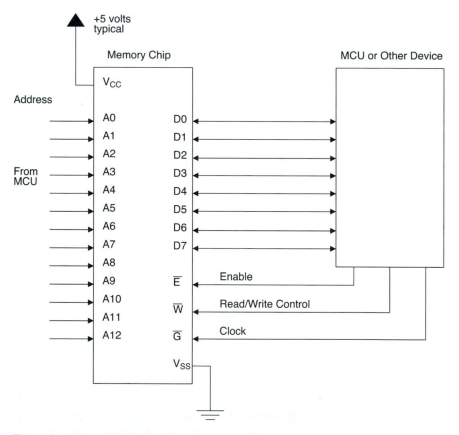

Figure 5.4   Generic Memory Chip

■ 5.4 ADDRESS DECODING

5.4.1 Address Decoding Circuit

Again, refer to Figure 5.4. For now, note that the MCU should enable the memory chip (make $\overline{E}$ low) only when it wants to address it. To do this, the MCU uses an *address decoding* circuit.

The signals on the address pins determine which memory location is selected or *addressed*. To review how this operates, refer to Figure 1.7. The example shows a 16-bit address of $2A73 selecting a memory location. The chip in Figure 5.4 has 13 address lines, so if the address is $2A73 as shown in Figure 1.7, it sees only the lower 13 bits. These select a byte location within the chip. However, there may be another chip that also has address lines in common. To select only the byte in this specific chip, it is necessary to enable it by driving $\overline{E}$ low only when bits A15, A14, and A13 are at logic levels 0, 0, and 1, respectively. Similarly, no other chips should be enabled when the address is $2A73.

A decoding circuit ensures that the chip is selected only when addressed. The address range of the chip is anything within the range shown in Table 5.1. Note that any address within the range $2000 to $3FFF has the same logic value for address bits A15, A14, and A13. It is a simple matter to build a logic circuit using NAND gates that has a low output only when the inputs are 0, 0, and 1. However, since other devices also have to be decoded for their respective address ranges, a more common approach is to use a decoder chip.

One such chip is the 74HC138 1-of-8 Decoder/Demultiplexer. Figure 5.5 shows its logic diagram and function table. It has enable lines, except that when they are not active, the outputs are forced to logic high. The active level of a signal is high if there is no inverting bubble shown, and it is low if an inverting bubble is shown. For example, CS2 is active low, meaning that it performs the function of selecting the chip when it is logic low. Note that either none of the outputs are low or only one of the outputs is low. If the chip is selected (chip select inputs active), the address input is a 3-bit binary number specifying which output is active (low). For example, if the input is %010 (the binary number for 2), Y2 goes low to logic zero.

This decoder chip can be connected to act as an address decoder for the generic memory chip example. The decoder chip selects are hardwired to their respective active logic levels (i.e., CS2, CS3 grounded, CS1 tied high). Do not confuse the signal labels for the decoder chip with those of the memory chip and address bus. Connect A13, A14, and A15 of the address bus to the decoder inputs A0, A1, and A2, respectively. Then connect Y1 to $\overline{E}$

TABLE 5.1 ADDRESS RANGE OF MEMORY CHIP OF FIGURES 5.4 AND 1.7

Memory chip	Hex address	Address bits															
		15	**14**	**13**	**12**	**11**	**10**	**9**	**8**	**7**	**6**	**5**	**4**	**3**	**2**	**1**	**0**
lower addr	2000	0	0	1	0	0	0	0	0	0	0	0	0	0	0	0	0
		0	0	1	x	x	x	x	x	x	x	x	x	x	x	x	x
selected addr	2A73	0	0	1	0	1	0	1	0	0	1	1	1	0	0	1	1
		0	0	1	x	x	x	x	x	x	x	x	x	x	x	x	x
upper addr	3FFF	0	0	1	1	1	1	1	1	1	1	1	1	1	1	1	1

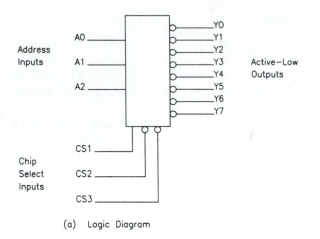

(a) Logic Diagram

Chip Select
Conditions
CS1 = 1
CS2 = 0
CS3 = 0

All outputs high
for other chip
select conditions

Inputs			Outputs							
A2	A1	A0	Y0	Y1	Y2	Y3	Y4	Y5	Y6	Y7
0	0	0	0	1	1	1	1	1	1	1
0	0	1	1	0	1	1	1	1	1	1
0	1	0	1	1	0	1	1	1	1	1
0	1	1	1	1	1	0	1	1	1	1
1	0	0	1	1	1	1	0	1	1	1
1	0	1	1	1	1	1	1	0	1	1
1	1	0	1	1	1	1	1	1	0	1
1	1	1	1	1	1	1	1	1	1	0

(b) Function Table

Figure 5.5 74HC138 1-of-8 Decoder

of the memory chip. With this circuit, when A15, A14, and A13 are %001, the decoder inputs A2, A1, and A0 are also %001. Hence, the decoder output Y1 goes low, which in turn drives the memory chips $\overline{\text{E}}$ input low to select the memory chip. Thus any address within the range $2000 to $3FFF will select a chip connected to Y1 for the circuit just described. Typically, the other outputs of the decoder would be connected to other chips. But none of these chips are selected whenever the address falls within the range $2000 to $3FFF.

You may have to read this section several times if you have not understood it the first time. Draw the circuit on paper and try different logic conditions if necessary. We have described address decoding for systems with a 16-bit address bus. The same applies for other bit sizes. But systems with an addressing range beyond several megabytes often use a *memory management unit* (MMU) that makes it possible to access an address range that exceeds the address bus size. For example, the Intel 80386 microprocessor has a built-in MMU that allows it to address a 64-terabyte range (46-bit) with only a 32-bit address bus (4-gigabyte). This extra nonphysical address space is called *virtual memory.*

5.4.2 An Address Decoding Design Example

We will now work through an address decoding design exercise. There are two memory chips, U1 and U2, with 2 kilobytes (KB) each. Each chip has an active-low enable signal. U1 has a start address of $5000 and U2 has one of $6800. To find the address range, add 2047 to each start address to obtain $57FF and $6FFF, respectively. Use a table similar to Table 5.1 as a template to determine the enable conditions. Fill in the logic levels for the lower and upper address for each chip. The result will be as shown in Table 5.2.

Figure 5.6 shows one possible solution. As we shall see in Chapter 6, it is also necessary for the clock signal (E) to be high when enabling a memory chip for a 68HC11 system.

PRACTICE

Design a decoder circuit for an 8KB (8-kilobyte) memory chip whose first address is $6000. First find the highest address in its range. Then fill in a table like Table 5.2. From this, derive decoder inputs. What is the address range for the chip?

Solution

```
Highest address = $6000 + (8K − 1) = $6000 + ($2000 − 1) = $7FFF
```

From the table, the first 3 address bits don't change: A15, A14, A13 = %011. So these are connected to decoder select inputs. Decoder output $\underline{Y3}$ (%011 = 3) is brought to memory chip enable.

Hence, use 74HC138 with the following connections:

$\overline{CS3}, \overline{CS2}$ to ground
CS1 to logic high
A0, A1, A2 to A13, A14, A15 of address bus
$\overline{Y3}$ to $\overline{E}$ (enable) of memory chip

The address range is $6000 to $7FFF.

TABLE 5.2   ADDRESS RANGE OF MEMORY CHIPS FOR ADDRESS DECODING EXAMPLE

Memory chip	Hex address	Address bits															
		15	14	13	12	11	10	9	8	7	6	5	4	3	2	1	0
U1 lower addr	5000	0	1	0	1	0	0	0	0	0	0	0	0	0	0	0	0
U1 upper addr	57FF	0	1	0	1	0	1	1	1	1	1	1	1	1	1	1	1
U2 lower addr	6800	0	1	1	0	1	0	0	0	0	0	0	0	0	0	0	0
U2 upper addr	6FFF	0	1	1	0	1	1	1	1	1	1	1	1	1	1	1	1

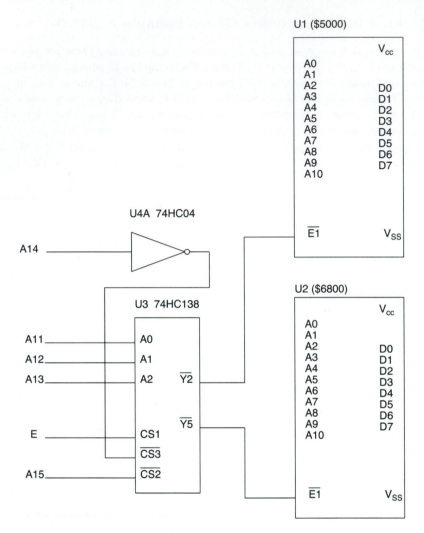

Figure 5.6 Solution to Address Decoding Exercise

5.4.3 Partially and Fully Decoded Systems

In the example seen so far, each memory location has a unique address. Certainly, it is not desirable for an address to select more than one data location, but often a location can have more than one address. This type of system uses what is known as *partial address decoding*.

Consider adding a 512-byte memory chip with one enable input to the system shown in Figure 5.6. A 512-byte chip has nine address inputs, A0 to A8. Let's say that we connected its enable or chip select to Y3. This chip would be selected whenever the address falls within the range $5800 to $5FFF. But this is greater than a 512-byte range. The chip is selected independent of address inputs A9 and A10. As a result, each location could then be accessed by four separate addresses whose only difference is the state of bits A9 and A10. For example, the first location can be addressed by $5800, $5A00, $5C00, or $5E00. The convention used by most programmers in a situation like this is to choose the lowest address.

It is not uncommon to use partial address decoding. Many programmable I/O chips require only 16 (or fewer) addresses. It is expensive to add hardware to fully decode a 16-bit range, particularly when most systems use only a few programmable I/O chips. For example, to fully decode an I/O chip that has 16 registers in a system with a 64KB (kilobyte) address space (16-bit) requires an address decoder with 12 inputs. On the other hand, if it was partially decoded such that the decoder used only six inputs, the I/O chip would be addressable by a 1KB range. Note that $2^{16-6} = 1024$. The decoder output selects the chip. The lowest-order address lines are used to select the register in the chip. For eight registers, three address lines are required. Hence, each register would have $2^{10-3} = 128$ possible addresses. Again, the standard convention is to use the lowest address possible.

■ 5.5 MODES OF OPERATION

In Section 1.3.2 we described the internal and external logical parts of an MCU. You may wish to refer now to Figures 1.4 and 1.5. Internally, the MCU has a CPU, memory, and registers. Externally, it has pins for I/O and bus signals. The I/O pins are grouped in sets of eight called ports. The Quick Reference (Appendix B) shows the pin assignments. When referring to the pin assignment figures, you should note that some pins have more than one function. For example, the mode A input pin (MODA) also has the function of the load instruction register output pin ($\overline{\text{LIR}}$). If you referred to the 68HC11 *Data Book* or *Reference Manual* (see Appendix C), you would find that the pin has an input function of MODA and an output function of $\overline{\text{LIR}}$. We describe the MODA function shortly and the $\overline{\text{LIR}}$ function later when we describe clocked operation in Chapter 7.

Multipurpose pins are common in programmable chips. One of the major problems in designing integrated circuits is the external connection. There is not enough room to provide a pin for each signal. To save on the *pin count,* semiconductor manufacturers have resorted to using some pins for more than one purpose. Software or other hardware connections determine what function a pin performs at a given time.

Many microcontrollers, such as the 68HC11, can be operated in different modes. The I/O and memory requirements of an application, or test requirements, determine in which mode to operate the MCU. The 68HC11 has four modes. The mode selected is determined by how pins MODA and MODB are connected at the time of reset. Usually, the mode is fixed for any application. A manufacturer of a microcontroller system can stock one standard part for a greater variety of systems. Two of the modes are operational: single-chip and expanded multiplexed. The other two are special bootstrap and special test. Table 5.3 shows the connections used to select the MCU mode.

■ 5.6 SINGLE-CHIP OPERATING MODE

Figure B.1 in the Quick Reference (Appendix B) shows the block diagram for the 68HC11. We have placed it there because we will be referring to it many times. The 68HC11 has five ports, A to E. In the rest of this book, we will examine how these ports are used. For now understand that I/O subsystems can control these ports to perform specialized I/O tasks. Control application programs require nonvolatile memory such as ROM for program code and writable memory such as RAM for temporary data storage. The block diagram shows that the 68HC11 chip has built-in memory as well as specialized subsystems to handle I/O.

TABLE 5.3   MODE SELECTION

Input pins		
MODA	**MODB**	**Mode selected**
0	1	Single-chip
1	1	Expanded multiplexed
0	0	Special bootstrap
1	0	Special test

If the built-in memory and I/O subsystems are sufficient, external chips for memory and I/O subsystems are not necessary. In this case the application can use the MCU in its single-chip mode.

As implied by the name *single-chip,* only one chip is required. There is no need for extra memory or I/O chips. Hence, an external bus is not required. When in single-chip mode, the pins on ports B and C are used for I/O. Using single-chip mode results in cost savings, and it improves reliability because it minimizes the use of external connections, which can be subject to failure. The EVB and EVBU boards mentioned in Chapter 4 can only be used to prototype single-chip applications.

PRACTICE

Is address $2A73 available as a RAM location when operating in single-chip mode?

Solution

No.

■ 5.7 EXPANDED MULTIPLEXED OPERATING MODE

Traditional microprocessor systems had connections to memory and peripheral control chips. These connections were known as a bus (address, data, and control). The 68HC11 can operate in an *expanded* mode to provide similar connections when it uses ports B and C as an address and data bus. The mode is also *multiplexed* because port C has a dual function. It carries address signals and data signals at alternating intervals.

Refer again to Figure B.1 in the Quick Reference (Appendix B). The block diagram shows ports B and C being used to provide external bus signals when operating in expanded multiplexed mode. The MCU chip does not use the I/O capability of these ports. Applications must use expanded multiplexed mode if they require more memory or I/O subsystems than are provided in the chip. To prototype an expanded mode system, you can use Motorola's EVM board or one of the development systems offered by other vendors. It is also possible to modify the EVBU board to operate in expanded multiplexed mode by cutting a trace to the MODA pin.

Note that port B carries only the upper address byte of a 16-bit address. By referring to the pin assignment drawings in the Quick Reference, you can note that pins PB0 to PB7 are also address pins, A8 to A15. Note that port C pins, PC0 to PC7, are also pins AD0 to AD7. In expanded multiplexed mode, port C is a bus that carries both the low byte of the address and the (single) data byte. The address and data signals cannot appear on the bus at the same time. They must be *multiplexed*. Multiplexed signals are different signals that share a common communication channel. In this case the channel is port C. Multiplexing address and data on the same pins is common practice. Many microprocessors and microcontrollers use this technique. Again, the reason is to keep the pin count down.

In addition to ports B and C, the MCU uses the strobe A/address strobe (STRA/AS) and the strobe B/read/write (STRB/R/$\overline{W}$) pins as address strobe (AS) and read/write (R/$\overline{W}$). These are part of the control bus. In single-chip mode they function as strobe A (STRA) and strobe B (STRB). We return to these signals again in Chapters 7 and 9. We'll say no more about them now.

When using a multiplexed bus, it is necessary to connect the bus to hardware to *demultiplex* it. Demultiplexing is the process of separating all the signals in a common communication channel into separate signals. This means that the demultiplexer separates the address and data components and places each on their own respective bus. In Chapter 7 we will examine how to connect the MCU for expanded multiplexed mode. It is necessary to cover clocked operation and timing diagrams first.

Nonmultiplexed Expanded Bus.   Some versions of the 68HC11 use a nonmultiplexed expanded bus. This leads to increased bus speeds and less hardware (less glue logic) since the requirement to demultiplex is eliminated. One example is the 68HC11 N-series. It has 80 pins, 62 of which are arranged into seven 8-bit ports: A, B, C, E, F, G, and H, and one 6-bit port, port D. When used in expanded mode, ports B and F act as the high- and low-order address lines of a 16-bit address bus. Port C acts as the data bus without multiplexing. Section 15.2.1 surveys the different variations of the 68HC11.

■ 5.8 SPECIAL BOOTSTRAP MODE

This is one of the test modes. Before we examine what this mode does, we'll look at the bootstrap (also known as boot) operation in general. This is an analogy to the advice to pull yourself by your own bootstraps—to help yourself out of difficulties. This may have been more appropriate in the days when high boots were in fashion and falling off a horse was more common. But the saying is still reasonably clear: Rely on yourself before relying on others.

Computers run a special program called an operating system that handles I/O, memory use, and execution of programs. The operating system is usually in RAM. When the computer is powered up (booted), it automatically loads the operating system into RAM. However, a program to load the operating system must already exist.

In Section 1.4.2 we mentioned briefly the operating system and the role of the basic input/output system (BIOS). The BIOS is a small program stored in ROM. When the computer is powered up, it executes this program. Recall that the power-on $\overline{\text{RESET}}$ vector specifies the start address. One of the functions of the BIOS is to load the operating system program code into RAM. Once the operating system is loaded in, program execution jumps to the operating system start address and executes its code. This is also commonly known as booting the system (not literally, please).

Figure 5.7 illustrates the operation. The $\overline{\text{RESET}}$ vector specifies the START address. The ROM code loads in memory from some storage media, usually a disk (boot disk). Then the program that was loaded in is in turn itself executed. For some operating systems the process requires several steps. The loaded-in program loads in more program code, which in turn loads in more program code, and so on. This step-by-step process came to be known as a *bootstrapping*.

What we have just described is the bootstrap process commonly used by computers. Although the bootstrap process used by the 68HC11 is similar, there are some differences. The manufacturer of a general-purpose computer does not know what the customer will use it for; hence, it must be able to load and run different programs as specified by the user. A microcontroller system does not normally run a variety of application programs that are loaded in from disk or tape. Its application is known beforehand. Generally, it uses bootstrapping only to load in a test program. This is necessary because the system supplier cannot predict what diagnostic tests will be necessary for each customer. Other uses are for EEPROM programming and for running a monitor program.

When the 68HC11 operates in bootstrap mode, it uses different vectors, known as bootstrap mode interrupt vectors. It also uses different ROM from that used for the two operating modes. The bootstrap ROM is a 192-byte block with an address range of $BF40 to $BFFF. This block is accessible only in bootstrap mode. This ROM has a bootloader program; hence, the code is a permanent part of the chip. The $\overline{\text{RESET}}$ vector address is $BFFE. After the system is reset with pins MODA and MODB grounded, the bootstrap ROM (also called bootloader) program executes. This program reads in more program code from the outside using the serial communications interface (SCI) subsystem with port

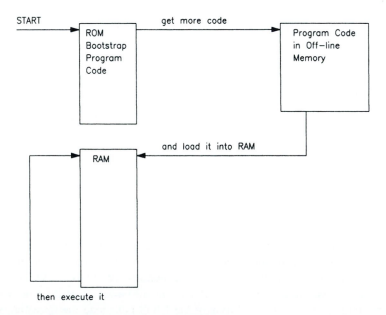

Figure 5.7 Computer Bootstrap Operation

D (see Figure B.1 in Appendix B). It loads a 256-byte program into the on-chip RAM at locations $0000 to $00FF. After the program is loaded, execution jumps to address $0000 to execute the program that was just loaded in. In Chapters 8 and 10 we will describe how the SCI inputs data into the MCU. Later types of the 68HC11, such as the 68HC11 N-series, use a larger ROM from $BE00 to $BFFF.

EEPROM Programming. The *Data Book* describes other options and considerations when using this mode. We will not detail these here. A Motorola application note (AN1010) describes how to use bootstrap mode to program (write to) the internal EEPROM or an external EEPROM using minimal extra hardware and a personal computer or terminal. To program an EEPROM, you have to execute a sequence of instructions, otherwise it is a type of read-only memory. The bootloader program loads this code into internal RAM. This loaded program in turn reads more data from the computer to load (program) it into EEPROM. In Chapter 6 we will cover the software to write data to EEPROM. If you are interested in knowing more about this application, order the document from your local Motorola supplier.

PCbug11. In Section 4.10 we mentioned the monitor program PCbug11. The personal computer contains the monitor program. It communicates with a program in the MCU called a *talker* using the serial communications interface (SCI). PCbug11 may use the talker in the *boot method* or the *ROMed method*. In the boot method it first loads the talker program from a computer disk into the internal RAM of the MCU. Then it runs the talker program for communication. In the ROMed method the talker program already resides in EEPROM or external memory. The monitor program in the personal computer synchronizes communication with the talker program.

■ 5.9 SPECIAL TEST OPERATING MODE

This mode is used by the chip manufacturer to test the chip in the factory. It allows the tester to change some register contents in ways that are not otherwise possible. Motorola provides very little documentation about this mode, for the simple reason that they discourage their customers from using it. It is meant to be used by Motorola only.

■ 5.10 SYSTEM CIRCUIT

Now it is time to look at a circuit using the 68HC11. Figure 5.8 shows a basic system that can be expanded to experiment with different types of interfaces. It closely resembles the circuit used in the EVBU and other development systems. The system is shown for single-chip mode. But it is possible to build upon the system by adding input/output (I/O) interfaces and additional memory when reconfiguring to expanded multiplexed mode. Subsequent chapters examine the use of additional external memory and I/O interfaces. You might also want to cross reference the schematic to the pin assignments shown in Appendix B, noting Table B.5 and the pin assignment diagrams in particular.

Following is an overview of a basic system.

V_{DD} and V_{SS} are for powering the microcontroller. The capacitors serve to filter some of the fluctuations in the power supply voltage.

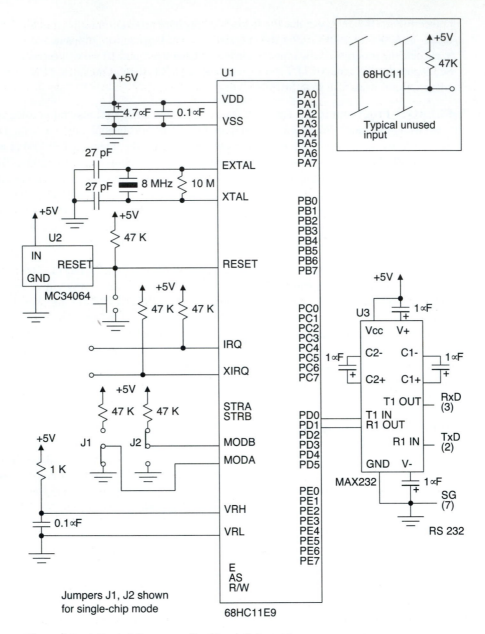

Figure 5.8 A Basic Microcontroller Circuit Schematic

EXTAL and **XTAL** are for connecting the crystal for clock signal generation (see Section 7.2).

RESET is the power-on reset signal. When the push button is not pressed, the 47K pull-up resistor keeps the signal high. Pressing the push button causes the pin to be pulled low, thus forcing a reset. When power is turned off, the supply voltage may drop very gradually. With marginal supply voltage, the MCU may *run away* by incorrectly reading RAM or ROM. This incorrect software operation may cause problems in some application situations. To prevent this, Motorola recommends the use of a low-voltage reset device such as the MC34064. The MC34064 forces a low RESET signal whenever V_{DD} drops below 4.65, thus preventing program execution at marginal supply voltage levels.

IRQ and **XIRQ** are interrupt signals pulled high through resistors. Hence, to trigger an interrupt, a device must *pull* these pins low.

MODA and **MODB** are connected to jumpers, shown for single-chip mode. For single-chip mode, MODA must be low and MODB must be high (see also Table 5.3).

V_{RH} and V_{RL} are connected to provide a 0- to 5-volt range for the analog-to-digital converter to be explained in Chapter 12.

PD0 and **PD1** are for a communications port in order to communicate with a personal computer. Section 10.1 will explain about serial communications. Maxim's MAX232 is a popular chip used for translating signals between a microcontroller and a standard computer serial port.

Other pins can be connected as required. With the system as shown, it is possible to download and run programs and connect some interface circuits. An input that is not connected may float to an arbitrary signal level which may be interpreted as a logic high or a logic low. It is best to connect them to pull-up resistors so as to define inputs as high by default.

■ 5.11 SUMMARY

In this chapter we have surveyed the basics of the system bus, tristate logic, address decoding, and the operating modes of the 68HC11. This is basic to understanding its hardware. These are fundamental when interfacing a microcontroller to peripheral circuits. Interfacing is the process of matching the output of one device to the input of another so that they can communicate. The process involves both hardware and software design and implementation.

A bus is an electronic pathway over which signals can travel between the devices connected to it. Normally, the extension of the MCU pins is known as the *system bus*. The system bus in turn is made up of the address, data, and control buses. To read or write, the 68HC11 first sends a 16-bit address word on the address bus to select/enable a peripheral chip. It controls the data transfer by asserting the appropriate control bus lines. For a read operation the MCU latches in the data bus contents. For a write operation it places the data onto the data bus. The convention is that the stated I/O direction is relative to the MCU (or main processor).

Devices using a bus must use tristate logic. The third state is high impedance. When one device tristates another, it effectively disconnects it from its output lines. The former device enables/selects the other device to connect it to its output lines. In a bus other devices share the same output lines; hence, all should have tristate capability.

A hardware address decoding circuit is often used to select (address) a device. The circuit inputs the address bus signals sent by the MCU to decode it to drive one output select line. This output select line is used to enable the "addressed" device. The address decoder asserts a select line within a specified address range. The high-order bits of an address specify this range.

A microcontroller can operate in several modes. A mode specifies the function of its pins. Other pins are hardwired to appropriate logic levels to select the mode. The 68HC11 has four modes. Two of these are operational and two are for special test and diagnostic purposes.

One of the operating modes is single-chip. In this mode the 68HC11 uses only its own built-in memory and I/O ports. It does not have an external system bus. The other operating mode is expanded multiplexed. In this case the port B and C pins are used as an external system bus. The 68HC11 can use external memory and I/O port chips. *Multiplexed* (in this case) refers to the fact that the port C pins alternate between sending the low address byte and sending or receiving the data byte. Although the terminology and specifications may differ, other microcontroller and microprocessor systems operate similarly.

■ EXERCISES

1. Describe the system bus in terms of its main components.

2. Explain why a block diagram is often used to document a microcontroller system.

3. What type of integrated circuits can have the outputs connected?

4. An 8-bit latch chip has an active-high enable pin. Its inputs are presently $5A. Its outputs are connected to 5-KΩ (pull-up) resistors, which in turn are connected to +5 V dc. What is the latch output when the enable signal is low? When it is high?

5. Repeat Question 4 with the following changes. The enable pin is active low. The input is $6B and the outputs are connected to 5-KΩ (pull-down) resistors, which in turn are connected to ground.

6. What hardware is required to enable a memory chip to use the data bus?

7. For the circuit shown in Figure 5.6, state the address range that will drive each decoder output low.

8. Describe how to modify the circuit of Figure 5.6 so that an address of $9205 drives Y2 low.

9. Design an address decoding circuit for the following 68HC11 system:

    8KB EPROM starting at address $E000
    8KB socket for RAM or EPROM starting at address $6000
    8KB RAM starting at address $C000
    I/O chip partially decoded to start at address $9800

    This happens to be the EVB board. If you have the EVB schematic, try this question without looking at it.

10. A system uses an 8KB EEPROM starting at address $C000 and a 2KB RAM starting at address $2000. It uses two I/O chips with start addresses of $4000 and $6000. Each one has eight addressable registers. Is it necessary to use a fully decoded addressing circuit? Explain why.

11. Why do some MCU pins have more than one function?

12. Explain how to connect the 68HC11 such that port B is an I/O port.

13. A toy manufacturer is designing a toy fire engine controlled by a 68HC11. It flashes light patterns, runs a pump to eject water, and sounds a siren. What operating mode should the designer choose? Why?

14. An assembly plant uses robots, each controlled by a 68HC11 with its program code in internal ROM. To troubleshoot robot failures, a technician has written a 200-byte program. Explain how the technician can run the program.

6 Microcontroller Hardware

OBJECTIVES

After completing this chapter, you should be able to

- Describe the basic physical characteristics and operations of semiconductor technology, and, in particular, the CMOS and HCMOS logic families.
- Determine considerations to avoid zap and latch-up.
- Identify the characteristics and limitations of chips used for interfacing.
- Use data sheets to determine chip specifications.
- Describe characteristics of semiconductor memory such as ROM, RAM, EPROM, flash, and EEPROM.
- Write programs to erase, program, and read EEPROM data.
- Describe the use of special configuration registers such as BPROT, CONFIG, and INIT.

In Chapter 5 we looked at how the 68HC11 uses its pins. In this chapter we examine the physical characteristics of the chip itself. Again, although the discussion is directed at the 68HC11, many of the points apply to other microcontrollers, microprocessors, memory chips, and other programmable integrated circuits.

■ 6.1 SEMICONDUCTOR TECHNOLOGY

6.1.1 The Integrated Circuit

In this section, we present a quick view of integrated circuit (IC) construction, fabrication, and operation. For some readers this will be review. For others, it will provide necessary background material. We will not cover everything there is to know about integrated cir-

214

cuits since there are many different types of integrated circuits. To cover it all would require several books. Also, there are always new developments in this field.

People often refer to integrated circuits as *chips*. Most integrated circuits are a small slice of silicon material containing a circuit pattern (sometimes known as a *die*). Usually, it is some type of digital circuit. The slice is packaged in a plastic or ceramic case. The circuit in the slice is connected to metal pins extending outside the case.

There are several types of cases. The pin assignment drawings in the Quick Reference (Appendix B) show the plastic-leaded chip carrier (PLCC) and windowed ceramic-leaded chip carrier (CLCC) pin assignments. These are also known as *cerquad* or simply *quad*. The plastic quad flat pack (QFP) is also available for some 68HC11 variations. These package options and the small outline integrated circuit (SOIC) are the most common types of IC packaging. For details, including exact package dimensions, refer to the technical data manual for the 68HC11 variation you are using. The metal pins are the power supply and signal connections for the IC. Most of the chip volume is used for packaging. The functional circuit is only a fraction of the IC. Some manufacturers produce a custom printed circuit board or chip with the slice built in without using the standard chip package. This saves space. For example, some automatic cameras use a printed circuit board that contains the silicon die for the 68HC11. The camera manufacturer bonds the die to the printed circuit board (PCB). Alternatively, the semiconductor manufacturer will fabricate a custom part if the quantities are large enough.

PRACTICE

What is the pin number for RESET?

Solution

Pin 17 for the quad package.

A digital circuit is made up of smaller switching circuits called gates (in Appendix F we review basic digital logic). These are the standard gates: AND, OR, NOT, NAND, and NOR. Gates, in turn, are made up of switching elements called *transistors*. As a result, a complex digital circuit is made up of thousands of transistors. Some microprocessors have more than 1 million transistors. The 68HC11 is not one of these, although it still has a high level of integration. Figure 6.1 shows a microscopic photograph of the silicon slice in a 68HC11 IC.

We said earlier that the IC has a slice of silicon material. Silicon is one type of semiconductor. Another type is germanium. Essentially, a semiconductor is a piece of crystalline material such as silicon with regions containing different impurities. Electrically, ICs can function as insulators, resistors, or (almost) perfect conductors by changing voltages applied to selected regions. The difference between a metal (conductor) and a semiconductor is that a metal conducts current by means of electrons. A semiconductor transports charges using two charge-carrying particles of opposite sign. The properties of semiconductors are used to make transistors. A transistor can also function as an amplifier, but this property is normally used in analog circuits instead of digital circuits.

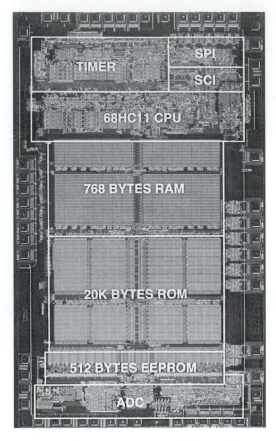

Figure 6.1 Motorola's MC68HC11E20 8-bit Microcontroller (Courtesy Motorola, Inc.)

6.1.2 Some Transistor Theory

We should look at some transistor theory first. This background is necessary in order to understand some of the characteristics we will examine later. Basically, a transistor can act like a switch that turns on and off, depending on the input signal. There are different types of transistors: bipolar, field effect (FET), and complementary metal-oxide semiconductor (CMOS). There are also variations of these.

The type that concerns us here is CMOS. The term *complementary* refers to the fact that CMOS chips use two types of metal-oxide semiconductor (MOS) transistors that behave oppositely. These are N-channel and P-channel MOS. Each MOS transistor has regions that are N-type or P-type. To make the different types of regions, the chip manufacturer adds impurities. This is also known as *doping*. An N-type region is doped with impurities that donate electrons. Some donors are antimony, phosphorus, and arsenic. A P-type region is doped with impurities that accept electrons. Some acceptors are boron, gallium, and indium.

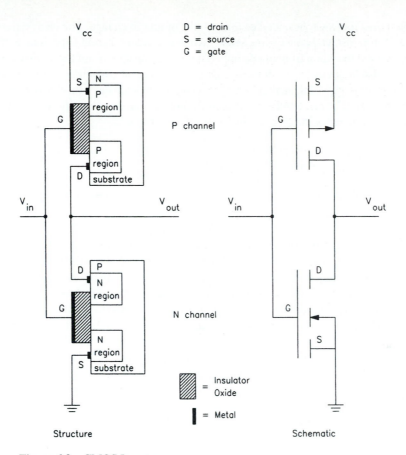

Figure 6.2 CMOS Inverter

Figure 6.2 illustrates an example of a CMOS logic gate, an inverter. The inverter output is always the opposite of its input. When the input is high, the N-channel MOS conducts (like a closed switch) and the P-channel MOS does not (like an open switch). This results in the output being switched to ground or logic low. The MOS transistors work oppositely for a logic-high input.

We look next at the operation of a MOS transistor in more detail. In an N-channel MOS current flows from the drain region to the source region through the substrate when the gate voltage is high enough. We use the convention that current direction goes from positive to negative, the opposite of physical electron flow. So, in physical reality, the impurities exchange electrons such that the net positive charge movement is from drain to source. Net negative charge movement is in the opposite direction.

Let's look at the physics of how current flow is created in an N-channel MOS transistor. A high gate voltage induces an electric field through the oxide layer. This creates a "channel" through the P-type substrate to connect the N regions by attracting electrons. The result is electron flow from source to drain (positive current from drain to source). The P-channel has current flow when its gate voltage is low.

When the input switches, it takes time for the output to change. The charge flow cannot start and stop instantaneously. In a sense each gate has an inherent capacitance. It takes time for an output to switch from low to high (charge the capacitor) and for it to switch from high to low (discharge the capacitor). The larger the inherent capacitance, the slower the rate of voltage change. A larger region means a larger capacitance; hence, a slower rate of change of the signal levels. Semiconductor manufacturers are continually seeking ways to reduce this inherent capacitance to improve speed performance.

6.1.3 Fabrication and Hardware Design

Of course, most microcontroller users do not fabricate their own chips. But some understanding of this process is useful for recognizing chip limitations. A chip manufacturer makes many copies of the silicon slices at the same time. Then each copy is mounted in a case with pins.

The fabrication steps are oxidation, photomasking, diffusion, and metallization. The *oxidation* step involves growing a silicon oxide layer on top of pure silicon. *Photomasking* is a photographic process to selectively remove areas of silicon oxide. This forms a pattern of exposed pure silicon. *Diffusion* is the process of diffusing (doping) the exposed silicon areas with N- or P-type impurities. *Metallization* is a process used to place a thin metal layer over the top of a slice to connect areas electrically. To specify the layout of the various regions, manufacturers use a set of *masks*.

It is possible for designers to specify customized chip layouts using design tools such as VHSIC Hardware Description Language (VHDL), where VHSIC means Very High Speed Integrated Circuits. VHDL is a digital hardware design language that defines register-transfer specifications, which in turn is translated to logic specifications for hardware. A VHDL program mimics the behavior of the digital system. It can incorporate timing specifications such as gate delays as well as other behaviors. The VHDL specification is then translated to mask specifications that can be used for manufacturing.

Instead of VHDL, some designers use the Verilog language, which is the other Hardware Description Language (HDL). Recently there has been a trend to use SystemC, a C++ language extension that enables hardware/software integration.

6.1.4 HCMOS and Other Logic Families

We have looked at one type of technology to fabricate chips using CMOS transistors. CMOS is an example of a logic family. There are other ways to fabricate logic gates using different types of transistors. In earlier days of digital IC technology, the two principal logic families were CMOS and transistor-transistor logic (TTL). TTL devices used a type of transistor known as bipolar instead of MOS. A short look at other logic families is needed to understand why the 68HC11 is built using a high-speed complementary metal-oxide semiconductor (HCMOS).

TTL has the advantage of switching logic states at a faster rate but CMOS consumes less power.[1] Also, CMOS has higher noise immunity. This means that it is less likely to

---

[1]Readers familiar with laptop and notebook personal computers should note that they often use CMOS chips. Reduced power consumption is a prime concern in their design.

change output voltages due to noise spikes induced from other sources. Low-level Schottky TTL (LSTTL) is a further development of TTL that reduces its power consumption. But LSTTL still consumes significantly more power than CMOS. For the most part, the choice of CMOS, TTL, and LSTTL is one of deciding to trade off speed for reduced power consumption. But this is a simplification since other factors have to be considered.

This brings us to HCMOS (or HSCMOS). HCMOS is an evolutionary development of CMOS. It is fabricated with smaller impurity regions. This reduces the inherent capacitance. Hence, it is a faster chip. We review the effects of capacitance later when we discuss timing specification. Also, we will look at some of its characteristics when we discuss chip specifications. Here we warn you of handling precautions that both HCMOS and CMOS require.

Supply Voltage.　The typical supply voltage for most logic families was 5 V. This became a standard for TTL and CMOS. Since the development of laptop computers and other portables, there is a trend to adopting a standard of 3.3 V. The lower operating voltage has advantages for portable applications, since power consumption can be reduced. When a CMOS transistor switches state, the voltage gap between source voltage and ground causes a surge of charge flow. This current causes power dissipation. Note that electrical power is proportional to the square of the voltage (or current). Thus, lowering voltage sounds like a simple answer to the problem of power consumption. However, transistors in logic circuits must distinguish between high and low states. When the voltage gap is reduced, the system becomes more sensitive to random noise. Noise can cause voltage levels to fluctuate. Sources of noise are supply-voltage variations, electromagnetic radiation, induction from neighboring lines, and small variations in ground voltage. When supply voltage is reduced, the ability to distinguish between high and low (also known as *noise immunity*) is reduced. The other problem is that switching speed is reduced. Semiconductor manufacturers overcome these problems by fabricating chips with greater precision and optimizing the physical layout of transistor interconnections.

Mixing 3-V and 5-V Devices.　Although the market for 3-V devices has grown, many 5-V systems remain in operation. The MC74LCX logic family works at both 3 and 5 V. It permits designers to mix 5- and 3-V components in the same system.

6.1.5　CMOS and HCMOS Handling Precautions

You have to be aware of a condition known as zap and another known as latch-up. We discuss zap first.

Zap.　CMOS and HCMOS devices can be damaged by electrostatic electricity. An electrostatic charge, picked up simply by walking across a carpet, can destroy a CMOS device. If a person with static charge touches a CMOS circuit, the charge can break down (destroy) the insulated gate of a CMOS input or output. Zap damage often appears as leakage or shorts because of the breakdown of the oxide layer.

Most semiconductor manufacturers build in some protection for their CMOS devices. Figure 6.3 illustrates the type of input protection built into HCMOS chips manufactured by Motorola. It can withstand a discharge of 2 kilovolts (2 kV = 2000 V). For further protection, connect unused inputs to logic high or low. Also, you can connect external protection networks as recommended by data books.

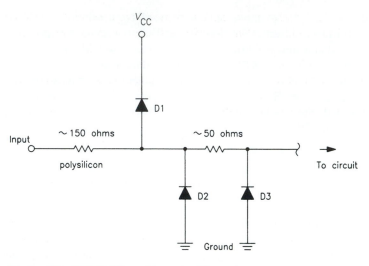

Figure 6.3 HCMOS Input Protection Network

We'll examine Figure 6.3 further. V_{CC} is the positive supply voltage. This is the standard label used by IC manufacturers to specify the functional range of a chip that is referenced to ground. Diode D1 limits the circuit voltage from going much higher than the supply voltage. If you are familiar with electronics, you may recall that the voltage drop across a forward-biased diode varies very little with increases in current (i.e., it is nonlinear).[2] The nominal voltage drop is 0.7 V. Increasing the current by 10 times may increase the voltage drop to only 0.8 V. The polysilicon resistor slows down the rate of change of voltage (slew rate).

Similarly, diodes D2 and D3 limit the voltage drop below ground. They become forward biased when the input drops below ground. Hence, these diodes limit the signal passed on to the rest of the circuit (transistor gates) to between V_{CC} +0.7 V and −0.7 V. Sometimes these diodes are known as clamping diodes.

However, this protection network will not protect the HCMOS circuit from large static voltage discharges. For example, a person walking across a waxed floor can generate 15 kV, depending on humidity and other factors. Manufacturers recommend precautions to avoid zap. One is to wear a special antistatic wrist strap. It connects the wearer to ground and will discharge any static charge that has been picked up. CMOS data books provide detailed procedures. These are particularly applicable in manufacturing workstations where there are more opportunities to generate static charges.

If it is not possible to employ these recommended precautions, you may still use CMOS devices safely by using common sense to avoid static. For example, touch ground

---

[2]In case you are not familiar with diodes, we present a quick explanation as follows. The pointed end with the bar (see Figure 6.3) is the cathode. The flat end is the anode. When the anode voltage is higher (more positive) than the cathode, the diode is forward biased and current flows through the diode (electrons flow from cathode to anode). If the anode voltage is lower than the cathode, the diode is reverse biased and no current flows through the diode. However, if the reverse bias voltage difference increases to the value of the diode's breakdown voltage limit, electrons will flow in the opposite direction (from anode to cathode).

before touching a CMOS device or printed circuit board. You will find that discrete chips are packaged in black conductive foam or other antistatic material. Do not store them in Styrofoam or plastic trays.

PRACTICE_____

In Figure 6.3 $V_{CC} = +3$ V. What is the maximum voltage that can be sent to the circuit?

Solution

D1 clamps voltage to $V_{CC} + 0.7$ V $= 3.7$ V.

_____

Latch-up.   In the condition known as latch-up, a gate stays (latches) in a permanent high or low state. This can occur when signal voltages and currents exceed recommended limits. We describe these limits in the next section. Because of the way that HCMOS is fabricated, the N and P regions effectively form two bipolar transistors. Consider them to be virtual bipolar transistors. Bipolar transistors have three terminals: the base, collector, and emitter. In this case each virtual bipolar transistor has its collector connected to the base of the other.

Latch-up occurs when high currents drive both transistors into saturation. *Saturation* refers to the condition of supplying enough base current to force a bipolar transistor to act like a closed switch. Due to their connections, the transistors continue to drive each other into saturation even after the initial high current is removed. This results in unlimited supply current, which will damage the chip or degrade its performance.

Latch-up can occur if an output is connected to a device with low impedance (or shorted accidentally). When the output pin attempts to supply a logic high voltage, the current it supplies is large enough to cause latch-up. This could occur if attempting to drive an inductive load such as a relay coil or motor. A load is any device connected to an output pin that draws current from or supplies current to the pin. Increased *loading* refers to increased current requirements. An inductive load is a device whose construction includes a wire winding (or coil). When a voltage is applied to the coil, the current is initially very large before it settles down to a lower value. This initial current is known as *inrush current*. Similarly, if voltage is suddenly removed from a coil, the initial current change is also very high but in the opposite direction. This is known as *outrush current*. Both inrush and outrush currents are known as *inductive transients*.

For inductive loads, one solution is to connect the output to a device called an *optoisolator*, which in turn connects to the coil. As implied by the name, there is no electrical connection between the chip and the coil. Another solution is to connect a suppresion diode (see Figure 12.3).

■ 6.2 CHIP SPECIFICATIONS

6.2.1   Data Sheets

The data sheets for any semiconductor integrated circuit contain tables listing physical specifications. A specification details the limits that a device can tolerate and those for proper functional operations. For example, there are temperature and power supply limits.

Systems designed to work in your living room may not operate without protection during midnight in the Antarctic or high noon in the Sahara.

Specifications are limits. The quality of the chip manufacturing process determines the "real" limits. It is not economical to test these limits for every chip. In fact, chips with better-stated specifications may come off the same assembly line as those with worse. The better ones have simply been tested more thoroughly by quality control. It is more likely that a chip will operate when the limits are exceeded, but this is not guaranteed. You may think of specifications as a way for semiconductor manufacturers to cover their rears!

Specifications for digital integrated circuits usually list several physical categories:

Maximum ratings

Recommended operating conditions

Direct current (dc) electrical characteristics

Alternating current (ac) electrical characteristics

Of course, they also describe logical characteristics such as truth tables, function tables, and instructions as required. The numerical examples we will use apply for the HCMOS family. The 68HC11 belongs to this family, which is why "HC" is included in its name.

The 68HC11 data sheets are available as part of the *Data Book* or a more general data book listing several products. One of these is Motorola's *Microprocessor, Microcontroller and Peripheral Data*, Volumes 1 and 2 (product code DL139). Data sheets for the 68HC11 can also be obtained separately. However, there are several types of 68HC11 ICs available with different specifications. Hence, you will have to ask for a product guide for the different types and then ask for the data book for the type in which you are interested. We survey some of these later in the book. An example of a specific product type is the MC68HC11E9. We have been lazy so far and have omitted the MC in the product name. We will continue to do so. You can probably guess that MC refers to "Motorola."

6.2.2 Maximum Ratings

Actually, the maximum ratings include minimum values as well. Sometimes, technical writers sacrifice precision in the use of language for the sake of brevity, the opposite of legal documents. Device damage may occur if the chip operates below the minimum or above the maximum values listed. These ratings concern protection from permanent damage and do not imply that the device will operate properly. These are dc supply voltages, input signal voltage and current levels, output signal voltage and current levels, power dissipation, and storage temperatures. The primary concerns here are that the system have the power supply operating within these specifications and that the chip not be overloaded.

The power supply should be able to maintain voltage within limits under minimum and maximum circuit loading conditions. A power supply has a specification that states the maximum current it can deliver for a given output voltage. For example, a 5-V dc supply rated for 1 ampere (A) is sufficient for many system circuit boards. To determine your supply requirements, calculate the current consumption by summing the consumption of all chips and components on the board.

Lower Operating Voltage. Like many other microcontrollers and microprocessors, the 68HC11 is available in two models. One normally operates with a 5-V supply. The other operates with a 3-V supply without a reduction in performance. There are more than 25 different low-voltage versions of the 68HC11. The 5-V version MCUs could operate at lower voltages but with a reduction in performance. The low-voltage versions still offer on-chip EEPROM (see Section 6.6) and analog-to-digital conversion (see Chapter 12). This is significant because both subsystems use a *charge pump* to increase the voltage internally. Motorola offers an evaluation kit called M68EBLP11KIT to learn low-power techniques for portable designs.

6.2.3   Recommended Operating Conditions

The recommended operating conditions are similar to the maximum ratings except that they state conditions to keep outputs within limits for proper functional operation. The operating conditions are more restrictive than the maximum ratings. Sometimes they include timing limits. Chips that depend on clock signals may require that the transitions occur within a maximum time limit, so any clock generating circuit must meet this specification.

6.2.4   Direct Current (dc) Electrical Characteristics

Generally, the operating conditions describe static logic-low and logic-high limits. Logic signals do not always have the same value. Induced noise from electromagnetic sources and circuit impedance (loading) will affect the actual signal values. Specifications state output and input limits based on stated operating conditions. The important ones with typical HCMOS figures are as follows ($V_{DD} = 5$ V and $V_{DD} = 3$ V):

		$V_{DD} =$	
		5 V	**3 V**
V_{OL}	Maximum low-level output voltage	0.1 V	0.1 V
V_{OH}	Minimum high-level output voltage	4.9 V	2.9 V
V_{IL}	Maximum low-level input voltage	1.0 V	0.6 V
V_{IH}	Minimum high-level input voltage	3.5 V	2.1 V

In general, the output and input limits are:

$V_{OL} = 0.1$ V

$V_{OH} = V_{DD} - 0.8$ V, not applicable to open-drain pins

$V_{IL} = 0.2 \times V_{DD}$

$V_{IH} = 0.7 \times V_{DD}$; $0.8 \times V_{DD}$ for RESET signal

In Section 9.8.2 we discuss *open-drain*. If the output voltage is between the output limits, there is no guarantee as to what logic level another input will consider it to be. These limits are stated for certain current loading conditions as stated by I_{LOAD}, the current limit. Similarly, if a device sees an input between the input limits, there is no guarantee as to whether it interprets it to be a logic high or a logic low. Hence, there is no guarantee as to what output it will produce. For example, if one gate outputs 0.5 V as logic low, electrical noise may cause the next gate to see an input above 1.0 V. It may interpret the input as a high instead of a low. If the system uses chips from the same logic family, the limits are met automatically as long as current limits are kept.

Typically, an HCMOS gate requires a current of 1 microampere (μA) per input. In the case of the 68HC11, V_{OH} is stated for a current of 0.8 milliampere (mA) flowing out of the pin. This is stated as $I_{LOAD} = -0.8$ mA. Note that current direction is positive when flowing into a pin. Current flowing out of a pin toward ground is known as *sourcing* and current flowing from the voltage supply into a pin is known as *sinking*.

Fan-in and *fan-out* indicate the number of inputs and outputs that can be connected to a logic gate safely without damaging the IC. Using the figures given in the preceding paragraph, a 68HC11 output can send a logic high to 800 HCMOS inputs. Also, it can send a logic low within specifications when sinking 1.6 mA. Since the high-output case is more restrictive, we can say that the fan-out is 800 for HCMOS.

Note the restriction to logic family type. An output is not always connected to an HCMOS gate. For example, would it be possible to drive a light-emitting diode (LED) as an indicating signal? Let's say that the LED cathode is connected to ground. An LED has a nominal voltage drop of 1.8 V. Normally, we would like to supply it with about 15 milliamperes (mA) in order to make the light visible. Therefore, it must also be connected in series with a resistor since the gate output will be greater than 1.8 V. In theory, we could drive it with a 68HC11 output since its maximum limit is 25 mA. But the output would drop below V_{OH}. Hence, you may not be able to use this output to send a signal to another logic gate as well. Actually, we do not recommend driving an LED directly using the 68HC11. This reduces its ability to drive other outputs due to total maximum power dissipation specification. It is much better practice to drive a buffer chip, which in turn drives the LED.

Note that experimentation may yield satisfactory results even when exceeding the stated specifications. But do not base a design on it. When manufacturing your microcontroller product, the chips that the manufacturer supplies may have variations in some of the unspecified electrical characteristics. However, all chips should meet the stated specifications.

Another important specification is the maximum quiescent current for a chip. This is the minimum current that a chip consumes when it does nothing. That is, no inputs or outputs cause the chip to supply more current. It is the current required to keep the chip operational. Part of it represents the bias currents of the internal transistors. For the 68HC11 this depends on the operating mode and whether it executes a wait for interrupt (WAI) or STOP. Recall that these were the low-power mode instructions. For example, the 68HC11 consumes only 100 microamperes (μA) during a STOP with no clock signal or outputs to drive. This makes it useful for portable battery-powered applications.

PRACTICE

Explain why it would be okay to connect a device with an equivalent resistance of 1 kΩ between an output pin and ground.

Solution

From Ohm's law (see the Glossary), the current is

$$I = \frac{V}{R} = \frac{5\ V}{1\ k\ \Omega} = 5\ mA$$

This is less than the maximum limit of 25 mA.

6.2.5   Alternating Current (ac) Electrical Characteristics

Generally, the ac electrical characteristics are timing parameters. We will only list these specifications for now. After we have covered timing diagrams and cycle-by-cycle operation, we examine some of the parameters more closely.

The most important speed specification for a microcontroller is the clock frequency. When the first edition was written, the allowable frequency range for the 68HC11 was 0 to 2 megahertz (MHz). Now some 68HC11 variations can operate up to 3 or 4 MHz. It is possible that these limits will increase again by the time you read this. Note that increasing the clock speed increases power consumption. Most of the power is consumed when the internal transistors switch states. In effect, a chip has internal capacitance. Every time there is a change in state, these capacitors are charged or discharged. Current supplies the charge, so higher clock speed means supplying this current more often, which means higher power dissipation. You can reduce power consumption by using a lower operating voltage. To avoid affecting other performance characteristics, use low-voltage variations of the 68HC11.

The 68HC11 is a static device. This means that it does not require a clock signal to keep its data. Therefore, there is no low limit for the clock frequency for the 68HC11 (and other static devices). Nonstatic devices require a series of repetitive pulses (clock signal) to refresh memory. Otherwise, they lose the stored data. You might not be terribly impressed with the clock speed of the 68HC11. Many other processors run at speeds from 5 to 20 times as high, although many of them internally *divide-down* the input clock for slower operation. However, the 68HC11 has a speed that is sufficient for many control applications.

Other timing parameters are various setup and hold times. A *setup time* is the time for a device to complete its output response to an input change. A *hold time* is the length of time a device will maintain its last output in response to a request to change it. The use of these specifications will become more apparent after looking at timing diagrams.

All logic gate inputs have parasitic capacitance connected to ground. The ac electrical characteristics state the maximum stated input capacitance. Nominally, the value is around 10 picofarads (pF). When an output is connected to an input of another device, there is also some built-in resistance. With the input capacitance there is an *RC* circuit. If you are familiar with basic circuit theory, you will remember that an *RC* circuit has a time constant of

$$T = RC \tag{6.1}$$

where T is the time constant in seconds, R the resistance in ohms, and C the capacitance in farads.

When a logic-low output is changed to a logic high, it takes time for the *RC* circuit capacitor to charge to the new value of the output signal. Similarly, for an output change from high to low, it takes time to discharge the input capacitor to its new value. Hence, the input pin will not see the change until after this time. The time constant is a measure of how quickly a signal changes from an initial value to a final value. It is the amount of time for a signal to change by approximately 63%. *Rise time* is defined to be the length of time it takes a signal to change from 10% to 90% of its final value.[3] A larger time constant means an increased rise time.

What happens if we connect the output to more inputs, each with its own input capacitance? Again, if you are familiar with circuit theory, you will realize that these capacitors are connected in parallel. Thus, the total capacitance seen by the output pin is the sum of all the input capacitances. Hence, the time constant and rise times increase proportionally. The result is that adding inputs limits the maximum frequency because it takes more time for inputs to change.

This covers the specifications for now. We address some other parts of a microcontroller chip next.

■ 6.3 MEMORY TECHNOLOGY

6.3.1 Review and Overview

Some semiconductor circuits are used to store data. These are known as *memory circuits*. Earlier we looked at semiconductor memory in Sections 1.4, 5.3, and 5.4. In particular, the 68HC11 has built-in memory blocks (see Figure B.1) for ROM, RAM, and EEPROM. The 68HC11 can also use external memory. These usually are random-access memory (RAM), erasable programmable read-only memory (EPROM), and electrically erasable programmable read-only memory (EEPROM). In Section 4.10 we also described how an evaluation board uses these external memory chips. In Chapter 5 we described how to interface these chips.

In this section we examined the internal ROM and RAM block. In the following section we will describe the EPROM type of memory. We cover it first to understand the final internal memory block, the EEPROM. Also, many microcontroller systems use external EPROM. Some models of the 68HC11 also use an internal EPROM instead of ROM. In the next section we will describe the EEPROM block. In the final section we will describe a special register called the configuration register (CONFIG). It determines how the internal memory resources are used.

6.3.2 Read-Only Memory (ROM)

Read-only memory (ROM) contains permanent data. It is nonvolatile, meaning that it does not lose data when power is shut off. It stores program code and constant data (such as the value of pi). When you order ROM chips from a semiconductor supplier, you must specify what data it contains. Alternatively, you can buy ROM chips with standard data in them.

---

[3]According to IEEE Standard 194, the preferred term for *rise time* is *first transition duration;* similarly, the preferred term for *fall time* is *last transition duration.*

The 1-of-8 decoder and similar decoder chips covered in Chapter 5 are sometimes disguised ROM chips. The input is an address that accesses a stored data value which is then sent to the output.[4]

When purchasing a 68HC11 with internal ROM, the customer must specify the data to be put into ROM. Note that the customer in this case is the system manufacturer (also known as OEM for *original equipment manufacturer*). The OEM buys the chip as one of the parts for the system it manufactures.

Recall that photomasking is part of the chip fabrication process. The system manufacturer specifies the ROM data so that the semiconductor manufacturer (Motorola, in the case of the 68HC11) can make up the custom photomask for the ROM. Specifying a custom mask for the 68HC11 internal ROM costs several thousand dollars. If you do not need many chips, it will be cheaper to use external EPROM chips or a 68HC11 model with internal EPROM to hold the nonvolatile data and program. It starts becoming economical to use a custom ROM mask when an OEM needs more than 2000 parts. This is only a rule of thumb. Consult Motorola for firm prices before committing yourself to purchasing quantities of chips with custom ROM.

One example would be a camera manufacturer. The company would expect to sell thousands of cameras worldwide. Hence, a custom ROM would be economical. Reducing chip count is another factor. Another example is a manufacturer of specialty drilling equipment for diamond mines. It is not likely that this company would sell enough equipment to justify the cost of using a custom ROM. Other considerations are whether you would expect to change system software during the life of your product.

If you decide to use a custom ROM, you can supply the ROM data as a hex file in a disk or send an EPROM chip that has been programmed with the data you want. We will discuss EPROM programming later. Recall that the 68HC11 also has a second ROM called the boot ROM. It is always preprogrammed for boot loading operations when the chip is wired for bootstrap mode.

6.3.3   Random-Access Memory (RAM)

Random-access memory (RAM) is memory a program can write to using an instruction such as STAA MEMORY_LOCATION. But it is volatile, meaning it loses data when power is turned off. None of this seems to explain why this memory is called RAM.

Why Is RAM Called RAM?   *Random access* means that each location in memory is accessible in the same amount of time, independent of its physical location. This differs from position-dependent storage such as disk, tape, or this book. For example, a disk *head* has to position itself to read or write data at any location. The time required depends on the original position of the head. By convention, any semiconductor memory that can be read and written to is called RAM.

There are two basic types of RAM, static and dynamic (SRAM and DRAM). Both types contain arrays of *cells*. A cell is an entity that stores 1 bit. Internal logic circuits decode the address bits to select the appropriate cells for each digital word.

---

[4]It is possible to buy BIOS chips to upgrade a personal computer. This is another example of ROM chips with standard data.

Dynamic RAM. Dynamic RAM uses a single transistor that acts like a charged capacitor to store each bit. But since the charge leaks, a DRAM must use a *refresh circuit*. It sends a signal periodically to recharge the capacitors. The refresh rate is typically every 2 milliseconds (ms).

The 68HC11 does not use dynamic RAM. It could use external DRAM with refresh circuitry, although it would not be common practice. Most DRAMs are bit addressable. That is, each address accesses a single bit. To store a block of bytes, a memory circuit would use eight DRAM chips connected to the same address lines. Each DRAM chip is connected to a single data line.[5]

DRAMs also provide an interesting lesson in government folly. Personal computers use DRAM for their main memory. During the mid-1980s, the semiconductor industry in the United States was worried about Japanese competition in supplying DRAM chips. They convinced the U.S. government to limit DRAM imports from overseas. But many U.S. domestic DRAM suppliers stopped manufacturing DRAMs despite the restrictions. The result was an artificial DRAM shortage that caused computer prices to rise. Interestingly, this also affected the software industry because some of the new software being marketed depended on using a lot of RAM. Time will tell whether history repeats itself.

Static RAM. Static RAM uses an internal *flip-flop* to store each binary bit. A flip-flop is a digital circuit requiring four or six transistors. A flip-flop latches in an input bit and stores it without requiring refresh pulses. The 68HC11 internal RAM is static. This means that it retains data even when the system clock stops.

Performance Differences. The two main specifications for RAM are memory size (or capacity) and access time (or speed). *Access time* refers to the length of time it takes to write or read from the time the chip is first addressed (address bus signals stabilize).

DRAM chips contain more data per chip size because they require only one transistor per cell, but SRAM is faster and simpler to use. Both types are being developed to improve their performance. As a result, it is difficult to state typical performance figures.

Since many microcontroller systems use less memory than do general-purpose computers, they use static RAM. In typical microcontroller applications, memory speed and simplicity are more important than storage capacity per chip.

◼ 6.4 ERASABLE PROGRAMMABLE READ-ONLY MEMORY (EPROM)

There is a distinction between *programming* and writing to a device. *Writing* means that an instruction can send data to an address or register for storage. *Programming* refers to a procedure of putting data into memory for nonvolatile or "permanent" (note the quotation marks) storage. When a memory chip is programmed, normal data storage instructions cannot write over the programmed data. Usually, a sequence of instructions is required to program each byte.

---

[5]Owners of personal computers may realize that each memory block uses nine DRAM chips instead of eight. The ninth chip stores parity bits. These are used to check for error detection. A parity bit indicates whether any of the other 8 bits in a byte changed accidentally.

Erasable programmable read-only memory (EPROM) is a user-programmable and erasable device. The system developer can program them using an EPROM *programmer*. In this case a programmer is the device that programs the chip and does not refer to a person who is paid to write software. An EPROM chip has a transparent quartz window. To erase the EPROM, expose the window to high-intensity ultraviolet (UV) light for a period of time.

In normal use, after it has been programmed, the EPROM chip acts like a ROM chip. Developers often use it to prototype a ROM chip because they can reprogram it when necessary. You cannot do that with a ROM. Also, EPROM chips are used for final application program storage when using a ROM is not economical (as discussed in Section 6.3.2).

Sometimes Motorola uses the same photomasks to produce both ROM and ROMless versions. For the ROMless version, they turn off a bit in a special register. This disables the internal ROM. The special register mentioned is the configuration register (CONFIG) we describe at the end of the chapter. There are two choices with the EVBU board. The first is to use the M68HC11E9 chip with the Buffalo monitor program stored in its internal ROM. The second is to use the M68HC711E9 chip with your own program stored in its internal EPROM.

Although many versions of external EPROM are available, we use the MC2764 as an example. It has a storage capacity of 8 kilobytes (8KB). The "64" in the part number refers to the number of kilobits. Note that $8K \times 8$ bits is 64 kilobits (64Kb).

When an EPROM is part of a final application's circuit board, it has a sticker to cover the quartz window. This prevents data erasure due to UV in sunlight. An EPROM without a sticker may not be a problem for developers since it will take several months to erase an EPROM using sunlight.

EPROM Cell.   Internally, each storage cell is a metal-oxide semiconductor field-effect transistor (MOSFET) with a floating gate embedded inside an oxide region. The logic level of a storage cell is determined by the condition of its floating gate. The nonprogrammed state when the logic gate is not charged is a logic 1. The programmed state when the gate is charged is a logic 0.

Thus an erased EPROM byte is all ones, $FF. When programming an EPROM, you are actually programming zeros into the device. To program a cell, the EPROM programmer must supply the chip's programming control pin (V_{PP}) with $+12$ to $+25$ V, depending on the type of EPROM. This higher voltage is required to charge the floating gate. The higher voltage provides enough force for electrons to *tunnel* through the oxide to reach the gate. For this reason, some people use the term *burn-in* to describe programming EPROM data. Similarly, UV light provides the energy for electrons trapped in the floating gate to tunnel through the oxide layer to escape.

Erasing the EPROM.   Before an EPROM can be programmed, it must be erased. To do this, remove the sticker to expose the quartz window. Place the EPROM chip in an EPROM eraser. The eraser exposes the window to UV light of the appropriate intensity. Erasing typically requires 20 to 40 minutes. Note that the entire EPROM is erased; it is not possible to erase a byte or group of bytes selectively.

Programming the EPROM.   Figure 6.4 shows the usual method for programming an EPROM chip. The computer communicates with an EPROM programmer that is cabled to it. The programming steps are

1. Plug in the "personality chip" for the specific type of EPROM you want to program.
2. Plug an erased EPROM chip into the programmer's EPROM socket.
3. Run the EPROM programming software and follow the steps specified by the software.

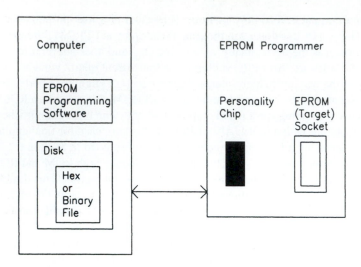

Figure 6.4   Programming an EPROM Chip

The EPROM programming software transfers the hex or binary file from the computer to the EPROM programmer. The EPROM programmer *burns* the data into the EPROM chip. When programming is completed, remove the EPROM chip and plug it into the circuit to test the program you just burned in.

Normal Operation.   Typically, an EPROM contains program code and constant data. When the EPROM is wired for normal mode, the V_{PP} pin is tied to $+5$ V. When wired this way, it operates like a ROM chip.

Internal EPROM.   Some versions of the 68HC11 have internal EPROM instead of ROM. For example,

> 68HC711D3 has 4KB EPROM.
>
> 68HC711E9 has 12KB EPROM.
>
> 68HC711G5 has 16KB EPROM.
>
> 68HC711K4 has 24KB EPROM.

Programming Internal EPROM.   We will describe briefly how to program an internal EPROM. Naturally, this is a concern only when you are using a 68HC11 version that includes an internal EPROM. With some exceptions, you can program it like a standard external EPROM with an EPROM programmer or use the CPU. Refer to the manuals and data sheets for further details, particularly since there are exceptions!

When programming the EPROM with an external programmer, you put the chip in PROG mode. To initiate PROG mode, pull pins $\overline{\text{RESET}}$, MODA, and $\overline{\text{MODB}}$ low. In PROG mode, pin $\overline{\text{XIRQ}}$ becomes the programming pulse pin (V_{PP}). The microcontroller pin *foot-print* becomes identical to that of a 27256 EPROM (256 kilobits). To program the

PROM using the CPU, the user must use a program sequence using bits in the programming register (PPROG). The user must also supply a high voltage (normally 12.25 V) to the programming pulse pin, V_{PP}.

PRACTICE _____

Would you use a 68HC711K4 in an orbiting satellite?

Solution

No. To use it, you would have to add excess weight to protect the EPROM from the sun's UV radiation. Note that there is no atmosphere to reduce the UV radiation.

■ 6.5 FLASH MEMORY

Flash memory is useful as a programmable nonvolatile memory. It offers some advantages over EPROM. It can be erased and programmed electrically, but unlike EEPROM, which we discuss next, it is only possible to change a sector of multiple bytes at a time. A flash memory cell is similar to an EPROM cell. The difference is its use of a control gate in addition to a floating gate. For information about interfacing, check the Motorola Engineering Bulletins, Application Notes, and other information posted on the Internet.

■ 6.6 ELECTRICALLY ERASABLE PROGRAMMABLE READ-ONLY MEMORY (EEPROM)

6.6.1 Organization and Applications

With electrically erasable programmable read-only memory (EEPROM) it is possible to program and erase an addressed byte electrically. Like an EPROM, it requires higher voltage to cause electrons to tunnel through an oxide layer to charge or discharge a floating gate. However, an external higher-voltage supply is not required to program or erase the internal EEPROM in the 68HC11. It has a built-in high-voltage *charge pump*. The 512 bytes of EEPROM have an address range of $B600 to $B7FF (see Figure 1.8). The EEPROM is further subdivided into *rows* of 16 bytes: $B600–$B60F, $B610–$B61F, . . . , $B7F0–$B7FF.

Typical applications are calibration and option information for a control system. For example, parameters for optimum engine control depend on where a car is sold. The parameters required for the prairies differ from those used in the mountains. The EEPROM can store small custom program codes that are different for each customer. It can also store diagnostic code and a serial number. Another application is to back up critical data. For example, the odometer reading of an automobile must be kept in nonvolatile memory.

Another version of the 68HC11 (68HC811E2) has a 2KB EEPROM. This version can hold larger programs and is good for small-volume applications because the ROM mask charge is avoided. But, at the time of this writing, EEPROM is expensive compared to other memory technology.

6.6.2   The Register Block

At this point the reader should refer to Table B.1 in the Quick Reference (Appendix B), which lists registers that control and monitor operations for the 68HC11. We introduce this block now because some of them control the operation of programming and erasing EEPROM bytes. Unlike the CPU registers (such as accumulators), these registers have an address. For example, to read the EEPROM programming register (PPROG at address $103B), use an instruction such as

```
LDAA    $103B
```

Many of the bits in the register block have specialized functions. They have mnemonic names to identify their functions. For example, to set (turn on) the EEPROM latch control bit (EELAT), use an instruction such as

```
LDX     #$1000      ;point to register block
BSET    $3B,X $02   ;set EELAT bit
```

The bit set instruction (BSET) turns on bit 1 of address $103B. This bit is EELAT (see Table B.1).

Some people refer to the register block as memory-mapped registers, I/O registers, or simply as registers. From now on, when we say registers, we mean those in the register block. CPU registers will be referred to individually or as a group called CPU registers. We refer to this block often in Part 4.

Using the Quick Reference.   The Quick Reference (Appendix B) provides a summary of facts and figures for the 68HC11. Table B.1, in particular, is very important. To follow programming examples given in this book, you will have to refer to it if you want to see where the address and bit position information comes from. If you have the *Data Book* or *Programming Reference Guide*, we advise you to use them since they provide explanations of register and bit functions for reference purposes. Since this is not a reference book, you will have to hunt through it to find the same information. Note that the "Cross References to Other Data" in the Quick Reference can assist you in finding this data.

Note on Slow Clock Speeds.   Motorola recommends that your software sets a bit called CSEL in the register called OPTION (bit 6 of address $1039) if the system clock (E) is less than 1 MHz. (In Chapter 12 we describe the OPTION register in more detail.) To do this, your program can execute the following instructions:

```
LDX     #$1000      ;point IX to address OPTION reg.
BSET    $39,X $40
*                   ;then CSEL = 1 (set bit 6 of $1039)
*                   ;note, doesn't affect other bits
```

6.6.3 EEPROM Operation and Specifications

Like an EPROM, a byte is erased when all bits are 1. To program it, turn a 1 bit into a 0 bit.

To Erase a Cell. The 68HC11 applies +20 V (internally) to a control gate. This causes electrons to tunnel through a thin oxide layer onto the floating gate. The *erased* state is a logic 1 because the floating gate is storing a charge.

To Program a Cell. The 68HC11 applies +20 V to a MOSFET drain to force electrons to tunnel away from the floating gate. The *programmed* state is a logic 0 since the gate is no longer charged.

Specifications. Motorola guarantees that the EEPROM will retain data for 10 years. They claim that the typical retention is 5 to 10 times longer. They also guarantee that the EEPROM will last for 10,000 programming and erase cycles. This is an industry standard. The limitation is due to depletion of electrons in the EEPROM cell P substrate.

6.6.4 A Programmer's Guide to Changing EEPROM

It is possible to program a 1 bit to zero, but it is not possible to do the reverse. To change a 0 bit to 1, you must erase the byte first. It is possible to erase a single byte, a single row, or the entire EEPROM. Programming involves one byte at a time. All programming and erasing operations use the PPROG register. Figure 6.5 illustrates the organization of the PPROG register. Note that the register contains a 0 byte after $\overline{\text{RESET}}$.

Erase Procedures. We consider erase procedures first.

1. Write to the PPROG register to select a byte, row, or entire EEPROM erase operation. That is, set or reset bits ERASE, BYTE, and ROW as required.
2. Write to an EEPROM address within the desired range. This latches the address(es) of the byte(s) to be erased.
3. Write to the PPROG register to turn on bit EEPGM to turn on the programming (erase) voltage.
4. Delay for 10 ms.
5. Write to the PPROG register to turn off all bits. This resets bit EEPGM to turn off the programming voltage. It also resets other bits to return the EEPROM to normal operation.

Listing 6.1 illustrates examples of subroutines to erase EEPROM bytes. Note that we do not show cross assembly pseudo-ops.

```
*Listing 6.1
*Subroutines to erase EEPROM bytes

*Subroutine BULKE
*Bulk erase 512-byte EEPROM
*No calling registers
*Return registers
*        ACCB $07, bulk erase flag
*        CCR affected, rest unaffected
```

PPROG

$103B	7	6	5	4	3	2	1	0
	ODD	EVEN	0	BYTE	ROW	ERASE	EELAT	EEPGM
RESET	0	0	0	0	0	0	0	0

ODD   Program Odd Rows (TEST)

EVEN   Program Even Rows (TEST)

Bit 5   Not Implemented, this bit always reads zero

BYTE   Byte Erase Select
               This bit overrides the ROW bit
               0 = Row or Bulk Erase
               1 = Erase Only One Byte

ROW   Row Erase Select
               0 = Bulk Erase
               1 = Row Erase

ERASE   Erase Mode Select
               0 = Normal Read or Program
               1 = Erase Mode

EELAT   EEPROM Latch Control
               0 = EEPROM Address and Data Configured for Read Mode
               1 = EEPROM Address and Data Configured for Programming/Erasing

EEPGM   EEPROM Programming Voltage Enable
               0 = Programming Voltage Switched Off
               1 = Programming Voltage Turned ON

Figure 6.5   EEPROM Programming Control Register

```
BULKE
        LDAB    #$06        ;ERASE, EELAT = 1 to
        STAB    $103B       ;set to bulk erase mode
        STAB    $B600       ;write any data to any EEPROM address
        LDAB    #$07        ;EEPGM = 1 to
        STAB    $103B       ;turn on programming voltage
        JSR     DLY10       ;delay 10 ms
        CLR     $103B       ;turn off high volt and set to READ mode
        RTS

*Subroutine ROWE
*Erases a row (16 bytes) of EEPROM
*Calling registers
*       IX      any address in row to be erased
*Return registers
*       ACCB    $0F, row erase flag
*       CCR affected, rest unaffected
```

```
ROWE
        LDAB    #$0E        ;ROW, ERASE, EELAT = 1 to
        STAB    $103B       ;set to row erase mode
        STAB    0,X         ;write any data to any address in row
        LDAB    #$0F        ;EEPGM = 1 to
        STAB    $103B       ;turn on high voltage
        JSR     DLY10       ;delay 10 ms
        CLR     $103B       ;turn off high volt and set to READ mode
        RTS

*Subroutine BYTEE
*Erases an EEPROM byte
*Calling registers
*       IX      address of byte to be erased
*Return registers
*       ACCB    $17,        byte erase flag
*       CCR affected, rest unaffected

BYTEE
        LDAB    #$16        ;BYTE, ERASE, EELAT = 1 to
        STAB    $103B       ;set to byte erase mode
        STAB    0,X         ;write any data to address to erase
        LDAB    #$17        ;EEPGM = 1 to
        STAB    $103B       ;turn on high voltage
        JSR     DLY10       ;delay 10 ms
        CLR     $103B       ;turn off high volt and set to READ mode
        RTS

*Subroutine DLY10
*Time delay of approx. 10 ms, assuming 2MHz clock
*really loops 3400 times for a delay of 6 × 3400 × E
*No calling registers
*No return registers
*CCR affected, rest unaffected

DLY10
        PSHX                ;preserve IX
        LDX     #3400       ;init loop
*run loop
LDLY10
        DEX
        BNE     LDLY10
        PULX                ;restore IX
        RTS
```

Programming Procedure. The procedure to program an EEPROM byte is as follows:

1. Write to the PPROG register to set the bit EELAT to configure for programming.
2. Write data to the EEPROM address selected. This latches in the address and data to be programmed.
3. Write to the PPROG register to set the bit EEPGM to turn on the programming voltage.

4. Delay for 10 ms.
5. Write to the PPROG register to turn off all bits.

Listing 6.2 shows a subroutine that can be used for programming.

```
*Listing 6.2

*Subroutine BPROG
*Programs a byte in EEPROM, assumes byte previously erased
*Calling registers
*     ACCA    data to be programmed
*     IX      valid EEPROM address
*Return registers
*     ACCB    $03
*     CCR affected, rest unaffected

BPROG
      LDAB    #$02
      STAB    $103B     ;set EELAT bit (EEPGM = 0)
      STAA    0,X       ;store data to EEPROM address
      LDAB    #$03      ;turn on programming voltage by
      STAB    $103B     ;setting EEPGM = 1 (and EELAT = 1)
      JSR     DLY10     ;delay 10 ms for charge pump
      CLR     $103B     ;turn off prog volt and set to READ mode
      RTS
```

PRACTICE

Field tests of a gas analyzer shows that it needs a 24-bit calibration setting of $321CB5.
Write the code to store the data.

Solution

```
*Solution
*Program a 24-bit (3-byte) calibration word in EEPROM
*Use subroutines of Listings 6.1 and 6.2
*Choose arbitrary row in EEPROM, address $B6A0

*Erase the row first
            LDX     #$B6A0     ;set up calling register
            JSR     BYTEE      ;call subroutine to erase 3 bytes
            INX
            JSR     BYTEE
            INX
            JSR     BYTEE
*Burn in 4 bytes
            LDX     #$B6A0     ;point to first EEPROM byte
            LDAA    CALIB      ;get first calibration byte
            JSR     BPROG      ;program it
            INX                ;point to next byte
```

```
*               ;and similarly program the rest
        LDAA    CALIB+1
        JSR     BPROG
        INX
        LDAA    CALIB+2
        JSR     BPROG
DONE
        BRA     DONE
*block of calibration data, recall FCB is the form constant byte
pseudo-op.
CALIB   FCB     $32, $1C, $B5
```

The BPROT Register. This is the Block Protection register that prevents a faulty program from accidentally modifying EEPROM data. To enable the ability to modify specified blocks of EEPROM, the software must clear specified bits within the first 64 clock cycles after reset. For example, the BUFFALO monitor program for the EVBU board does this. To protect a block from accidental modification, the software can set the applicable bit high at any time. However, once set, it is not possible to clear bits again without a reset in normal modes. In bootstrap and test modes a program can modify the BPROT register at any time. We can also use the BPROT register to protect the CONFIG register (which we will discuss next). For more information about the BPROT register, refer to the Technical Data manual for any 68HC11 type that uses it.

■ 6.7 CONFIGURATION CONTROL REGISTER (CONFIG)

We introduced the configuration control register (CONFIG) in Section 3.9.2 when we described computer operating properly (COP) failure reset. Unlike other registers in the register block, this one is actually an EEPROM byte. To change its contents, you must use EEPROM programming procedures with the 68HC11 operating in the bootstrap or test mode. But changing the CONFIG register contents has no effect until the MCU is reset.

If you intend to change the CONFIG register contents, be very cautious. If bit ROMON is 0, the internal ROM is disabled. If bit EEON is 0, the internal EEPROM is disabled. Accidentally resetting these bits for single-chip mode could make the microcontroller system inoperative.

The 68HC11 has a security feature. If the security mode disable bit (NOSEC) is 0, a user will not be able to read the program code in ROM, although it will still execute. This prevents the customer or competitor of an OEM from copying the code that resides inside the device. To make use of the NOSEC bit, the supplier has to request this option when ordering the 68HC11 chip from Motorola since it is implemented as a *mask* option. Motorola also uses this register to supply different versions of the 68HC11 using the same chip photomask.

■ 6.8 RAM AND I/O MAPPING REGISTER (INIT)

The RAM and I/O mapping register (INIT) is an example of a time-protected register. Writing to the register has an effect only if it occurs within the first 64 clock cycles after a reset ($\overline{RESET}$ active). This prevents accidental changes to this register due to software problems.

Figure 1.8 shows the default memory map for the 68HC11E9. The default map is in effect when register INIT is $01. Bits RAM3,2,1,0 make up a 4-bit number specifying a 4KB boundary for the RAM block. In other words, it is the first hex digit of the start address. For example, if it were %0110, the RAM block would start at $6000. Similarly, bits REG3,2,1,0 make up a 4-bit number specifying the start address of the register block. To start the register block at zero, clear these bits.

You may wonder why one would want to remap RAM or registers. Some applications make little use of RAM but use registers very often. Programs for such applications would run faster if they could use the direct addressing mode for registers.

■ 6.9 SUMMARY

An integrated circuit (IC) is a slice of silicon containing doped regions that form transistors. The slice is connected to I/O pins and is mounted in a plastic or ceramic package. The 68HC11 IC is available as a quad package. There are several steps in fabricating an IC: oxide layer growth, photomasking, diffusing, and metallization. The photomask defines the circuit pattern. The 68HC11 is fabricated using HCMOS technology. This is a method of fabricating logic gates. Like CMOS, its main features are low power consumption and high noise immunity. However, HCMOS offers faster performance. When handling HCMOS, one must be careful to avoid zap (electrostatic charge) and latch-up. Latch-up is the condition in which a gate latches into a permanent logic state that results in the gate drawing excessively large currents.

The specifications of an IC include maximum ratings, recommended operating conditions, dc electrical characteristics, and ac electrical characteristics. These define when damage occurs and when performance suffers. It also states what signal levels and timing to expect when the chip is operating normally.

Semiconductor memory is generally classified as ROM, RAM, EPROM, and EEPROM. The 68HC11(A8) has 8KB of ROM, 512 bytes of EEPROM, and 256 bytes of RAM. Normally, ROM contains program code, and EEPROM contains calibration and critical data. RAM stores changing data such as sensor measurements. Programming and erasing EEPROM involves a sequence of instructions using the PROG register. To enable or disable internal ROM or EEPROM, reprogram the CONFIG register. To change the start address of RAM or the register block, write to the INIT register.

■ EXERCISES

1. Describe how you could modify the CMOS inverter circuit to be a two-input NAND gate.

2. Describe how you could modify the CMOS inverter circuit to be a two-input NOR gate.

3. The first microprocessors were packaged as a Dual Inline Pin (DIP) integrated circuit. Would a DIP and quad version have the same signals?

4. The first microprocessors were packaged as DIP ICs. Explain why manufacturers switched to quad packaging.

5. The manufacturer of VCRs uses an HCMOS microcontroller in the remote controller. Explain why.

6. The same manufacturer wants to use an HCMOS MCU to control a servomotor for a satellite dish tracking system. Explain why the manufacturer would use an optoisolator to interface the MCU to the motor.

7. Give instructions to avoid zap damage to workers assembling the controller board for the satellite dish system.

8. The following shows some typical voltage specifications:

	LSTTL (V)	HCMOS (V)
V_{CC}	5.0	5.0
V_{OL} max.	0.5	0.1
V_{OH} min.	2.7	4.9
V_{IL} max.	0.8	1.0
V_{IH} min.	2.0	3.5

Would you recommend connecting an LSTTL output to an HCMOS input? Would you recommend connecting an HCMOS output to an LSTTL input? Why?

9. The remote controller described in Question 5 executes the STOP instruction when it is not being used. This stops the system clock. Explain why this reduces power consumption.

10. How many bits and bytes are there in an MC2732 EPROM?

11. What is the logic level of an erased bit in EPROM?

12. What determines the logic level of an EEPROM storage cell?

13. What signals should be applied to the following pins in an EPROM when using it as a read-only chip?
    (a) V_{CC}
    (b) V_{PP}
    (c) V_{SS}

14. Name two basic types of RAM.

15. What advantages does SRAM have over DRAM?

16. Why is it not necessary to use an external +12-volt supply when programming the internal EEPROM for the 68HC11?

17. Indicate whether it is possible to program an EEPROM byte with the data $84 without erasing the byte for the following initial EEPROM data in hex:

    A6   A5   9C

18. Is it possible to program 10 bytes in 20 ms?

19. Write a subroutine to program a byte. It uses indexed X addressing and bit manipulation instructions. The calling registers are

    1Y = address of EEPROM byte
    1X = $1000
    ACCA = data to be programmed

20. An automobile uses a 68HC11 for some applications. It uses the internal EEPROM to store the car's odometer reading for every 100 m. Some EEPROM addresses are reserved for the following functions:

Address	Label	Function
$B602	ODOM2	High byte of odometer reading, $01 = 6553.6 km, $FF = 1,671,168 km
$B603	ODOM1	Middle byte of odometer, $01 = 25.6 km, $FF = 6528 km
$B604	ODOM0	Low byte of odometer, $01 = 0.1 km, $FF = 25.5 km

Write a subroutine that when called (every 0.1 km) increments the odometer count by one. Assume that the car will break down before it has traveled 1,000,000 km. Also, ignore the life expectancy of an EEPROM.

7 Clocked Operation

OBJECTIVES

After completing this chapter, you should be able to

- Describe the clocked operation of a bus and the circuit it connects to.
- Interpret timing diagrams to determine control sequencing and timing considerations.
- Monitor and interpret memory access operations at the signal level.
- Describe direct memory access (DMA) operations.
- Use oscilloscope test loops to interpret cycle-by-cycle operation of instruction execution.
- Determine basic operations for using a logic analyzer.
- Take into account transmission line effects for circuit design.

A microcontroller is a dynamic device that is clock driven. It changes state with every clock pulse. In this chapter we will examine how to document (describe) the signal changes by using timing diagrams. We will also examine how the bus changes during execution of instructions. From this we can examine the operation of microcontroller circuits operating in expanded multiplexed mode. After understanding clocked operation, you can use commonly available test instruments to monitor operation of the system.

■ 7.1 TIMING DIAGRAMS

Timing diagrams show how signals change over time. They can show one signal or bus signals (collection of signals). Time is shown on the horizontal axis, increasing from left to right; Figure 7.1 shows the format of many timing diagrams. Figure 7.2 shows an example of a timing diagram. For now, consider Figure 7.1. We will interpret Figure 7.2 later.

Part (a) of Figure 7.1 is reasonably simple. The rising edge and falling edge times are part of the ac characteristics (see Section 6.2.5). If an output signal also has tristate

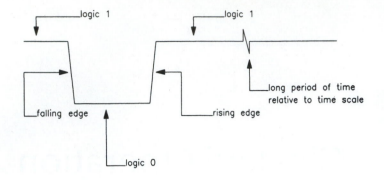

(a) Single Bit (Line) Timing Diagram

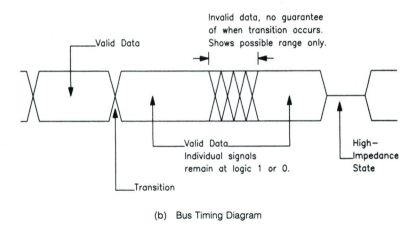

(b) Bus Timing Diagram

Figure 7.1 Timing Diagram Formats

(high-impedance) capability, the diagram would show this as a horizontal line between logic 0 and logic 1.

Part (b) illustrates bus signal timing. It is impractical to show every bus line separately since some buses may have 32 lines. Since a timing diagram must show many lines, it shows stable bus signals as two lines representing logic low and high. Each individual bus line may be low or high. During a change in a clock signal, the bus signals change. The timing diagram also indicates how long it takes for the transition to occur. Sometimes there is no definite time for a transition to occur. In this case the diagram shows a range of possible times. The circuit designer needs to consider that transitions could occur any time in the possible range. Since many buses are tristate, this condition needs to be shown as well.

In microcontroller systems bus transitions occur in relation to a *clock* signal as shown in Figure 7.2. Sometimes, a timing diagram shows a specific sequence, such as the behavior of the data bus during an interrupt. Recall that the MCU stacks the registers when an interrupt occurs. This means that it writes the register contents to memory (the stack). A

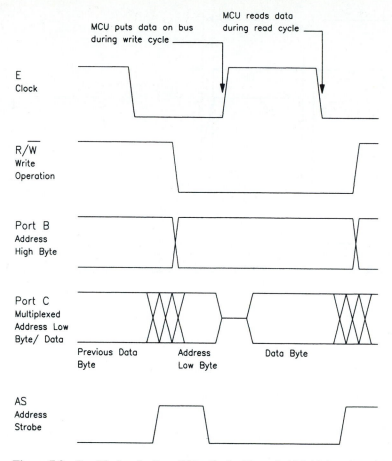

Figure 7.2 Bus Timing for Data Write Cycle (Expanded Multiplexed Mode)

timing diagram showing this activity would label each stable data bus state as the register data. For interrupt stacking, the sequence order is PC low, PC high, IY low, and so on, as shown in Figure B.2 of the Quick Reference (Appendix B).

■ 7.2 SYSTEM CLOCK

The clock signal is a periodic sequence of pulses. It is not necessarily a square wave, although it resembles one. The 68HC11 can generate its own internal clock signal. To do this, connect a crystal to the crystal pins (XTAL and EXTAL). Also connect a resistor in parallel and a capacitor to ground for each pin as shown in Figure 5.8. For more detailed instructions, refer to the Reference Manual.

The internal clock frequency is one-fourth of that supplied to the crystal pins. A typical system designed for maximum clock frequency uses an 8-MHz crystal. Hence, the clock speed (frequency) is 2 MHz. This clock is often referred to as the *system clock* or the *E clock*. When the E clock output is low, an internal process is taking place. When it is high, the MCU is writing or reading data.

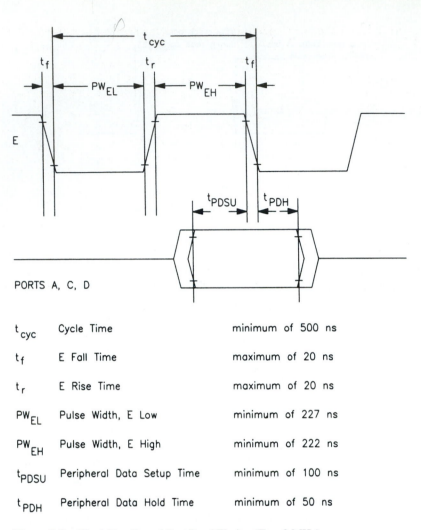

t_{cyc}	Cycle Time	minimum of 500 ns
t_f	E Fall Time	maximum of 20 ns
t_r	E Rise Time	maximum of 20 ns
PW_{EL}	Pulse Width, E Low	minimum of 227 ns
PW_{EH}	Pulse Width, E High	minimum of 222 ns
t_{PDSU}	Peripheral Data Setup Time	minimum of 100 ns
t_{PDH}	Peripheral Data Hold Time	minimum of 50 ns

Figure 7.3   Clock Details and Port Read Timing (E = 2 MHz)

Refer to Figure 7.2. Recall that port C is the data bus when the 68HC11 operates in expanded multiplexed mode. Note that port C contains a data byte when E is high. You may also note that the previous explanation was simplified. The MCU writes data onto the bus after E goes high and retains the data for a short period after E goes low again.

Figure 7.3 shows the E clock in more detail. You may wish to refer back to Section 2.4 to review how the central processing unit (CPU) executes instructions. When E falls (transition from logic 1 to logic 0) to begin a new cycle, the CPU puts the contents of the program counter (PC) onto the address bus. When it rises, the CPU increments the PC (pointing to the next instruction). Also, the CPU or another device sends data to the I/O port (or data bus). For a read cycle, the addressed location sends data to the port. For a write cycle, the CPU sends the data to the port. At the end of a cycle, when E falls, the CPU or another device latches in the signals at the port pins, depending on whether it is a read or a write cycle. The diagram shows a read cycle. Note that an external device connected to an input port should output data when E is high. A system operating in expanded multiplexed mode executes similar cycles. Data is sent to or read from the data bus instead of an I/O port. (This is what is described in Section 2.4, albeit in less detail.)

Figure 7.3 also serves to illustrate timing specifications. In this case it shows what happens when the MCU reads data from an I/O port. Recall that port C is an I/O port when the MCU operates in single-chip mode. During each clock cycle, E goes low first, then high. An instruction to read data from an input data takes several cycles, depending on the type of instruction and its addressing mode. Figure 7.3 shows the last cycle during which MCU latches in the data at the port pins. The timing parameters are important in circuit design. The peripheral device must provide stable data within the time limits shown relative to the clock. The MCU latches in the data when the clock falls. Later we will show how to monitor timing activity.

Note that timing diagrams shown in manufacturers' data sheets are more complete and complicated. In Chapter 6 we covered IC specifications. The ac characteristics state limits for rise and fall times. Other limits are hold and setup times and pulse widths. A typical device sends or reads data when it detects a rising or falling edge on one of its pins. This signal is a clock or a latch enable signal. The data must be sent before the clock or latch enable signal to ensure proper recognition of the data. The amount of time required for this is the setup time. Similarly, data must be maintained for a minimum time period after the clock or latch enable to ensure proper recognition of data. A pulse width is the time interval between 50% levels of an input pulse. This guarantees proper operation of a logic device.

PRACTICE

The MCU executes the instruction

```
LDAA $1003  ;read port C
```

Explain what the MCU actually reads if the signals at port C change during execution.

Solution

It will read the port C signals during the trailing edge of the last E cycle (see Figure 7.3).

■ 7.3 BUS OPERATIONS

Larger microcontroller applications use circuits with peripheral chips connected to the microcontroller chip. This means that the microcontroller must use some pins for the external bus (address, data, and control). Recall from Section 5.7 that the 68HC11 can operate in an expanded multiplexed mode. Other microprocessors and microcontrollers may operate using a similar mode.

In this section we cover how expanded mode operation provides external bus signals. We also look at direct memory access (DMA) operations and the first in, first out (FIFO) queue operation. These two techniques are not normally employed in applications run by the 68HC11. However, we cover them since you will see references to DMA and FIFO in other technical literature. All three operations involve exchanging information using the bus.

Expanded Mode Operation. It is important to understand timing when building or monitoring a circuit using the microcontroller unit (MCU) in expanded multiplexed mode. Recall that port B is replaced by a bus that carries the high byte of the address. Port C is a bus that carries both the low byte of the address and the data byte. Since these two signals cannot appear on the bus at the same time, this bus is multiplexed.

Refer to Figure 7.2. The MCU puts the high byte of the next address on port B after the clock falls. The port C pins send the address low byte at the same time, but it holds it there only until the clock rises. It sends or receives a data byte when the clock is high. Note that Figure 7.2 shows the case when the MCU sends data (write). Note that the address strobe signal (AS) is asserted high when port C has an address byte. External hardware can use the address strobe signal (AS) to *demultiplex* the address and data signals. Demultiplexing is the process of separating all the signals in a common channel into the separate signals.

Figures 7.4 and 7.5 show how an MCU uses tristate to access an external memory device. The output of an address decoder (see Section 5.4) connects to pin $\overline{E}$. As shown in Figure 7.4, when the decoder output to $\overline{E}$ is high (not selected), the data pins in the memory chip are in the high impedance state. This effectively disconnects the data pins from the address bus, so the memory chip is not accessed. When the address decoder selects the memory chip by driving $\overline{E}$ low (selected), the data pins are connected to bus. Then the MCU signals to pins $R/\overline{W}$ and $\overline{G}$ determine when data is read from the memory chip or written to the memory chip.

Figure 7.5 shows read and write operations when the memory chip is selected ($\overline{E}$ is low). During each clock cycle (see Figure 7.2), the MCU drives the address bus to select a location and the read/write ($R/\overline{W}$) line to tell the addressed memory device whether a read or write operation is occurring. The low-order address lines determine which location to access in the memory chip. Remember that the high-order address lines control the address decoder, which in turn determines which memory chip to select. During a read operation, the MCU drives the read/write line ($R/\overline{W}$) high. When the MCU drives the clock (E), the NOT gate drives the memory gate-enable pin ($\overline{G}$) low. The memory chip puts the addressed data on the data bus when it sees a low signal on pin $\overline{G}$ and a high on pin $R/\overline{W}$. When the MCU drives the clock low, the MCU latches in the data (see Figure 7.2). This also disables the memory chip. During a write operation, the MCU drives $R/\overline{W}$ low. When the MCU drives the clock high, it drives the data bus with the data (see Figure 7.2). The memory chip will latch in the data bus signals when it sees a rising edge on its gate enable pin $\overline{G}$ (NOT-gate output of falling clock E) while $R/\overline{W}$ is low. This occurs when the MCU drives the clock low.

Now we will consider an actual system using the expanded multiplexed bus. Figure 7.6 shows typical connections for expanded mode operation. The 74HC138 decoder chip selects external devices, which in this case is the MCM6164 8KB RAM chip. The 74HC373 latch demultiplexes port C so that the RAM chip receives the address low byte for an entire clock cycle. The NAND and NOT gates ensure that the RAM reads data only when data exists on port C.

To study the circuit operation in more detail, refer to both Figures 7.2 and 7.6. At the beginning of a clock cycle, the MCU puts the address high byte on port B, the address low byte on port C, and pulses AS high for address strobe. This forces latch enable (LE) high, which causes the address low byte to be latched. The output of the 74HC373 is now the cur-

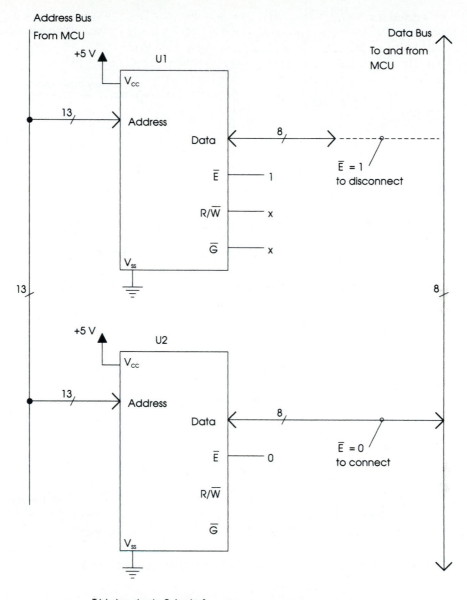

Tristate: logic 0, logic 1
high impedance (i.e., disconnect)

Figure 7.4 Tristate Operation for Memory Access

rent low address byte. The address pins of the RAM (or any other external device) will see both the high and low address bytes for the entire clock cycle.

In Figure 7.5 we saw that the MCU reads or writes data when the clock goes high. Consider a read cycle. The read/write signal (R/$\overline{\text{W}}$) remains high. A NOT gate forces inverted R/$\overline{\text{W}}$ to remain low. When the MCU drives E high, another NOT gate forces the RAM

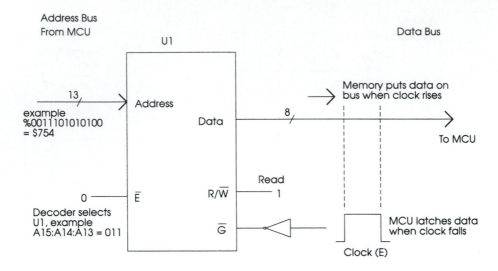

(a)   Read Operation   MCU read from selected address, example $6754

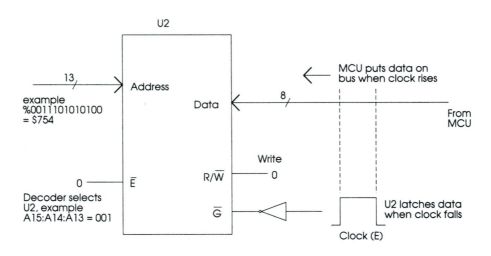

(b)   Write Operation   MCU write to selected address, example $2754

Note   Data Bus is bidirectional.
        MCU writes data when Clock goes high.
        MCU reads data when Clock falls from high to low.

Figure 7.5   Memory-Access Operations

gate-enable pin $\overline{G}$ low. Therefore an external device such as RAM will put data on the port C lines only when E is high. The NAND gate forces write enable ($\overline{WE}$) high when inverted R/$\overline{W}$ is low. This causes a high signal at RAM pin $\overline{W}$, which tells it to send out data to port C. The MCU latches in the data at port C when the clock goes low again (i.e., at the end of the cycle).

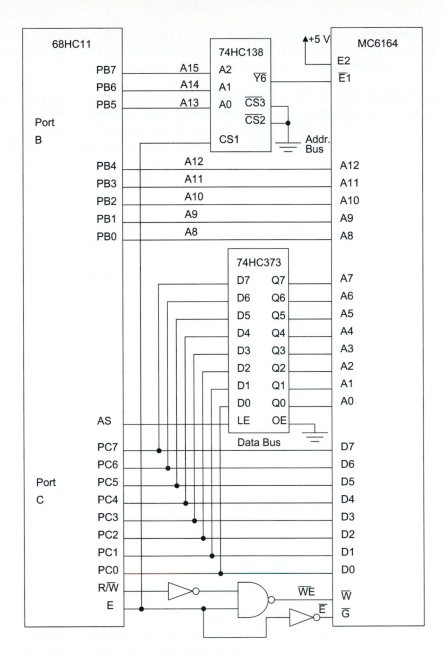

Figure 7.6 Address/Data Demultiplexing

If it is a write cycle (the RAM inputs data from the bus), the MCU sends out data on port C. Also, it forces R/$\overline{\text{W}}$ low, which in turn forces inverted R/$\overline{\text{W}}$ high. When the MCU drives E high, the NAND gate output is forced low. This drives write enable ($\overline{\text{WE}}$) low. This causes the RAM to read in data from port C. An external device (e.g., RAM) must latch in the data before E goes low again (before $\overline{\text{G}}$ goes high again). At the end of the cycle, R/$\overline{\text{W}}$

goes high again. Note that write enable ($\overline{WE}$) is low only when E is high and R/$\overline{W}$ is low. If $\overline{WE}$ is low, RAM inputs data. If it is high, RAM outputs data. The NOT and NAND gates ensure that the RAM will read in data only when it is on the data bus.

Direct Memory Access (DMA) Operation.   The usual method to access a memory chip is to execute a load or store instruction. These instructions require several clock cycles to execute each time. In the next section, cycle-by-cycle operation, we will examine these operations further. For now accept the fact that each memory access requires several clock cycles of the CPU's time.

There are situations when a processor must store a sequence of data very quickly. For example, disk drives and communications systems require fast transfers. A fast data transfer may occur in a system that must take measurement samples of a rapidly varying signal during a short time period. During this time, the only thing the system is doing is moving data from one place to another; for example, moving sensor data to RAM. Alternatively, data in RAM is moved to an external device. The RAM functions as a *buffer.* A buffer is a temporary storage location to hold data while the processor is busy. After the processor is free, it reads the buffer for a read operation. For a write operation, it writes data to it (fills it).

There is no need for the bus to carry opcodes and other information. In this case the CPU simply wants to select an address, move data to the location, increment the address, move the next data to it, and so on. The process should be automatic without having to write a program loop requiring a separate instruction for each step.

The direct memory access (DMA) controller is a device that performs this operation. It allows the CPU to wait or do other tasks while the data transfer is taking place. The DMA controller can be a separate chip. Alternatively, it can be the part of the MCU chip designed for specialized input/output operations. The latter type may control data transfer operations for an external bus or the chip's internal bus.

The basic CPU with DMA operation is

CPU sends start address to DMA controller.

CPU sends block length and other control information to DMA controller.

DMA controller takes over the system bus and does (a) or (b).

    (a) Read data and store it in RAM.

    (b) Read RAM data and send it out.

CPU waits for DMA controller to hand back control of the bus.

DMA controller hands back control of the bus.

CPU resumes normal operation.

While the DMA controller uses the bus, the CPU must tristate its bus lines. Alternatively, it can continually read the data bus while tristating other bus lines using an external circuit. This is possible for the 68HC11 because it continually reads the data bus after executing a wait for interrupt (WAI). The CPU does this until it is interrupted.

A DMA controller may have several channels. An I/O device or memory is connected to each one. With the DMA controller (it may be built in), the microprocessor can write or read a sequence of bytes with little intervention by the CPU. These are known as *burst transfers.* A DMA controller operating this way is said to be operating in *burst mode.* Another is *alternate cycle mode.* A DMA cycle and instruction can occur alternately on the bus. This is useful when CPU processing must occur during the DMA transfers. DMA controllers often use FIFOs, the next topic.

The First In, First Out (FIFO) Queue.   The FIFO is a small block of memory used by some processors. Often, the term *queue* is used, although it is sometimes called a *stack*. A FIFO queue can be implemented in hardware or software. As the name implies, data is removed in the order in which it was put in. This is the opposite of the "main" stack, which is a LIFO (last in, first out). Recall that the LIFO stack operates by using push and pull operations. It uses one pointer, the stack pointer (SP). The FIFO, in contrast, uses two pointers. The input pointer indicates the next location where data can be stored. The output pointer indicates the next address where data can be removed.

Hardware FIFO queues are commonly used in communications systems. While data is being received, it is put into the FIFO as a DMA operation. During this time, the processor can continue with its current task. When enough bytes are written into the FIFO (fills up), it reaches a trigger level. The processor detects the trigger (it may be an interrupt) and reads the data from the FIFO. Without the FIFO the processor would have to handle the reception of each byte individually. With the FIFO it can handle them in blocks.

A transmit operation to send out bytes is similar to a receive operation. The processor fills up the FIFO and then works on another task. During this time, the transmitter hardware removes data from the FIFO and sends it out. When enough FIFO bytes are removed (emptied), it informs the processor that the FIFO is empty (it may be an interrupt). The processor can then write more bytes to the FIFO.

■ 7.4 CYCLE-BY-CYCLE OPERATION

The *Data Book* includes a table that shows cycle-by-cycle operation. We show a sample segment of this table since we chose not to include the entire table in this book. If you look at the Instruction Set Summary (Appendix A), you will note a column labeled "Cycle by Cycle." The entries in this column are a cross-reference to the cycle-by-cycle operation for the instruction. For example, STAA (operand) extended has the reference number 5-3. For illustration purposes, we show this segment as Table 7.1.

The address bus is port B and demultiplexed port C (after the latch, such as the one in Figure 7.6) for a total of 16 bits. The data bus is port C when the clock is high. This table is useful when monitoring bus activity using an oscilloscope or logic analyzer. We will examine the use of these instruments later in the chapter.

Next, let's look at a specific example. Consider the following section of a program:

```
                        ORG     $F800
F800      B7 C1 5A      STAA    $C15A
```

TABLE 7.1   CYCLE-BY-CYCLE OPERATION FOR STAA EXTENDED

Cycles	Cycle no.	Address bus	R/W	Data bus
4	1	Opcode address	1	Opcode
	2	Opcode address +1	1	Operand address (high byte)
	3	Opcode address +2	1	Operand address (low byte)
	4	Operand address	0	Accumulator data

Cycles	Cycle no.	Address bus	R/W	Data bus
4	1	$F800	1	$B7
	2	$F801	1	$C1
	3	$F802	1	$5A
	4	$C15A	0	$75

Table 7.2 illustrates the bus activity if accumulator A contains $75.

This is a good opportunity to review address decoding and the operation of the circuit shown in Figure 7.6. Assume that the system also uses an external EPROM instead of internal ROM. The EPROM select line is connected to Y7. During the first three cycles, address lines A15, 14, 13 are high, which causes Y7 of the 74HC138 to go low to select the EPROM. The first cycle is the instruction fetch cycle. The MCU fetches the opcode $B7 from the EPROM and latches it in at the end of the cycle. Remember that the latch circuit output is the low byte of the address for the duration of an entire cycle. During the next two cycles, the MCU fetches the address specifying where to store data. During the last cycle address bits A15, 14, 13 form the binary number %110, which causes the address decoder to drive Y6 low to select the RAM. When E goes high during this cycle, the MCU drives R/W low and puts the accumulator data on the bus. This also causes $\overline{G}$ to go low. The RAM latches in the signals at its data pins when $\overline{E1}$ is low and $\overline{G}$ rises.

PRACTICE

What are the 68HC11 signals during every half clock cycle when it executes the following instruction?

```
STAA $2A73 ;ACCA = $17, this instruction at address $E145
```

Solution

E	Port B (hex)	Port C (hex)	R/$\overline{W}$	AS
0	E1	45	1	Pulse high
1	E1	B7	1	0
0	E1	46	1	Pulse high
1	E1	2A	1	0
0	E1	47	1	Pulse high
1	E1	73	1	0
0	2A	73	Goes low	Pulse high
1	2A	17	0	0

Remember: Port C = low address byte when E = 0
            Port C = data byte when E = 1.
See also Figure 7.2 for last clock cycle.

The load instruction register $(\overline{\text{LIR}})$ pin is also driven as an output. Back in Section 5.5 we mentioned pin MODA/$\overline{\text{LIR}}$. To connect the 68HC11 for expanded mode operation, this pin must be tied high through a pull-up resistor. When in expanded mode, the pin is driven low for the first cycle of each new instruction. The signal is intended to help debug the system during development of an application. This signal can be monitored by an oscilloscope or logic analyzer. In the following sections we will discuss the use of these instruments.

■ 7.5 OSCILLOSCOPE TEST LOOPS

You may have to monitor bus signals for test purposes. There may be a fault or you might wish to verify that the circuit operates as designed. One of the most common instruments used to measure electronic signals is the oscilloscope. However, it does have some limitations when used to monitor signals in any computer system. Multimeters are even more limited. They can only display dc signals and show the root-mean-square (rms) value of a 60-Hz sine wave. (It may be 50 Hz, depending on where you live.)

The oscilloscope has a screen that shows how a signal changes voltage over time. With the exception of storage oscilloscopes, it can only display repetitive signals. These are periodic signals such as sine waves or square waves (see Appendix G). Therefore, it can display the clock signal E. But typical bus activity is nonrepetitive or repeats after many clock cycles.

The oscilloscope operates by waiting for a trigger signal. A trigger is a signal that rises or falls below a threshold voltage. When it is received, the oscilloscope does what is called a *sweep*. It draws the signal shape across the screen. After it completes the sweep, the oscilloscope waits for another trigger signal before beginning the next sweep. Note that a trigger occurring during a sweep does not affect the sweep. When a signal is repetitive, each sweep draws the same shape. Hence, the screen shows a stable display. If the signal is nonrepetitive (or triggering controls are not set up properly), the screen shows a display that overlaps, flickers, or moves—or all of these.

Unless you use a digital storage scope, the only way to monitor bus activity is to make it repeat in as few cycles as possible. To do this, write a program that loops in a few cycles. Also, it has to have one unique rising/falling edge to use as a trigger. This program is know as an *oscilloscope test loop*. Any program can be an oscilloscope test loop as long as it has the properties mentioned. Ideally, the loop should execute in 10 or fewer cycles in order to fit the display on the screen.

The program of Listing 7.1 is a sample oscilloscope test loop. It can be used to test output Y3 of the address decoder shown in Figure 7.6. Many problems are caused by faulty address decoding. Note that the program code itself is put in RAM (Figure 7.6). It would not be part of application code ROM, although it could be part of EPROM during development.

```
*Listing 7.1
*Oscilloscope Test Loop
*used to check Y3 of address decoder

    ORG    $D100
```

```
            LDAA    #$A5    ;any data
LOOP
            STAA    $6105   ;any address within range of device
            BRA     LOOP
```

The program executes the first instruction once and the last two continuously. The loop requires only seven clock cycles to execute. Note that it performs one write operation during the loop. Thus R/$\overline{\text{W}}$ can be used as a trigger for this program. It goes low during the last cycle of STAA.

An oscilloscope is still awkward to use since most oscilloscopes have only two channels. This limits the number of signals you can see at one time. However, most labs and electronic repair shops have one. The oscilloscope is still a valuable tool if you wish to repair circuit boards instead of throwing them away. This may be awkward since most chips are soldered in, so it is difficult to swap chips to locate a problem. A resourceful person can sometimes invent suitable test programs to locate a fault. If the system uses the MCU in single-chip mode, you will have to load in the test loop program using the bootstrap mode.

PRACTICE

Write the cycle-by-cycle operation for the LOOP part of the oscilloscope test loop shown in Listing 7.1 Also refer to Question 8 in the end-of-chapter exercises.

Solution

Instruction	Cycle no.	Address bus	R/$\overline{\text{W}}$	Data bus
STAA	1	$D102	1	$B7
	2	$D103	1	$61
	3	$D104	1	$05
	4	$6105	0	$A5
BRA	5	$D105	1	$20
	6	$D106	1	$FB
	7	$FFFF	1	xx

Note that R/$\overline{\text{W}}$ is low once during each loop iteration. This is the reason for using it as a trigger. Note that xx represents a logic don't care condition.

■ 7.6 LOGIC ANALYZER

A logic analyzer is an instrument more suitable for monitoring computer bus signals. It is primarily a development tool. The logic analyzer can display many channels, typically multiples of 8. In operation it differs from an oscilloscope. It samples a sequence of bus signals and stores the captured sequence in memory. The analyzer displays the stored sequence and not *real-time* bus activity. In contrast, an oscilloscope displays the current signals. Because

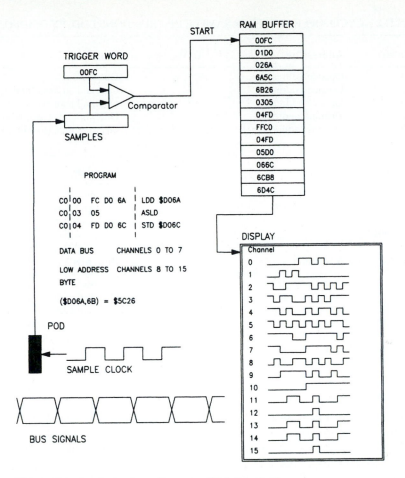

Figure 7.7   Logic Analyzer Operation With Sample Program

the logic analyzer captures a sequence, it can store the bus cycles for many instructions. Perhaps *analyzer* is not a good word since the basic instrument itself displays data but does not analyze it for you. Another important difference between an oscilloscope and a logic analyzer is that the analyzer displays logic states, 0 or 1. The oscilloscope displays the actual voltage values.

Figure 7.7 shows the operation of a typical logic analyzer. An operator sets the sampling rate. Normally, the system clock is used to determine the sampling times. Since many instructions execute at very high rates, it is necessary to tell the analyzer when to store the samples for display. This is determined by specifying a *trigger* word. When the analyzer sees the trigger word, it displays data relative to the trigger.

In the sample setup shown in Figure 7.7, a 16-channel logic analyzer monitors the data bus and low address byte. The analyzer samples the bus continually. Let's say that the operator is interested in seeing bus activity when the indicated instructions are executed.

TABLE 7.3   CYCLE-BY-CYCLE OPERATION FOR SAMPLE INSTRUCTIONS LDD, EXTENDED MODE

Cycles	Cycle no.	Address bus	R/$\overline{\text{W}}$	Data bus
5	1	Opcode address	1	Opcode
	2	Opcode address + 1	1	Operand address (high byte)
	3	Opcode address + 2	1	Operand address (low byte)
	4	Operand address	1	Operand data (high byte)
	5	Operand address + 1	1	Operand data (low byte)

ASLD, Inherent Mode

Cycles	Cycle no.	Address bus	R/$\overline{\text{W}}$	Data bus
3	1	Opcode address	1	Opcode
	2	Opcode address + 1	1	Irrelevant data
	3	$FFF	1	Irrelevant data

STD, Extended Mode

Cycles	Cycle no.	Address bus	R/$\overline{\text{W}}$	Data bus
5	1	Opcode address	1	Opcode
	2	Opcode address + 1	1	Operand address (high byte)
	3	Opcode address + 2	1	Operand address (low byte)
	4	Operand address	0	Register data (high byte)
	5	Operand address + 1	0	Register data (low byte)

The operator sets the trigger for the LDD instruction. This is the low byte of the instruction address and its opcode. The operator arms the analyzer by pressing a start button. Meanwhile, the program is executing. When the analyzer sees the trigger data, it stores this cycle and following cycles in memory. After the data capture, the analyzer displays the bus activity. It can display it as a timing diagram or as a sequence of hex and/or binary numbers. Some analyzers have pods that make it possible to disassemble the bus data so that it can display the source code (without symbols) assuming that the operator connected the pod correctly. It can use the $\overline{\text{LIR}}$ signal (Section 7.4) to determine the beginning of an instruction cycle.

The sample program is shown executing for the case when the address $D06A,6B contains the data $5C and $26. The result of the arithmetic shift left (ASLD) instruction is $B84C, as can be seen by the low byte of the last two samples in the RAM buffer. The interested reader can use Table 7.3 to confirm that Figure 7.7 shows the RAM buffer data and timing diagram for the sample program. In this case the address $FFFF is an address containing the data $C0. As shown in the cycle-by-cycle operation, this data is irrelevant and the MCU ignores it.

Logic analyzers are available in models that can store from 256 up to 8192 samples. To capture all of the bus signals, it should have channels for the data bus, address bus, and control bus. For the situation shown in Figure 7.7, a compromise had to be made. The system can show only the data bus and part of the address bus. This could make triggering difficult. For example, the system executes a main program continuously. Also, we are unlucky enough to have the instruction ADDA #$FC at address $C0FF. This means that address $C100 has the instruction operand $FC. The analyzer will see this as a valid trigger. If the operator happens to press the Start button at the wrong time, the analyzer uses the byte at address $C100 as the trigger instead of what was wanted. Note that this situation can be avoided in the 68HC11 by using an active load instruction register ($\overline{\text{LIR}}$) signal as a trigger condition.

Some logic analyzers allow you to use a short sequence to specify a trigger. This reduces the chance of triggering at an unwanted time. When the trigger specifies the start of a sample, it is known as *pretriggering*. This is the situation shown in Figure 7.7. If the trigger specifies the end of a sample, it is known as *posttriggering*. It is also possible to use the trigger as an offset to specify a sample *window*. The trigger is somewhere in the middle of the display sample sequence. For example, you might want to see what happens 100 instructions before the trigger and what follows the trigger.

The logic analyzer can also monitor I/O port lines and other signal lines in the system circuit board. The testing application determines what connections to make. To interpret the analyzer display, you have to have detailed knowledge of system operation and cycle-by-cycle operation of the processor used (e.g., 68HC11). It is not normally used for field service work. We repeat: It is primarily a development tool. Use it to verify system operation and to debug flaws in system design and prototype construction. A typical microcontroller application does not use a computer terminal to monitor execution. The only way to check operation during system development may be to measure bus signals when a prototype program executes. As an alternative or in addition to a logic analyzer, you may use an emulator (Section 4.10).

PRACTICE

What is the trigger word to capture the execution of the program of Listing 7.1? We are using a 24-channel logic analyzer.

Solution

Using pretriggering, the trigger word is: $D10086. This assumes that pod channel 0 is connected to data line zero (D0), and so on.

■ 7.7 TRANSMISSION LINE EFFECTS

You may wonder what transmission lines have to do with microcontrollers. Generally, one would not consider how a wire or circuit board trace connecting devices together would affect a signal. It turns out that when data changes at a fast rate for long lines, the line itself affects the signal. Such a line is considered to behave like a transmission line. The relationship between frequency, how fast data changes, and physical line length is relative. The line affects the signal significantly if the line is long relative to the wavelength of the signal propagating through it. Real transmission lines are physically many kilometers long; hence, a 60-Hz power signal is affected by the line itself. In the following discussion we call any line that is long relative to the signal wavelength (reciprocal of frequency) a transmission line.

All signal lines store energy when transmitting voltage and current. This is because inductance and capacitance are natural properties of wire. When a signal is sent, the energy stored in an electric field (as capacitance) produces a current. The current generates a magnetic

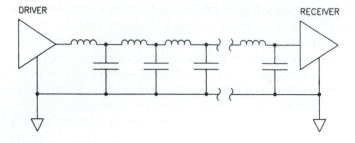

(a)   Model of any digital signal line

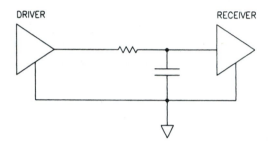

(b)   Simplified model for shorter line or lower frequency

Figure 7.8   Transmission Line Models

field (as inductance) to store energy. This current, in turn, induces another electric field. This process continues. The result is that electromagnetic energy propagates along the wire. Figure 7.8(a) shows a model of a signal line. This line is known as a *transmission line.*

For low-frequency (large-wavelength) signals, the effects of the capacitors and inductors are negligible. In the extreme case of dc signals, the capacitors act like an open circuit and the inductors act like a short circuit. When the frequency becomes very high (short wavelength), the capacitors tend to act like short circuits and inductors tend to act like open circuits. When the signal line is many times shorter than the signal wavelength, transmission line effects are negligible and can be ignored.

A connecting line within a printed circuit board is short enough to be modeled as shown in Figure 7.8(b). Most of the energy is stored as capacitance (inductance ignored). The inherent capacitance prevents signals from changing instantaneously. A rising edge must first charge the capacitor and a falling edge must discharge it. Data sheets indicate what capacitive loading is used to state the ac electrical characteristics. The loading effects signal rise and fall times. If you use a system with more capacitive loading than that stated in the data sheets, the design should be based on rise and fall times longer than those stated in the sheets.

Refer back to Figure 7.8(a). At any point along the line, a signal sees an impedance caused by the inherent capacitance and inductance. This impedance, known as the *characteristic impedance,* is commonly labeled as Z_0.

When the line is long relative to the signal wavelength/frequency, oscillations may occur on the line when a device changes its output. This oscillation is also known as *ringing.* You have probably observed ringing when measuring bus signals with an oscilloscope. Ringing is most apparent for the system clock (E) because it uses the highest frequency. The oscillations are caused by signal reflections at both the driver and receiver ends. We can consider digital signals to be *steps,* either transitions to low or high. When the driver sends the signal voltage step (wave), the wave travels along the line. Eventually, it reaches the receiving-end termination. A wave approaching a termination is the incident wave. When it hits the termination, the termination reflects the wave back as a reflected wave. This action is similar to that of ocean waves hitting a concrete pier. At the time of initial reflection, the reflected and incident waves add to cause an overshoot for a rising edge (undershoot for a falling edge). The reflected wave travels back to the driver and will again be reflected. Each time a reflected signal arrives at any termination, its magnitude is smaller than that of the incident wave. Eventually, the reflections die down and steady state is achieved.

The amount of reflection is determined by the reflection coefficient of the termination. The reflection coefficient is

$$\rho = \frac{Z/Z_0 - 1}{Z/Z_0 + 1} \tag{7.1}$$

where ρ is the reflection coefficient, Z the termination impedance, and Z_0 the characteristic line impedance. Note that the reflection coefficient can be negative if Z is less than Z_0. The reflection coefficient lies in the range -1 to $+1$. The reflected voltage is

$$V_r = \rho V_i \tag{7.2}$$

where V_r is the reflected voltage and V_i is the incident voltage.

Impedance Matching. There is no reflection, hence no oscillation, if the reflection coefficient is zero. This occurs when the termination impedance equals the characteristic impedance. Figure 7.9(a) shows a termination impedance of 600 Ω ; 1.0 kΩ in parallel with 1.5 kΩ is 600 Ω. It is also in parallel with receiver input impedance. But since this is very high, it can be neglected. The resistor values shown in Figure 7.9 are the values recommended by Motorola. It is a trade-off between matching the signal line impedance and reducing power consumption. For example, a ribbon cable has a nominal impedance of 200 Ω. However, using a 200-Ω termination increases power consumption. To reduce ringing, the termination resistors should be as close as possible to the receiver. Figure 7.9(b) shows an alternate circuit terminating both driver and receiver.

Considering Transmission Line Effects. One question may be: When is it necessary to consider transmission line effects? You must consider the effects when the line distance is long relative to the data rate. Electromagnetic energy travels at the speed of light, approximately 3×10^8 m per second. This is equivalent to 3 m every 10 ns. If a driver sends a 100-ns pulse along a 3-m line, there will be five reflections before the driver ends the pulse. If the termination impedance is not matched to the line's characteristic impedance, the pulse will be distorted. On the other hand, if the driver sends a 1-ms pulse along the same line, there are

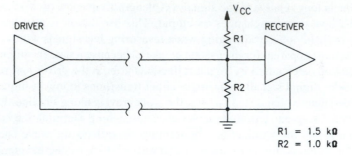

(a) Termination Resistors at the Receiver

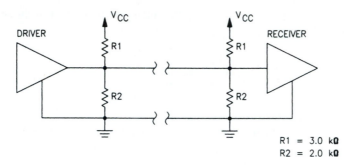

(b) Termination Resistors at both the Driver and Receiver

Figure 7.9 Transmission Line Termination for HCMOS

theoretically 50,000 reflections before the driver ends the signal. Of these reflections, only a fraction would be significant.

From these examples you should be able to see why transmission line effects are noticeable in very high speed microprocessors. The system clock for a 50-MHz microprocessor has very fast rise and fall times; thus the physical dimensions of connecting cable approaches the system bus signal wavelengths (6 m).

■ 7.8 SUMMARY

Semiconductor vendors often use timing diagrams to illustrate timing specifications. They indicate how fast a device must respond to a change in input. They indicate the time period during which the bus holds data. This time period is relative to the edge of a control bus signal.

The clock (E) is one of the control bus signals. It supplies the timing to the microcontroller to perform its tasks serially, one after another. The central processing unit (CPU) fetches program code and data stored in memory, word by word. For an 8-bit MCU such as the 68HC11, a word is a byte. The CPU processes the data and sends the result to an I/O device or to memory. Any processor that operates in this serial fashion is known as a *von Neumann machine* (see Chapter 1).

Two pins, XTAL and EXTAL, provide the interface for either a crystal or a CMOS digital clock. The signals must be four times higher than the desired clock rate. These signals control the internal clock generator circuitry. The 68HC11 generates bus cycles.

When operating in expanded multiplexed mode, the bus cycles are visible to the port B and C pins. During each cycle, the MCU puts the high address byte to the port B pins. During the first half of a cycle, it puts the low address byte to port C and asserts an address strobe (AS) pulse. During the second half, it uses port C as a data bus. It asserts the read/write ($R/\overline{W}$) line as required, depending on whether the cycle is a read or write cycle. An external latch circuit can use the AS signal to demultiplex the low address and data bus on port C.

Data books document what is on the address bus, data bus, and R/W line for each clock cycle for each instruction. This information is useful when interpreting the results of oscilloscope or logic analyzer measurements. They also aid in interface design.

An oscilloscope is a versatile instrument used by many electronic technicians and engineers. However, it has limitations when used to measure microcontroller system signals. Typically, you would write a program that loops. The loop is a few cycles long and generates one unique pulse to use as a trigger.

The logic analyzer can store bus activity data for many clock cycles. However, it is primarily a laboratory instrument instead of a field instrument. Interpreting its data requires detailed knowledge of system hardware and software.

As systems with higher clock speeds are introduced, transmission line effects will become more important. Any line whose length is long relative to the signal pulse width or wavelength (period) acts like a transmission line. Transmission line effects can cause signal distortion and ringing. To avoid this, device inputs and outputs should be connected to appropriately sized resistor networks. These networks match termination impedance to be equal or close to the signal line characteristic impedance.

■ EXERCISES

1. Draw a bus timing diagram that shows valid data, a transition, and a high-impedance state.

2. Is it possible to capture a 100-ns pulse input at an I/O port?

3. Complete the table to show signal levels at the 6164 RAM when connected for typical 68HC11 expanded multiplexed mode operation.

68HC11		MCM 6164	
$R/\overline{W}$	E	W	$\overline{G}$
0	0		
0	1		
1	0		
1	1		

4. For Question 3, during which of the conditions can the RAM output data assuming that E1 is low?

5. Explain how an MCU can use 16 pins to output a 16-bit address and 1 data byte.

6. Refer to Figure 7.6. The multiplexed address hold time is at least 33 ns. This is the time the address byte stays on the bus after AS falls. The 74HC373 latch has a minimum hold time of 10 ns from the time its LE input falls. Explain why it can demultiplex the bus.

7. Address $00 contains the word $6C54. Predict the bus activity for the following program segment.

```
ORG     $E000
LDD     $0000
ASLD
STD     $0005
```

8. A 68HC11 system runs at 2 MHz. It executes the following oscilloscope loop:

```
        ORG     $D100
        LDAA    #$A5
LOOP
        STAA    $6105
        BRA     LOOP
```

The cycle-by-cycle operation for branch instructions is the following (with the exception of BSR, BRCLR, and BRSET):

Cycle	Address bus	R/W	Data bus
1	Opcode address	1	Opcode
2	Opcode address + 1	1	Branch offset
3	$FFFF	1	Irrelevant data

An oscilloscope measures address lines A0 and A1, triggered by an appropriate external trigger edge. Show the oscilloscope readings at 500-ns intervals after the trigger edge. Show the readings as logic 0 or 1. When can the accumulator data be observed?

9. Refer to Figure 7.7. Given only the program information and the contents of $FFFF = $C0, derive a RAM buffer data and timing display similar to that shown in Figure 7.7.

10. For the program described in Question 8, indicate the logic analyzer display as hex numbers. The analyzer is triggered by the STAA instruction and has 24 channels. It can display some channels as single bits. Also, the reset routine starts at $E000.

11. An 8-channel logic analyzer samples the data bus for the following program. The trigger word is $00. Which instructions will it capture?

```
PORTB    EQU      $04
         ORG      $C000
         LDX      #$1000
         BSET     PORTB,X  $5C
         BCLR     PORTB,X  $48
```

12. Repeat Question 11 for a trigger word of $1D.

13. Would there be a problem with connecting a 68HC11 to a device a few hundred meters away?

14. Derive the equation for the reflection coefficient for transmission lines. Note that this question requires basic electric circuit theory knowledge.

8 Interfacing Concepts

OBJECTIVES

After completing this chapter, you should be able to

- Describe key interfacing concepts and conventions such as drivers, input/output (I/O) direction, subsystems, I/O registers, ports, and mapping.
- Write programs that perform bit manipulation to monitor binary inputs and drive binary outputs.
- Identify memory-mapped register and bit addresses using the data book.
- Write fundamental programs using polled I/O and interrupt-driven I/O.
- Describe fundamental characteristics, operations, and applications for parallel, synchronous serial, asynchronous serial, programmable timer, and analog/digital I/O subsystems.
- Describe fundamental configuration operations at the bit level with control, status, and data registers.
- Identify some interfacing standards.

In this chapter we will first examine what is meant by *interfacing* and some of the terminology encountered in this subject. Next, we will present the concept of the generic subsystem and its component parts. Then the two basic methods of interfacing, polling and interrupts, are explained. We will then proceed to preview the different subsystems used by most microcontrollers. Later we will examine how the microcontroller generally uses I/O registers for interfacing operations. Finally, we will introduce the existence and need for interface standards.

■ 8.1 INTRODUCTION

Interfacing is one of those words that is becoming more and more commonly used in the English language. You may have heard expressions such as "That computer program has a good user interface," which means that the program is easy to learn and use.

Formally, an interface is a device and/or set of rules to match the output of one device to send information to the input of another device. Interfacing is the process of connecting devices together so that they can exchange information. Note that an interface includes the physical connection, the hardware, and a set of rules or procedures, the software. We have seen examples already. The connection between a microcontroller and external memory is an interface. There are physical connections for addressing, data flow, and control signals. The software consists of instructions that read from or write to an addressed location.

There are hierarchies of interfacing. Connecting two supercomputers to work together on the same program is quite different from connecting a memory chip to a microcontroller chip. There are also interfacing *standards*. A standard is an accepted way of interfacing agreed to by different people, companies, and industry organizations. When connecting devices built by different manufacturers, there has to be an agreement as to how this should be done. A telephone is a good example of a standard. All telephones are built to conform to an accepted method of connecting to the telephone line. They also conform to accepted procedures to make calls and to receive calls. Later we will cover some of the standards applicable to microcontroller interfacing.

Among interface standards are those used by the personal computer. It is a system with many interfaces. These interfaces are standardized because many manufacturers supply components and others supply programs. One such interface is that between the main board (motherboard) and the keyboard. The keyboard has physical cabling and a mechanical connector with specific pin assignments. The data signals have specified logic levels and timing. The data is coded according to the standard. Software (sub)routines get keyboard data or check its status. Application programs use the subroutines to check which key a user has pressed. For example, pressing the A key in an IBM or compatible personal computer keyboard sends a data byte of hex 1F to the microprocessor.

The subroutine just mentioned is an example of a *software driver* or *device driver*.[1] A computer or microcontroller system typically has subroutines or software interrupts that read inputs from or send outputs to standard devices such as keyboards and monitor screens. At the user or application level, it may be necessary to develop a program that among other tasks needs to read keyboard information and send output to a screen. The person writing the program needs only to know how to use the software drivers for the input/output tasks. It is not necessary for the programmer to know the details of the interface itself. In a custom system the developers have to invent their own software drivers. These should be developed in such a way that reuse by the programmer and others is easy.

This might be a good time to review the automotive application described in Section 1.1. Since microcontrollers are used in control applications, they include many of the peripheral features needed to communicate with the outside world. These are the built-in subsystems for interfacing.

The process of reading input signals and sending output signals is called *I/O*, which is the short form for *input/output*. The subsystems are known as I/O subsystems. Some MCU interfaces are referred to as MCU I/O. Generally, a memory interface is not classified

---

[1]DOS uses software interrupt routines called *interrupts* (INT) for interfacing requirements. Some of the INT functions handle standard devices such as the keyboard, screen, printer, and disk drives. Note that these devices are standard in the sense that the hardware and software is designed for use with a DOS computer.

as I/O since memory is considered to be an internal part of the system, even if the chips are located on another board. Memory contains information already in the system. I/O are signals from or to the outside. In summary, the term *I/O* refers to communication with the world outside a computer or microcontroller system.

I/O Conventions. I/O direction is relative to the MCU. Input is data read by the MCU. Output is data sent out by the MCU. Recall the read operation described in Section 7.3. When $R/\overline{W}$ is high, data is *input* to the MCU. It is sent out (output) by the peripheral chip. When $R/\overline{W}$ is low, data is *output* by the MCU and is sent to the peripheral chip (peripheral's input).

The words *input* and *output* can also be verbs instead of nouns. For example, the MCU inputs data from sensors and outputs commands to the ignition system. Other terms for the verb *input* are *read* and *receive*. Other terms for the verb *output* are *write* and *transmit*.

◼ 8.2 INPUT/OUTPUT SUBSYSTEMS AND REGISTERS

8.2.1 Different Subsystems

Different types of I/O have developed and become accepted in industry. An *I/O subsystem* is a convenient way to categorize the part of the microcontroller responsible for using an I/O interfacing technique. In this chapter we describe them conceptually. In later chapters we will describe each one in detail for the 68HC11. Most computer systems and microcontrollers use similar I/O subsystems. Microcontrollers generally have built-in I/O subsystems because they are designed to handle the different types of interface requirements commonly found in industry. Some applications require more I/O than can be handled by the MCU's built-in I/O subsystems. In this case the MCU board could include programmable I/O peripheral chips. Figure B.1 in the Quick Reference (Appendix B) shows the 68HC11 I/O subsystems. These are the blocks connected to the port blocks. In later chapters we will look at what is inside these blocks.

Historically, I/O processing was controlled by discrete logic or programmable I/O chips. The development of microcontrollers has reduced the need to use this extra hardware in many applications. It is part of the trend for further system integration. An I/O subsystem is sometimes called an *interface*. This is further evidence that the meaning of the word *interface* can be broad. Terminology can be a problem in computer technology; this technology develops so quickly that terms are used differently by experts before they can be standardized.

8.2.2 I/O Registers

Both microcontrollers and programmable I/O chips handle I/O processing using registers. You may wish to refer back to Figures 1.4 and 1.5, where we introduced the I/O registers. Again, refer to Table B.1 in the Quick Reference (Appendix B). This documents the I/O registers for the 68HC11. In Section 6.6 we examined some of these for EEPROM programming. Most of the registers belong to one of the I/O subsystems. This may not be evident initially when examining Table B.1. It simply lists the registers numerically by address.

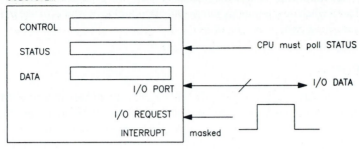

Case 1
Polled I/O, no interrupt
CPU must check a status bit to determine whether an I/O request occurred.

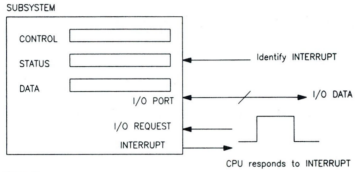

Case 2
Interrupt−driven I/O
I/O request causes interrupt and CPU responds to interrupt.

Figure 8.1   Generic I/O Subsystems

We can categorize each subsystem's registers as control, status, and data. This is a generalization that is more or less true. It provides a useful model for a *generic* subsystem. A generic subsystem has features that are common for most real subsystems. We look at a generic subsystem in order to understand features shared by the other subsystems. After understanding the generic system, we can look at the specific features of the other subsystems.

Figure 8.1 shows a generic subsystem. For now ignore the details that explain the difference between polled I/O and interrupt-driven I/O. We will come back to these later. Recall that the central processing unit (CPU) is the part of the microcontroller that controls data movement within the MCU and execution of programs. It contains its own registers, such as accumulator A and index register X. Note the line for I/O data has a slash across it. This is one convention for indicating a multiline bus.

The 68HC11 Registers.   For the most part the 68HC11 subsystems follow the generic model. The subsystems use different names for the registers. Some registers contain bits used by different subsystems. Unfortunately, some of the terminology and register organization sometimes differs between the different subsystems. You can note this by referring to Table B.1 in Appendix B. Presumably, Motorola used this organization because of design decisions faced when developing an integrated circuit. One reason is to use existing industry conventions for a particular type of I/O.

8.2.3 I/O Ports

Although we have already discussed ports, we will describe them here in terms of I/O. This will define what we mean by I/O ports. Again, refer to Figure 1.4 and to Figure B.1 in Appendix B. Each subsystem has chip pins or external lines for I/O data bits associated with it. Each section of lines is called an *I/O port*. An I/O data register holds the most recently read data from an input port. For an output port the I/O data register (when written to) stores the next data to be sent to the port. When the 68HC11 operates in single-chip mode, all the ports are I/O ports. For expanded multiplexed mode, ports B and C are used for addresses and data buses.

PRACTICE

Find the address for the port A register.

Solution

$1000. You can find this by referring to Table B.1 in Appendix B.

8.2.4 Port Replacement Unit

When the 68HC11 operates in expanded multiplexed mode, it loses the I/O port functions for ports B and C. By connecting a special chip called a *port replacement unit* (PRU), the MCU recovers the use of the ports B and C for I/O. In addition, the MCU still has the external address and data bus.

Refer back to Figure 5.1. Note that the EVB board uses the PRU. The MC68HC24 PRU IC was specially designed by Motorola for 68HC11 systems operating in expanded multiplexed mode. We can use the evaluation (EVB) board to develop applications running the 68HC11 in single-chip mode, but the 68HC11 in the EVB board actually runs in expanded multiplexed mode. The MCU can run programs from EPROM or RAM. This allows the developer to test programs using RAM first. The EVB board has an I/O port connector similar to the EVBU shown in Figure 1.2. It replaces the MCU pins to prototype circuits using a 68HC11 chip. In the EVB board the PRU is used to provide port B and C signals to the I/O port connector. The I/O port connector pins only replace the pin functions for single-chip mode. There are no external connections for the address and data bus signals. Another development system is required to develop expanded multiplexed mode applications. The evaluation multiplexed (EVM) board is one such system. It also provides external connections for operating in either mode. The universal evaluation board (EVBU) does not have a PRU, although it is possible to add one if desired. (It would also be necessary to cut a trace in order to operate the 68HC11 in expanded-multiplexed mode.)

Since other microcontrollers use equivalent modes, they too use a chip to recover I/O ports. Intel microcontroller systems typically use the Intel 8243 input/output expander or the Intel 87C75 microcontroller peripheral I/O port expander with 32K 8 EPROM.

■ 8.3 MEMORY OR INPUT/OUTPUT MAPPING

At this point we will look at two ways to identify and select an external device. These are memory mapping and I/O mapping. In this section we will also illustrate a simple control program to give you a flavor of interfacing software. In memory-mapped I/O each device has an address just like a memory location. Recall that the I/O registers in the 68HC11 have addresses. The following illustrates how this technique is used in some instructions. These instructions are part of a program to control an automobile turn indicator.

The input is address $1003. In our sample automobile, bit 0 is on only when the turn lever is pressed down to signal a left turn. Bit 1 is on only when the lever is up to signal a right turn. The output address is $1004. To turn on the left-turn indicator, turn on bit 6. To turn on the right-turn indicator, turn on bit 7. Hence, the following instruction inputs the lever status into accumulator A:

```
LDAA $1003 ;read turn lever status
```

The next two instructions check if a left turn is to be signaled.

```
ANDA    #%00000001    ;left turn signaled?
BEQ    NO_LEFT        ;if no, do something else
```

Note that the logical AND results in a nonzero value only if bit 0 is set, that is, if the driver moves the lever down. To turn on the left-turn indicator light, the following instructions can be used:

```
LDAA  $1004       ;get old output data
ANDA  #%01111111  ;turn off right signal indicator
ORAA  #%01000000  ;turn on left signal indicator
STAA  $1004       ;this sends out the new data
```

Note that this example also illustrates turning on one output only. This enables us to keep the other outputs in their present condition. Later we look at how bit manipulation instructions such as BSET and BCLR and the bit checking branches BRSET and BRCLR can be used.

The other method to identify external devices is input/output mapping, a technique used by some microprocessors, such as certain Intel and Zilog models. They have separate memory and I/O instructions. The I/O ports have separate addresses from the memory map.

The following illustrates how an Intel 8086 might handle the same turn indicator application.[2]

---

[2]An automobile controller would more likely use a microcontroller such as the Intel 8051 instead of the 8086 microprocessor. We use the 8086 as an example because the 8051 microcontroller uses memory-mapped I/O.

```
IN    AL, 03H          ;read turn lever status
AND   AL, 00000001B    ;left turn signaled?
JE    NO_LEFT          ;if no, do something else
IN    AL, 04H          ;get old output status
AND   AL, 01111111B    ;turn off right signal indicator
OR    AL, 01000000B    ;turn on left signal indicator
OUT   04H, AL          ;this sends out the new data
```

Note that the program uses the special instructions IN and OUT. It also illustrates other differences between Intel- and Motorola-type assembly languages. An Intel instruction indicates data movement from right to left. It uses postfix notation to indicate the numbering system. A pound sign (#) is not used to indicate immediate addressing as shown by the AND and OR instructions.

The IN instruction moves data from port 03 hex into register AL. The AND instruction performs a logical AND of register AL with a binary number. Intel does not use the term *branch*. The JE instruction is a relative jump. The next IN instruction loads the output port 04 hex into register AL. Then the OR instruction sets bit 6 of register AL. The OUT instruction transfers the contents of register AL to port 04 hex. To keep things simple, we have used two-hex-digit (8-bit) port addresses for the Intel example. The Intel 8086 can use four-hex-digit (16-bit) addresses as well with some extra instructions.

The bit manipulation instructions of the 68HC11 are useful for I/O control. However, these instructions use only direct or indexed addressing. A common technique is to use an index register to point to the memory-mapped registers. Consider Listing 8.1.

```
*Listing 8.1
*Demonstrate automobile turn signal control using
*bit manipulation instructions
*Note: Listing doesn't show all necessary pseudo-ops
*Assumes right signal already off


*Some useful equates

BIT0    EQU    $01        ;bit 0
BIT6    EQU    $40        ;bit 6
REGBAS  EQU    $1000      ;start address of register block
PORTC   EQU    $03        ;port C offset
PORTB   EQU    $04        ;port B offset

        LDX    #REGBAS    ;point to register block
*                         ;if left turn signaled
        BRCLR  PORTC,X  BIT0 NO_LEFT
*                         ;then turn on left signal indicator
        BSET   PORTB,X  BIT6
```

Note that these examples do not take into account causing the indicators to flash on and off.

C Code Examples. There are C compilers for most types of microprocessors. The C language is a *standard*, however, microprocessor-specific features such as their registers are not part of the C language standard. A C compiler for a specific microprocessor or microcontroller typically includes definitions to handle certain processor-specific features. The ICC 11 compiler includes data declarations that correspond to the I/O registers. To access these registers using C, it will be necessary to *include* the HC11.H header file that is part of ICC 11. To do this, use the directive statement #include <hc11.h>. C Listings 8.1a and 8.1b illustrate C programs to access I/O registers. Hence it is possible to access the registers by name, such as PORTB and PORTE.

```
/* C Listing 8.1a */

#include <hc11.h>

/* Declarations */
unsigned char In, Out;

void main(void)
{
    PORTB = Out;
    In = PORTE;
}
```

This program outputs data to port B and reads data from port E. The next program does the same thing as the turn signal program of Listing 8.1.

```
/* C Listing 8.1b */

#include <hc11.h>

#define BIT0 0x01
#define BIT6 0x40

void main(void)

{
/* if PC0 == 0 then PB6 = 1 */
    if ((PORTC & BIT0)==0)
            PORTB = PORTB | BIT6;
}
```

PRACTICE

Write software to turn on the rear brake lights when the driver presses the brake pedal. Use port A for I/O.

Solution

```
*Solution
*Turn on rear brake lights if brake pedal pressed
*Choose PA0 as input from pedal, active high
*Choose PA3,4,5 as output to lights, active high
```

```
REGBAS    EQU    $1000         ;register block address
PORTA     EQU    0
BPEDAL    EQU    %00000001     ;mask for brake pedal bit
BLIGHT    EQU    %00111000     ;mask for brake light bits


          LDX    #REGBAS       ;point to register block
*                              ;if brake pedal pressed
          BRCLR  PORTA,X BPEDAL NO_BRAKE
*                                ;then turn on brake lights
          BSET   PORTA,X BLIGHT
*                                  ;other instructions follow if
*                                  ;necessary
*                              ;else don't turn them on
NO_BRAKE
```

8.4 INTERFACING USING POLLING OR INTERRUPTS

Interfacing involves handling I/O, that is, reading inputs or sending outputs to peripheral devices. A peripheral device is connected to an I/O port using data lines. As shown in Figure 8.1, an MCU (subsystem) may use either polled I/O or interrupt-driven I/O when using an I/O port.

8.4.1 Polled I/O

Configuration.   Each subsystem has its own control register. It is the control register that determines whether the subsystem uses polled or interrupt-driven I/O. A bit in the control register selects polled I/O mode, usually by masking a specific interrupt. Recall (see Section 3.8) that an interrupt is masked by setting or clearing a corresponding mask or enable bit. Another bit may determine how the subsystem recognizes a request. Some subsystems may require additional control bit settings to specify further variations. *Configuration* refers to the step to set up a subsystem initially. It comprises instructions that tell the subsystem how to behave. You can also think of configuration as programming the subsystem.

Description.   The I/O transfer operation is as follows. The MCU checks periodically whether a peripheral requests servicing. This is known as *polling*. When it finds that a peripheral has requested service, the MCU performs the data transfer operation, either a read or write. An analogy to polling is the situation of checking whether anybody telephoned you while you were out. If you did not have a portable phone or answering machine, after returning you would have to telephone all your friends to check if any of them tried to call you. Next, we will examine the steps in a polling procedure.

Request.   An external peripheral may request service by pulsing an I/O request line (see Figure 8.1). The I/O request may be for a read or write operation. As shown in Figure 8.1, either a low-to-high (rising edge) transition or a high-to-low (falling edge) transition on the I/O request line causes a bit (called a REQUEST flag) in the status register to set (or reset). Also, it may be possible to program one of the control register bits to determine whether a rising or falling edge causes this action. This would have been done in the configuration stage.

Polling.   To determine whether or not a peripheral actually requires service, the MCU reads (polls) a status register periodically to check the corresponding REQUEST flag. If it indicates that an I/O request occurs, the MCU clears (or sets) the bit. This rearms the status register to detect the next request. The MCU then proceeds to handle the request as required.

Servicing.   Before we examine the service routine, let's see how the data register is used. For an input or read operation, the peripheral sent data to the I/O port. This data is passed on to the data register. When an output or write request occurred, the MCU will eventually write data to the data register. The data register contents are then sent to the I/O port. The MCU executes the I/O service routine. For a read operation, it could be something like "read the data register and pass the value to the main program in an accumulator." For a write operation, it might be an output routine that passes an accumulator value and writes it to the data register. The service routine may be an assembly language, C, or other language subroutine.

Further Notes.   The polling process is usually a loop. The program reads the status register. If it does not detect a set REQUEST flag, it repeats the loop. The program exits the loop only when a request has been detected. We can call this a *polling loop*. Another term is a *gadfly loop*. The gadfly is an insect that is continually bothersome. A person who continually bothers someone else until he or she gets the wanted answer is also know as a gadfly. A gadfly loop does not exit until it sees a request.

8.4.2   Interrupt-Driven I/O

Configuration.   A control register bit selects the interrupt I/O mode. This is usually done by enabling a specific interrupt. Another bit may specify whether a falling or rising edge triggers the interrupt (request). Other subsystems may require other control bit settings to be specified due to other options.

Description.   When a peripheral requests service, an interrupt is generated. The MCU suspends current processing to service the interrupt. The interrupt service routine handles the required I/O processing. It may read data from the peripheral or write data to it. After it is done, the MCU resumes execution of the interrupted program. Using the telephone analogy, it is like being at home when the telephone rings. You can respond immediately and talk to the caller, returning to your previous task once the telephone conversation is complete.

Request.   Like polling, an active transition on the request line sets the REQUEST flag. In this case it also asserts an interrupt.

Interrupt.   The MCU stacks the CPU registers and sets the I bit in the condition code register. If required, it resolves any interrupt priorities between multiple interrupts and fetches the vector for the highest-priority interrupt. Typically, each peripheral may have an associated interrupt vector. You may wish to refer back to Section 3.8 to review interrupt operation. In addition, you may wish to refer to Table B.2 and Figure B.2 in the Quick Reference (Appendix B).

Servicing.   The MCU executes the service routine addressed by the vector. If more than one device could cause the same interrupt, the MCU will poll other status bits to determine which device caused the interrupt. The MCU clears the REQUEST flag. It then completes the I/O operation, depending on the requirements. After completion, the MCU executes the return from interrupt (RTI) instruction and returns to the interrupted program.

Vector Jump Table.   When developing an application prototype, it is necessary to use an interrupt (and reset) jump table so that you can choose the start address of the service routines. To review how this is done, refer to Figure 3.8. One way to set up the jump table is to use a standard label for the contents of each vector. Header file hc11vec.h (Appendix E) shows a vector segment that reserves each label as an address pointer for each vector. A jump segment reserves 3 bytes for each pseudovector in the jump table. Also refer to the cross-assembly section in Appendix D. It illustrates the inclusion of hc11vec.h and a sample method of setting up the IRQ pseudovector. Note that in this case the IRQ service routine must begin with the label RIRQ.

■ 8.5 THE PARALLEL I/O SUBSYSTEM

This part of the chapter surveys different types of subsystems. The first one covered is the parallel I/O subsystem. Figure 8.2 illustrates the parallel I/O subsystem. Each line carries a bit of the data word. The data word is a byte (8 bits) for most ports. The diagram shows two of the parallel I/O ports for the 68HC11. Other ports can also be configured to be parallel ports if they are not used for another subsystem. Other microcontrollers would use similar techniques.

Port B is an output port. The data register address is $1004. Port C is shown as an input port. It is one of the ports that can be configured to be either input or output. Its data register has the address, $1003. In case 1 the MCU sends the data byte $A5 to the corresponding data register (address $1004). This sets the logic level of the corresponding port lines to high or low, depending on the state of each line's corresponding data register bit. In case 2 the MCU reads the port C data register (address $1003).

Applications.   Typical applications include programmable controllers. These are specialized industrial computers designed for a manufacturing plant environment. They are programmed for tasks such as reading sensor switch inputs. Based on the input status, they switch actuators such as motors and valves on and off.

Another application is a security system. It is similar to a programmable controller because it reads alarm sensor data to determine which (if any) alarms to turn on. A keyboard is another application. The fact that the subsystem reads the status of all keys may seem obvious. However, a keyboard may also use outputs from an I/O port. (We will examine keyboard systems in more detail in Chapter 9.)

A printer is another application because many printers use a microcontroller to read input from a computer and drive the motors and print head in the printer. (Printer interfacing is also described in Chapter 9.) The turn indicator program described in Section 8.3 is a parallel I/O application. Note how the example used the port B and C addresses.

■ 8.6 SERIAL SYSTEMS

Parallel I/O requires many data lines, one for each bit. This can increase system expense. Serial systems use a single line to transfer data bits one after the other. This reduces the number of lines but slows down the rate of data transfer. There are two basic types of serial systems: *synchronous* and *asynchronous*.

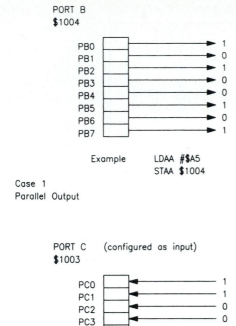

PORT B
$1004

PB0	→ 1
PB1	→ 0
PB2	→ 1
PB3	→ 0
PB4	→ 0
PB5	→ 1
PB6	→ 0
PB7	→ 1

Example   LDAA #$A5
            STAA $1004

Case 1
Parallel Output

PORT C   (configured as input)
$1003

PC0	← 1
PC1	← 1
PC2	← 0
PC3	← 0
PC4	← 0
PC5	← 1
PC6	← 1
PC7	← 0

Example   LDAA $1003
Then ACCA
has the value $63

Case 2
Parallel Input

Figure 8.2   Parallel I/O Subsystem

8.6.1   Synchronous Serial I/O Subsystem

Figure 8.3 illustrates the basic operation of synchronous serial data transfer. One device controls the data transfer. This is called the *master*. It communicates with other devices called *slaves*. Figure 8.3 shows only one slave device. This is a three-wire system in which each wire is used, respectively, for master-to-slave data, slave-to-master data, and synchronizing clock.

This example illustrates a *full-duplex* serial transfer since each device receives a new byte at the same time as transmitting a different byte. The principle of operation could be similar to a revolving luggage carousel in an airport arrival section. Two people standing at opposite ends could exchange luggage. Each person deposits his or her luggage onto the carousel. The rotating carousel carries the luggage to the other person for pickup. In the synchronous serial system each data transfer operation is a cycle to exchange data between the shift registers of two devices. This involves one device sending a byte to the other at the same time as receiving a byte from it. To output data, each device writes a byte to the shift

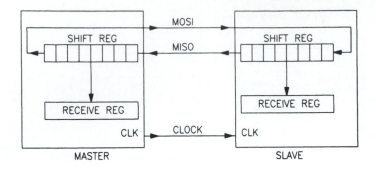

MASTER initiates data transfer

MASTER drives serial data clock to synchronize transfer

MASTER and SLAVE(s) typically are local

MISO = Master In, Slave Out

MOSI = Master Out, Slave In

Figure 8.3 Synchronous Serial I/O Subsystem

register. Then the master drives a clock for eight pulses. This shifts the byte from one device into the shift register of the other. With each clock pulse, one bit is shifted. After eight clock pulses, each device transfers the contents of their respective shift registers into their receive registers. The net result is that each device has transmitted and received. Note that the slave cannot initiate any transfers. (In Chapter 10 we will examine how more than two devices can communicate this way.)

Synchronous serial systems are typically used when all devices are local, such as in the same circuit board or the same machine. An automobile is an example of a machine that may use chips that communicate with each other using synchronous I/O.

More Terminology. We use the term synchronous serial I/O because all devices use the same clock signal. The data transfer is synchronized to the same clock reference. Recall that the system shown in Figure 8.3 is a type of communications system known as *full-duplex*. In full-duplex systems data is received and transmitted at the same time. Related terms are *half-duplex* and *simplex*. For half-duplex, a data line can either transmit or receive at any one time. For simplex, the data line is used only for transmitting or receiving.

Applications. Many chips use a standardized interface for synchronous serial I/O data transfer. Typical applications include display driving chips, data converter chips, and real-time clock chips. This serial interface can also be used as a control bus linking many devices together to control the same process or machine.

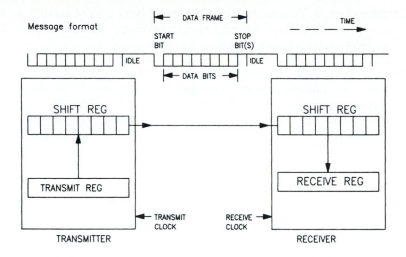

TRANSMITTER and RECEIVER typically are remote

TRANSMITTER and RECEIVER use separate clocks

RECEIVER samples bits at same rate that TRANSMITTER sends bits

START and STOP bits used to synchronize data transfer

Figure 8.4   Asynchronous Serial I/O Subsystem

8.6.2   Asynchronous Serial I/O Subsystem

Asynchronous serial is commonly used for computer-to-computer or computer-to-peripheral communications. Typically, the devices are remote, meaning that there is a considerable distance between them. For example, two devices may communicate over a telephone line. The term *asynchronous* refers to the fact that each device uses its own clock.

Figure 8.4 illustrates the operation of this system. Although the transmitter and receiver use separate clocks, the two clocks must run at the same rate. The transmitter can send data whenever it feels like it. But there has to be a way to inform the receiver so that it knows when data bits arrive. By convention, an idle line is always high. For each byte (or frame), the transmitter always sends a *start* bit (low) before the data bits. This wakes up the receiver, which has previously been sensing an idle line. The first low it detects tells it to expect data bits to arrive at the agreed-upon clock rate. For example, if the clock rate is 1 bit per millisecond, the receiver will sample the line every millisecond to input 8 bits after detecting the start bit.

The transmitter also ends each transmission with *stop* bits, which by convention are high. This ensures that the receiver can always detect the next start bit. When the receiver has shifted in 8 data bits, it detects the stop bits and waits for the next start bit before it repeats the process. Since an idle line is also high, it does not matter how much time there is between each framed byte. The use of start and stop bits is referred to as *framing* the data. The rate of bit transfer is known as the *baud rate*.

Since there is not one standard for the baud rate and the number of data bits per frame, a protocol has to be specified. Users of serial systems (transmitter and receiver) have to agree ahead of time which protocol to use. If they are not in agreement, errors will occur.[3]

Applications.   Typical applications include computer communication to peripherals such as modems, mice, instruments, printers, or to other computers. One of the most common standards is asynchronous serial. It is called the RS-232 interface. Virtually all computer-related equipment uses it. If you want to connect a factory controller to a telephone line so that it can send a report to management, there is a good chance that it will use RS-232. The EVB, EVBU, and EVM development boards use an RS-232 interface so that you can connect a terminal or computer to it.

■ 8.7 PROGRAMMABLE TIMER I/O SUBSYSTEM

A timer subsystem can perform a variety of functions, as shown in Figure 8.5. It handles some timing tasks so that the CPU is free to do other things. By using a timer the CPU does not have to execute software time-delay loops to keep track of time. This also allows it to measure time more accurately because interrupts will have no effect. Recall how a software time delay measures time. It simply loops for a set number of times before exiting. An interrupt will affect a software time delay because it will still loop the same number of times no matter how long the interrupt duration.

The tasks may involve timed outputs or measuring input times. The timer subsystem can generate a stream of pulses at different frequencies and pulse widths. It can also generate single pulses. The system can act as an internal timer without an output to start and stop a task at set times. The timer inputs are used to measure period (or frequency) and pulse widths of signals. (In Chapter 11 we will examine how to configure the control registers, preset time delays by writing to data registers, and read timing information from other data registers.)

Applications.   The automotive engine control system described in Chapter 1 used the programmable timer subsystem. It can measure the car's speed by counting speedometer pulses within a time period. A similar sensor measures crankshaft position. One of the outputs sends timed pulses to the ignition system.

Another application is to control the speed of a dc motor using a technique called pulse-width modulation (PWM). It effectively regulates the average power supplied by varying the percentage of time the motor is turned on. Programmable timer outputs can generate simple audio signals for speakers to make a variety of sounds. It can generate a reference timing signal for an instrument. A process controller needs to sample data at periodic intervals.

■ 8.8 ANALOG/DIGITAL I/O SUBSYSTEM

Many control systems use *analog* signals. That is, they use voltage signals that can be of any value within a specified range. An analog signal represents some physical quantity instead of a logic value. For example, 32 mV may represent the temperature 52 degrees Cel-

---

[3]Computer users may have experienced the need to make some settings when using a device with a serial interface. The settings specify baud rate, number of data bits and stop bits, and something else called parity. We cover parity in Chapter 10.

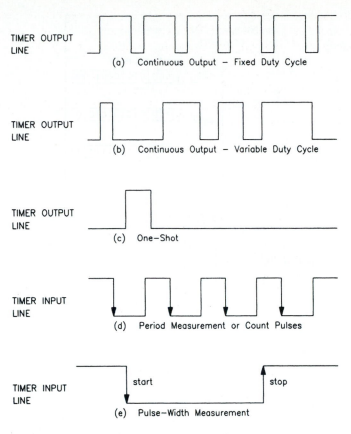

TIMER OUTPUT LINE

(a)   Continuous Output – Fixed Duty Cycle

TIMER OUTPUT LINE

(b)   Continuous Output – Variable Duty Cycle

TIMER OUTPUT LINE

(c)   One–Shot

TIMER INPUT LINE

(d)   Period Measurement or Count Pulses

TIMER INPUT LINE

start   stop

(e)   Pulse–Width Measurement

Figure 8.5   Timer I/O Subsystem

sius. Since the "outside world" is analog, there are many requirements when using analog signals. Figure 8.6 shows both an analog-to-digital converter (A/D) and a digital-to-analog converter (D/A). The A/D reads an analog voltage or current and converts it to a binary number to represent its value in volts. A D/A converts a binary number to an analog voltage or current.

In Chapter 12 we will look at how a binary number can represent analog information. For now we consider the simplest method. Note that the smallest binary number represents the minimum analog value. The largest binary number represents the maximum analog value. The other binary numbers proportionally represent the analog values within the range. The following shows the relationship for an 8-bit number representing the range 0 to +5 V:

Analog (V)	8-Bit binary
0	0000 0000
2.5	1000 0000
5.0	1111 1111

Despite the fact that conversions take time, the software to control A/D converters is usually quite simple. Configure the control register(s), then wait for a status bit to set and read the data register. The 68HC11 has eight built-in A/D channels, but it does not

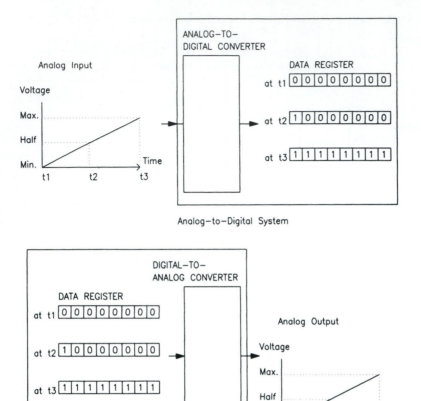

Figure 8.6 Analog/Digital I/O Subsystem

have a built-in D/A. It has to use peripheral chips or pulse-width modulation (see Section 12.3) for this purpose. Chapter 12 covers typical conversion processes for both types of converters.

A/D Applications. A *sensor* converts a physical quantity such as pressure to another quantity, usually voltage. The microcontroller can read analog sensors used to measure pressure, temperature, light intensity, and flow rate. It can check battery voltage for a portable unit. An alternative way to measure speed is to read voltage from a tachometer.

A microcontroller using an A/D can be a truly versatile voltmeter. It can measure an alternating wave at regular time intervals and calculate the root-mean-square (rms) magnitude. The voltmeter does not have to assume that it is measuring a sine wave at a set frequency. Note that the timer subsystem can trigger the MCU to take measurement samples during periodic time intervals.

Another A/D application is to monitor audio and video signals. This is not a typical application for a microcontroller such as the 68HC11. Specialized embedded controllers or digital signal processings (DSP) are more likely to be used (see the following section on embedded controllers).

Sophisticated telephone systems with programmable features may use a microcontroller with specialized chips for voice applications known as *codecs* (coders-decoders). A codec combines an A/D and D/A in one chip. Many telephone networks use digital signals for transmission because of reduced noise and other reasons. Since the human voice is analog, there is a need to perform both conversions.

D/A Applications.   Variable resistors are used to control the volume (sound intensity) of audio amplifiers. One way to control resistance is to change the voltage at the gate of a field-effect transistor (FET). An alternative way to generate variable-frequency square waves is to send a variable voltage to a voltage-to-frequency converter (V/F). Servomotors are dc motors used for positional control of disk drive heads, robot arms, rudders, and machining tools. The rotational position is controlled by an analog signal. A digital system can control a servomotor using a D/A.

Embedded Controller or Microcontroller?   At this point we will mention more terminology. There may be confusion when some people talk about embedded controllers or microcontrollers. Sometimes they might be talking about the same thing. But most often, an embedded controller is any kind of programmable processor used for real-time control. We will use the term *microcontroller* to mean a general-purpose microcomputer system on a chip that can be used for a variety of applications. It is one type of embedded controller. An embedded controller can also be a specialized device designed for a specific category of applications. Video displays, hard disk drives, and signal processors often use embedded controllers.

In Chapter 1 we called a microcontroller a computerized Swiss army knife, since it is a general-purpose tool. You can consider special-purpose embedded controllers to be like tools for specific types of work such as a knife set, screwdriver set, and so on. The digital signal processor (DSP) is a special type of embedded controller more appropriate for applications using varying-frequency signals. Audio and video processing systems use DSPs. Other examples are diagnostic systems that monitor vibrations to analyze performance of machinery.

■ 8.9 THE I/O SUBSYSTEM REGISTERS

Earlier we introduced the general categories of registers used by a subsystem. We will now return to this topic. The intention of this section is to provide the groundwork for the material in later chapters. This section does not refer directly to the 68HC11, but the concepts should help you to interpret data books for other microcontrollers. You may find it confusing at first. There is a possibility that it will make more sense after examining the examples in the later chapters, so please be patient.

For a quick review, recall that a subsystem uses registers for control, status, and data. This is a very broad statement. As mentioned earlier, each subsystem may use different terminology and organization. This is partly because each subsystem has some unique requirements. Our quick survey of the subsystems should make this clear. For example, an asynchronous serial subsystem's framing requirement has no equivalent in an analog converter subsystem. The change in terminology becomes greater when dealing with other chips, particularly those supplied by other manufacturers. Many do not use the word *subsystem*.

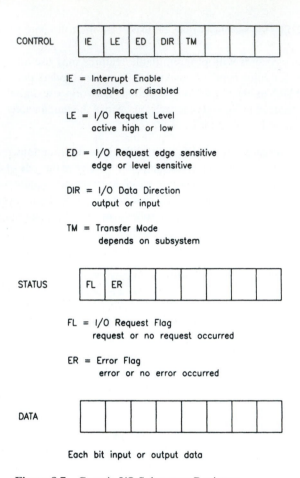

Figure 8.7   Generic I/O Subsystem Registers

Since the electronics industry is developing continually, there will be different techniques to control I/O subsystems. To sum it all up, refer to Figure 8.7. It attempts to show the typical usage of registers. Basically, each subsystem uses these registers in one form or another.

For specific information concerning the 68HC11, refer to Table B.1 in the Quick Reference (Appendix B). It lists all the registers and bits for the I/O subsystems. Table B.1 also lists other registers and bits to control the CPU.

8.9.1   Control Register and System Configuration

The control register may have a bit to *enable* an interrupt. In some systems it must be a logic 1 to enable an interrupt. In others, it must be a logic 0. In this case the bit may be called a *mask* bit. To mask an interrupt, set the mask bit. Interrupt masking and enabling are opposites. Some systems use one term and some use the other. In our generic control register, this is an interrupt enable bit (IE). Generally, a bit label describes what it does when the bit is high (when it is active). To allow (enable) interrupts, a system may set an enable bit or clear a mask bit. To prevent (mask) interrupts, a system sets a mask bit or clears an enable bit.

TABLE 8.1 ACTIVE REQUEST CONFIGURATION

LE	ED	Action causing request
0	0	Low level
0	1	Falling edge
1	0	High level
1	1	Rising edge

Subsystems with multiple sources of interrupts such as timer subsystems may have extra registers known as *mask registers*. They contain only mask bits. In the case of the 68HC11 timer subsystem, the status registers are known as *flag registers*.

Some systems allow you to define what is considered to be an active transition or event. Does it react to a rising edge or a falling edge? Perhaps it is level sensitive? When the defined transition occurs, a *flag* bit sets in the status register. The generic control register uses the edge-sensitive (ED) bit and the level (LE) bit for this purpose. Table 8.1 shows how these bits define an active signal.

In our case the generic control register uses a single bit (DIR) for direction. Some microcontrollers employ a second "control" register known as a *data direction register.* Each bit specifies the direction of a corresponding data line in a port. If a bit is 0, the data line is an input. If it is a 1, the data line is configured to be an output. Parallel I/O subsystems may have a data direction register.

The transfer mode (TM) bit merely identifies the fact that each type of subsystem has its own techniques and characteristics. Typically, it specifies alternatives for subsystem behavior. For example, it may specify whether a device initiates a transfer or must wait for another device to initiate the transfer. Admittedly, this is vague for a generic subsystem. A specific subsystem uses it for a more definite purpose. Also, it would probably have a different name for the bit (or group of bits).

Configuration.   Before a program uses polling or interrupts to handle an I/O request, the subsystem must be initialized (configured or set up). Initially, it runs a section of code or subroutine that specifies what system options to use. This section of the program performs what is called configuration. If a configuration routine is not executed, the system uses what is called the *default* values. Default values are what the registers contain when the MCU is first powered up. Every time the 68HC11 is $\overline{\text{RESET}}$, it loads most of the registers with default values. The *Data Book* specifies the default values. If you wish to use other values, you have to reconfigure the subsystem every time the system is $\overline{\text{RESET}}$.

To configure a subsystem, the program writes data to the appropriate control register(s), data direction register, or mask register to set or clear the different control bits as required. We can call this data a "magic number." Some may call the configuration procedure a "ritual."

Let's look at an example of configuring the generic system. Let's say that we want to make the port an output. An interrupt will occur when the I/O REQUEST line is high. Then we can set up the control register as follows:

Control Register Configuration

B7	B6	B5	B4	B3	B2	B1	B0
IE	LE	ED	DIR	TM	.	.	.
1	1	0	1	x	x	x	x

If we make the logical don't cares (x) zero, to configure this subsystem we can use the magic number $D0 as follows:

```
LDAA    #$D0
STAA    CONTROL_REG
```

If you are using C to write the application program, you will probably use a library of subroutines (functions). Some of the subroutines are used to input or output data using a particular subsystem. Before these subroutines can be used, the subsystem has to be configured. The manual supplied with the C compiler should document how to do this for each subroutine. Other C libraries will contain a selection of configuration subroutines for different requirements. The application program needs to call these subroutines only once before executing the I/O subroutines as often as necessary.

8.9.2   Status Registers

Many status register bits are called *flag* bits. A flag indicates the status. Before radio and telegraph, people used flags for communication beyond voice range. The messages were simple and direct, such as "It is safe to approach."

Software performs two basic operations with the status register. It reads the register to check which flag bits have set (active-high flag). Then it must clear the flag bits so that it can detect the next time it sets. Refer back to Figures 8.1 and 8.7. The I/O request flag, FL, sets (changes from 0 to 1) if an I/O request has occurred. Remember that the control register defines what type of input is considered to be a request. As noted earlier, the I/O service routine must clear this bit so that a future request can be detected.

In some 68HC11 subsystems writing or reading a data register clears a flag. In others it involves reading the status register, followed by accessing a data register. In the timer subsystem the service routine must write a 1 in order to clear the flag bit. This sounds opposite to what you would normally expect, but this is how it works. If you look at the *Data Book* for details on some registers, such as TFLG1 and TFLG2, it will state:

```
(Write with bit[s] set to clear corresponding flag[s].)
```

The generic status register uses an error bit, ER. The two serial I/O subsystems have several error flags. Each indicates a particular type of error.

There are other variations. We mentioned that the timer subsystem uses flag registers. A parallel I/O subsystem may not use a separate status register. Some of the control register bits may indicate status. Let's say that the generic subsystem clears a status flag when the status register is read followed by any access of the data register (read or write). Let's continue the generic example begun in Section 8.9.1. Consider what the service routine could do after a high input to the request line caused an interrupt. The status register would be as follows:

Status Register

B7	B6	B5	B4	B3	B2	B1	B0
FL	ER	.	.	.	.	.	.
1	x	x	x	x	x	x	x

To begin the flag-clearing sequence, the program can do the following:

```
LDAA STATUS_REG
```

We show the rest of the sequence in the next section. After the sequence is completed, the subsystem can detect another request in the future. The service routine, whether it is a C library or assembly language subroutine, will contain the code necessary to handle the status register.

8.9.3   Data Registers

This register holds the input data that has been received or the output data that was most recently sent out. Subsystems may have several data registers. The 68HC11 A/D subsystem calls them *result registers*. The timer subsystem calls them *capture registers* for inputs and *compare registers* for outputs. The parallel I/O subsystem simply calls them *port registers*. Table B.1 in the Quick Reference (Appendix B) lists them all.

Let's say that in our generic example, for some reason or other, the service routine needs to drive port lines 3 and 4 high and the rest low. We then have the following:

Data Register

B7	B6	B5	B4	B3	B2	B1	B0
0	0	0	1	1	0	0	0

The service routine could use the following instructions:

```
LDAA    #$18
STAA    DATA_REG
```

Remember that the system is an output in this case. For an input port an I/O service routine would load in the data register.

Note that our service routine clears the status flag FL as a result of performing actual I/O. In general, it is important that any service routine clears the flag sometime before returning control back to a higher level of the program. Remember that other subsystems may have to use different status flag clearing sequences.

The same port may have both input and output lines. In this case writing to the data register affects only those bits that correspond to outputs. A write to a physical input has no effect. However, with the 68HC11 this write information is latched for output at a later time (if that pin's data direction control bit is changed to an output). The other data register bits still indicate the physical input line logic. Reading this data register results in reading the physical inputs and what was sent out as output.

PRACTICE

For the generic subsystem, write software to read the input if the request line is high and no error occurs. It will also handle an error condition if one is detected.

Solution

```
*Solution
*Choose polling

*Configuration step
    LDAA    #$40          ;LE = 1, rest = 0
    STAA    CONTROL_REG

*Polling loop (request status check)
POLL
    LDAA    STATUS_REG
    BITA    #$80          ;if FL == 0 then continue loop
    BEQ     POLL

*Error status check
    BITA    #$40          ;if ER == 0 then no error occurred
    BEQ     GET_DATA      ;read data (if no error)
    JMP     HANDLE_ERR    ;otherwise handle error condition
*                         ;which should also clear error flag, EL

*handle I/O
GET_DATA
    LDAA    DATA_REG      ;read operation
*                         ;also clears I/O request flag, FL
```

■ 8.10 INTERFACE STANDARDS

Many manufacturers build different computers, controllers, and support peripherals. To interconnect all of these different types of equipment, it was necessary to agree on interface details. Otherwise, a user (customer) would have to rely solely on one manufacturer.

An interface standard is a document or agreement specifying some interface details. They typically specify mechanical connections, pin assignments, signal levels, signal timing, and framing formats. Manufacturers of equipment using interfaces agree to follow the specification of a standard. It may happen that one manufacturer has developed an interface that has become popular with others. In this case a technical association may decide to adopt this as a standard for others to follow. Alternatively, it may not be adopted by a technical association, but the standard is still widely used. This type of standard is known as a *de facto* standard. At other times a technical association develops a standard. Other companies choose to follow the standard.

There are too many standards to list in this book. Also, it is a dynamic field. Every year sees the introduction or adoption of new standards. Others fall out of use. Some standards are intended to be used within a single organization. However, most are intended to be used by every organization involved in building or using a type of interface. Interface standards typically restrict themselves to a narrow category. For example, there is a standard that is used solely for interfacing a hard disk drive to a specific class of personal computers.[4] We will introduce some of the more common standards as the need arises.

The purpose and benefits of a standard are to ensure that equipment from one manufacturer can communicate with that of another. A technical association is made up of representatives from government, universities, and different companies in industry. Of course, there are no police to force manufacturers to comply with these standards. Sometimes a large company tries to impose its own standard, with the aim of increasing market share.

Technical associations developed to avoid problems of "commercial bullying" and to coordinate the efforts of the various parties, to avoid duplication of effort. This is particularly necessary at the international level when dealing with different sovereign states. These technical associations have become standards organizations.

We will introduce some standards organizations as the need arises. For now we mention three of them: the Institute of Electrical and Electronics Engineers (IEEE), the American National Standards Institute (ANSI), and the International Standards Organization (ISO). We will have a bias toward standards commonly used in North America.

■ 8.11 SUMMARY

An interface matches the output of one device to the input of another. The interface itself consists of the hardware (circuit and connections) and the software (device service/handler routine) to send information from one device to another. Microcontroller units (MCUs) process information from the outside world and control things by sending commands to them. The information comprises input signals. The commands are output signals. An MCU may also convert input signals to display the information for human use. These signals are commonly known as input/output (I/O).

In this chapter we discussed interfaces used for I/O as opposed to those used for memory (Chapters 5 and 7). There are several categories of I/O interfaces that *we* call subsystems. These may be logical parts in an MCU or individual chips. A subsystem can handle I/O using polling or interrupts. Each subsystem has one or more ports. A port is a collection of pins used to input or output data. Also, subsystems use registers to perform the functions of control, status checking, and data transfer. A data register stores the data for a corresponding port (or port line).

The I/O subsystems can be configured. In other words, they are programmable. The software must first configure the subsystem by writing a configuration word (magic number) to a (some) control register(s). The logic state (1 or 0) of each bit selects or deselects a subsystem option. Generally, the software writes to the control register once after reset. After this, it usually uses only the status and data registers to handle I/O with a peripheral device connected to the port.

When an I/O peripheral device requests service, the subsystem hardware automatically sets a corresponding bit, called a flag, in a status register. If the subsystem was configured for interrupts, an interrupt is generated automatically each time the flag sets. If it was configured for polling, the software must read the status register periodically (poll it)

---

[4]This is the *intelligent drive electronics* (IDE) interface. It is used for computers with an AT bus. Yes, "AT" is another standard. This does not mean that there are no other hard-disk interface standards.

to find out whether or not the flag has set. Although interrupt-driven I/O and polling detect a request differently, they both handle the request the same way. When a request is detected, the service routine clears the flag by writing to the status register before or after using the data register. It reads the data register for an input operation or writes to the data register for an output operation. In some subsystems the flag is cleared differently.

A typical MCU such as the 68HC11 has built-in subsystems for parallel, serial, programmable timer, and analog I/O. Since many computer peripherals use microcontrollers, we will examine how they use the various subsystems. A keyboard uses parallel I/O to check which key has been pressed. The communications board uses asynchronous serial I/O to transmit data to a remote device or to receive data from it. A programmable timer can schedule events the computer should perform periodically. A computer mouse uses it to count encoder pulses to keep track of its position. A joystick uses analog-to-digital converters to measure potentiometer signals, which in turn indicate stick position. A data acquisition system may use synchronous serial I/O to communicate with the chips in its circuit. Note that both the mouse and joystick use parallel I/O to sense button position. A mouse may use serial I/O to communicate with the computer.

Due to the variety of possible mechanical connections, signal functions, and pin assignments, it is necessary to standardize some interfaces. A standards organization typically defines the standard. Sometimes one manufacturer's interface becomes adopted as a *de facto* standard. In this industry it is necessary to be familiar with the use of the standards applicable in your line of work.

The chapters that follow in Part 4 cover the subsystems in more depth.

■ EXERCISES

1. In an MCU system, what do we call data that is written to a peripheral device?

2. What are other terms for the verb *input*? What are other terms for the verb *output*?

3. What is an I/O subsystem?

4. Name three types of hardware used to build an I/O subsystem or interface.

5. Draw a block diagram of a typical I/O subsystem.

6. List the addresses for the following registers.
   (a) PORTD
   (b) TOC4
   (c) SPCR
   (d) ADR2

7. List the 68HC11 ports and identify whether the lines are input only, output only, or either.

8. Do you think a typical MCU application requires more inputs than outputs? Explain why.

9. An MCU controls a vending machine to dispense hot drinks. Can it use polling to handle I/O?

10. Your company is designing a new "energy-conscious" vending machine to be sold in remote areas without electricity. It will run on batteries. Would you use polling in the MCU program to handle I/O?

11. Port A has the following pins:

PA0	Input
PA1	Input
PA2	Input
PA3	Output
PA4	Output
PA5	Output
PA6	Output
PA7	Configured as output

Presently, PORTA has the value $EE. What is the value of accumulator A and PORTA after the following instructions are executed?

(a) LDAA #$29
(b) STAA $1000
(c) LDAA $1000

12. What are the advantages and disadvantages of serial I/O compared to parallel I/O?

13. Using Figure 8.3 as a guide, describe how 3 bytes are transmitted.

14. Which serial I/O method has a potentially higher data transfer rate?

15. An MCU uses a 2-MHz clock. If a timer register is a 16-bit register and it increments (counts up) every E cycle, what is the longest period of time that it can count before resetting to zero?

16. Name two methods to write a program to run a 24-hour clock.

17. In one MCU system it takes 15 E cycles to send a new output to an 8-bit D/A. Conversion requires 15 ns. What is the maximum frequency for a sawtooth wave that can be generated using a 68HC11 running with a 2-MHz clock?

18. An input of $+5$ V to an A/D produces an output of \$FF (digital). An input of -5 V produces an output of \$00. What will be the digital output if the input is 0 V? Questions 19 and 20 both use the generic subsystem described in Chapter 8.

19. The MCU will read the port when it detects the falling edge of a positive pulse on the I/O request line. It will use polling. Use pseudocode and stepwise refinement to describe a program to do this.

20. The leading edge of a positive pulse causes the MCU to write data to the port. It will use an interrupt. Use pseudocode and stepwise refinement to describe a program to do this.

9 Parallel Input/Output

OBJECTIVES

After completing this chapter, you should be able to

- Design the hardware and software interface for discrete switch inputs and LED outputs.
- Design the hardware and software interface for a seven-segment display.
- Design software for LCD interfacing.
- Design the hardware and software interface for a matrix keyboard, including the handling of debouncing.
- Describe the operation for strobed and handshake I/O.
- Write programs using strobed and handshake I/O.
- Write programs to implement some standards such as Centronics parallel interface, IEEE-488, and SCSI.
- Design circuits for HCMOS-TTL interfacing and wired-AND/OR with open-drain outputs.

Two themes present themselves throughout most of this chapter. One is to present the specific implementation of parallel input/output (I/O) as handled by the 68HC11 IC. The other is to examine some common parallel I/O applications. These can be implemented differently, depending on the microcontroller or other types of digital ICs. Note that other types of microcontrollers have similar I/O. We present the two together to show the reasoning behind the design of the 68HC11 parallel I/O subsystem.

In the first half we will examine how the microcontroller can be interfaced to some common devices. These are the seven-segment light-emitting diode (LED) display and liquid-crystal display (LCD) to provide visual output for a person to read. Next, we see how input and output lines are used for a keyboard interface that allows users to enter data into an MCU system.

The second half explores how a parallel interface can be used for communication between computers and peripherals. Strobed I/O is a simple technique for communication involving one control line. More often, handshaking is used. The devices use control lines to ensure that one device has received the data sent by the other.

The 68HC11 has one type of built-in handshaking system. We will look at how to use the applicable registers and ports. We will also examine the signal sequence. There are other implementations of a handshaking interface. In the section on parallel interface standards, we will examine the timing and logic for some of the commonly used implementations. It illustrates the handshaking sequence used in the Centronics parallel printer interface, IEEE-488 General Purpose Instrumentation Bus (GPIB), and the Small Computer Systems Interface (SCSI).

■ 9.1 INTRODUCTION TO THE SUBSYSTEM

The 68HC11 has five 8-bit ports. Any of these can be used as parallel I/O ports, depending on how their respective control registers are configured. Of these, ports B and C can only be used for parallel I/O. The others can be configured for use by other subsystems if desired.

Any of the port C and D lines can be used as input or output, depending on how their respective data direction bits are configured. Pin 7 of port A (PA7) is also bidirectional. Its data direction bit (DDRA7) is bit 7 of the pulse accumulator control register (PACTL). When port A is used for parallel I/O, lines 0, 1, and 2 are input only and lines 3, 4, 5, and 6 are output only. Port B is used for output only. Port E is an input port when used for parallel I/O.

The term *general-purpose I/O* is used to describe digital I/O handled simply by reading or writing to parallel I/O ports. Hence we can also say that all ports can be used as general-purpose I/O. Although ports B and C are used only for parallel I/O, we shall see that they can be configured for specialized parallel I/O functions. We will cover these functions— strobed I/O and handshaking—later in the chapter. First, we will cover general-purpose I/O using two applications, output to a digital display and keyboard input.

To illustrate the use of general-purpose I/O, consider the use of port C and its associated data direction register as shown in Figure 9.1. The circuit shows light-emitting diode (LED) outputs and switch inputs connected to the port C pins. Each associated bit in the port C register will indicate the corresponding switch status or drive the corresponding LED. Note that driving an LED high will turn it off (reverse bias to prevent current from flowing) and driving it low will turn on the LED (forward bias to permit current to flow). Also note that each switch has a pull-up resistor so that the voltage at a port C input is 5 V when the switch is open. When a switch is closed, the corresponding port C input is pulled low to ground, so the voltage is 0 V.

Each bit in the data direction register (DDRC) specifies the direction of the corresponding pin in port C. To set up the data direction as shown in Figure 9.1, we can use the following instructions based on using the library file HC11REG.ASM listed in Appendix E:

```
LDAA   #$0F   ;Configure port C PC7,6,5,4 as inputs, PC3,2,1,0 as outputs
STAA   REGBAS+DDRC   ;by writing %00001111 to data direction register
```

To drive PC0 and PC1 high and PC2 and PC3 low, we could use the following instruction:

```
LDAA   #$03   ;Write %00000011 to drive PC0, 1 high and PC2,3 low
STAA   REGBAS+PORTC
```

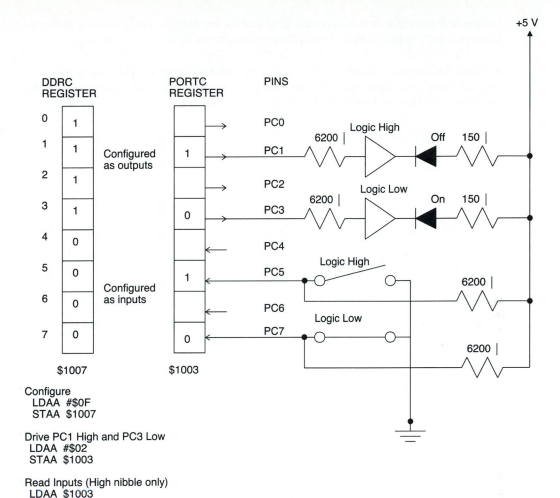

Figure 9.1 General Parallel I/O Example

Note that the high nibble of the data sent to the port C register (PORTC) is a logic don't care. This is because bits in PORTC for pins configured as inputs are read only. Now let's say we want the program to branch somewhere if the input to PC5 is high. We could use the following instructions to branch to location PC5HI:

```
        LDX        #REGBAS       ;Point to register block
        BRSET      PORTC,X $20 PC5HI
        ;other instructions
PC5HI
        BCLR       PORTC,X $02
```

Since bit checking and manipulation instructions such as BRSET use either direct or indexed addressing, we can use an index register to point to the block of registers. You might want to refer to Table B.1 in Appendix B to see how the registers are addressed and to the listing for HC11REG.ASM in Appendix E to see how the symbols are defined.

C Code Examples.   As described earlier in Section 8.3, it is possible to access the registers by name when developing C code using ICC11. The following is an example. Remember to *include* the header file `hc11.h`.

```
/*C Listing 9.1 */

#include <hc11.h>

void main(void)
{
    DDRC = 0x0f; /* configure data direction */
    PORTC = 0x03; /* drive PC0, 1 high, PC2,3 low */
    /* if PC5 == 1 then PC1 = 0 */
    if ((~PORTC & 0X20) == 0)
            PORTC = PORTC & ~0x02;
}
```

It so happens that ICC11 also provides a convenient way to access bits. The header file `hc11.h` provides a definition for bits 0 to 7 as `bit(0)` to `bit(7)`. C Listing 9.2 illustrates this format.

```
/*C Listing 9.2 */

#include <hc11.h>

void main(void)
{
    DDRC = 0x0f; /* configure data direction */
    PORTC = 0x03; /* drive PC0, 1 high, PC2,3 low */
    /* if PC5 == 1 then PC1 = 0 */
    if ((~PORTC & bit(5)) == 0)
            PORTC = PORTC & ~bit(1);
}
```

■ 9.2 SEVEN-SEGMENT (LED) DISPLAY OUTPUT

9.2.1   The Seven-Segment Display

Many systems have display requirements for people to get alphanumeric information. These include instrument panels, calculator displays, cockpit instruments, and various clock and watch displays. Many of these use liquid-crystal displays (LCDs) and light-emitting diodes (LEDs). The LCD is more versatile and is more commonly used. We cover the LED type first and in more depth because it is easier and cheaper to use. A later section will cover LCD interfacing.

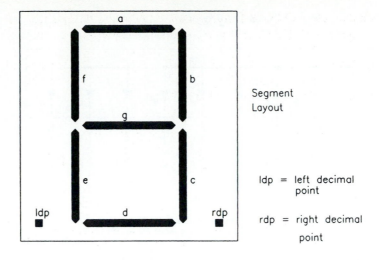

Segment
Layout

ldp = left decimal
point

rdp = right decimal
point

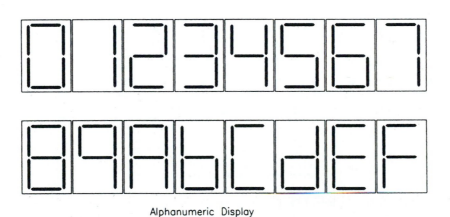

Alphanumeric Display

Figure 9.2 Seven-Segment Display

A seven-segment display is a common name for a type of LED. For example, the channel selector display in your television set may use this type of display. Some LCD displays are also known as seven-segment type. Figure 9.2 shows the layout of the seven-segment display. Each segment is an LED. Seven segments form a pattern that looks like the digit 8. A character can be displayed by illuminating some of the segments in the group of seven. In reality, most displays include extra segments to be used as decimal points. However, they are still known as seven-segment displays. Other displays have a different segment layout. For example, one type has a plus (+) and minus (−) sign layout. Even though it is not always accurate, the name "seven segment" has remained in use.

There are two types of seven-segment LEDs: the common anode and the common cathode (Figure 9.3). Recall that an LED is illuminated by applying a forward voltage across it from anode to cathode. This is called forward biasing the LED. (You may wish to refer back

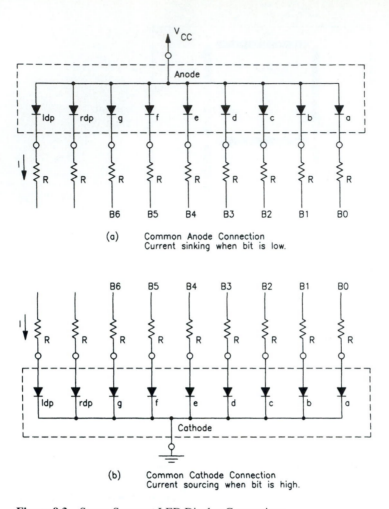

Figure 9.3 Seven-Segment LED Display Connections

to the discussion of LEDs in Section 6.2.4.) Note that each LED segment is connected to a current-limiting resistor. To display a character, forward-bias the appropriate segments. For example, forward-bias segments b and c to display the digit 1. Each character can be represented by a binary code. This code depends on the type of seven-segment display and its connections.

We consider the common anode type first. If a logic low occurs at any segment pin, the corresponding segment is illuminated since it is forward biased. If a logic high occurs at any segment pin, the corresponding segment is not illuminated because the segment LED is reverse biased. Common cathode operation is opposite to common anode operation.

9.2.2   Seven-Segment Code

Each display shown in Figure 9.2 is generated when a specific bit pattern is sent to the segment pins. Figure 9.3 shows where the output lines (B0, B1, . . .) are connected for each

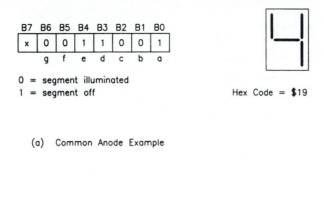

0 = segment illuminated
1 = segment off

Hex Code = $19

(a)   Common Anode Example

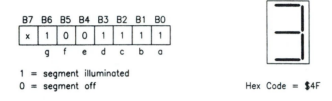

1 = segment illuminated
0 = segment off

Hex Code = $4F

(b)   Common Cathode Example

Figure 9.4   Seven-Segment Code Examples

type. Each character has a corresponding binary (or hex) code. Figure 9.4 shows some code examples for both types of displays. These examples apply for the connections shown in Figure 9.3. Note that only 7 bits are used. The most significant bit (MSB) is a logical don't care (x) because it is not connected.

9.2.3   Driving the Seven-Segment Display

In this section we will remind you that current-limiting resistors must be included. If they are omitted, a forward-biased condition will short-circuit the LED and cause it to draw too much current. This will damage the display and the output driving it. Common cathode seven-segment displays are driven by current sourcing (Figure 9.3[b]). Common anode types are driven by current sinking. Figure 9.5 shows a common cathode display driven by current sourcing. The attentive reader may ask: "Aha, why are there no current-limiting resistors?" The answer is that the MC14495-1 IC (Hexadecimal-to-Seven Segment Latch/Decoder ROM/Driver) has built-in current-limiting resistors.

Two of the important LED parameters are forward voltage (V_F) and forward current (I_F). Nominal values are

$$V_F = 1.7 \text{ V}$$
$$I_F = 15 \text{ mA}$$

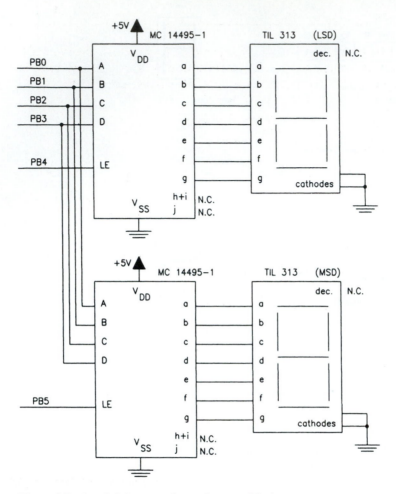

Figure 9.5 Parallel Output to Seven-Segment Displays

With these values there is sufficient illumination for visibility. A display driver chip such as the MC14499 can be used to supply the required source current. From this information we can determine appropriate current-limiting resistor values.

The driver chip logic high level, V_{OH}, is nominally 4.2 V. From this we can determine the resistor value using Ohm's law:

$$R = \frac{V_{OH} - V_F}{I_F} = \frac{2.5\ V}{15\ mA} = 170\ \Omega$$

Many LED circuits use 180-Ω resistors since these are commonly stocked by electronic suppliers.

PRACTICE

Write software to display the digit 5 when port C is connected to a noninverting buffer. The buffer drives a common anode seven-segment display. The data lines are connected as shown in Figure 9.4(a) except that one decimal point is also connected.

Solution

```
*Solution
*Display the digit '5' using seven-segment display

*Define data and register addresses

*seven-segment code for '5' without decimal point
*dp g f e d c b a
*1 0 0 1 0 0 1 0 = $92

*0 = segment on, 1 = segment off

DISP5     EQU     $92

*define registers used in this program

REGBAS    EQU     $1000     ;register block pointer
PORTC     EQU     $03
DDRC      EQU     $07

BEGIN
          LDX     #REGBAS   ;point to registers
*                           ;configuration, make port C an output
          BSET    DDRC,X$FF
          LDAA    #DISP5    ;send code to port C
          STAA    PORTC,X   ;to drive display
```

9.2.4   Software Seven-Segment Code Look-Up

If your interface circuit uses only buffers and resistors to connect an output port to a seven-segment display, you will have to use software to generate the character codes. The program can do this by using a look-up table. The contents of the table depend on the application and type of display. For example, to display hex digits for a common anode display, entry 4 would have the byte $19 (Figure 9.4). Listing 9.1 shows program code to send output to a display.

```
*Listing 9.1
*Display one digit, assume that DATA high nibble is zero

LDX     #TABLE    ;point to table
LDAB    DATA      ;get data in hex
ABX               ;point to its seven-segment code
LDAA    0,X       ;get seven-segment code
STAA    OUTPUT    ;send it to output
```

OUTPUT can be a parallel I/O port, such as port B at address $1004. TABLE can be an EEPROM location to mark the beginning of a 16-byte block.

9.2.5   Hardware Decoder

Some ICs are specially designed to drive seven-segment displays. They contain buffers to supply the required drive currents and latches to store data. Some include a decoder. In this case the IC accepts a binary number, decodes it, and drives the appropriate segments to display the number. The MC14495-1 is one example (Figure 9.5). Although decoders remove some of the burden from the software, there is some loss of flexibility. It will not be possible to display custom characters. For example, the MC14495-1 does not support a blank display or the use of decimal points.

9.2.6   Multiplexed Displays

A straightforward way to connect multidigit displays is to dedicate port pins for each display. To save pins, one can use *multiplexed* displays. Recall that *multiplexing* refers to using the same signal channel for more than one signal.

In Figure 9.5 the low nibble of port B outputs the hex character for both displays. The lines PB4 and PB5 control which display is updated. The MC14495-1 has a 4-bit latch. When its latch enable (LE) is high, the IC decodes what is currently in the latch to drive the display. Inputs at pins A, B, C, and D are ignored while LE is high. When LE is low, these inputs are loaded into the chip's latch. When LE rises, the data is latched in and the display shows the new data. The data remains stored in the latch as long as LE is high.

In the circuit shown in Figure 9.5, lines PB4 and PB5 are normally high. To display a hex byte, the software must send the two nibbles (4 bits) separately. To send the least significant digit (LSD), it masks out the upper nibble. It writes the nibble to port B. Then it pulses PB4 low, then high again. To display the upper nibble, it shifts these 4 bits into the lower 4-bit position. It writes these 4 bits to port B, except that PB5 is momentarily pulsed low. If both PB4 and PB5 are pulsed low, the nibble is sent to both displays. Listing 9.2 shows the pseudocode.

```
*Listing 9.2
*Pseudocode to display a byte to seven-segment display

Set PB4 = 1, PB5 = 1
Set up port B to display the MSD as follows
    Shift high nibble of byte into low nibble of port B

Toggle PB5 low then high again, the MSD is latched in

Set up port B to display LSD as follows
    Shift low nibble of byte into low nibble of port B

Toggle PB4 low then high again

*end of pseudocode
```

Listing 9.3 shows a sample software driver for this display interface. Again (see Listing 8.1), we illustrate the use of the bit manipulation instructions using indexed addressing mode. Index register X (IX) is preloaded with $1000 to point to the register block before the subroutine is called. A program can use subroutine OUTSEG to display a hex digit.[1]

```
*Listing 9.3
*Subroutine OUTSEG
*Displays a hex byte on a multiplexed
*two-digit seven-segment display
*driven by two MC14495-1 ICs connected to port B
*Calling Registers
*          IX = register pointer
*          ACCA = hex byte to be displayed
*Return Registers
*          none, only CCR affected

*NOTE: The following segment has already been defined.
*          See "Cross Assembly," Appendix D.
*          PORTB label defined in header file HC11REG.ASM.
*---------------------------------------------------
UTILITY    EQU   $0180
           ORG   UTILITY
OUTSEG

           PSHB              ;preserve ACCB
           TAB               ;set up to display MSD first
           LSRB              ;shift high digit into low nibble
           LSRB
           LSRB
           LSRB
           ORAB  #$D0        ;disable LSD, enable MSD
*                            ;i.e., PB4 = 1, PB5 = 0
           STAB  PORTB,X     ;and display MSD
           BSET  PORTB,X $20
*                            ;latch in MSD, i.e., PB5 = 1
           TAB               ;set up to display LSD next
           ANDB  #$0F        ;mask out and disable MSD and
*                            ;enable LSD
           ORAB  #$E0        ;i.e., PB4 = 0, PB5 = 1
           STAB  PORTB,X     ;and display LSD
           BSET  PORTB,X $10
*                            ;latch in LSD, i.e., PB4 = 1
           PULB              ;recover old ACCB
           RTS               ;return
```

---

[1]The remainder of the assembly language programming examples in this book will assume that library files HC11REG.ASM and HC11VEC.ASM are implicitly included (see Appendix E). Most will omit essential pseudo-ops that are required to assemble a source code. They will also omit some standard initialization, such as the stack pointer value. This will shorten the code of the example program listings. For more information, refer to the section on cross assembly in Appendix D.

You can expand the multiplexing principle to use four displays using the remaining port B pins PB6 and PB7. Hence, the high nibble of port B selects a display, and the low nibble outputs the data to be displayed. Of course, you can design a circuit to multiplex more than four displays by using an extra port. To multiplex more displays without using additional port lines, you can connect the high nibble lines to a decoder IC such as the 74HC138 (discussed in Section 5.4). Each of its outputs can be connected to a separate LE pin of the driver IC.

■ 9.3 LIQUID-CRYSTAL DISPLAYS (LCDs)

A liquid-crystal display (LCD) is another type of display. You have probably seen them on watches and calculators. They are also commonly used in instrument panels, consumer electronic displays (e.g., a videocassette recorder panel), and diagnostic displays for sophisticated printers. Portable (laptop) computers have flat screens using LCDs. Flat-screen LCD panels are controlled by special embedded controllers and are not commonly part of 8-bit microcontroller systems. An 8-bit microcontroller system would use the smaller LCD panels. The chief advantage of an LCD is its low power consumption and the variety of display formats available.

9.3.1 Operation

First, we will cover the structure of the LCD. Figure 9.6(a) illustrates its basic internal structure. An LCD display has two plates separated by crystal material. Thus its structure is like that of a capacitor. The polarizer plates are used to pass light. As we shall see, the liquid crystal in the middle can be made to pass or stop light.

The LCD display requires an alternating *excitation* wave applied to selected electrodes to charge selected areas. The excitation wave develops an electrostatic field to align the liquid-crystal molecules in these selected areas. When the crystals are aligned, they allow light to pass through to the mirror. In the charged areas the mirror reflects more light than the uncharged area. This makes the charged area appear to light up. This is the *reflective* type of LCD. An *absorption* LCD works oppositely. Instead of a mirror, it uses a black surface. The charged area absorbs light to make it appear darker than the uncharged areas.

Note that excitation is supplied by an alternating wave. This could be a square wave. But more often it is a waveform with several logic levels (see Figure 9.8 in Section 9.3.4). A constant (dc) excitation signal will polarize and destroy the crystals. To supply the excitation wave, a square wave is supplied to an IC designed to drive an LCD or an LCD with built-in drive circuits.

There is no universal standard for LCDs. They differ in segment layout, excitation requirements, and methods of connecting frontplanes and backplanes. Some use a layout similar to seven-segment LEDs. Others use a matrix of dots (Figure 9.7).

9.3.2 Seven-Segment LCD

Refer to Figure 9.6(b). The driver IC, an MC14543B (BCD-to-Seven Segment Latch/Decoder/Driver for Liquid Crystals) is designed to drive a seven-segment LCD. It requires a square wave for excitation. An external source is required to supply the square wave. You may recall from Chapter 8 that the 68HC11 can supply a square wave. This square wave is

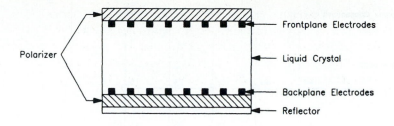

(a)   Internal Structure

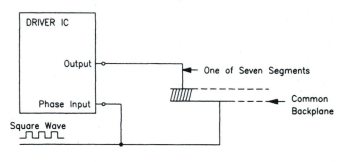

(b)   Driver Connection to Seven–Segment LCD

Figure 9.6   Liquid-Crystal Display

supplied to the backplane electrodes of the LCD display and to the driver IC. The driver IC modifies the phase input, depending on the input data. To turn on a segment, it inverts the square wave and then sends it to the segment frontplane. With the frontplane and backplane waves out of phase, a logic-high rms (see the Glossary) voltage turns on the segment. If the input data specifies to turn off the segment, the driver outputs the square wave to the segment frontplane. With the frontplane and backplane waves in phase, the rms voltage between the electrodes is zero. This driver chip has a parallel interface with four data input lines. Like the LED driver described earlier, it has a latch input (LD) to update the display. However, it decodes only the 4-bit nibbles, %0000 to %1001, into decimal digits, 0 to 9, respectively. Other binary inputs result in a blank display.

9.3.3   Dot Matrix LCD

An LCD using segments arranged as a matrix of dots can display a variety of symbols. An LCD display with a large matrix of dots can be used to display graphical information as well as *alphanumeric* (letters and digits) information. One format is a 5×8 dot matrix. Figure 9.7 shows one way to connect it to an MC145000 (Serial Input Multiplexed LCD Drivers) LCD driver. This driver has 12 frontplane (FP) outputs and four backplane (BP) outputs. Not all outputs are used in this case.

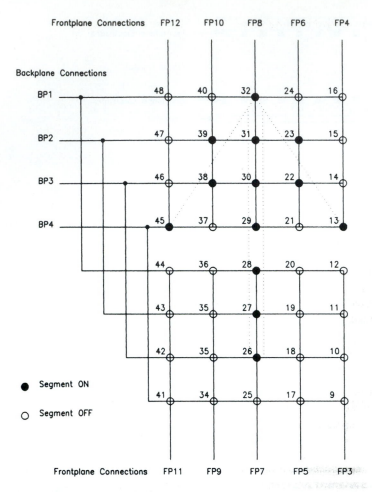

Figure 9.7 Example of a 5 × 8 Dot Matrix LCD Display Controlled by an MC145000 LCD Driver

This is an example of *matrix addressing* an element. The display element is turned on or off by controlling the corresponding row (BP) and column (FP) lines. The BP waveforms are constant. The driver has a latch to store 48 bits of data to control each display segment. With the BP waveform constant, the form of the FP waveform determines whether a segment is on or off. We will look at the waveforms later.

Each data bit determines the frontplane signals to be sent to control an LCD segment. A high bit turns on the corresponding segment. A low bit turns off a corresponding segment. Table 9.1 shows the bit locations to control the segment located at each frontplane and backplane intersection. Interestingly, the data sheet for the MC145000 uses a different bit-numbering convention. The least significant bit is number 48 and the most significant bit is number 1.

Figure 9.7 also shows the binary data required to display an "up arrow." The dark dots indicate which bits should be set in the 48-bit data sent to the driver. This will

TABLE 9.1 LCD SEGMENT CONTROL BIT LOCATIONS[a]

	FP1	FP2	FP3	FP4	FP5	FP6	FP7	FP8	FP9	FP10	FP11	FP12
BP1	4	8	12	16	20	24	28	32	36	40	44	48
BP2	3	7	11	15	19	23	27	31	35	39	43	47
BP3	2	6	10	14	18	22	26	30	34	38	42	46
BP4	1	5	9	13	17	21	25	29	33	37	41	45

[a]MSB = 1, LSB = 48.

give you an idea of the coding requirements to create a display. In this case the display data is

Column	Hex byte
FP12,11	08
FP10,9	06
FP8,7	7F
FP6,5	06
FP4,3	08
FP2,1	xx

The MC145000 is versatile enough to be connected to LCD displays with different segment layouts. The corresponding coding will differ. A circuit using several LCD drivers, such as a combination of MC145000 and MC145001 chips, can drive an LCD display with more than 48 segments. The data is sent to the drivers using a serial interface (see Section 8.6.1). This type of display is analogous to cathode ray tube (CRT) displays used in computers. Each uses an array of picture elements (pixel). The display is bit mapped. A block of memory represents the display image. The block in memory stores the bit patterns that determine whether a pixel is on or off. This type of display is known as *bit mapped*. You may note that each byte corresponds to a column in the LCD panel if it is connected as shown in Figure 9.7.

For an application that uses only certain characters, such as alphanumeric, a look-up table can determine which bits to set to display the desired character. This table codes the dot matrix pattern for each character. The software references the table to place the character dot matrix pattern in the bit map. For example, to display the message "BEGIN," it will have to place the pattern for B in the first section of the bit map, E in the second, and so on. Many vendors supply LCD panels with the alphanumeric look-up table already supplied in a ROM.

PRACTICE

Write software to store the bit map for the letter B as the fourth character in a bit map. Assume that there are only 50 alphanumeric characters in the look-up table. Use a 5×8 dot matrix pattern. Each character in the bit map should be separated by two columns.

Solution

```
*Solution
*Test subroutine FILLMAP
*Main code puts dot matrix pattern for character 2
*as fourth letter (position 3) in bit map.
```

```
NCOL    EQU    5              ;number of columns
*                             ;for each character
POSN    EQU    3              ;char position in bit map

        ORG    0

BITMAP         RMB    70      ;assumed maximum of 10 characters

*bit map locations should normally be initialized
*to zero since 2 bytes (columns) separate each
*letter

        ORG    $100

        LDS    #$FF           ;init stack
*                             ;set up calling registers
        LDAA   #2             ;char code
*                             ;and bit map position
        LDX    #BITMAP+(7*POSN)
        JSR    FILLMAP
DONE
        BRA    DONE           ;stop for now

*Subroutine FILLMAP
*Puts the dot matrix codes for a character in the bit
*map. Each character requires 5 bytes. After return
*the bit map pointer references next position,
*two columns after current character.

*Calling registers
*      IX = bit map pointer
*      ACCA = character code from $0 to $31
*Return result to bit map
*      *IX to *(IX + 6) = char dot matrix code
*      (Appendix D describes the pointer operator [*])
*Return registers
*      IX = IX + 7 if no error
*      IX = IX if error
*      ACCA = ACCA
*      CCR affected
*Error code
*      ACCA > $31

FILLMAP
        CMPA   #$31           ;exit if char code invalid
        BGT    EXIT
        PSHA                  ;preserve registers
        PSHB
        PSHY
```

```
*                              ;look up pattern
*                              ;char start address is
*                              ;DOTMAT + NCOL × CODE
        LDY     #DOTMAT
        LDAB    #NCOL
        MUL                    ;should get ACCA = 0
        ABY
*                              ;now copy pattern to bit map
        LDD     0,Y
        STD     0,X
        LDD     2,Y
        STD     2,X
        LDAA    4,Y
        STAA    4,X
        LDAB    #NCOL + 2  ;point to next bit map position
        ABX
        PULY                   ;recover registers
        PULB
        PULA
EXIT
        RTS


*look-up table for dot matrix pattern of characters
DOTMAT
*<space> is char 0
        FCB     0, 0, 0, 0, 0
*A is char 1
        FCB     $F0, $16, $11, $16, $F0
*B is char 2
        FCB     $FF, $89, $89, $56, $20
*rest of table continues
```

9.3.4   LCD Excitation Waveforms

Most LCD driver ICs or LCD displays with built-in drivers require a square-wave input.
The driver converts the square wave into appropriate frontplane and backplane waveforms.
The driver data sheets specify the input square-wave requirements. With some displays,
these waveforms can be complicated, as shown in Figure 9.8. The MC145000 Driver IC
supplies frontplane and backplane waveforms depending on the input data specifying
which segments to turn on. Recall that this IC can supply 4 backplane signals and 12 frontplane signals. The display circuit designer can choose different combinations to connect
these signals to a display.

Normally, we do not concern ourselves with what waveforms to use for generating a
desired display; we let the driver IC handle this for us. However, to use it, it is helpful to
have an idea of what a driver IC does. Recall that the backplane waveforms are constant.
They do not depend on the input data, but specifying the state of an input data bit determines the frontplane waveform. It decides whether to turn a segment on or off. For the

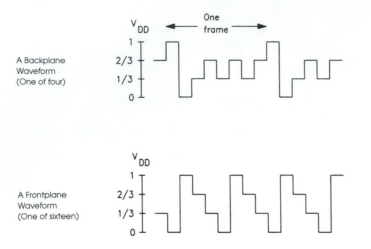

Figure 9.8 Sample LCD Excitation Waveforms Based on MC145000 Driver IC

MC145000 driver, each frontplane has 16 possible waveforms. Normally, each LCD segment is connected to one frontplane and one backplane electrode (Figure 9.7). Thus each frontplane controls four segments. There are 16 possibilities, ranging from turning off all four segments, to turning some on and some off, to turning on all four. Again, we can leave it to the driver IC to determine which waves to supply.

The following equations show the relationship between segment state and plane voltages:

$$V_{ON} = \frac{V_{DD}}{\sqrt{3}} \tag{9.1}$$

$$V_{OFF} = \frac{V_{DD}}{3} \tag{9.2}$$

V_{ON} and V_{OFF} are the rms voltages between the frontplane and backplane to turn a segment on and off, respectively. V_{DD} is the logic supply voltage.

■ 9.4 KEYBOARD INTERFACING

9.4.1 The Matrix Keyboard

Most of us realize that a keyboard is used by a person to enter input into a computer (or control panel). You may think of it as being only an input device. Actually, it turns out that most keyboard interfaces require both input and output. A common type of keyboard is the matrix type. This type saves on the amount of I/O wiring because the keys share wires. Every key does not require its own individual wire. Instead, each has its own combination of row and column wires.

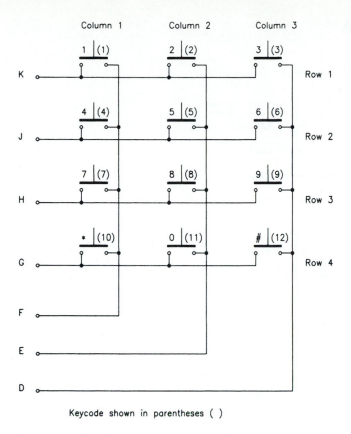

Figure 9.9　3 × 4 Matrix Keyboard (Telephone Format)

Consider Figure 9.9. It shows a 12-key matrix keyboard that is used in telephones. Each key has a momentary contact switch that is connected to an intersection of row and column wires. When a key is released (break), an open circuit exists between all wires/ terminals. When a key is pressed (make), the contact closure connects the row and column wire. Thus, a short circuit exists between row and column wires when any key is pressed. To determine which key is pressed, a microcontroller must *scan* the rows and columns to identify the row and column intersection of the short circuit. The terminal labels correspond to those used by the Grayhill 88AB2 keyboard.

Each key has an identifying number (key code or scan code) as well as a character or function associated with it. For the telephone format keyboard connected as shown in Figure 9.9, key number 11 (key code 11) has the character zero (0). In this case, when the MCU checks terminals G and E, it will find a short circuit. Thus, it will identify the key code to be 11. The IBM PC uses the term *scan code* to identify each key on its keyboard. The matrix keyboard principle can be extended to keyboards with many keys. However, other modifications are required. If more than one key is pressed, there is no way to state for certain which ones were pressed. Some keyboards require a multikey pressing sequence, such as CONTROL or SHIFT key sequences for computers. In this case some keys have their own dedicated input lines.

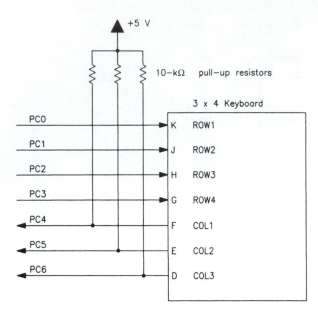

Figure 9.10 Parallel I/O With Keyboard. Also connect PC7 high using a pull-up resistor.

The key code numbering starts with a column and then adds the row number before continuing with the next column. With this numbering sequence, a keyboard decoder can drive a signal to a column and sense the row lines to determine which key is pressed, if any. Figure 9.10 shows a connection to a matrix keyboard. Port C is a convenient port to use because its lines can be configured as input or output. However, other factors in the overall system design may lead you to choose other I/O lines.

To identify the key code, the MCU scans each contact in sequence. It can increment a counter every time it scans the next key. When the MCU detects a closed contact, it stops the count. The count then corresponds to the key code. If no closed contacts are found, the key code is reset to zero. The system uses a key code of zero to represent the condition of no keys pressed. This method will vary for different types of keyboards and for different connections. The process of determining which key is pressed is called *keyboard decoding*.

Refer to both Figures 9.9 and 9.10 to follow the next explanation. To detect a short circuit, the MCU drives one of the output lines low and checks a corresponding input line. If it is also low, the key was pressed. If the key was not pressed, the open circuit allows the resistor to pull up the input line to logic high. For example, to check key code 5, the MCU drives PC1 (terminal J) low and checks the input at PC5 (terminal E).

9.4.2 Keyboard Decoding for 2 out of 7

Some keyboards have an extra common terminal. If a key is pressed, a short circuit occurs between the common and the key's row and column line. Thus there are two possible short circuits when a key is pressed. For a 12-key matrix, this is known as a 2-out-of-7 style. There are two choices when using this method:

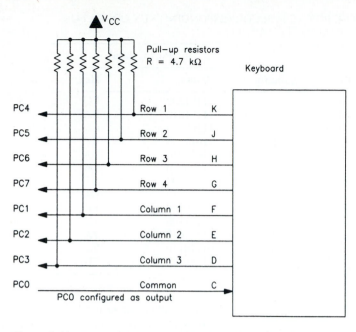

Figure 9.11 2-out-of-7 Keyboard Decoding Using Port C

1. The common is output and the rest are inputs.
2. The common is input and the rest are outputs.

Figure 9.11 shows a connection for the first choice. Using the data direction register (DDRC), port lines PC1 to PC7 are configured as inputs and PC0 is configured as output. The pull-up resistor ensures that any input is logic high when a switch is open. When PC0 is driven low and one key is pressed, one of the row inputs and one of the column inputs will be low. The keyboard software driver routine checks which inputs are low and determines the key code.

A look-up table can be used to determine the key code. But in this case, with seven inputs, the table would require 128 entries, most of which will not correspond to any key. Alternatively, the driver routine can find the key code, as shown in the pseudocode of Listing 9.4.

```
*Listing 9.4
*Decode 2-out-of-7 matrix keyboard

    Drive PC0 low
    Read port C
    Identify low bit for COL (PC1,2,3)
    Identify low bit for ROW (PC4,5,6,7)
    KEY = COL + (3 × ROW) - 3
```

To refine this software technique further, we can make use of the fact that only 12 input conditions are possible when one key is pressed. Table 9.2 shows the possible inputs for the circuit shown in Figure 9.11.

TABLE 9.2   PORT C INPUTS FOR CORRESPONDING KEY CODES FOR 2-OUT-OF-7 KEYBOARD OF FIGURE 9.11

Key code	Row	Column	Port C data Binary	Port C data Hex
1	1	1	11101100	EC
2	1	2	11101010	EA
3	1	3	11100110	E6
4	2	1	11011100	DC
5	2	2	11011010	DA
6	2	3	11010110	D6
7	3	1	10111100	BC
8	3	2	10111010	BA
9	3	3	10110110	B6
10	4	1	01111100	7C
11	4	2	01111010	7A
12	4	3	01110110	76
None			11111110	FE

We can use Table 9.2 to write some subroutines that can be used by the keyboard software driver. These are shown in Listing 9.5. The main program would call INITKYBD once near the beginning to ensure that the data direction of port C is set up. Note that the reset condition of data direction register DDRD is $00. An application program calls FROW-COL, then IDKEY, to find the key code of the key pressed. For these subroutines, a zero or negative key code implies that no key was pressed (or the key was released while the subroutines were executing).

```
*Listing 9.5
*Some subroutines that can be used for the 2-out-of-7
*keyboard interface of Figure 9.11.

*--------------------------------

        ORG     $180


*Subroutine INITKYBD
*Initialize keyboard interface
*No calling or return registers
*REMINDER - PORTC, DDRD labels defined in HC11REG.ASM
*     See Appendices D and E

INITKYBD
        LDX     #$1000    ;point to registers
        BSET    DDRC,X$01
*                         ;make PC0 output
        RTS               ;return

*Subroutine FROWCOL
*finds row and column of key pressed
*No calling registers
```

```
*Return registers
*       ACCA = ROW if found, else 0
*       ACCB = COL if found, else 0
*       CCR affected

FROWCOL
        CLRA                    ;initialize ROW, COL to 0
        CLRB
        LDX     #$1000          ;point to registers
        BCLR    PORTC,X $01     ;drive PC0 low
*                               ;find ROW number
        BRSET   PORTC,X $E0 ROW1
        BRSET   PORTC,X $D0 ROW2
        BRSET   PORTC,X $B0 ROW3
        BRSET   PORTC,X $70 ROW4
        BRA     EXIT1           ;exit if no key pressed
ROW4    INCA
ROW3    INCA
ROW2    INCA
ROW1    INCA
*                               ;find COL number
        BRSET   PORTC,X $0C COL1
        BRSET   PORTC,X $0A COL2
        BRSET   PORTC,X $06 COL3
        BRA     EXIT1           ;exits here if key released
COL3    INCB
COL2    INCB
COL1    INCB
EXIT1   RTS                     ;then return


*Subroutine IDKEY
*Calculates key code of detected key
*using formula KEY = COL + (3 × ROW) − 3
*Calling registers
*       ACCA = ROW
*       ACCB = COL
*Return registers
*       ACCA = keycode
*       CCR affected

IDKEY
        PSHB                    ;save COL
        TAB                     ;use ACCB as temp. reg.
        ASLB                    ;ROW = (3 × ROW) − 3
        ABA
        SUBA    #3
        PULB                    ;restore COL
        ABA                     ;KEY = COL + (3 × ROW) − 3
        RTS                     ;return with keycode
```

However, there are more requirements for a keyboard interface than those just covered.

9.4.3 Debouncing

One important mechanical consideration is contact bounce. When a mechanical switch is pressed or released, it bounces microscopically for a period of a few milliseconds. Hence, when you press or release a key, the MCU will see many occurrences of make and break instead of one occurrence. A mechanical key requires a debouncer when used with logic circuits or microcontroller hardware.

The MCU should see *only one* break-to-make transition when a key is pressed and *only one* make-to-break transition when a key is released. A hardware debouncing circuit can be built using a latch or a low-pass (*RC*) filter. Alternatively, software can provide debouncing. A software debouncer uses a time delay after detecting a transition to check if the key condition has stayed the same. Listing 9.6 shows pseudocode that can be used for debouncing.

```
*Listing 9.6
*Keyboard debouncer pseudocode

*debounce make

repeatmake  ;repeat make loop
                            ;until make detected twice
    if break then repeatmake
        else time delay     ;make detected
                            ;this is the debounce delay
    if break then repeatmake
        else continue       ;stable make condition occurred

*debounce break

repeatbreak ;repeat break loop
                            ;until break detected twice
    if make then repeatbreak
        else time delay     ;break detected

    if make then repeatbreak
        else continue       ;stable break condition occurred
```

Basically, the key is debounced if the key codes before and after the time delay are identical. The pseudocode can be modified such that a loop repeats (after a time delay) if the key codes differ.

9.4.4 Software Drivers for Matrix Keyboard

Now we consider keyboard decoding using inputs (or outputs) for the columns and outputs (or inputs) for the rows. We examine the case for a 12-key keyboard. However, the same techniques can be used for larger keyboards. In fact, this may be the only way to handle larger keyboards. You can refer to Figure 9.10 for this entire section.

For a keyboard scan, the MCU sequentially drives each output low and reads the inputs to check if any inputs are low. We will develop two subroutines, called `getkey` and `breakkey`. Subroutine `getkey` waits for a key to be pressed and then returns the identity of the key pressed. Subroutine `breakkey` waits for the key to be released. Also, we want to test the subroutines. This can be done by writing a main program to call them and display the identity of the key pressed. How this information is displayed depends on the development system and other hardware you are using. This exercise also serves to review the use of stepwise refinement to develop an algorithm. Listing 9.7 illustrates the first level of refinement.

```
Listing 9.7
*Level 1 pseudocode for matrix keyboard driver

main            ;this tests the subroutines
    call initkybd
main 1
    call getkey
    display keychar
    call breakkey
    repeat main 1
```

The next step is to refine the two subroutines as shown in Listing 9.8. We also define a key code of zero to be the condition when no keys are pressed.

```
Listing 9.8
*Level 2 pseudocode for matrix keyboard driver
*This shows the basic steps executed by the subroutines

getkey

scanon
    call idkey                ;idkey returns the key code
    if key == 0         then repeat scanon
                        else call debounce   ;debounce delay
    save key                  ;save key code found
                              ;prior to debounce
```

TABLE 9.3   PORT C INPUTS FOR CORRESPONDING KEY CODES FOR MATRIX KEYBOARD OF FIGURE 9.10

Key code	Row	Column	Port C data	
			Binary	Hex
1	1	1	11101110	EE
2	1	2	11011110	DE
3	1	3	10111110	BE
4	2	1	11101101	ED
5	2	2	11011101	DD
6	2	3	10111101	BD
7	3	1	11101011	EB
8	3	2	11011011	DB
9	3	3	10111011	BB
10	4	1	11100111	E7
11	4	2	11010111	D7
12	4	3	10110111	B7
None			xxxx1111	0xF

```
        call idkey
        if key ! = key - 1      ;key - 1 is the previous key code
                                then repeat scanon
                                else return


breakkey


scanoff
        call idkey
        if key ! = 0               then repeat scanoff
                                   else call debounce
        call idkey
        if key ! = 0               then repeat scanoff
                                   else return
```

Note that we have now identified two more subroutines, idkey and debounce. A further level of refinement is needed to develop these subroutines. One strategy is to pre-set a counter. Scan each key in sequence and increment the counter sequentially. Compare the port C data with the data expected if the key were pressed. If port C equals the data expected, the counter value is the key code. If no successful compares occurred, reset the counter to zero to return a key code of zero. The expected data for a pressed key can be derived by checking which row and column bits should be low. This is similar to the procedure for deriving Table 9.2. A look-up table can be used to reference the counter. Table 9.3 shows the data for the look-up table. Listing 9.9 shows the pseudocode.

```
*Listing 9.9
*Level 3 pseudocode for matrix keyboard driver
*This shows details of nested subroutines

*Subroutine idkey
idkey
        init keypt, key = 0   ;initialize key count
idkey1
        keypt++, key++       ;increment for scan
        port c = *keypt      ;drive port C outputs
        if portc == *keypt   ;compare port C input
                             ;with look-up code
                then return
                             ;exit subroutine with key code
                             ;if found, otherwise continue
                             ;until last key checked
        if key < lastkey
                then repeat idkey 1
        key = 0                ;reset key count if no key found
        return

*Subroutine debounce
debounce
        delay 10 ms
        return
```

In this case, `lastkey` is 12. The pseudocode of Listing 9.9 uses some C language *syntax*. The asterisk (*) means that the variable is pointing to another value. In fact, `keypt` is the look-up table pointer. From here we can assign CPU registers and memory locations to implement the pseudocode.

The source code listing for assembly language is shown as Listing 9.10. In this case we do not display the key code. Note also that writing to a port with some lines configured as input does not affect the corresponding data register bits. The source code uses pseudo-ops for the Avocet cross assembler and header files (Chapter 4). Appendix E has the source listing of the header files.

```
*Listing 9.10
*Test keyboard driver subroutines for
*matrix keyboard interface of Figure 9.10


*---------------------------------------
        ORG     $100

MAIN    LDS     #$FF            ;initialize stack
        LDX     #REGBAS         ;point to registers
        JSR     INITKYBD
*                               ;initialize keyboard interface
MAIN1
        JSR     GETKEY          ;wait for key to be pressed
        JSR     BREAKKEY
*                               ;then wait for it to be
*                               ;released
        BRA     MAIN1           ;and repeat loop


*---------------------------------------


*Subroutine INITKYBD
*Initializes the keyboard interface
*Calling Register
*       IX = register block address
*Return Registers
*       CCR affected

INITKYBD
        PSHA                    ;preserve registers
        LDAA    #%00001111
        STAA    DDRC,X          ;PC0-3 output, PC4-7 input
        PULA                    ;restore registers
        RTS                     ;return


*---------------------------------------


*Subroutine GETKEY
*Waits for key to be pressed and returns its key code
*Calling Registers
```

```
*               IX = register block address
*Return Registers
*               ACCA = key code
*               CCR affected

GETKEY
        PSHB                    ;preserve registers
GETKEY1
        JSR  IDKEY              ;return key code in ACCA
        CMPA #0                 ;if key == 0 then repeat scan
        BEQ  GETKEY1
        JSR  DEBOUNCE
*                               ;debounce delay
        PSHA                    ;save key found prior to debounce
        JSR     IDKEY           ;return new key code in ACCA
        PULB                    ;if key != key-1 then repeat scan
        CBA
        BNE     GETKEY1         ;else return
        PULB                    ;restore registers
        RTS


*----------------------------------------

*Süubroutine BREAKKEY
*Waits for key to be released (break condition)
*Calling Registers
*               IX = register block address
*No return registers except that CCR affected

BREAKKEY
        PSHA
BREAK1
        JSR     IDKEY           ;key code of zero
        CMPA    #0              ;means break occurred
        BNE     BREAK1          ;if break detected then debounce
        JSR     DEBOUNCE
        JSR     IDKEY           ;and check for break again
        CMPA    #0
        BNE     BREAK1          ;if no break then repeat
        PULA                    ;else return
        RTS
*----------------------------------------

*Subroutine IDKEY
*Returns key code from keyboard. A key code of zero
*means that no key was pressed.
*Calling Registers
*               IX = register block address
*Return Registers
*               ACCA = key code
```

```
IDKEY
        PSHY                    ;preserve registers
        PSHB
        LDY     #KYTAB          ;point to table, init keypt
        CLRA                    ;key = 0
IDKEY1
        INCA                    ;key++
        INY                     ;keypt++
        LDAB    0,Y             ;drive port C outputs
        STAB    PORTC,X
        PSHA                    ;preserve key
        LDAA    PORTC,X
        CBA                     ;if portc == *keypt then return
        PULA                    ;note CCR unaffected
        BEQ     IDKEY2          ;if key < lastkey (12)
        CMPA    #12             ;then repeat idkey1
        BLO     IDKEY1
        CLRA                    ;else key = 0 if no key pressed
IDKEY2
        PULB                    ;restore registers
        PULY
        RTS                     ;return(key)


*---------------------------------------
*Subroutine DEBOUNCE
*Delay 10 ms at E = 2 MHz
*No calling or return registers
DEBOUNCE
        PSHX                    ;preserve register
        LDX     #$0D06          ;init loop counter
*this is the delay loop
DEBOUNCE1
        DEX
        BNE DEBOUNCE1
        PULX                    ;restore register
        RTS                     ;return


*---------------------------------------


*Look-up table to map port C data to each key code
*Note interface has PC7 tied high
KYTAB
        FCB     $FF             ;key code of zero, don't care
        FCB     $EE, $DE, $BE
        FCB     $ED, $DD, $BD
        FCB     $EB, $DB, $BB
        FCB     $E7, $D7, $B7
*---------------------------------------
```

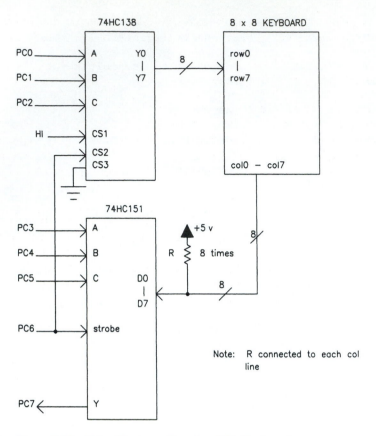

Figure 9.12 Using Hardware Decoders With Keyboard

9.4.5   Using Hardware Decoding Chips

As with most interfaces, there is a trade-off between hardware and software. The decoding software can be simplified by using additional hardware. Consider Figure 9.12, which shows an 8×8 matrix keyboard with 64 keys. We have already used the 74HC138 (3-to-8 Decoder) IC when discussing address decoding in Section 5.4. The 74HC151 is an 8-input data selector/multiplexer. The selector address signals A, B, and C select which data input, D0 to D7, is transferred to the chip's output Y. For example, if the address signal is %010 (binary 2), output Y equals data input D2. This transfer occurs only when the strobe signal is low.

If we define the key code to start at zero (first row, first column), the 6-bit number in the lower 6 bits of port C can be used as a key code. The key code sequence continues down the column to row 1, row 2, and so on to row 7. It then continues starting at the top of the next column. Output PC6 must be low to enable the decoder chips.

Example.   Sending an output of %00010011 will drive row 3 low and select input from column 2. If key 19 was pressed, a logic low will be sensed at input PC7. Note that the least significant 6 bits have the binary pattern for decimal 19 and writing to PC7 has no effect because it is an input. Thus scanning the keys involves incrementing the 6-bit count in port C and checking whether PC7 is high or low to determine if and which key is pressed.

■ 9.5 OTHER USER INPUT/OUTPUT

The keyboard is a common input device. For industrial applications, an operator controls the process from a control panel. The control panel typically uses push buttons, selector switches, or thumbwheel switches. Other inputs are a mouse, trackball (for computers), voice recognition system, image scanner (for text), or vision system (to recognize parts in a manufacturing system). We have discussed the use of LEDs and LCDs as display devices. Another output used in control panels is the low-powered lamp. It typically indicates the on or off status of a machine such as a pump.

The computer uses a cathode ray tube (CRT) or an LCD flat-screen panel as a display device, but typical 8-bit microcontroller applications do not use them. Many computers use an embedded controller to control a CRT screen. The computer's CPU sends the binary data to be displayed to the CRT controller. The controller translates this data to the video timing signals for the CRT screen. An example of an embedded controller for the CRT is the MC6845 CRT Controller (CRTC).

Vacuum Fluorescent Display (VFD). This type of display looks similar to an LED display and is found in many consumer electronic devices such as compact disc players. Vacuum fluorescent displays (VFDs) emit a higher-intensity light and are available with more color variety. The basic parts of a VFD are the anode segments, filaments, and a wire mesh grid. The anodes nominally require 15 to 30 V to switch on. The grid is supplied with -5 V. Some VFDs have a seven-segment layout and others have a matrix layout. To control a VFD, an MCU drives data lines to turn anode segments on and off while switching the grid voltage from negative to slightly positive. It does this by using a multiplexing technique.

■ 9.6 STROBED INPUT/OUTPUT

An I/O port may have to read a sequence of bytes or transmit a sequence. This situation is common for parallel communication interfaces used in printers, floppy disks, and parallel buses for peripheral communication. These are examples of devices that transfer data in parallel for fast I/O. Their interfaces use control signals to coordinate the transfer of data. The simpler controlled data transfer technique is strobed I/O. Later we will look at a more comprehensive technique known as handshaking.

Basic strobed I/O operates as follows. When an MCU sends data to a peripheral, it tells the peripheral that data is available. For example, if the peripheral uses flip-flops to latch in data from the MCU, it will need an extra *strobe* signal to clock or latch in data. Similarly, if the MCU is reading data from a peripheral, it has to know when the next data is sent. Otherwise, how would it know whether the peripheral is merely sending the same data repeatedly or simply has no new data to send? The peripheral can send a strobe signal that tells the MCU when new data is available.

Refer to Figure B.1 in the Quick Reference (Appendix B). The parallel I/O subsystem also uses two control pins, strobe A (STRA) and strobe B (STRB). Also refer to Figure 9.13.

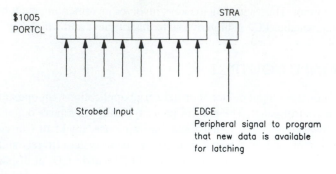

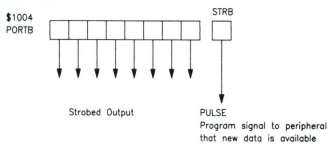

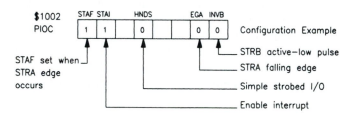

Figure 9.13 Strobed I/O

Port C has two data registers: the port C data register (PORTC at $1003) and the port C latched data register (PORTCL at $1005). Figure 9.13 shows only the latched data register, PORTCL. Register PORTCL is used for strobed and handshake I/O because it latches input data. The parallel I/O control register (PIOC at $1002) is the control register for this system. It determines how the MCU will behave for parallel I/O operations. If the mode select bit (HNDS, which is bit 4 in $1002) is low, the MCU will operate ports B and C in simple strobe mode. Again, refer to Table B.1 in Appendix B to note the 68HC11 I/O registers and bits. Figure 9.14 provides more detailed explanation about register PIOC.

9.6.1 Input Strobe

A peripheral has its output data lines connected to port C and an output control line (strobe) connected to strobe A (STRA). Using the configuration example shown in Figure 9.13, if the peripheral keeps pin STRA high, changes in the port C lines will also change the contents of data register PORTC. However, the changes will not show up in data latch register PORTCL.

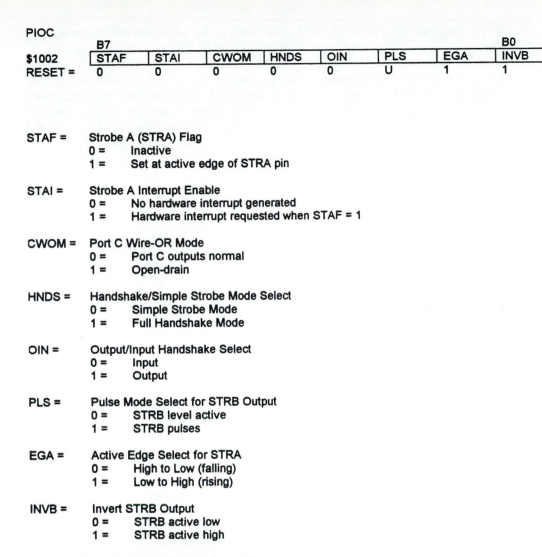

PIOC

	B7							B0
$1002	STAF	STAI	CWOM	HNDS	OIN	PLS	EGA	INVB
RESET =	0	0	0	0	0	U	1	1

STAF =   Strobe A (STRA) Flag
           0 =   Inactive
           1 =   Set at active edge of STRA pin

STAI =   Strobe A Interrupt Enable
           0 =   No hardware interrupt generated
           1 =   Hardware interrupt requested when STAF = 1

CWOM =   Port C Wire-OR Mode
           0 =   Port C outputs normal
           1 =   Open-drain

HNDS =   Handshake/Simple Strobe Mode Select
           0 =   Simple Strobe Mode
           1 =   Full Handshake Mode

OIN =   Output/Input Handshake Select
           0 =   Input
           1 =   Output

PLS =   Pulse Mode Select for STRB Output
           0 =   STRB level active
           1 =   STRB pulses

EGA =   Active Edge Select for STRA
           0 =   High to Low (falling)
           1 =   Low to High (rising)

INVB =   Invert STRB Output
           0 =   STRB active low
           1 =   STRB active high

Figure 9.14   Parallel I/O Control Register

When the peripheral drives pin STRA high to low (falling edge), the data on port C pins will be latched into the data latch register, PORTCL. This causes the strobe A flag (STAF) in the control register PIOC to set and an interrupt will occur. If strobe A goes high and stays high and the data input at port C changes, the contents of data register PORTC will also change. However, the contents of data latch register PORTCL will not change because it latches in new data only during an active strobe. In this case a falling edge is the active strobe (since bit EGA = 0) and the data is input to port C.

When the MCU reads control register PIOC and then reads data latch register PORTCL, the strobe A flag (STAF) will clear (reset to zero). The latching sequence will repeat when strobe A goes low again. If a second strobe A edge occurs before the MCU reads PORTCL,

Parallel Input/Output   323

the previous contents of PORTCL will be lost. Note that this subsystem does not have a separate status register. The control register (PIOC) has one bit (STAF) reserved for status.

The example shown in Figure 9.13 shows control register PIOC configured for interrupts. Alternatively, a driver program can use polling. The program of Listing 9.11 uses polling to input 10 bytes sent by a peripheral to port C. It latches in data every time a rising edge occurs at pin STRA. Remember: Most of the assembly language programming examples will assume that library files HC11REG.ASM and HV11VEC.ASM are included (see "Cross Assembly" in Appendix D).

```
*Listing 9.11
*Strobed I/O Demo
*Section of program to input 10 bytes
*from port C and store them


*--------------------------------------

        ORG       $100

        LDX       #REGBAS
        CLR       DDRC,X              ;configure port C as input
        LDAA      #02                 ;configure PIOC register
        STAA      PIOC,X              ;active STRA is rising edge
        LDY       #PTR                ;initialize storage pointer
*poll STAF for rising edge
CIN
        BRCLR     PIOC,X $80 CIN
        LDAA      PORTCL,X
*                                     ;latch in port C input
*                                     ;when STRA edge detected
        STAA      0,Y                 ;and store it
        INY                           ;repeat 10 times
        CPY       #PTR+10
        BNE       CIN
        STOP                          ;then stop

PTR     EQU       $180
```

The program operates as follows: When pin STRA sees a rising edge, flag STAF sets in control register PIOC. This is detected by the branch if bits clear (BRCLR) instruction. The flag is cleared afterward because the BRCLR instruction also reads register PIOC and the following LDAA instruction reads register PORTCL.

9.6.2 Output Strobe

Refer to Figure 9.13 again. A peripheral has its input lines connected to port B. Its strobe input line is connected to pin STRB. Using the configuration example shown, strobe B (STRB) will normally be high. When the MCU writes to port B, strobe B will pulse low for two clock cycles and then go high again. It is up to the peripheral to respond to this signal.

Consider the case when a peripheral outputs a signal to pin STRA to tell the MCU that it wants data. The MCU can poll register PIOC to determine when a request occurs. It responds by writing a new byte to port B. This also pulses the strobe B line to tell the peripheral that it has sent the requested data. Also, the MCU must read register PORTCL to clear flag STAF even though it may not use the data latched in there. Listing 9.12, which illustrates this type of sequence, shows only the relevant part of the program code for this example.

```
*Listing 9.12
*Demonstrate output strobe
*PIOC configured as shown in Figure 9.13
*and all other required initialization has been done.


*-------------------------------------

        ORG     $100

        LDY     #PTR        ;initialize data pointer
        CLI                 ;enable interrupts
REPEAT
        WAI                 ;wait for falling STRA
*                           ;to cause interrupt
        LDAA    0,Y         ;send out data when it occurs
        STAA    PORTB,X
*                           ;note, MCU also pulses STRB low
*                           ;for two E cycles
        INY                 ;repeat for next data transfer
        BRA     REPEAT


*-------------------------------------


RIRQ
*This is the interrupt handler routine for parallel I/O
*Note it has the same vector ($FFF2,F3) as IRQ

        LDAA    PIOC,X      ;these two instructions clear STAF
        LDAA    PORTCL,X
        RTI

PTR     EQU     $180
```

In this case the interrupt service routine only clears flag STAF and returns control to the interrupted program. Refer to Table B.2 in the Quick Reference (Appendix B) and you will note that the parallel I/O interrupt uses the same vector as does the external pin IRQ. Also recall that the interrupt mask bit (I) in the condition code register has to be clear for any "maskable" interrupt to occur. In addition to this, the local interrupt enable bit, strobe A interrupt enable (STAI), must be set as shown in Table B.4 in Appendix B.

9.6.3 Configuration

We have already covered aspects of configuration for strobed I/O. This section will cover them comprehensively as well as review them. Figure 9.14 shows the bit functions of control register PIOC. It also shows their functions for handshake I/O. These will be examined later in the chapter.

The strobe A interrupt enable bit (STAI, which is bit 6) determines whether the strobe A flag (STAF) being set will also cause an interrupt. If bit STAI is clear, the MCU will have to poll register PIOC to determine if an edge occurred at pin STRA. If bit STAI is set, an active edge on pin STRA will also cause an interrupt. The active edge select bit (EGA, which is bit 1) defines the "active" edge for STRA. If bit EGA is logic 1, then a rising edge on pin STRA causes flag STAF to set. If it is a logic 0, a falling edge causes flag STAF to set.

To detect additional STRA edges, a program must clear the STAF flag before the next edge occurs. When configured for simple strobe mode (HNDS = 0), the MCU must read register PIOC first and then read register PORTCL. This is the flag-clearing sequence for STAF when the MCU is in simple strobe mode. The invert STRB output bit (INVB, which is bit 0) determines whether pin STRB will pulse low (normally high) or pulse high (normally low) whenever the program writes data to register PORTB. When bit INVB is 0, the strobe B pulse will be low. If it is 1, the pulse will be high. The pulse width is always two clock (E) cycles for simple strobed I/O.

Note that the reset condition for register PIOC is for simple strobed I/O with no interrupts. When we examined parallel I/O applications for seven-segment LEDs, LCDs, and keyboards, we were in fact using register PIOC configured for simple strobed I/O. We simply have not used pins STRA and STRB and register PORTCL because their use was not required for these applications.

We cover the uses of bits OIN and PLS later since they apply only for handshake modes (HNDS = 1). Bit CWOM will be discussed when we cover an application requiring open-drain outputs.

PRACTICE

Configure the parallel subsystem for a simple strobe with open-drain output. The output strobe signal should be active high. The input strobe is a don't care and no interrupt is generated.

Solution

```
LDX      #REGBAS     ;point to registers
LDAA     #$21        ;CWOM, INVB = 1, rest = 0
STAA     PIOC,X      ;write magic number to control register
```

■ 9.7 FULL HANDSHAKE INPUT/OUTPUT

We have seen how strobed I/O can be used to transfer bytes between an MCU and a peripheral. Sometimes more control is required. Strobed input includes a signal from the peripheral telling the MCU that data is available. But there is no signal from the MCU to the peripheral telling it that the MCU is ready to receive data. Strobed output includes a signal for the MCU to tell the peripheral that data is available. But there is no signal from the peripheral to the MCU to tell the MCU that the peripheral is ready to receive data.

When there is a transfer of data, there may have to be a set of rules to define when and how to transfer each byte. This set of rules is known as a *protocol*. A protocol is a set of standard procedures used in data communications that coordinates the transmitting and receiving of information. For example, the rulers of two countries have a protocol to establish a meeting between them. This may be a process of having lower-ranking officials set up a meeting place, a time, and facilities.

The handshake protocol is an agreement whereby the receiver acknowledges each unit of data it receives. The transmitter waits for this acknowledgment before it sends the next unit. A unit can be a block of bytes or a single byte. A parallel I/O subsystem handshake protocol transfers each byte as a unit. We will see that communications systems use protocols for data units that are single bytes and blocks of bytes.

The 68HC11 supports automatic handshaking for parallel I/O for port C. This is determined by configuring control register PIOC. For a full handshake the configuration routine must set the handshake mode select bit (HNDS, which is bit 4) to a 1. Other bits in register PIOC specify the type of handshake. An automatic handshake means that the hardware responds to certain signals automatically without using program instructions. Simple strobed input is an example. The 68HC11 automatically latches in data into data latch register PORTCL when an active edge occurs at pin STRA. No instructions are required to poll pin STRA and then to read port C.

Software can facilitate the use of handshake for other ports or types not supported automatically. Software handshaking means that each action is performed using program instructions: for example, to latch in data at port E using pin $\overline{\text{IRQ}}$ as a strobe signal, the IRQ service routine must execute a load instruction to store the port E value somewhere in memory before its signal changes. We will look at the different types of handshake; then we will look at how to program the 68HC11 to handle a common handshaking protocol used for printers.

9.7.1   Input Handshake

Refer to Figure 9.15 for this entire section. When configured for input handshake, strobe A is an input strobe line and strobe B is an output strobe line. To configure port C for input strobe, reset the output/input handshake select bit (OIN, which is bit 3) to 0. With input handshake (and output handshake) there are two output strobe options; pulsed operation and interlocked operation. The pulse mode select bit (PLS, which is bit 2) in control register PIOC selects which is used.

Pulsed Operation.   The peripheral will generate a pulse on strobe A to indicate that it has sent data. The 68HC11 responds to the pulse. It should have a program that responds by reading in data. When the 68HC11 reads the data, it automatically acknowledges by pulsing strobe B for two clock cycles. This is known as *pulsed handshake operation*. When configured for pulsed operation, the 68HC11 hardware detects the input "request" strobe and automatically acknowledges after its software has read the data from register PORTCL. To configure for pulse handshaking operation, bit PLS must be high. Figure 9.16 shows the timing diagram for the case when the 68HC11 is configured for strobe A active falling and strobe B active low. Remember, to cause the output acknowledge pulse in input handshake mode, the PORTCL register must be read by 68HC11 software after reading PIOC.

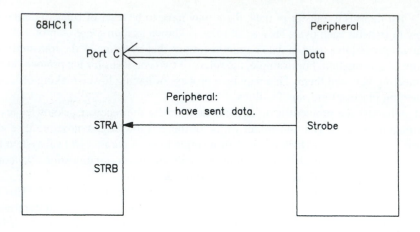

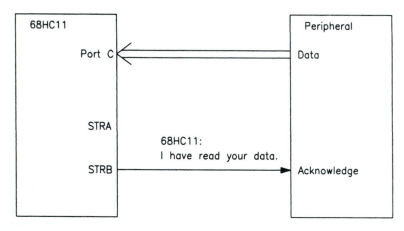

Figure 9.15   Input Handshaking

Note that setting bit PLS in configuration register PIOC selects pulsed operation. Also, the configuration information shown in Figure 9.16 is for the case of polled I/O (because bit STAI = 0). Note that the timing diagrams show control register PIOC with strobe A flag bit STAF clear. The flag clearing sequence for pulsed and interlocked input handshake is to read control register PIOC first, then read data latch register PORTCL.

Interlocked Operation.   Another handshake operation is called *interlocked*. This mode of operation is selected when the configuration routine resets bit PLS in register PIOC. In this case strobe B acts like a ready signal. It is asserted (active) only when the MCU is ready to read data. This is its normal condition. When it is busy with the process of reading in data, the ready line is deasserted. It asserts again after the MCU has read the data. Figure 9.16 shows interlocked operation when strobe B, the ready line, is active low.

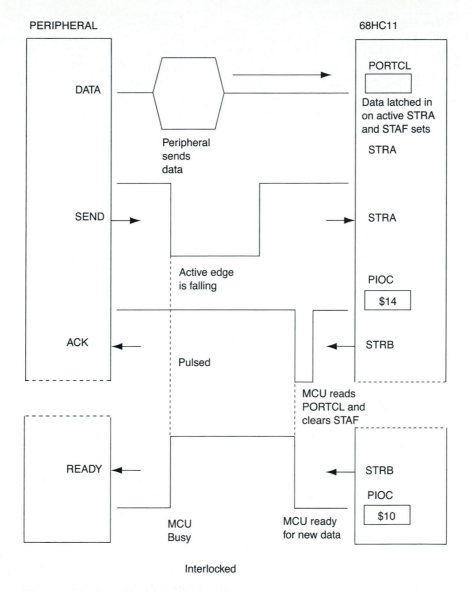

Figure 9.16 Input Handshake Timing Diagrams

When the peripheral sees the ready line deasserted, it should follow the rules by not sending new data at that time. When the 68HC11 reads the contents of data latch register PORTCL, strobe B asserts automatically to indicate that the MCU is ready for new data. The peripheral should not send new data until it sees the strobe B line low.

Listing 9.13 demonstrates some subroutines to implement input handshaking. It can be modified to do something with the data it reads in. Note that this routine can also be modified to use interrupts instead of polling. The configuration data can also be changed.

```
*Listing 9.13
*Demonstrate input handshake subroutines
*Program simply reads in data as it comes
*but does nothing with it. Subsystem configuration
*identical to that of Figure 9.16, interlocked.


*----------------------------------------


      ORG        $100

DIR   EQU        $00             ;data direction stored here
CONF  EQU        $01             ;configuration stored here

DEMOIN
      LDX        #REGBAS         ;point to registers
      CLR        DIR             ;configure for input handshake
      LDAA       #$10            ;HNDS=1: OIN,PLS,EGA,INVB=0
      STAA       CONF
      JSR        INITHNDS
*Read input data continously
REPEATIN
      JSR        INHNDS          ;does not return back to demo main
      NOP                        ;until data received
      BRA        REPEATIN


*----------------------------------------


*Subroutine INITHNDS
*Initializes port C for handshake mode
*Modify addresses IODIR, IOCONF, to change configuration.
*Calling Registers and Addresses
*     IX = address of register block
*     DIR ($00) = data direction of port C
*     CONF ($01) = configuration byte for PIOC
*No Return Registers except CCR affected
INITHNDS
      PSHA                       ;preserve registers
      BCLR       PIOC,X $10      ;put in strobe mode first
      LDAA       PIOC,X          ;then clear STAF if set
      LDAA       PORTCL,X
      LDAA       DIR             ;set up port C direction
      STAA       DDRC,X
      LDAA       CONF            ;configure with magic number
      STAA       PIOC,X
      PULA                       ;restore registers
      RTS                        ;and return

*Subroutine INHNDS
*Uses input handshake to read port C
*Before calling this subroutine, must call
```

```
*subroutine INITHNDS to configure for handshake
*Calling Registers
*     IX = address of register block
*Return Registers
*     ACCA = data read from port C
*     Others unaffected except for CCR

INHNDS
*                              ;poll for STRA transition
     BRCLR    PIOC,X $80 INHNDS
     LDAA     PORTCL,X     ;input strobed data and clear STAF
     RTS                   ;return(data)
```

9.7.2   Output Handshake

Figure 9.17 shows the basic operation for output handshake. When the MCU is configured for output handshake, strobe A is an input ready or busy line and strobe B is an output strobe line. To configure port C for output handshake, set bit OIN to 1.

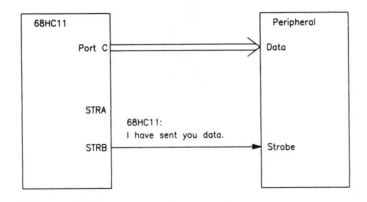

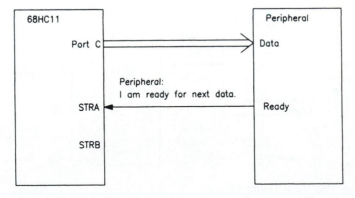

Figure 9.17   Output Handshaking

Pulsed Operation. The 68HC11 writes data to port C (PORTCL) and automatically sends a strobe signal by pulsing strobe B for two clock cycles. This tells the peripheral that data is available for it to read. The peripheral reads the data and responds by asserting a ready signal. This tells the MCU that the peripheral has read in the data and is ready for new data. The 68HC11 sees the ready signal as a transition on pin STRA. It sees it because the active edge of STRA causes flag bit STAF to set. The MCU can now write the next byte to register PORTCL.

Again, to configure for pulsed operation, pulse mode select bit PLS must be high. Figure 9.18 shows the timing diagram when strobes A (edge) and B (pulse) are configured to be active low for polled I/O operation. The flag-clearing sequence for bit STAF is to read control register PIOC first and then write to data latch register PORTCL. This is true for both pulsed and interlocked operation when control register PIOC is configured for output handshake.

Interlocked Operation. In the case of interlocked output handshake, writing to PORTCL asserts strobe B. When the 68HC11 receives the acknowledgment (or ready), it

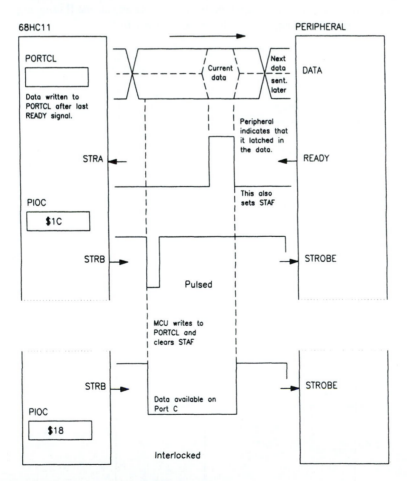

Figure 9.18 Output Handshake Timing Diagrams

deasserts strobe B. Writing new data to PORTCL will assert strobe B again. For interlocked output handshake the configuration routine must reset bit PLS and set bit OIN in control register PIOC.

Listing 9.14 shows program code that can be used to implement output handshaking. It can be modified to send a sequence of bytes. Note that this routine can also be modified to use interrupts instead of polling.

```
*Listing 9.14
*Output handshake demonstration.
*Program simply outputs same data when requested by
*peripheral. Subsystem configuration identical to
*that of Figure 9.18, interlocked.

*-------------------------------------


        ORG     $100
DIR     EQU     $00
CONF    EQU     $01


OUTDATA EQU     $02             ;data stored here


DEMOOUT
        LDX     #REGBAS         ;point to registers
        LDAA    #$FF            ;configure for output direction
        STAA    DIR
        LDAA    #$18            ;and output handshake, interlocked
        STAA    CONF
        JSR     INITHNDS
REPEATOUT
        LDAA    OUTDATA
        JSR     OUTHNDS
        BRA     REPEATOUT


*-------------------------------------


*Subroutine INITHNDS, see Listing 9.13

*Subroutine OUTHNDS
*Uses output handshake to send data to port C
*and waits for peripheral ready before returning
*Calling Registers
*    IX = address of register block
*    ACCA = data to send out
*No Return Registers except CCR affected

OUTHNDS
    STAA     PORTCL,X       ;output data, clears STAF also
*poll for STRA transition
```

```
OUTHNDS1
*                                   ;by checking if STAF high
       BRCLR    PIOC,X $80 OUTHNDS1
       RTS                          ;and return, can now
*                                   ;send next byte if desired
```

Note that subroutine OUTHNDS exits with flag STAF set because it waits for a strobe A transition. When the subroutine is called again, it clears flag STAF when writing the next data byte. Then it waits for it to set again before exiting.

Three-State Variation. Unlike simple strobed I/O or input handshake mode, it is still possible to have output handshake with port C even while some bits are configured to be inputs.

To do this, the subsystem uses the three-state variation of output handshake. As long as the strobe A input is at its inactive level, all port C pins obey their corresponding data direction bit as specified by data direction register DDRC (i.e., driven for outputs and high impedance for inputs). When strobe A goes to its active level, all port C pins act as driven outputs. In the case of the system shown in Figure 9.18 (if DDRC = 00), the port C pins are driven as outputs only when strobe A is high. Note that the ready signal is asserted on the trailing edge of the strobe A pulse. Bit EGA defines the active *edge*. The active *level* is the level before the active edge.

PRACTICE

Configure the parallel subsystem for a pulsed input handshake that is serviced by an interrupt when a rising edge occurs on the request line. The acknowledge should be active low. Also show a service routine that stores the input data in address $0.

Solution

```
*Configuration part of program

       LDX      #REGBAS   ;point to registers
       LDAA     #$56      ;STAI, HNDS, PLS, EGA = 1, rest = 0
       STAA     PIOC,X
       CLI                ;enable interrupt


*Service routine part of code

RIRQ
*      ;assume input handshake is the only source of IRQ
       interrupt
       LDAA     PIOC,X    ;sequence to clear STAF
       LDAA     PORTCL,X  ;and read data
       STAA     $0        ;then store data
       RTI                ;and return
```

■ 9.8 PARALLEL INTERFACE STANDARDS

Recall that we introduced standards in Section 8.10. In this section we will examine how some commonly used interface standards use handshaking for parallel data transfer.

9.8.1 Centronics Parallel Interface

The Centronics Parallel Interface is commonly used for printers. Printers typically have either a serial or a parallel interface. If the printer specification states "parallel" interface, it usually implies the Centronics standard. It defines how to connect a printer to a computer and the form of the data sent to it. Microcontrollers often use this standard because many printers use a microcontroller. Also, some other microcontroller systems send output to a printer. *Centronics* is the printer manufacturer that originally developed the predecessor to this standard. Although others have built upon their work, the name *Centronics* has stuck. This is an example of a *de facto* standard.

Handshaking is required because a computer sends a sequence of bytes to the printer. The printer must receive them, store them in a buffer memory, and then use the data to drive the print mechanics to produce an image on paper (hard copy). We cover printers in more detail in Chapter 14 since a microcontroller will use other I/O subsystems to control other parts of a printer. It will be useful to become more familiar with the other subsystems before considering the entire printer system. In this section we will deal specifically with the printer data interface to a computer.

Figure 9.19 shows the handshake protocol defined by this standard. It shows how data is transferred from a computer port to the printer. The computer sends a strobe pulse ($\overline{STB}$) every time it sends data to the printer. Handshaking is accomplished by two printer signals, acknowledge ($\overline{ACK}$) and busy (BUSY), instead of one. When the printer receives a falling $\overline{STB}$ signal, it responds by setting the BUSY line high. While $\overline{STB}$ is low, the printer reads in the data. The $\overline{STB}$ pulse must be at least 1 µs wide. When the printer is ready to receive the next byte, it sends BUSY low (it is no longer busy) and pulses $\overline{ACK}$ low for approximately 5 µs.

The timing diagram shows the timing constraints. Some computer ports will not tri-state the data lines but keep the data constant until it sends out the next byte.

Figure 9.20 shows the pin assignments and mechanical connection part of the standard. The signal functions are described below.

Centronic Parallel Interface Signal Description

$\overline{STB}$	Strobe pulse to read data in.
DATA	First to eighth bit of parallel data (note that numbering is from 1 instead of 0)
$\overline{ACK}$	Acknowledge. An approximately 5-µs-wide low pulse to indicate that data has been received and printer is ready to accept new data.
BUSY	A high signal indicates that the printer cannot receive data (NOT READY). It becomes high in the following cases: 1. During data entry 2. During printing operation

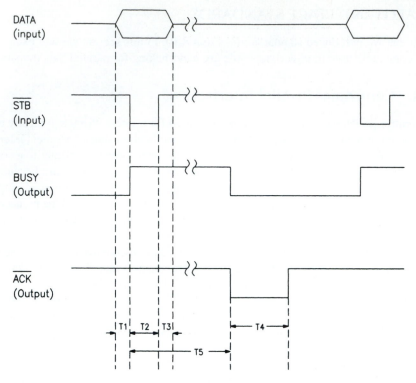

DATA (input)						
$\overline{STB}$ (Input)						
BUSY (Output)						
$\overline{ACK}$ (Output)						

T1 0.5 μs min.

T2 1 μs min.

T3 0.5 μs min.

T4 5 μs max. (approx.)

T5 1 ms or less when buffer is not full.

Input/Output data direction is relative to printer.

Figure 9.19 Centronics Parallel Interface Timing Diagram

3. In "off-line" state

4. During printer error status

PO — Paper out. Goes high when printer detects absence of paper.

SLCT — Select state. High when on-line. *On-line state* means that printer can communicate with computer. Low when off-line. *Off-line state* means printer is effectively disconnected from the computer.

$\overline{\text{AUTO FEED}}$ — If low, a linefeed is automatically added to a carriage return.

$\overline{\text{PRIME}}$ — Used to initialize printer. When low, the printer resets.

$\overline{\text{ERROR}}$ — Low when printer detects a fault. It can be caused by:

1. Paper-out condition

2. Off-line state

SG — Signal ground.

FG — Frame ground.

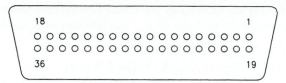

Parallel Interface Connector
Printer Side, looking away from printer.
(Amphenol 57-40360, wiring side)

Signal Pin	Return Pin	Signal	Direction with Respect to Printer
1	19	$\overline{\text{STB}}$	INPUT
2	20	DATA 1	INPUT
3	21	DATA 2	INPUT
4	22	DATA 3	INPUT
5	23	DATA 4	INPUT
6	24	DATA 5	INPUT
7	25	DATA 6	INPUT
8	26	DATA 7	INPUT
9	27	DATA 8	INPUT
10	28	$\overline{\text{ACK}}$	OUTPUT
11	29	BUSY	OUTPUT
12		PO	OUTPUT
13		SLCT	
14		$\overline{\text{AUTO FEED}}$	INPUT
16		SG	
17		FG	
18		+5V	OUTPUT
31	30	$\overline{\text{PRIME}}$	INPUT
32		$\overline{\text{ERROR}}$	OUTPUT
33		SG	

Pin Configuration

Figure 9.20 Centronics Parallel Interface Pin Assignments

+5 V TTL power supply. Note that it is the responsibility of the printer to supply the power.

Since any error condition causes BUSY to go high, in practice, most computers monitor only the BUSY and $\overline{\text{ACK}}$ lines to determine whether a printer can receive data. Many computer printer driver routines will exit (time out) if BUSY is high longer than a specified period of time. In this case the printer service routine returns an error code to the calling application program to tell it that a printer error has occurred.

You may wonder, "Surely, a programmer who wants to send output to a printer doesn't need to know all this detail." In a typical computer system, the software drivers already exist. Recall that a software driver handles the details. In this case the programmer does not have to know anything about printer interfacing. He or she simply uses a printer function. On the other hand, a system developer for a microcontroller system usually has to develop a custom interface. A technician may have to prototype or test it. In this case detailed knowledge of the interface is necessary.

We will look at using the 68HC11 as the device sending data to a printer and as the device controlling the printer. In the latter case, it receives data from another device that is typically a computer.

68HC11 As Computer.   Refer to Listing 9.15. The ASCII data to be printed is stored in a RAM buffer. An application program using the printer calls INITPRN to initialize the printer. It then sets the Y index register to point to the buffer and calls PRINT_BYTE to print each byte. The service routine PRINT_BYTE follows the handshaking protocol. It gets each byte to transmit from the RAM buffer and sends it to port C, the data port connected to the printer. When it detects the end of file (EOF) character, it sends a formfeed (FMFD). The calling program could check the contents of accumulator A for the FMFD character to determine when printing has stopped. The program uses a polling style instead of an interrupt-driven style.

```
*Listing 9.15
*I/O service routines to send output to "Centronics"
*Printer Interface. Also demonstrates peripheral
*interlocked output handshake. Calling program uses
*subroutines to print data stored in a print spool buffer.
*


*INTERFACE CONNECTIONS
*PRINTER                 MCU as computer
*DATA          <---------        Port C
*STB           <---------        STRB
*ACK           ---------->       STRA
*BUSY          ---------->       PD1
*PRIME         <---------        PD0

*PIOC CONFIGURATION DETAILS
*INVB=0      STRB/STB active low
*EGA=0       STRA/ACK active low
*PLS=0       Interlock
*OIN=1       Output handshake
*HNDS=1      Full handshake mode
*CWOM=0      Normal CMOS outputs
*    (LS compatible for Centronics standard)
*STAI=0      Disable interrupt
*STAF=x      Sets on falling STRA line
*To clear STAF, read PIOC, then write PORTCL
*(when configured)


*ASCII control characters

FMFD  EQU    $0C         ;ASCII FF character - formfeed, Ctrl-L
EOF   EQU    $1A         ;ASCII EOF character - end of file, Ctrl-Z
*-------------------------------------

      ORG    $100
```

```
*Subroutine INITPRN
*Initializes printer interface
*Calling Registers
*     IX = address of register block
*No Return Registers except CCR affected

INITPRN
      PSHA                    ;preserve registers
      BSET    DDRD,X $01      ;PD0 output, PRIME (reset) input
      BCLR    PORTD,X $01     ;PRIME low to initiate (reset) printer
      BSET    PORTD,X $01     ;PRIME high to enable printer
      LDAA    PIOC,X          ;clear STAF if set
      LDAA    PORTCL,X
      BSET    DDRC,X $FF      ;configure port C output
      LDAA    #$18            ;configure PIOC with
*                             ;magic number %00011000
      STAA    PIOC,X
      PULA                    ;restore registers
      RTS                     ;return


*Subroutine PRINT_BYTE
*Go here after ACK sets STAF during
*calling program and makes STRB/STB active
*Transmits addressed data byte to printer
*unless end of file encountered, in which case
*it sends a formfeed to printer.
*Calling Registers
*     IX = address of register block
*     IY = address of data byte
*Return Registers
*     ACCA = printed character or $0C (formfeed)
*     if end of file encountered
*     CCR affected

PRINT_BYTE
BUSY
      BRSET   PORTD,X $02 BUSY
*                             ;wait here if printer busy
      LDAA    0,Y             ;get character to print
      CMPA    #EOF            ;if EOF character then
*                             ;stop printing
      BEQ     SFMFD
      STAA    PORTCL,X        ;else send current char
*                             ;to printer
*                             ;also clears STAF
NOACK
      BRCLR   PIOC,X $80 NOACK
*                             ;if no ACK then wait here
*                             ;(until ACK sets STAF)
      RTS                     ;otherwise return
```

```
*send formfeed to printer
SFMFD
*                              ;and return
     LDAA    #FMFD
     STAA    PORTCL,X
     RTS
```

68HC11 as Printer.   An 8-bit microcontroller is more likely to be a controlling device for
the printer itself since most computers are 16- or 32-bit devices. In this section we will concern ourselves only with the printer's data interface. Chapter 14 covers other parts of a
printer control program.

The 68HC11 handshaking subsystem can be configured for automatic control of two
handshaking signals, but the Centronics standard uses three. Listing 9.16 illustrates the connections used in our program example. The design is based on using port C for a simple input strobe. The printer output handshaking signals are driven using bit manipulation
instructions in the software. Remember that Listing 9.16 only demonstrates the portion of
a 68HC11 program that inputs printer data.

```
*Listing 9.16
*I/O service routines to control "Centronics" interface
*part of printer control system. Also demonstrates
*software handshaking because port C configured
*for simple input strobe. Main program would call
*service routines to store input data in a RAM print
*buffer for later output to print mechanism.


*PIOC CONFIGURATION DETAILS

*CONNECTIONS
*Computer Printer Port     MCU as Printer Controller
*DATA        -------------->     Port C
*STB         -------------->     STRA
*ACK         <--------------     PD2
*BUSY        <--------------     PD3


*PIOC CONFIGURATION DETAILS
*INVB=x    STRB/STR active low
*EGA=0     STRA/ACK active low
*PLS=x     Interlock handshake
*OIN=x     Output handshake
*HNDS=0    Simple strobe mode
*CWOM=0    Normal CMOS outputs
*    (LS compatible for Centronics standard)
*STAI=0    Disable interrupt
*STAF=x    Sets on falling STRA line
*To clear STAF, read PIOC, then read PORTCL


*---------------------------------------
```

```
        ORG     $100

*Subroutine INIT_INTRF
*Initializes parallel interface part
*of printer upon power-up reset.
*Calling Registers
*    IX = address of register block
*No Return Registers except CCR affected

INIT_INTRF

        PSHA                    ;preserve registers
        LDAA    PIOC,X          ;clear STAF if set
        LDAA    PORTCL,X
        BSET    PORTD,X $0C     ;PD2,3 output and
        BSET    DDRD,X $0C      ;BUSY, ACK high
        LDAA    #00             ;configure PIOC
        STAA    PIOC,X
        PULA                    ;restore registers
        RTS                     ;and return
*Subroutine INPUT
*Reads parallel port to get byte sent
*by an external device
*Calling Registers
*    IX = address of register block
*Return Registers
*    ACCA = input data byte
*    CCR affected

INPUT
        BCLR    PORTD,X $0C     ;BUSY low and pulse ACK
        NOP                     ;for approx. 5 us (E = 2 MHz)
        BSET    PORTD,X $04
CIN
        BRCLR   PIOC,X $80 CIN
*                               ;wait for STB pulse
        LDAA    PORTCL,X        ;get input and clear STAF
        BSET    PORTD,X $08     ;set BUSY high
        RTS                     ;return
```

The observant reader may notice that the source code does not use an overscore for active-low signals. For example, it shows STB instead of $\overline{STB}$. The source code has to restrict itself to using standard ASCII characters; otherwise, it cannot be assembled. Hence, overscores or other special markings cannot be used.

HCMOS-TTL Interfacing.   The Centronics standard states that the signal levels should be TTL or LSTTL (see Section 6.1.4). Since the 68HC11 uses HCMOS, not TTL, it is necessary to interface two different logic families. It turns out that HCMOS outputs can be connected directly to TTL inputs, but a TTL output connection to an HCMOS input should

TABLE 9.4   ELECTRICAL SIGNAL CHARACTERISTICS FOR TTL AND HCMOS

Characteristic	TTL	HCMOS
V_{IH}	2.0 V	3.5 V
V_{IL}	0.8 V	1.0 V
V_{OH}	2.4 V	$V_{CC} - 0.1$ V
V_{OL}	0.5 V	0.1 V

use a pull-up resistor because the two logic families have different signal threshold levels. For a review of signal limits, refer back to Section 6.2. Note that the keyboard interfaces described earlier also used pull-up resistors.

Table 9.4 lists some typical values of signal levels for the two families. The table assumes a power supply of +5 V. For HCMOS, the input characteristics are proportional to the supply voltage at pin V_{CC}. Note that a TTL high output could be as low as 2.4 V. But an HCMOS input recognizes only a signal greater than 3.5 V as being a valid high input. It is necessary to connect a pull-up resistor to ensure that a TTL logic-high output is greater than 3.5 V. Connecting HCMOS outputs to TTL inputs can be direct since the HCMOS output tolerances fall within the TTL input tolerances.

Choosing a value for the pull-up resistor is a matter of making compromises. A larger resistor value will reduce power consumption for logic-low outputs. The resistors will draw less current for the same voltage drop. But every logic input has associated with it a parasitic capacitance (C_{IN}). Hence, a larger resistor value also slows down signal transition speeds. It will take longer to charge up or discharge the parasitic capacitance when higher resistance is used.

For illustration, consider input capacitance for the port C and strobe A pins. They are listed in the *Data Book* to be 12 pF. If a resistor of 10 kΩ is used, then the time constant (RC) is 120 ns. The rise time will still be less than the minimum time durations specified in the standard. Also, experience has shown 10-kΩ resistors to be suitable.

9.8.2   The IEEE-488 General-Purpose Instrumentation Bus

The *Hewlett-Packard Corporation* (now *Agilent Technologies*), a major instrument manufacturer, developed a parallel data transfer standard in the early 1980s. It is used for communication between a computer and instruments. For example, a computer tells a digital storage scope to sample data and to send the data to the computer for analysis. Later, the IEEE organization adopted this standard and called it the IEEE-488 General-Purpose Instrumentation Bus (GPIB) standard. As a result, other manufacturers can purchase a license to use this standard. Since its beginning in 1975, the standard has been revised. IEEE-488 specifies the physical characteristics and the protocol for transferring bytes of data. Like many other standards, it uses its own terminology. The IEEE-488 Interface is also an example of an interface in which open-collector (TTL) or open-drain (HCMOS) outputs are used.

Devices.   Each device is defined to be a *talker,* a *listener,* or a *controller.* A device can change its function. The listener receives data. The talker transmits data. The controller sends commands to talkers and listeners and conducts polls to see which devices are active.

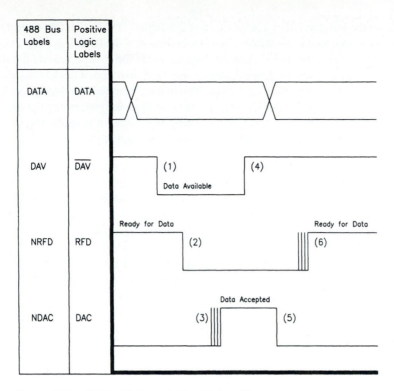

Figure 9.21 IEEE-488 Handshaking Timing Diagram

Bus Lines. These are:

Eight bidirectional data lines to carry commands, address, and data information

Five general control lines to manage the bus

Three handshaking lines to control each byte transfer

Handshaking. The only part of the standard we examine in further detail is the handshaking protocol. The protocol is fully asynchronous, meaning that it does not depend on timing. Each time a byte is transferred, the bus goes through a handshake cycle. Figure 9.21 shows the timing diagram. Officially, the bus uses negative logic (logic 1 is low voltage), but some manufacturers prefer to use terminology based on positive logic. We will use the positive logic terminology. The handshake control lines are

$\overline{\text{DAV}}$ *Data available* output signal from the talker, active low.

RFD   *Ready for data* output signal from the listener, active high. NRFD means low when not ready for data.

DAC   *Data accepted* output signal from the listener, active high. NDAC means low when data not accepted.

The listener asserts the ready for data (RFD) line high when it is ready to accept data. The talker checks that the RFD line is high before it starts transmission of data. When it has placed data on the bus, the talker asserts the data available $(\overline{\text{DAV}})$ line low (step 1) to notify other devices that data is available. The listener deasserts the RFD line (step 2) because it has begun processing the data. When the listener reads the data, it asserts the data accepted (DAC) line high (step 3). When the talker sees DAC asserted, it deasserts $\overline{\text{DAV}}$ high (step 4). Then it may put new data on the bus. The listener responds to DAV going high by deasserting DAC low (step 5). The listener may have to continue processing the data it read in, perhaps to store it, before it is ready to read new data. When it is ready, the listener asserts RFD high (step 6) to indicate that a talker may send data to it.

Wired-AND/OR Logic With Open-Drain Outputs.   There can be up to 15 devices on the bus. Only one device can be a talker at any time. But more than one device can be a listener. Hence, the talker has to ensure that all listeners received the data byte before initiating a new handshake cycle or before it relinquishes control to a new talker. To accomplish this, the handshaking lines should be open-collector or open-drain, thus implementing wired-OR and wired-AND logic.

Figure 9.22 shows several HCMOS open-drain outputs connected to a common line. The outputs of an open-drain are a virtual short when the output is low and are high impedance for a logic high. In other words, the output behaves like a switch. A low output "closes the switch" to ground the output line. A high output "opens the switch" and the external pull-up resistor pulls the line level high. Section 6.2.4 mentions open-drain briefly.

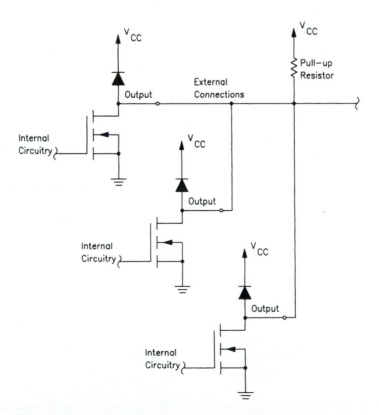

Figure 9.22   Open-Drain Outputs for Wired-AND Application

There we mentioned that the specification for minimum high-level output voltage (V_{OH}) did not apply. This is because the level is determined by voltage supplied to the pull-up resistor and the resistance of the loads.[2] Since outputs can be connected together, the output is high only when *all* the gate outputs are high impedance (logic high). If you think in terms of positive logic, the connection is like an AND gate. All inputs to the connection must be high for the line to remain high. If any drain is turned on to provide a logic zero input, the line is pulled low despite what the other outputs are doing. If you think in terms of negative logic, meaning that a logic zero is defined to be active, asserting any signal (low) causes the output (the line) to be asserted (also low). Hence, you will also see the term *wired-OR* applied to this connection.

In the case of the IEEE-488, $\overline{\text{DAV}}$ is a wired-OR. If any talker asserts $\overline{\text{DAV}}$, the $\overline{\text{DAV}}$ line will be driven low. The RFD and DAC lines are wired-AND. All listeners have to assert these lines before the lines stay high. The timing diagram of Figure 9.21 shows spikes for RFD and DAC to indicate other listeners asserting these lines. The port C and port D outputs of the 68HC11 can be configured to be open-drain. For port C, set the port C wire-OR mode bit (CWOM) in control register PIOC. For port D, set the port D wire-OR mode bit (DWOM) in another control register (SPCR).

Other Open-Drain Pins. We should mention that $\overline{\text{RESET}}$ and MODA are also open-drain pins. However, they are open-drain because they can also act as outputs. A COP or Clock Monitor Fail Reset causes $\overline{\text{RESET}}$ to pulse low (see Section 3.9.2). Section 7.4 already mentioned that the microcontroller drives pin MODA low when fetching an opcode.

GPIB Interface ICs. Some ICs are designed for interfacing a device to an IEEE-488 bus. Some of these are the Intel 8292, the Motorola MC68488, and the NEC μPD7210.

9.8.3 Small Computer Systems Interface (SCSI)

The small computer systems interface has the acronym *SCSI,* pronounced "scuzzy." This is an interface protocol designed for communications between personal computers and intelligent I/O devices such as disk drives, tape drives, and remote printers. SCSI has been an ANSI standard since 1986 (X3.131-1986). We will present a summary of the standard first; then we will concentrate on a detail of the standard that specifies a handshake sequence.

This interface uses a 50-pin connector with nine data lines and nine control lines for a single-ended system (referenced to ground). A differential system uses more of the available 50 connections. The ninth data bit is used for parity. A parity bit is used to check if an error has occurred in the data byte. We cover parity further in Chapter 10. The data lines carry other information, such as addresses and commands. Perhaps it would have been more appropriate to call them information lines.

---

[2]According to circuit theory, the voltage-divider rule determines the output voltage:

$$V_{\text{output}} = \frac{V_{CC}R_{\text{load}}}{R_{\text{load}} + R_{\text{pull-up}}}$$

Note that R_{load} is the load resistance of all high outputs connected in parallel.

TABLE 9.5 SCSI INFORMATION TRANSFER PHASES AND CONTROL SIGNALS

Signal				
MSG	C/D	I/O	Phase name	Description
0	0	0	DATA OUT	Initiator sends data to target
0	0	1	DATA IN	Target sends data to initiator
0	1	0	COMMAND	Initiator sends command to target
0	1	1	STATUS	Target sends status to initiator (reserved)
1	0	x		
1	1	0	MESSAGE OUT	Initiator sends message to target
1	1	1	MESSAGE IN	Target sends message to initiator

The SCSI bus can support up to 8 devices, each of which can have 8 *logical units*. Each logical unit can have 256 *logical subunits*. Theoretically, it is possible to have up to 14,000 peripherals on the SCSI bus. A device is either an *initiator* or a *target*. The initiator issues commands and is sometimes known as the host. A target performs commands. A typical SCSI system is a computer (the initiator) that reads data from or writes it to a hard disk drive (the target). In any SCSI system at least one device is an initiator and at least one other device is a target.

The SCSI bus makes transitions between eight possible states that are known as *phases*. A phase determines the direction of the data lines and the type of information put on the data lines. Four of the phases are information transfer phases. Each of these phases is identified by three of the control signals: message (MSG), control/data (C/$\overline{\text{D}}$), and input/output (I/$\overline{\text{O}}$). Information can be command, status, or data. Information transfers can be synchronous or asynchronous. Table 9.5 shows the phases and control signals. The other four phases are used to negotiate the use of the bus prior to information transfer. It is during these phases when devices are addressed.

The message signal (MSG) is controlled by the target. It indicates when a message is on the bus. Control/data (C/$\overline{\text{D}}$) is a signal asserted by the target during a transaction. It indicates whether control or data information is on the bus. The input/output (I/$\overline{\text{O}}$) signal indicates the direction of a data transfer relative to the initiator (host). Figure 9.23 shows part of a SCSI transaction when an initiator reads data from a target. After the initiator gains control of the bus and selects the target, it issues commands. The target responds by sending data. The transaction ends with the transfer of status and message bytes. Figure 9.23 shows the command and data part of the transaction. In this case the handshake signals are active low. A DATA OUT phase would be similar to a COMMAND phase except that C/$\overline{\text{D}}$ is zero.

Note that the use of request ($\overline{\text{REQ}}$) and acknowledge ($\overline{\text{ACK}}$) pulses differ depending on the direction of data transfer. When data flows from the initiator to the target, data is considered to be valid from the start of the $\overline{\text{ACK}}$ pulse and held until at least the end of the $\overline{\text{REQ}}$ pulse. When data flows from the target to the initiator, data is valid from the start of the $\overline{\text{REQ}}$ pulse and held until at least the start of the $\overline{\text{ACK}}$ pulse.

An asynchronous transfer cycle uses the $\overline{\text{REQ}}$/$\overline{\text{ACK}}$ handshake sequence as follows. Interestingly enough, the target begins the handshake sequence.

Case 1: Initiator writes data to target (I/$\overline{\text{O}}$ = 0).

The target asserts $\overline{\text{REQ}}$ to indicate ready-to-read data.

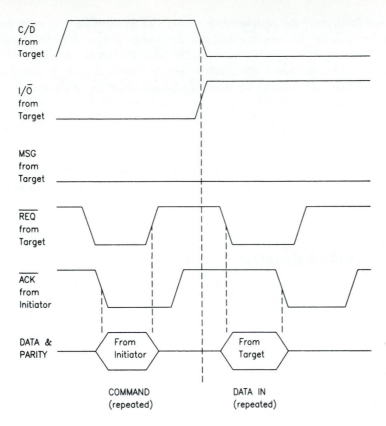

Figure 9.23 Command and Data Phases of a SCSI Bus

The initiator responds by sending the data and asserts $\overline{\text{ACK}}$ to acknowledge the request.

The target responds to the $\overline{\text{ACK}}$ signal by latching in the data and deasserts $\overline{\text{REQ}}$.

The initiator responds to the deasserted $\overline{\text{REQ}}$ line by deasserting $\overline{\text{ACK}}$ to enable the next transfer.

Case 2:   Initiator reads data from the target ($\text{I}/\overline{\text{O}} = 1$).

The target puts data on the bus and asserts $\overline{\text{REQ}}$.

The initiator latches in data and asserts $\overline{\text{ACK}}$.

The target responds by deasserting $\overline{\text{REQ}}$.

The initiator responds by deasserting $\overline{\text{ACK}}$ to enable the next transfer.

If the target and initiator negotiate an agreement, they can use the faster synchronous transfer to avoid waiting for handshake signals. A synchronous data transfer also uses the same handshake signals except that the target pulses $\overline{\text{REQ}}$ for each byte of data, and the initiator pulses $\overline{\text{ACK}}$ the same number of times. It is possible for both devices to agree to transfer bytes at rates greater than 4 megabytes per second.

More and more computer peripherals are using a SCSI interface. It is more practical to use ICs designed to use this interface than to build your own. The MC68HC99 hard disk controller chip has a SCSI compatible interface. The SN75C091A is a SCSI bus controller chip offered by Texas Instruments, which can be used in system designs. You may wish to consult reference 6 (in Appendix C) for more information about the SCSI bus.

■ 9.9 SUMMARY

Section 9.9.1 of the summary reviews the parallel I/O subsystem as implemented by the 68HC11. Section 9.9.2 reviews some of the applications that can interface to almost any parallel I/O subsystem.

9.9.1 Parallel I/O Subsystems Summary

Ports B and C can be used for parallel I/O only when in single-chip mode or as part of the port replacement unit (PRU). Pins in other ports can be used for parallel I/O if other I/O subsystems do not use them. The control register for the parallel I/O subsystem is PIOC ($1002). It also has a status bit that is the strobe A flag, STAF (bit 7). The following summarizes the use of 68HC11 ports for parallel I/O and parallel I/O operations.

Port A:

Data register is PORTA ($1000).

Three input pins (PA0,1,2), four output pins (PA3,4,5,6), one bidirectional pin (PA7).

Bit DDRA7 in register PACTL (B7 in $1026) sets the direction for PA7.

Port B:

Data register is PORTB ($1004).

Parallel outputs only when in single-chip mode. Its pins are the high address byte in expanded mode.

The 68HC24 Port Replacement Unit restores port B parallel output functions for expanded mode operation.

Port C:

Data register is PORTC ($1003).

Parallel inputs or outputs when in single-chip mode. Direction is configured by register DDRC ($1007). Its pins are the multiplexed low address byte and data byte when in expanded mode.

The 68HC24 Port Replacement Unit restores port C parallel I/O functions for expanded mode operation.

Port D:

Data register is PORTD ($1008).

Only pins PD0 to PD5 are available for input or output.

PD6 is reserved for STRA/AS.

PD7 is reserved for STRB/R/$\overline{\text{W}}$.

Data direction register is DDRD ($1009, B0 to B5).

Port E:

Data register is PORTE ($100A).

Input only.

Reset condition:

Data direction bits are zero.

Output port data bits are zero.

Input port data bits high impedance.

Strobed I/O:

Input. Upon detecting an edge on STRA, the 68HC11

> Sets a status flag, STAF in PIOC (bit 7 in $1002).
>
> Has data available to it in PORTCL ($1005).
>
> Data is latched in to PORTCL by the active STRA edge.

If enabled by bit STAI in PIOC (bit 6), an interrupt occurs with the setting of flag STAF.

Output. When the MCU writes a byte to register PORTB ($1004), a pulse is generated on pin STRB. No interrupts or status flags are available for strobed output on port B.

Registers used:

DDRC	Select desired port C pins as inputs or outputs.
PORTCL	Latch port C input data of STRA edge, used to read latched data.
PORTC	Input pin values (not latched) on port C pins, used to read current data.
PORTB	Latch port B output data (outputs data using STRB).
PIOC	Parallel I/O control register.

Pins:

Port B	PB0 to PB7 as outputs.
Port C	Any of PC0 to PC7 as inputs.
Port D	PD6 as STRA input, PD7 as STRB output.

Handshake:

Uses port C only for either input or output handshake operation.

STRB acts as a READY line to indicate to a peripheral that the 68HC11 is ready to input or output data.

The peripheral pulses STRA when it has provided data to, or taken data from, port C. The following also occurs:

> Set status flag, STAF, in PIOC (B7 in $1002).
>
> For input, data is latched in PORTCL ($1005).
>
> For output, data from PORTC ($1003) is latched in the peripheral.

The READY line, STRB, is negated.

If enabled by bit STAI in PIOC (B6), an interrupt occurs with the setting of flag STAF.

Registers used:

DDRC	Select desired port C pins as inputs or outputs.
PORTCL	Latch port C input data on STRA edge for input handshake mode (used to read latched data). Output data for output handshake mode.
PORTC	Output pin values on port C pins for direct output without STRB assertion or read current (not latched) port C pin data.
PIOC	Parallel I/O control register.

Pins:

Port C	PC0 to PC7 as inputs or outputs.
Port D	PD6 as STRA input, PD7 as STRB output.

9.9.2 Parallel I/O Applications Summary

Seven-Segment Display. The character displayed depends on which segments are on and which are off. Each character has a corresponding binary code whose bits specify the state of the segments. The code can be part of the display software in the form of a look-up table. Other interfaces use driver ICs with built-in decoding. In a common cathode display a high digital output turns on the segment and a low output turns it off. The common anode display functions in the opposite manner. To turn on a segment requires more current than normally available at a digital output. Hence, most display interfaces use driver ICs to supply the current necessary.

Liquid-Crystal Displays (LCDs). An LCD has the primary advantage of requiring very little power. However, they require specialty driver ICs. LCD interfacing is not standardized. Methods of interfacing different LCDs vary widely because there are many different types of LCD segment layouts. The most common are seven-segment and dot matrix. The binary codes required to drive selected segments vary.

All LCDs require some form of alternating waveform to turn the segments on or off. These can be provided by an external IC. Alternatively, it is built into an LCD panel. A simple LCD interface would use a parallel port to send data to an LCD driver and display. Other drivers may use a serial interface for connection to a microcontroller. But the interface between the LCD and the driver is parallel.

Keyboard Interface. Most keyboards have their individual keys wired in a matrix of rows and columns. Pressing a key causes a unique connection of a row line to a column line. A typical interface drives a column (or row) line low and checks a row (or column) line to see if it, too, is driven low. When a low line is detected, the pressed key is identified by the binary code representing the columns and rows. A keyboard software driver typically returns a key code (or scan code) to indicate which key has been pressed. It does this by running a scan sequence to check row and column combinations sequentially.

Because keys bounce mechanically, debouncing is also required. Hardware (filters or latch circuits) or a software time delay can provide debouncing. Other variations in keyboard interface design are due to keyboard size, keyboard layout, and use of decoding ICs.

Handshaking.   Many peripherals use parallel interfaces. These require handshaking. A handshake sequence is used to confirm transfer of data from one device to another. A device will not send another unit of data (normally, a byte) until it has received an acknowledgment that the other has read the previous unit and is now ready to receive the next one. The acknowledgment and request signals are usually logic edges.

Parallel Interface Standards.   Due to the vast number of possible implementations, standardized parallel interfaces had to be agreed upon. Among other things, these standards specify handshake sequences. Most computers with a parallel printer port use the Centronics standard. Hence, most application programs use a printer driver that uses the Centronics handshaking sequence. The IEEE-488 GPIB is used in instrumentation systems. The instruments send or receive data from a computer controlling the bus. The SCSI interface is a standard for connecting peripherals to a computer. This allows any type of peripheral to communicate with any type of computer as long as they both conform to the standard.

■ EXERCISES

1. Write program instructions to configure port C so that the low nibble is an input and the high nibble is an output.

2. Write program instructions to read the port C inputs when it is configured as in Question 1.

3. For port C configured as in Question 1, write instructions to turn off PC4, PC5 and turn on PC6, PC7.

4. Refer to Figure 9.5. What should be displayed after RESET? Explain why.

5. The display of Figure 9.5 is connected to an EVBU board running the Buffalo monitor. One of the monitor commands is "mm" (memory modify). To use it:

>   Enter mm space and the address to modify.
>   System displays existing data at address.
>   Enter new data.

   For example, to change the contents of $0100 to $55,

```
mm 0100 xx 55
```

   where xx is the existing data. Note that Buffalo assumes that all numbers are hex. Design a way to test the interface of Figure 9.5 using the memory modify command.

6. The connections to a common anode seven-segment display are as follows:

rdp	a	b	c	d	e	f	g
B7	B6	B5	B4	B3	B2	B1	B0

Complete the following table to show the seven-segment codes.

Character	Hex code
0	
1.	
2	
3.	
4	
5.	
6	
7.	
8	
9.	
blank	

7. What has to be accounted for when building an interface to read switches from a factory control panel?

8. Is it possible to interface an LCD display directly to the 68HC11? Why?

9. For an LCD panel connected as shown in Figure 9.7, write instructions to display a large zero. Use the subroutine OUTLCD.

```
*Subroutine OUTLCD
*Calling registers (pseudo)
*        Address $00 = 48-bit segment data
*Return registers none
```

10. For the system of Question 9, write instructions to test the LCD panel. Also use subroutine DLYHS. It is a 1/2-second time delay with no calling or return registers.

11. Write the subroutine GETKEY for the hardware shown in Figure 9.12.

12. Describe how you could build a keyboard interface for a 16×16 matrix keyboard. You may use Figure 9.12 as a design concept.

13. If you are developing a keyboard controller for a 16×16 matrix keyboard, would you use the 68HC11 ports only or include hardware decoders? Explain why.

14. A 68HC11 is used to program a 2716 EPROM. It is inserted in a socket so that it is connected for programming mode. The connections are as follows:

Pin	Connection
V_{pp}	+25 V
V_{cc}	+5 V
$\overline{G}$	Through a 10-kΩ pull-up resistor to +5 V

(a) Indicate what connections to make to the EPROM's data pins (D0 to D7) and address pins (A0 to A10). Also, there should be a connection to the EPROM's pin $\overline{E}$/Prog.

(b) *Note:* The 2716 EPROM is organized as 2K × 8 bits. You can also think of it as having eight 256-byte pages. Each *page* has 256 *offset* addresses. For example, address $23F is in page 2 with an offset of $3F. The system programs each byte as follows: Set up the address lines and data lines. Apply a 50-ms high pulse at $\overline{E}$/Prog. Repeat the procedure for each address.

15. For the system of Question 14, write a program to program the EPROM with the data stored in RAM whose start address is the label DATABUF.

16. Design a system to generate an XIRQ interrupt every time a user writes to port B.

17. A peripheral sends data. It also pulses a control line high for 2 μs every time it sends the data. This also initiates an interrupt service routine in the 68HC11 to read data from the peripheral. Design the hardware and software for this interface.

18. Write the instructions required to configure the parallel I/O subsystem. In other words, find the "magic" numbers to put in the appropriate registers. Also write sample instructions to handle I/O.

(a) Configure the subsystem for an output handshake to an HCMOS device where port C can still be used for input. The handshake is interlocked. The MCU polls for a rising request signal and acknowledges the request with a high signal.

(b) Configure the subsystem for a simple strobe with open-drain output. Port D also has open-drain output. The output strobe signal should be active low. The input strobe is a don't care and no interrupt is generated.

(c) Configure the subsystem for a pulsed input handshake that is serviced by an interrupt when a falling edge occurs on the request line. The acknowledge should be active high. The MCU can also output to HCMOS devices connected to the data port when it is not being used for the input handshake.

19. Write a program that reads data from an IEEE-488 bus. The connections are as follows:

IEEE-488 bus	68HC11
Data	Port C
DAV	IRQ, STRA
DAC	STRB
RFD	PD0

All handshake signals are connected to noninverting buffers with open-collector or open-drain outputs. Therefore, it is not necessary to configure ports C (an input) and D to be open-drain.

20. (a) Invent a protocol for handshake communication between two 68HC11 systems.

    (b) Design the hardware and pseudocode for the system.

10 The Serial Subsystems

OBJECTIVES

After completing this chapter, you should be able to

- Describe settings used for asynchronous serial character transmission.
- Design an RS-232 circuit.
- Describe the operations (such as double buffering), signals, and registers used in the serial communications interface (SCI).
- Write software using the SCI, including configuration, I/O, and the handling of errors.
- Explain differences between asynchronous and synchronous serial communication.
- Describe the operations, signals, and registers used in the serial peripheral interface (SPI).
- Design I/O circuits using SPI for bus and cascade topology.
- Write software using the SPI including configuration and I/O for master and slave operations.
- Describe some basic principles of network communications such as the open systems interconnection model and some protocols.

The 68HC11 has two subsystems for serial interfacing. The serial communications interface (SCI) can be used to connect a terminal or personal computer to the microcontroller. It can also connect several widely distributed microcontrollers to form a communication network. The serial peripheral interface (SPI) can provide high-speed serial communication to peripherals or other microcontroller units.

In this chapter we will initially cover concepts of asynchronous communications. These apply to most asynchronous serial interfaces. In particular we will examine the commonly used standard RS-232. In the next section we will detail how the subsystem is implemented in the 68HC11 microcontroller. We will then cover uses and the basic operation of the synchronous serial interface used by many digital peripheral chips to load in or send out data. The final part of the chapter details how this peripheral subsystem is implemented in the 68HC11.

355

■ 10.1 ASYNCHRONOUS COMMUNICATIONS SYSTEMS

10.1.1 A Serial Communications Primer

Recall that in Section 8.6.2 and Figure 8.4 we examined the concepts of asynchronous I/O. Each device uses its own clock. The clocks must run at the same rate but do not need to be synchronized. The asynchronous I/O subsystem for the 68HC11 is called the serial communications interface (SCI). As implied by the word *communications* in its name, the SCI subsystem is used primarily for data communications. This usually implies that *remote* devices are exchanging information. *Remote* typically means a distance greater than a few meters. An asynchronous communications system (or network) must have at least one transmitter and one or more receivers. A transmitter sends data to be read by the receivers. For some asynchronous networks, devices may be a transmitter, a receiver, or both.

Start and Stop Framing. When serial transmission is used between computers, all of the data bits travel in a single line, one after the other. The basic unit (group of bits) of information is the *character* or data frame. It is normally a byte. The transmitter can send characters at any rate, so there may be time delays between the transmission of each character. For asynchronous communication the receiver needs to know when a character starts and when it stops, so each character is *framed* by start and stop bits. Figure 10.1 shows some of the possible framing configurations. Note that a constant high signal is considered to be idle with no data being transmitted. A start bit is always a zero (0) and a stop bit is always a one (1). Also note that data bits are transmitted with the LSB first. There is no level transition between bits of the same value. For example, if all data bits were 1, the transmitted frame would be low only during the start bit. This format is known as "non return to zero" (NRZ). This means that the voltage does not return to zero between adjacent 1 bits. Other systems may use 2 or $1\frac{1}{2}$ stop bits. Typically, the transmitter hardware automatically creates the framing bits and the receiver hardware automatically removes them.

Parity. *Parity* is used to detect single-bit errors. It refers to the quantity of 1 bits in a binary number. If that quantity is even, the number has even parity. If it is odd, the number has odd parity. A parity bit may be used to detect errors. Often, it is the eighth bit in a data byte. Note that the standard ASCII code uses only 7 bits.[1] Hence, the eighth bit may be used as a parity bit.

The transmitter is responsible for creating the parity bit and the receiver deciphers it. In some serial hardware, the transmitter creates it automatically. In others, software must be used to create the parity bit. The parity bit can be a 1 or 0, depending on two things:

1. The type of parity selected, even or odd
2. The quantity of 1 bits in the data byte to be transferred

---

[1] Character sets vary among computer manufacturers. Most follow a standard definition for the first 128 characters, which is ASCII (see the Quick Reference [Appendix B]). However, they may differ in their definitions of characters 128 to 255. Another standard for coding characters as binary numbers is the ANSI character set. For example, the U.S. version of the IBM PC uses character 156 as the British pound symbol (£), whereas it is character 163 in ANSI.

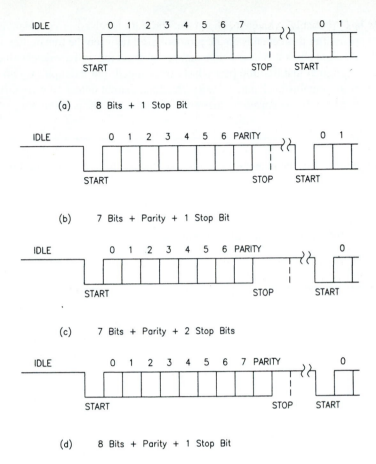

Figure 10.1 Asynchronous Serial Transmission and Framing

To illustrate the use of parity, consider the case when the transmitter and receiver agree to use 7 bits plus even parity plus one stop bit. Let's say that the transmitter will be sending the character ASCII 8 (binary 111000). Since this character has three 1s in the binary code, the transmitter must set the parity bit to 1 to make the total amount of 1s an even number (4). Then the transmitter sends the following framed bit pattern, shown in order of transmission (left to right):

0	0001110	1	1
Start	Data	Parity	Stop
bit	bits	bit	bit

There is now an even number of 1s in the data byte (parity bit is included as part of parity determination). Note that the start and stop bits are not included in the parity determination.

As we said, parity is used for error detection. Let's say that one of the bits switched logic levels due to noise or synchronization errors: For example, bit 2 switched to logic 1. Thus the receiver detects two 0 bits, four 1 bits, a 0 bit, and then a 1 parity bit. It will note that although the parity is supposed to be even, in fact it is detected as odd (five 1s). It will conclude that one of the bits has "probably" changed. Therefore, the data received is considered to be wrong. What happens after this depends on the protocol used between the transmitter and receiver. (We will talk more about this later.) Note that parity checking will not detect all errors. Two bits may change such that the true parity has not changed. For example, if two 0 bits switch to 1, the parity remains unchanged. Parity checking cannot detect all errors. Also, it can only note that an error has occurred. It does not indicate which bit is in error.

PRACTICE

Show the framing bits when the character B is sent at 7 data bits, 2 stop bits, odd parity.

Solution

Start bit   0
Data bits   0100001 (shown in order of transmission)
Parity bit   1
Stop bits   11

Data Speed, Baud.   Serial data speed is the number of bits transmitted per second (BPS). *Baud* is the rate at which the signal changes. It is the rate of change of the signal (reciprocal of shortest signal element). Sometimes each signal element represents one data bit, so data speed and baud are the same. In some cases communication requires extra signal bits that are not data. In other cases, a signal element may encode several data bits, which will be described in Section 10.1.4. In the MCU's serial communications interface (SCI), each signal element encodes 1 bit.

Baud rate includes the start, parity, and stop bits. Often the term *baud* is used to express the number of bits transferred per second. Thus 9600 baud means that serial data is transmitted and received at the rate of 9600 bits per second. The number of BPS is the number of "data" bits sent per second. This is confusing. Perhaps less confusion would result if BPS was called DBPS, where D stands for data. We can express BPS as follows:

$$\text{BPS} = \frac{\text{number of data bits}}{\text{number of frame bits}} \times \text{baud} \qquad (10.1)$$

Also,

$$\text{characters per second} = \frac{\text{baud}}{\text{number of frame bits}} \qquad (10.2)$$

Equation (10.2) assumes ASCII. Note that one frame contains one character whether it is standard 7-bit ASCII or an extended 8-bit character code such as ANSI.

The following illustrates the use of baud, BPS, and characters per second. A system uses 9600 baud, 8 data bits, no parity, and 1 stop bit. Each 8-bit data frame has a total of 10 bits (start, data, and stop). Then

$$\text{data bit rate} = 8/10 \times 9600 \text{ baud}$$
$$= 7680 \text{ BPS}$$

and

$$\text{char/sec} = 9600 \text{ baud}/10 \text{ bits per character}$$
$$= 960 \text{ characters per second}$$

We will return to BPS in Section 10.1.4 when we look at modems.

RS-232 Interface Standard.   When equipment communicates using asynchronous serial, their connection is usually made with an RS-232 interface. The standard defines signal levels, connectors, and pin assignments, among other things. Just about every computer uses this standard for serial communication. Most equipment has a *serial port* that follows a subset of this standard. Since it is an early standard (1971) that has been revised several times, it is probably one of the most confusing. Many still use the term RS-232, although the official name is now EIA 232. A serial port is the serial interface connector plus all the electronics required to make the interface fully functional.

The logic levels used for RS-232 signals differ from the usual digital circuit logic. The nominal RS-232 data signals are $+12$ V (space) for logic 0 and -12 V (mark) for logic 1. These allow signals to be transmitted over greater distances. The standard specifies the minimum and maximum voltages allowed for both logic levels for transmitter and receiver. Note that the signal levels seem backward to normal logic polarity. Also, the standard specifies that no logic signal can be between $+$ or -3 V. Hence, no logic signal is 0 V. This is a bipolar form of the "non return to zero" (NRZ) format.[2] During a sequence of data bits that are all logic 1s (or 0s), the signal is constant. It does not change. There is no return to a reference for each data bit.

The standard defines 25 different signals, although in practice few of these are used. Most RS-232 interfaces use a mechanical connector with 25 pins known as a DB-25. The shape of the connector is similar to that used for the Centronics interface (see Figure 9.20) except that it uses 25 pins. The shape resembles the letter D. These connectors are available as plugs (DB-25P) or as sockets (DB-25S). Since few of the 25 signals are ever used, many serial ports also use a DB-9 connector. As you may guess, it uses only nine plug pins (or socket holes). Very often, only three signals are required for a connection. For example, the evaluation board (EVBU) uses only three signals: transmit data, receive data, and signal ground. These three signals are shown in Table 10.1.

Standards organizations such as ISO and CCITT have categorized equipment for data communications into two types: data terminal equipment (DTE) and data communications equipment (DCE). The RS-232 standard also uses these categories. The two types use their transmit (TxD) and receive (RxD) lines differently, as follows:

DTE: TxD transmits, RxD receives

DCE: TxD receives, RxD transmits

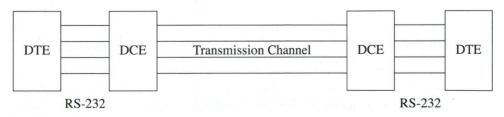

<hr>

[2]The term *NRZ* also refers to the technique used to store data on a magnetic (floppy or hard) disk. The direction of current in the recording write head determines the logic bit value. The current is never zero.

TABLE 10.1 COMMONLY USED RS-232 SIGNALS

Name	Common abbreviation[a]	DB25 Pin/EIA circuit label	DB9/EIA-574 Pin/EIA circuit label
Transmit data	TxD or TD	2/BA	3/103
Receive data	RxD or RD	3/BB	2/104
Request to send	TSR	4/CA	7/105/133
Clear to send	CTS	5/CB	8/106
Data set ready	DSR	6/CC	6/107
Signal ground	SG	7/AB	5/102
Data carrier detect	DCD	8/CF	1/109
Data terminal ready	DTR	20/CD	4/108
Ring indicator	RI	22/CE	9/125

[a]Many people use the common abbreviations instead of the standard EIA circuit label.

Normally, a DTE interfaces with a DCE in order to communicate with another DTE that is also interfaced to a DCE. A DTE is the initial source of data or the final recipient. It usually provides the control logic for data transfer. Examples are terminals and computers. The DCE provides the functions to establish, maintain, and terminate a connection. It also codes or decodes signals as required. Examples are line drivers and modems for telephone communications. According to the standard, a DTE uses a plug and a DCE uses a socket (although there are exceptions).

The idea here is that one can make a direct connection between a DCE and a DTE without crossing wires. Since the conception of RS-232, the distinction between DCE and DTE has become less clear. Manufacturers do not always categorize the same equipment the same way. For example, how does one classify equipment type for a personal computer and printer connection? There are often situations when one must connect a DCE to a DCE or a DTE to a DTE. This situation requires switching some connections in the interface. For the simple system with only three connections, wires 2 and 3 have to be crossed. This switched connection is often called a *null modem*.

Because the RS-232 signal levels differ from normal HCMOS or TTL levels, it is necessary to use special driver chips for signal conversion. Normally, this would require a power supply that can provide ± 12 V. Fortunately, semiconductor manufacturers have developed ICs that can supply RS-232 signal levels using only a $+5$-V supply. This is done by using an internal charge pump. All that is required is to connect some capacitors to the IC. The MC145407 5-Volt-Only Driver/Receiver is an example of this type of chip. Other chips are the MAX232 series from Maxim and the LT1081 from Linear Technology. Figure 10.2 illustrates its use to build an RS-232 port for a 68HC11. Figure 5.8 illustrates using a MAX232. Note that PD0 is the SCI's RxD line and PD1 is its TxD line. In this case the 68HC11 is connected as a DCE device. If more RS-232 signals are required, the MC145407 can also supply ± 10 V to a sister driver chip, the MC145406. According to the RS-232 standard, ± 10 V is sufficient.

10.1.2 Protocols and Flow Control

Basically, a protocol is a set of rules for making connections and transferring messages. Flow control refers to methods of stopping data transfers and restarting them. When trans-

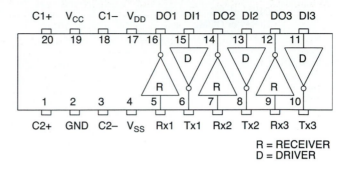

C1+   V_{CC}   C1−   V_{DD}   DO1   DI1   DO2   DI2   DO3   DI3

R = RECEIVER
D = DRIVER

(a)   Pin Assignment of the MC145407
5-Volt Only Driver/Receiver

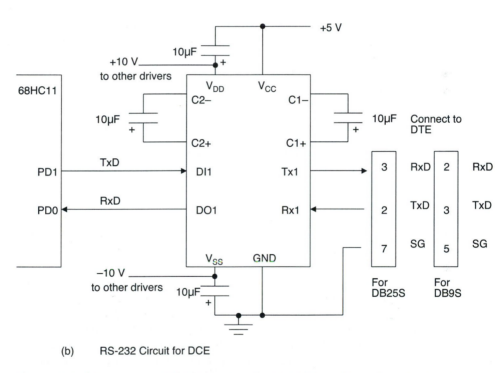

(b)   RS-232 Circuit for DCE

Figure 10.2   Microcontroller RS-232 Interface. For MAX232, see Figure 5.8. *(Continued)*

ferring data, there have to be agreed-upon rules by both the transmitter and receiver. In some cases a program can initiate a transmission to another device. The other device is running a program that is doing nothing else but waiting for data and reading it when it comes. Note that we are referring to communication flow control, not program flow control (Section 2.8).

There are several protocols used in the communications industry. We will describe some of them briefly because you will certainly read about them in some manuals, and many communications software packages use them.

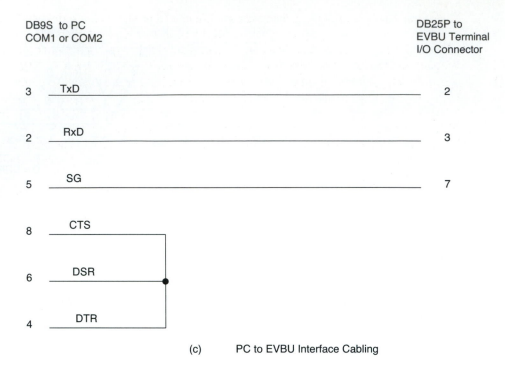

DB9S to PC
COM1 or COM2

DB25P to
EVBU Terminal
I/O Connector

3   TxD   2

2   RxD   3

5   SG   7

8   CTS

6   DSR

4   DTR

(c)   PC to EVBU Interface Cabling

Figure 10.2   Microcontroller RS-232 Interface. For MAX232, see Figure 5.8. *(Continued)*

Acknowledge/Not Acknowledge (ACK/NAK) Flow Control.   Some protocols, such as XMODEM, Kermit, and Bisynch, use a transmitter that waits for the receiver to acknowledge each block of data before the next block can be sent. After a block of data is sent, the receiver checks that the data is valid. If the data is valid, the receiver sends an acknowledge (ACK) signal. The transmitter uses the ACK signal as permission to send the next block. If the data is invalid, the receiver sends a negative acknowledgment (NAK). The transmitter will retransmit the previous block.

Cyclic Redundancy Check (CRC).   Data integrity can be ensured by calculating a special code based on the bit pattern in the data block. The cyclic redundancy check (CRC) is one of these codes. This can be used in addition to or instead of parity check. CRC uses a feedback mechanism to make a code word depend on the previous history of the message. The transmitter calculates the code for the message block it sent and adds it to the end of the block. The receiver also calculates the code based on the data it receives. It includes the CRC character sent by the transmitter in its calculations. If the result is zero, there is a very high probability that the message was received error free.

CRC refers to a technique. A communications protocol specifies the type of CRC used. CRC-16 is a common type. It has a greater than 99% chance of detecting error bursts that can be up to 16 bits in length. In a communications system noise tends to cause bursts of errors.

XON/XOFF Flow Control.   Another common method for flow control is XON/XOFF, which works as follows: For some reason the receiver needs to tell the transmitter to stop sending data. For example, a computer sends data to a printer with a serial interface. If the printer buffer is full, the printer has to tell the computer to stop transmitting; otherwise, transmitted data will be lost. The receiver sends an XOFF signal, which tells the transmitter to stop transmission. Afterward, when the receiver is ready, it sends an XON signal to tell the transmitter to resume transmission. XON and XOFF are the ASCII control characters DC1 (CTRL-Q) and DC3 (CTRL-S), respectively.

The Break Signal.   Sometimes it is necessary to get a device's attention to abort a transfer in midstream. It may be necessary to wake up a *sleeping* receiver. The *break signal* is a signal that drives the line low for longer than a frame time. Recall that each transmitted character has at least one bit high, the *stop* bit. Also, an idle line is always high. Often, the break signal is used to get a device's attention. It may be interpreted as a command instead of data. It may tell a device to disconnect.

10.1.3   Communication Channel Operation

One definition of a *communication channel* is a signal path that connects two stations/devices. There are three basic ways (operations, modes, protocols) to transmit data along a communication channel:

1. Simplex
2. Half-duplex
3. Full-duplex

Figure 10.3 illustrates these operations. We will use roads and highways as an analogy. *Simplex* is like a one-way street. *Half-duplex* is a two-lane road with one lane under repair. A flag person directs traffic from one lane to stop while allowing the other to travel. *Full-duplex* is a fully operational multilane road.

For simplex operation transmission can occur in one direction only. For half-duplex operation each device can transmit at one time and receive at the other. Simultaneous transmission and receive cannot occur. Figure 10.3(b) illustrates half-duplex being used for an ACK/NAK protocol. For full-duplex operation, both devices can transmit and receive simultaneously. In Figure 10.3(c) each device also echoes back an acknowledgment block that duplicates the block previously received. This verifies that all data has been transferred.

10.1.4   Modems

A modem is used to send and receive serial digital data over a telephone line. A microcontroller system may use a modem to send data to a remote location using the existing telephone lines. One such system is a security alarm panel. When its sensors detect an intruder, the microcontroller sends an alarm message to the telephone operator of a security company that specializes in monitoring security systems. The operator can then call the police in the hope of catching the intruder. Modems may be used simply to send data from one control system to another or a computer using the telephone network. Many remote file transfers are done using modems.

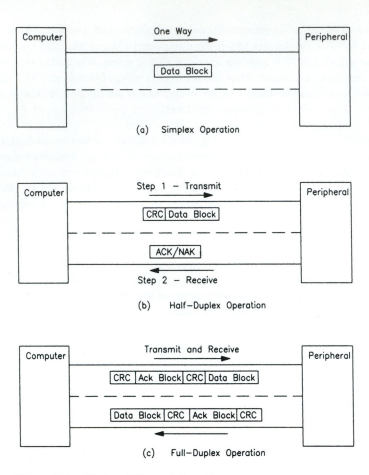

Figure 10.3 Modes of Channel Operation

Basics. Typically, a modem is connected to a serial port. This may be a dedicated circuit instead of the physical port using the DB-25 connector. The serial port, the RS-232 data terminal equipment (DTE), is connected to a modem, a data communications equipment (DCE). The modem is connected to a telephone line.

The name *modem* is a contraction of *modulator-demodulator.* For transmission it converts digital signals to sound (audio) signals for transmission along the telephone line (modulates). For receiving the modem converts the audio signals back to digital (demodulates). The audio signal is known as a carrier signal. You may know that audio signals are made up of sine waves of different frequencies. A typical waveform of the human voice does not look like sine waves on an oscilloscope. This is because it is a complex waveform containing many sine waves. Digital transmission with modems uses only a few sine-wave frequencies.

Low-speed modems use a technique known as frequency shift keying (FSK). One carrier frequency represents a logic 1 (mark) and another represents a logic 0 (space). Higher-speed modems use variations. One is phase shift keying (PSK). It makes use of the fact that a sine wave can be divided into phases. A phase corresponds to a sector of a circle or angular range of rotation. In PSK a phase is used to represent unique logic levels. Using four phases (each of 90°), PSK can encode 2 bits with each phase. Eight-phase PSK can encode 3 bits at a time. Another variant, differential phase shift keying (DPSK), uses relative phase shift instead of absolute. Higher speeds are obtained by varying both phase and amplitude. This is known as quadrature amplitude modulation (QAM).

Recall that a baud is defined as a unit of signaling information. In the case of modems, a phase is the unit of signaling information. Hence, a phase represents 1 baud. But since a phase may encode several bits, modem speeds are often stated in bits per second (BPS) instead of baud.

There are several modem standards. They define a set of operating characteristics. These include bit speeds, carrier frequencies, and modulation method. Some modem chips are the Am 79191, 7911, and 7910 by Advanced Micro Devices. They use FSK and can be configured for most modem standards.

Modem Handshaking Control.   In any telephone conversation, one person must call the other. We can reword this as "one person originates the call and the other answers." Modem handshaking uses the same terminology. The modem that starts transmission is known as the *originate modem*. The one that is being called is known as the *answer modem*. Hence, for a modem to communicate with another, they must establish a connection. They must also maintain the connection and later terminate it.

For modem-to-modem communications, the serial interface port (DTE) and the modem (DCE) use RS-232 *control* signals. The nominal RS-232 control signals are $+12$ V for an ON condition and -12 V for an OFF condition. These are shown in Table 10.2. They are part of the RS-232 standard. Table 10.2 does not show the transmit, receive, and ground signals. For these, see Table 10.1.

A serial communications IC typically has pins for at least some of the modem control signals. Software control of general-purpose I/O pins is required for any that are missing. This is not a problem since not all signals have to be controlled. Note that the 68HC11 does not have any modem control signals. Some modem chips also have a data carrier detect (DCD) output. Some serial communication chips have a DCD input. This signal can be used to assert an interrupt if the carrier is lost. For the 68HC11, this signal can be connected to an interrupt pin (IRQ or XIRQ).

TABLE 10.2   MODEM CONTROL SIGNALS[a]

Signal name	DTE (port)	DCE (modem)	Function	Pin/EIA circuit label
$\overline{\text{RTS}}$	Output	Input	*Request to send* tells the DCE that the DTE wishes to transmit.	4/CA
$\overline{\text{CTS}}$	Input	Output	*Clear to send* tells the DTE that the DCE is ready to receive; DCE transmits any data sent by DTE.	5/CB
$\overline{\text{DTR}}$	Output	Input	*Data terminal ready* prepares the DCE for connection to the communications channel.	20/CD
$\overline{\text{DSR}}$	Input	Output	*Data set ready* indicates that the DCE is ready to operate the communications channel.	6/CC
$\overline{\text{RI}}$	Input	Output	*Ring indicator* indicates that the DCE received a ring signal on the communications channel.	22/CE

[a] Control signals are active low when converted to HCMOS levels.

Handshake control is as follows. In simplex mode, transmission or reception can occur at any time. To send data, the serial port asserts the ready to send ($\overline{\text{RTS}}$) signal. It waits for the modem to respond with a clear to send ($\overline{\text{CTS}}$) before sending data to it. In full-duplex mode, these signals can be kept low. Often they are simply hardwired together. But before data transmission can occur, the connection must be established. This requires the use of the data terminal ready ($\overline{\text{DTR}}$) and clear to send ($\overline{\text{CTS}}$) lines. The modem that initiates the connection assumes the role of the originate mode. This is done when the serial port for the originate modem asserts the data terminal ready ($\overline{\text{DTR}}$) line. The originate modem sends the dial signal (address). The answer modem responds and asserts a ring indicator ($\overline{\text{RI}}$) for its serial port. This port responds to $\overline{\text{RI}}$ by asserting its $\overline{\text{DTR}}$. The answer modem responds by sending an answer signal, the answer mode mark. The originate modem responds to the answer mode mark by sending an acknowledge signal, the origination mode mark. The frequency and time duration of the answer and origination mode marks are defined by the appropriate modem standard.

When the answer modem recognizes the origination mode mark, it asserts clear to send ($\overline{\text{CTS}}$) for its serial port. The originator modem also asserts $\overline{\text{CTS}}$. Full-duplex communication can now occur. Of course, a call or communication session must also be terminated. This can be done by sending a break signal. Other systems terminate simply by stopping transmission.

■ 10.2 THE SERIAL COMMUNICATIONS INTERFACE (SCI)

Basic and Enhanced Versions.   In this section we will restrict the discussion to the basic version used by 68HC11 models, such as the 68HC11A8. Some models, such as the 68HC11N series, use an enhanced serial communications interface. They are similar, but some details of the registers and corresponding bit assignments differ.

10.2.1 The SCI Subsystem in the 68HC11

It is at this point that we direct our attention to the basic serial communications interface (SCI) used by the 68HC11 microcontroller. With additional conversion circuits the SCI can be used to communicate with remote devices using industry standard cables, telephone lines, or radio transmitters. The SCI subsystem uses port D pin PD1 as a transmit line (TxD) and it uses port D pin PD0 as a receive line (RxD). These lines can be enabled or disabled by one of the SCI control registers (SCCR2). When disabled, the port D line can be used as a general parallel I/O line, as determined by its data direction register (DDRD). If a transmitter and/or receiver is enabled, the SCI subsystem has control of the respective port D line and overrides the DDRD setting.

For the 68HC11 software routine, transmitting a byte is a simple matter of writing bytes to a data register (SCDR). The SCI subsystem handles the data framing requirements (no parity). The MCU reads serial input by reading the data register to get each character received. The SCI receiver automatically converts each framed serial character into a parallel byte. This is a simplified view and we have ignored timing and coordination requirements for now. Later we will look at how the system handles these requirements.

The SCI system uses an internal clock tapped off the main MCU clock for serial transfer or sampling of data. By configuring another control register (BAUD), a program can set the baud rate to available integral fractions of the system clock frequency (E). The SCI system does not have a built-in parity check or generation. It does not automatically insert a parity bit when transmitting and it does not check for parity when receiving. However, software can perform this function quite easily. Other communication chips, such as the 6850 Asynchronous Communications Interface Adapter (ACIA), have built-in parity generation/detection systems.

The SCI transmitter can send break signals. A program sets or toggles the send break bit (SBK) in SCI control register 2 (SCCR2). The SCI receiver can be put to sleep by setting another bit, the receiver wake-up control bit (RWU), in the same register. Another bit, the wake-up bit (WAKE), in another register, SCI control register 1 (SCCR1), determines how to wake up the receiver.

Like most serial communication systems or programmable chips, the SCI subsystem uses *double buffering*. Figure 10.4 shows the double-buffering operations of both the transmitter and receiver. In the following sections we describe these operations. The SCI subsystem uses one data register address ($102F) and name (SCDR) for both registers. An instruction to write data to SCDR causes data to be sent to the transmit data register (TDR). An instruction to read SCDR causes data to be read physically from the receive data register (RDR).

Some specialized communications chips or subsystems use a first in, first out (FIFO) queue to provide a buffer of several bytes. This allows considerably faster communications since the software does not have to service every time a byte is received. You may wish to refer back to the discussion of the FIFO queue in Section 7.3.

10.2.2 Transmit Operation

To transmit, the system writes a byte to the transmit data register (TDR) buffer, which in turn is transferred to the output shift register. If enabled, it also transfers a ninth bit as determined by transmit bit 8 (bit T8 in register SCCR1) to the shift register. For example, if T8 was a logic 0, the ninth bit shifted out will be a logic 0. The shift register automatically shifts out a start bit and the data bits and automatically adds a stop bit at the end. The setting of a baud rate register (BAUD) determines the shift rate.

The system also sets a flag, the transmit data register empty flag (TDRE), every time it transfers data from the buffer to the shift register. The software detects that TDRE has set (polling or interrupt-driven). The transmit driver routine may respond by writing another byte to the TDR buffer. Writing to the TDR buffer clears the TDRE flag. This can occur while the previous data byte is being shifted out. The term *double buffering* is used to refer to the fact that while the shift register is shifting out a byte, the data buffer is storing the next byte to be transmitted. Hence, the system can write the second byte while the first one is still being transmitted.

When the shift register has completed shifting out the byte, the SCI system automatically transfers the contents of the TDR buffer to the shift register. It starts to shift it out immediately. This also causes the system to set the TDRE flag to indicate that the software may write another byte to the TDR buffer.

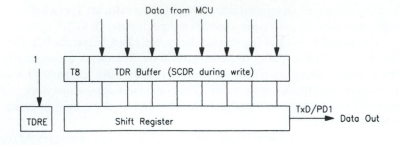

Transmitter

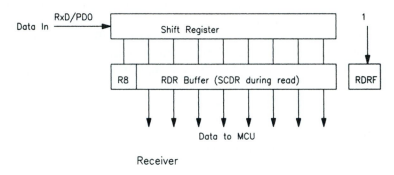

Receiver

Figure 10.4   Double Buffering in the SCI System

10.2.3 Receive Operation

If the receiver is awake, it will do nothing while there is an idle condition. The first low signal, a start bit, triggers the receiver to shift in the next 8 bits. If configured for 9 bits, it will also shift in a ninth bit. The setting of register BAUD determines the rate of sampling and shifting in. When all bits are shifted in, the system transfers the contents of the shift register to the receive data register (RDR) buffer. If configured for 9 bits, it also transfers the ninth bit to bit R8 in control register SCCR1. Transferring this data also sets the receive data ready flag (RDRF) to tell the software that data is ready to be read. The receiver samples the stop bit and waits for another start bit before it shifts in the next group of data bits.

A receiver driver routine may read the RDR buffer (read data register SCDR) any time while the next data is being shifted in. This is also double buffering. In this case the buffer stores the most recent received byte. It stores the currently received byte while the next byte is shifting in. The software should read the buffer before the next byte has shifted in completely. Otherwise, it will miss reading the next data byte. Reading the RDR buffer resets flag RDRF and enables the next received byte to be transferred to it.

10.2.4 SCI Software

Figure 10.5 shows the SCI registers. We will need to refer to them when discussing software to handle I/O. As usual, the software should configure (initialize) the SCI subsystem first. This could be a subroutine that is executed after a reset. Other SCI I/O handler subroutines could be responsible for transmitting a byte or receiving a byte. We will outline the procedure for each of these routines. Further details about the SCI register bits are presented in Section 10.3.

Configuration Procedure

1. Select the baud rate. Write to the BAUD register (SCP0-1, SCR0-2).
2. Select the word length and wake up. Write to the SCCR1 register (M, WAKE).
3. Enable interrupts, transmit, receive, and wake up as required. Write to the SCCR2 register (TIE, TCIE, RIE, ILIE, TE, RE, RWU).

Transmit Procedure for Each Byte

1. Poll the SCSR register or respond to the interrupt (read SCSR).
2. If applicable (M = 1), write to T8 in the SCCR1 register.
3. If TDRE == 1, write to the SCDR register.

Receive Procedure for Each Byte

1. Poll the SCSR register or respond to the interrupt (read SCSR).
2. If RDRF == 1, then read SCDR register.
3. Option: If there is an error (OR|NF|FE == 1), handle the error.
4. If applicable (M = 1), read R8 in the SCCR1 register.

We look at the receive errors in more detail later.

Sample Programs.   In this section we look at two sample programs, which are testing programs. They can be modified to handle blocks of data for data transfer between two devices.

The first sample program illustrates a polling technique to transmit and receive. It is a *loopback* testing program. Its purpose is to test the serial interface, such as an RS-232 port. The transmit pin is connected back to the receive pin. The program sends a byte of data to the SCI transmitter and then reads it back again from the SCI receiver. Listing 10.1 shows this program. If the byte received equals the transmitted byte, the interface operates correctly. The program can be modified to transfer blocks of data through the SCI loopback connection.

```
*Listing 10.1
*This program transmits the data stored at $00 continuously
*to the SCI transmitter and reads it back from the SCI
*receiver. It stores the received byte at $01.

*User can set configuration registers BAUD, SCCR1, and SCCR2
*as desired by setting addresses $02, $03, and $04.
*Warning: program set up only to handle I/O polling,
```

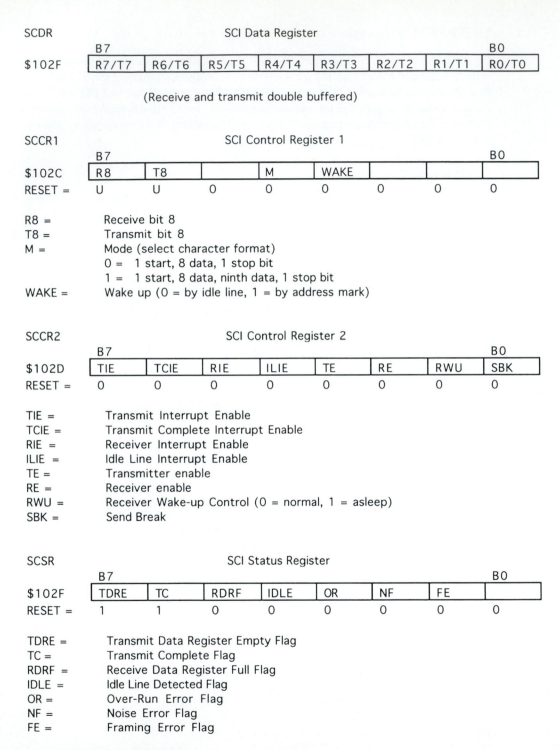

SCDR SCI Data Register

B7							B0
R7/T7	R6/T6	R5/T5	R4/T4	R3/T3	R2/T2	R1/T1	R0/T0

$102F

(Receive and transmit double buffered)

SCCR1 SCI Control Register 1

B7							B0
R8	T8		M	WAKE			

$102C

RESET = U U 0 0 0 0 0 0

R8 =   Receive bit 8
T8 =   Transmit bit 8
M =   Mode (select character format)
            0 = 1 start, 8 data, 1 stop bit
            1 = 1 start, 8 data, ninth data, 1 stop bit
WAKE =   Wake up (0 = by idle line, 1 = by address mark)

SCCR2 SCI Control Register 2

B7							B0
TIE	TCIE	RIE	ILIE	TE	RE	RWU	SBK

$102D

RESET = 0 0 0 0 0 0 0 0

TIE =   Transmit Interrupt Enable
TCIE =   Transmit Complete Interrupt Enable
RIE =   Receiver Interrupt Enable
ILIE =   Idle Line Interrupt Enable
TE =   Transmitter enable
RE =   Receiver enable
RWU =   Receiver Wake-up Control (0 = normal, 1 = asleep)
SBK =   Send Break

SCSR SCI Status Register

B7							B0
TDRE	TC	RDRF	IDLE	OR	NF	FE	

$102F

RESET = 1 1 0 0 0 0 0 0

TDRE =   Transmit Data Register Empty Flag
TC =   Transmit Complete Flag
RDRF =   Receive Data Register Full Flag
IDLE =   Idle Line Detected Flag
OR =   Over-Run Error Flag
NF =   Noise Error Flag
FE =   Framing Error Flag

Figure 10.5 SCI Registers

BAUD SCI Baud Rate Control Register

	B7							B0
$102B	TCLR		SCP1	SCP0	RCKB	SCR2	SCR1	SCR0
RESET =	0	0	0	0	0	U	U	U

TCLR, RCKB are for tests mode only

SCP1,0 = Serial Prescaler Selects

SCP1	SCP0	Divide E by	Highest Baud Rate, E = 2 MHz
0	0	1	125.00K
0	1	3	41.666K
1	0	4	31.250K
1	1	13	9600 (+0.16%)

SCR2, 1, 0 = SCI Rate Select

SCR2	SCR1	SCR0	Prescaler Output Divide-by Factor	Highest Baud Rate, 9600 (SCP1 = 1) (SCP0 = 1)
0	0	0	1	9600
0	0	1	2	4800
0	1	0	4	2400
0	1	1	8	1200
1	0	0	16	600
1	0	1	32	300
1	1	0	64	150
1	1	1	128	75

Note: U = undefined

Figure 10.5   SCI Registers *(Continued)*

```
*not interrupts.
*Note: Xtal = 8 MHz, E = 2 MHz
*-------------------------------------
*can be modified by user
      ORG    0
TRDAT    FCB    $A5    ;data to be transmitted
RECDAT   RMB    1      ;received data
BAUDDAT  FCB    $30    ;baud rate, initially 9600
CONF1    FCB    $00    ;SCCR1 config. data
CONF2    FCB    $0C    ;SCCR2 config. data

    *-------------------------------------
```

```
        ORG     $100

        LDX     #REGBAS         ;point to registers
        LDAA    BAUDDAT         ;set up baud rate
        STAA    BAUD,X

        LDAA    CONF1           ;configure SCI control registers
        STAA    SCCR1,X
        LDAA    CONF2
        STAA    SCCR2,X

        LDAA    SCSR,X          ;clear RDRF if set
        LDAA    SCDR,X
AGN
        LDAA    TRDAT           ;get data
        STAA    SCDR,X          ;and transmit, also clears TDRE
*wait for empty transmit register
TR
        BRCLR   SCSR,X $80 TR
*                               ;continue when TDRE sets
*wait for full receive register
REC
        BRCLR   SCSR,X $20 REC
*                               ;continue when RDRF sets
        LDAA    SCDR,X          ;read received data,
*                               ;also clears RDRF
        STAA    RECDAT          ;and store it
        BRA     AGN             ;and repeat transmit loop
```

You should be able to verify that the configuration data sets up the SCI system to enable both receiver and transmitter at 9600 baud (see also Figure 10.5). To set up for 9600 baud, the prescaler bits SCP1 and SCP0 are set in register BAUD. To enable the transmitter and receiver, set bits TE and RE in register SCCR2. The program can easily be modified to transmit 9 bits by setting the mode bit (M) in control register SCCR2. The logic level of bit T8 determines the state of the ninth data bit. For example, changing the contents of address CONF1 to $50 results in sending 9 data bits with the ninth bit set. More program modifications are required to read 9 bits.

Two load instructions are used to clear the receive data register full flag (RDRF) in case it was set previously. The clearing sequence for RDRF is to read status register SCSR and then to read data register SCDR. Reading the status register also arms the clearing mechanism for the transmit register empty flag (TDRE). The byte is transmitted by writing it to data register SCDR. This also clears flag TDRE. The program now loops at the TR loop.

When the byte is transferred to the transmit shift register, the flag TDRE sets. This allows the program to continue past the TR loop to the REC loop. Here it waits for the receive buffer to fill by waiting for the flag RDRF to set. When it does set, program execution continues past the REC loop to read data register SCDR. This also clears the flag RDRF. The program stores the received byte and repeats execution. A user running the program on a development system would check the contents of addresses TRDAT and RECDAT after halting execution. If the two are equal, the interface works.

Note the flag-clearing sequences for transmit buffer empty and receive buffer full. For transmit (clear TDRE), the sequence is to read status register SCSR and then write to data register SCDR. For receive (clear RDRF), the sequence is to read status register SCSR and then to read data register SCDR.

Alternatively, this program can be used with a cable tester, a device that you can plug into a serial port. Often, it has a switch to allow loopback testing. The switch connects the transmit line to the receive line. The tester uses two LEDs (red and green). One illuminates during a logic 0 condition and the other during a logic 1. If a program like that of Listing 10.1 transmits continually, both LEDs will turn on while the program is executing. For idle and break conditions, one or the other LED will turn off.

C Code Examples.   C Listing 10.1 shows a C program equivalent to that of Listing 10.1. SCI subsystem registers and bits are defined in the header file, but it will be necessary to use the statement `#define _SCI` for the compiler to understand the bit definitions such as TDRE. The program also uses the built-in library routine for setting the baud rate with predefined baud rate data. In this case, there are user-defined functions to handle reading and writing data from and to SCI. It is also possible to use built-in library functions, which use the integer data type, as shown in C Listing 10.1a.

```c
/* C Listing 10.1 */

#define _SCI */ Tell compiler to use SCI bit definitions in
hc11.h */
#include <hc11.h>      /* Include register definitions */

/* Declarations */
unsigned char TrDat = 0xa5;
unsigned char RecDat; /* also used as temporary */
unsigned char BaudDat = BAUD9600; /* Defined in hc11.h */
unsigned char Conf1 = 0x00;
unsigned char Conf2 = 0x0c;

/* Prototypes - using character data type
instead of the int data type specified in hc11.h */
void writechar_sci(unsigned char);
unsigned char readchar_sci(void);

void main(void)
{
        /* Configure SCI */
        setBaud(BaudDat);
        SCCR1 = Conf1;
        SCCR2 = Conf2;
        /* Clear RDRF if set */
        RecDat = SCSR;
        RecDat = SCDR;
        while (1)          /* Loop forever */
        {
                writechar_sci(TrDat);
```

```
                    RecDat = readchar_sci();
        } /* end while loop */
}

void writechar_sci(unsigned char Data)
{
        SCDR = Data;      /* Get and transmit data */
        while ((SCSR & TDRE) == 0); /* continue when TDRE sets */
}

unsigned char readchar_sci(void)
{
        while ((SCSR & RDRF) == 0); /* wait until RDRF sets */
        return (SCDR);   /* then return SCDR data */
}

/* C Listing 10.1a */
/* Use ICC11 Library for SCI */

#define _SCI */ Tell compiler to use SCI bit definitions in
hc11.h */
#include <hc11.h>         /* Include register definitions */

/* Declarations */
unsigned int TrDat = 0xa5;
unsigned int RecDat;     /* also used as temporary */
unsigned char BaudDat = BAUD9600; /* Defined in hc11.h */
unsigned char Conf1 = 0x00;
unsigned char Conf2 = 0x0c;

void main(void)
{
        /* Configure SCI */
        setBaud(BaudDat);
        SCCR1 = Conf1;
        SCCR2 = Conf2;
        /* Clear RDRF if set */
        RecDat = SCSR;
        RecDat = SCDR;
        while (1)         /* Loop forever */
        {
                write_sci(TrDat);
                RecDat = read_sci();
        }/* end while loop */
}
```

Second Sample. The second sample assembly language program illustrates an interrupt-driven I/O technique. It transmits the same byte continuously without receiving. The program also generates a strobe pulse after every transmit instruction. This is another oscilloscope loop program. A user can monitor the serial data framing by using the strobe B line as a trigger. Listing 10.2 shows this program.

```
*Listing 10.2
*This program transmits the data stored at $00 continuously
*to the SCI transmitter. User can set configuration
*registers BAUD, SCCR1, and SCCR2 as desired by, respectively,
*setting addresses $01, $02, and $03.
*Warning: program set up only to handle interrupts,
*not I/O polling.
*Observe data framing with oscilloscope channel connected
*to PD1/TxD and trigger channel connected to STRB.


            ORG     0

TRDAT       FCB     $A5      ;data to be transmitted
BAUDDAT     FCB     $30      ;baud rate, initially 9600
CONF1       FCB     $00      ;SCCR1 config. data
CONF2       FCB     $8C      ;SCCR2 config. data


*-------------------------------------

            ORG             $100

*REMINDER: Interrupt vectors should be already set up
*See "Cross Assembly," Appendix D.

    LDX     #REGBAS         ;point to registers
    SEI
    LDAA    BAUDDAT         ;set up baud rate
    STAA    BAUD,X

    LDAA    CONF1           ;configure SCI control registers
    STAA    SCCR1,X
    LDAA    CONF2
    STAA    SCCR2,X

    LDAA    #$06            ;configure PIOC
    STAA    PIOC,X          ;to define STRB active low pulse

    LDAA    SCSR,X          ;clear TDRE if set
    LDAA    TRDAT           ;get data
    STAA    SCDR,X          ;and transmit
    CLI
AGN
    WAI                     ;and wait for empty
*                          ;transmit register
    BRA     AGN

*-------------------------------------

*SCI interrupt handler to transmit
RSCI
    LDAB    SCSR,X          ;clear TDRE
```

```
        STAA    SCDR,X          ;and transmit
        STAA    PORTB,X         ;write anything to PORTB
*                               ;to pulse STRB
        RTI                     ;return
```

PRACTICE

Write instructions to transmit the message "good" at 2400 baud, 8 data bits, no parity, and 1 stop bit. Assume an 8-MHz crystal.

Solution

```
*Solution, transmit "good"

    LDX     #REGBAS         ;point to registers
    LDAA    #$32            ;set up baud rate
    STAA    BAUD,X
    CLR     SCCR1,X         ;set up framing
    LDAA    #$08            ;set TE to enable transmitter
    STAA    SCCR2,X
    LDY     #CHARBLK        ;initialize pointer/counter
NEXT

    LDAA    0,Y             ;get character
    STAA    SCDR,X          ;send it to transmit buffer
* wait here until char transmitted
POLL
    BRCLR   SCSR,X $80 POLL
    INY                     ;point to next char
    CPY     #CHARBLK+4      ;until all transmitted
    BLT     NEXT
DONE
    BRA     DONE            ;stop here for now
*the message to send
CHARBLK
    FCC     "good"
```

■ 10.3 SCI REGISTERS

We have already examined the registers and seen how they can be used for basic transmit and receive operations. Now we will present a more formal coverage.

10.3.1 Serial Communication Data Register (SCDR)

Refer back to Figure 10.4. Both the transmit data register (TDR buffer) and the receive data register (RDR buffer) have the address of the SCI data register (SCDR). If data is written to data register SCDR, data goes to the TDR. If data register SCDR is read, data comes from the RDR. Thus the MCU software cannot read back the data it has already sent.

10.3.2 The Baud Rate Register (BAUD)

The baud rate register (BAUD) selects the various baud rates. They control the shift rates for the transmitter and receiver. The SCP0 and SCP1 bits function as a prescaler for the SCR0,1,2 bits. Together, these 5 bits provide a variety of baud rates for a given crystal frequency. Figure 10.5 shows the baud rates for an 8-MHz crystal. Remember that the system clock (E) has a frequency one-fourth that of the crystal frequency (E = 2 MHz).

The 68HC11 has a rate generator that divides the system clock by 16. Hence, the highest possible baud rate is E/16 (or crystal frequency/64). The bits in register BAUD provide further division of these frequencies. Figure 10.5 shows the settings used to obtain the baud rates used most often by other equipment. Another common baud rate is 19.2 kilobaud. It cannot be obtained with an 8-MHz crystal. An SCI can use a baud rate of 19.2 kilobaud by using a 4.9152-MHz crystal. Then the configuration data for register BAUD has to be $20 or $02.

Actually, we have presented a simplified picture. The system divides the crystal frequency by 4 and then divides this by the prescale factors as specified by register BAUD. This frequency is used to drive the sampler for the receiver. The system further divides this sampler frequency by 16 for the transmit clock. The result is that the receiver uses a sample clock that is 16 times the baud rate. We can divide each bit into 16 equal intervals. Once the receiver has detected a start bit, it samples each data bit and stop bit at the eighth, ninth, and tenth intervals. The value of the bit is taken to be the value of the majority of samples. You might want to look ahead to study Figure 10.7 in Section 10.3.6 to see how the receiver samples a bit.

For a device to receive data transmitted by another, they should operate at the same baud rate or almost the same rate. The trick is for the receiver to determine the boundaries of each start bit. It is only necessary for the receiver samples to fall within each data bit boundary to determine the correct logic value for each bit. The transmitter and receiver baud rates may differ by up to 4% without the receiver missing any bits. The *Reference Manual* (see Appendix C) shows detailed calculations for baud rate frequency tolerance. In Figure 10.5 note that the system clock divided by 13 is not exactly 9600 but it is within tolerance limits. We will not detail how the receiver determines the beginning of a start bit during any frame. Motorola's documentation provides enough explanation for those who wish to know more.

Configuration Examples

Example 1. With a 4-MHz crystal, the highest possible baud rate is 4 MHz/64 = 62.5 kilobaud. If BAUD = %00010011, the baud rate selected is

$$\frac{62.5 \text{ kilobaud}}{3 \times 8} = 2604 \text{ baud}$$

Example 2. With an 8-MHz crystal, the highest possible baud rate is 8 MHz/64 = 125 kilobaud. If BAUD = %00110010, the baud rate selected is

$$\frac{125 \text{ kilobaud}}{13 \times 4} = 2404 \text{ baud}$$

This is approximately 2400 baud. It is within the tolerance specified.

10.3.3 Serial Communication Control Register 1 (SCCR1)

This register provides the control bits that determine word length and select the method used to wake up a sleeping receiver.

Multiple Receivers. The wake-up control is used primarily in systems with multiple receivers, such as a communications network. There may be active signals at the receive pin (RxD) intended for another receiver connected to the same (common) signal line. In this case we would want our SCI subsystem to ignore the message intended for or addressed to another device. We can do this by putting the receiver to *sleep*. When the SCI is in this mode, it does not shift in bits into the receive shift register. To put the receiver to sleep, the software sets bit RWU in the second control register, SCCR2.

Wake-Up. Once a receiver is asleep, there has to be a procedure to allow it to respond to messages addressed to it. This is the *wake-up* process. There are two ways to wake up a sleeping receiver. One option is to wake it up as soon as the line goes idle. This is idle line wake-up. To select this method, the software should clear bit WAKE in the first control register, SCCR1. The other option is address-mark wake-up. The software selects this by setting bit WAKE in control register SCCR1.

If the wake-up process is automatic, how does the application program know that the receiver has been awakened? When awakened, the receive buffer will fill and cause the status flag, RDRF, to set. Depending on the configuration, this will be detected by an interrupt (typically) or by polling for the receive buffer full.

Idle-Line Wake-Up. If the SCI system is using idle-line wake-up, a protocol should be established. For example, the first few bytes indicate the address of the intended target receiver. When the receiver software establishes that the message is addressed to another device, it puts the receiver to sleep again. It does this by setting bit RWU in register SCCR2. The receiver will not wake up again until the receive line again goes idle. Idle-line wake-up is used when messages are relatively long (more than 10 bytes). Note that one idle time (one bit longer than a frame time) separates each message.

If the application program does not detect an idle line, it may not realize that the receiver has awakened as long as the line remains idle. It can detect an idle line by polling or responding to an interrupt depending on configuration. In any case it will know when the idle line goes active again and when a byte is shifted into the receive shift register. We will return to idle line detection later.

Address-Mark Wake-Up. A character whose most significant bit (MSB) is 1 will wake up all receivers enabled for address-mark wake-up. This character is considered to be an address character. Hence, data characters should have their MSB at zero. The following type of protocol can be used. Each receiver's software reads the address character when awakened and compares it to its own address. If they match, the receiver will read the rest of the message bytes. If the address does not match, the receiver can go to sleep. Address-mark wake-up is used when messages are short since the separation of messages requires 1 bit for each data byte transferred.

Mode Select (Character Format). Mode bit M in control register SCCR1 selects the character format mode. This can be 8 or 9 bits (M = 0 or 1). This is useful if a framing format for using 2 stop bits or 8 data bits plus parity is desired. Note that there are no control bits to select automatic parity generation or detection. Subroutines can modify transmit bit T8 and check receive bit R8 to handle these cases. These bits can also be used to distinguish 8-bit data from 8-bit address bytes for an address-mark wake-up protocol.

If M is 1, the software reads the ninth bit by reading bit R8 in the control register. Then it reads the remaining bits from the data register, SCDR. To transmit the ninth bit, a program writes to bit T8 in register SCCR1 and then writes the remaining bits to register SCDR. We will see an example of this in Listing 10.3, which follows after the next section.

10.3.4 Serial Communication Control Register 2 (SCCR2)

This register provides the control bits that enable or disable individual functions for the serial communications interface. We present each of the bits in order of most to least significant position (or left to right).

Transmit Interrupt Enable (TIE). If bit TIE is 1, the SCI subsystem generates an interrupt when the transmit data register SCDR is empty. This occurs when the TDR buffer is transferred to shift register, as shown in Figure 10.4. The MCU can respond by writing another byte to data register SCDR.

Transmit Complete Interrupt Enable (TCIE). If bit TCIE is 1, the SCI subsystem generates an interrupt when transmission is complete. This differs from the TIE interrupt. It is asserted when the contents of the shift register have been shifted out with the stop bit. In addition, there is no new data in the transmit buffer (TDR) that is waiting for transfer to the shift register. In other words, no data is pending. This interrupt can be used to establish the end of message (separated by idles) for some protocols.

Receive Interrupt Enable (RIE). If bit RIE is 1, the system generates an interrupt when a byte is transferred from the shift register to the receive data buffer. The MCU program should respond by reading the buffer (register SCDR) before another byte is received.

Idle Line Interrupt Enable (ILIE). If bit ILIE is 1, the system generates an interrupt whenever the receive line (RxD) is detected as being idle. This can be used to end communications, depending on the protocol. An idle condition is defined as a logic high for one frame time as defined by mode bit M in control register SCCR1. We will note later that setting the receiver wake-up control bit (RWU) inhibits the idle-line function.

Transmitter/Receiver Enable (TE/RE). If bit TE is 1, pin PD1 is an SCI transmit pin. Otherwise, it is a general-purpose I/O pin as defined by the port D data direction register (DDRD). If bit RE is 1, pin PD0 is an SCI receive pin. Otherwise, it is a general-purpose I/O pin as defined by the port D data direction register.

Receiver Wake-Up Control (RWU). If bit RWU is 1, the receiver is put to sleep and ignores any activity on the receiver until it is awakened. You may wish to review Section 10.3.3 about wake-up. Note that bit WAKE in register SCCR1 selects the wake-up method.

Send Break (SBK). If bit SBK is 0, the transmitter is operating normally. If bit SBK is 1, the transmitter will continually send bits of zeros until bit SBK is cleared. When bit SBK is cleared, the transmitter sends at least one high bit to guarantee recognition of a valid start bit. Another method is to toggle bit SBK on and off. The transmitter sends one break character and then reverts to idle or sending data. For the 68HCII, a break character is 10 or 11 zeros.

Sample Program. The program of Listing 10.3 uses both control registers, SCCR1 and SCCR2, to frame data based on a variety of configurations. In this case it mimics the configuration set up by another communications IC. The configuration code is used to calculate a jump address. Each jump address destination has code applicable to its corresponding configuration code. We show only the program code for configuration codes (FRAME data) 4 and 5.

The program begins by performing a basic configuration common to all selections. The code applicable for each frame type modifies this configuration. As shown in the code for frame FOUR, the program sets bits T8 and M in SCI control register 1 (SCCR1) to send 2 stop bits. The program takes similar action when other frames are selected. It would use a utility subroutine to determine the actual parity of the data. Based on this information, the code for frames 6 and 7 would set or reset bit T8 as required. The others would set or reset the MSB of the data byte as required.

```
*Listing 10.3
*Transmit a block of bytes starting at address BUFF
*and ending at address ENDBUFF
*For brevity, not all source code is shown.
*Configures different framing formats using a software
*"pseudo" configuration register FRAME at address $02
*as follows:

*FRAME Word Select Bits Table, similar to 6850 ACIA
*2,1,0      Function
*000        7 bits + even parity + 2 stop bits
*001        7 bits + odd parity + 2 stop bits
*010        7 bits + even parity + 1 stop bit
*011        7 bits + odd parity + 1 stop bit
*100        8 bits + 2 stop bits
*101        8 bits + 1 stop bit
*110        8 bits + even parity + 1 stop bit
*111        8 bits + odd parity + 1 stop bit

*data definitions can be modified by user

           ORG      0

BAUDDAT    FCB      $30           ;baud rate, initially 9600
FRAME      FCB      $00           ;pseudo config. register
BUFF       RMB      32
ENDBUF

*-------------------------------------

           ORG      $100
```

```
*Stack and vectors should be already set up
*See "Cross Assembly" in Appendix D.
          LDX     #REGBAS        ;point to registers
          SEI                    ;disable interrupts for now
          LDAA    BAUDDAT        ;set up baud rate
          STAA    BAUD,X
          CLR     SCCR1,X        ;init. T8 = 0,M = 0,
*                                ;default configuration
          LDAA    #$88           ;enable transmitter and interrupt
          STAA    SCCR2,X


*Jump to transmit code as specified by FRAME data

          LDY     #FRTAB
          LDAB    FRAME
          ANDB    #$07           ;mask bits 3 to 7 of FRAME
          ASLB                   ;multiplies by 2
          ABY
          LDY     0,Y            ;jump address = FRTAB+2xFRAME
          JMP     0,Y            ;jump to selected transmit code


*{This section contains sections of code
*to handle each type of FRAME configuration}

*{code for FRAME 0}
ZERO
*{code for FRAME 1}
ONE
*{code for FRAME 2}
TWO
*{code for FRAME 3}
THREE


*Code for FRAMEs 4 and 5
*This is a sample of one section of code
*to show 8 data bits plus 2 stop bits or 1 stop bit
*-------------------------------------
*no parity, 2 stop bits
FOUR
          BSET    SCCR1,X $50
*no parity, 1 stop bit
FIVE
          LDY     #BUFF
FIVE1
          LDAA    0,Y            ;get data
          STAA    SCDR,X         ;transmit it
          CLI                    ;enable interrupt
          WAI
          INY                    ;get next data byte
          CPY     #ENDBUF        ;until all bytes
```

```
          BNE     FIVE1          ;transmitted
          BRA     DONE
*{code for FRAME 6}
SIX
*{code for FRAME 7}
SEVEN
*{applicable code when transmission complete}
DONE


*---------------------------------------


*Service routine RSCI
*SCI interrupt handler to transmit byte (This is complete)

RSCI
          LDAB    SCSR,X         ;clear TDRE
          STAA    SCDR,X         ;transmit byte
          RTI


*---------------------------------------


*Frame Look-up Table
*to specify jump addresses to execute code selected
*by FRAME data

FRTAB
          FDB     ZERO, ONE, TWO, THREE
          FDB     FOUR, FIVE, SIX, SEVEN
```

Vectors and Local Masks. Table B.2 in the Quick Reference (Appendix B) shows that all of the SCI interrupts share the common vector address $FFD6,D7. Table B.3 shows the local masks just described. Hence, if more than one of the enable bits are set, the interrupt service routine will have to check status register SCSR to find the cause of the interrupt. In the following section we will describe the status register flags.

10.3.5 Serial Communication Status Register (SCSR)

This register provides inputs to the MCU internal interrupt logic circuits for generation of the SCI subsystem interrupt. Alternatively, polled I/O techniques may be used to check the status of these bits. The flag-clearing procedures require that the status register be read. The second step is to read or write to the SCDR data register, depending on the particular flag. The order of operations is important. It is permissible to execute other instructions between the two steps as long as it does not compromise the handling of I/O due to timing constraints.

Transmit Data Register Empty Flag (TDRE). Bit TDRE sets when the transmit buffer (data register SCDR during a write) has transferred its contents to the transmit shift register. To clear bit TDRE:

Read status register SCSR.

Then write to data register SCDR.

Transmit Complete Flag (TC).   Bit TC sets when the transmitter has reached an idle state, meaning that there is no pending data and the last byte has been shifted out completely. To clear bit TC:

Read status register SCSR.

Then write to data register SCDR.

Receive Data Register Full Flag (RDRF).   Bit RDRF sets when the receive shift register transfers its data to the data buffer, register SCDR. To clear bit RDRF:

Read status register SCSR.

Then read data register SCDR.

Idle Line Detected Flag (IDLE).   Bit IDLE sets when the active receive line RxD becomes idle. To clear bit IDLE:

Read status register SCSR.

Then read data register SCDR.

Receive Error Flags (OR, NF, FE).   These flags set when their respective errors occur. To clear any of these flags:

Read status register SCSR.

Then read data register SCDR.

It is up to the software to check if any of these flags have set and how to respond to the error. Typically, this could be for the receiver device to send an NAK character to request retransmission of the block of data. In the next section we describe how these errors occur.

10.3.6   Receive Errors

Overrun Error Flag (OR).   Overrun occurs when a newly received character has been shifted completely into the receiver before the MCU software has read the receive data register (and hence cleared RDRF). A high RDRF bit inhibits any writes to the buffer until the flag is cleared so that the previous received data byte is available. An occurrence of this error sets flag OR and any data currently in the shift register will be lost. In the case of Figure 10.6, byte 1 will be lost because the CPU has not read byte 0.

Noise Error Flag (NF).   The receiver uses a sampling clock that has a frequency of 16 times the baud frequency (see Section 10.3.2). Once the receiver has established the bit boundaries, it samples the bits during the eighth, ninth, and tenth cycles of the sampling clock. The system sets flag NF if these samples are not identical. Figure 10.7 shows a high noise spike during a low bit.

Framing Error Flag (FE).   A framing error occurs if an invalid stop bit is detected. In other words, it sets if the system detects a low bit after all data (and parity) bits are shifted in. This can be caused by baud rate mismatch between the transmitter and receiver. Another cause

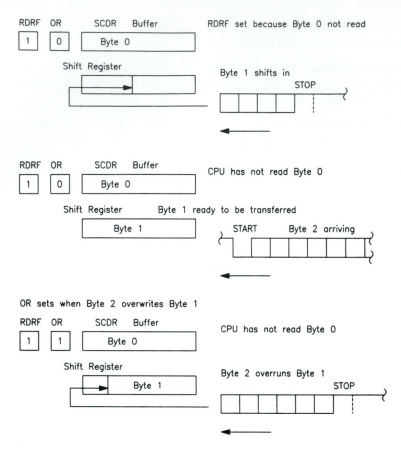

Figure 10.6 Overrun Error

can be a mismatch in the framing protocol between the two. The situation shown in Figure 10.8 can conceivably happen if the transmitter uses a frame of 9 data bits and the receiver assumes (is configured for) a frame of 8 data bits. The receiver detects the last bit (the transmitter's ninth bit) as low and flags an error because it thinks it should be high to indicate stop.

Parity Error. The SCI subsystem has no built-in logic to detect parity error, unlike some other programmable communication subsystems. Most computer communication protocols do not require the use of parity checking, so this should not be a problem. However, software can be used if it is necessary. A receiver handler routine can use a pseudoparity error flag.

Error Detection. Listing 10.4 shows the application of conditional branch instructions to detect errors. We do not show how the errors are handled since this depends on the requirements of a particular situation.

```
*Listing 10.4
*Demonstrates error detection using polling
*---------------------------------------

        ORG       $100
```

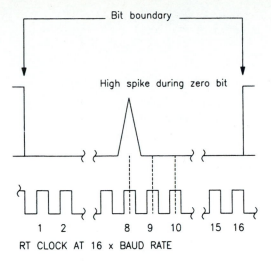

Receiver samples bit at 8th, 9th, and 10th pulse

In this case, one sample is high due to noise.

Figure 10.7 Noise Error

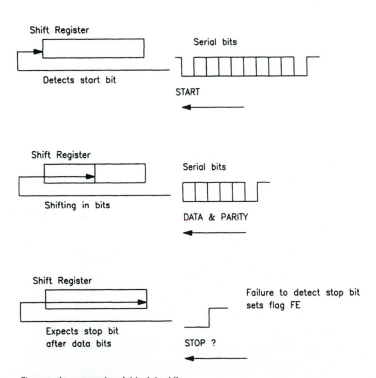

The receiver expects eight data bits.

Figure 10.8 Framing Error Example

```
        ONSCI
        *initialize SCI for 9600 baud at 8-MHz Xtal
                LDAA    #$30
                STAA    BAUD,X
                LDAA    #$00
                STAA    SCCR1,X
                LDAA    #$0C
                STAA    SCCR2,X             ;enable
        INSCI
                LDAA    SCSR,X
                BITA    #$20
                BEQ     INSCI
                BRSET   SCSR,X $10 ID       ;handle idle condition
        CHK1    BRSET   SCSR,X $08 ORF      ;handle overflow error
        CHK2    BRSET   SCSR,X $04 NF       ;handle noise error
        CHK3    BRSET   SCSR,X $02 FE       ;handle frame error
        INSCI1
                LDAA    SCDR,X              ;reads error-free data

        *Rest of code stores error-free data.
        *ORF, NF, and FE are labels pointing
        *to sections of code to handle their
        *applicable error. These routines may
        *simply be identical. They can inform the
        *main program to send an NAK after the current
        *block transmission is complete

        ID
        *Idle condition code as required
        ORF
        *Overflow code as required
        NF
        *Noise code as required
        FE
        *Frame code as required
```

PRACTICE

Write an SCI interrupt service routine that receives bytes and stores them in a RAM buffer.
If any receive error is detected, the routine writes a null character ($00) to the buffer loca-
tion for the erroneous character. The SCI has already been configured.

Solution

```
        *Service routine RSCI
        *Assumes interrupt caused only by receive operation
        RSCI
                LDY     BUFFPTR             ;buffer pointer stored as
                                            static variable
```

```
                LDX       #REGBAS                 ;point to registers
*                                                 ;if no error then char =
                                                  char
                BRCLR     SCSR,X $0E NO_ERROR
                CLRA                              ;else char = null
                STAA      0,Y                     ;and store it
                LDAA      SCDR,X                  ;and clear flags
                BRA       SCI_END                 ;and exit
NO_ERROR
                LDAA      SCDR,X                  ;get char and clear flags
                STAA      0,Y                     ;store char to buffer
SCI_END
                INY                               ;increment buffer pointer
                STY       BUFFPTR
                RTI                               ;return
```

Closing Notes.   This ends our look at the asynchronous serial subsystem. The remainder of the chapter deals with the synchronous subsystem.

■ 10.4 SYNCHRONOUS SERIAL INPUT/OUTPUT

In Section 8.6.1 we introduced the general principles of synchronous serial I/O. *Synchronous* means that transmitter and receivers all use the same clock signal. The devices in an asynchronous communication network have to use clocks with identical frequencies, but they do not have to be physically the same signal. Although of identical frequencies, their respective clocks may be out of phase. The start and stop framing bits allow receivers in an asynchronous network to lock into data bits in a byte. You may say that the network is synchronized for the duration of a frame.

On the other hand, the transmitter in a synchronous communication system also sends a clock signal for the receivers. Synchronous serial is faster than asynchronous serial because there is no need for framing bits such as start and stop. Also, synchronous systems can use a faster clock because the receiver does not have to sample several times for each bit due to the uncertainty of bit boundaries. Since all devices use the same clock, they shift bits at exactly the same rate.

We can classify synchronous serial systems into two broad categories. One is for handling local transfers with other ICs in a system. The other is for linking computers and associated peripheral devices over long distances. The former is the type built into a general-purpose microcontroller. The second has special coding requirements as specified by a standard. It uses high speeds. The interface for this type of synchronous communication system is handled by special-purpose ICs. This type is often called a *network*. Normally, a general-purpose microcontroller needs to interface with the special-purpose IC in order to use a network.

In the former category, data transfer is made as simple transfers of bytes without any special handshaking protocols. This is the type illustrated in Figure 8.3. Motorola calls this type of subsystem the *serial peripheral interface* (SPI). Others may call it the *synchronous communication interface* or *serial control port.* Note the use of the word *peripheral.* This type of system is more or less a standard. Motorola and others build chips that can use serial lines for processor-to-chip communications. These are local (i.e., peripheral) devices. The SPI is typically used for local devices instead of remote devices because of the extra line required for the clock. For example, local can mean anywhere inside a machine, robot, or automobile. *Remote* typically means distances of greater than several meters. This can range from distances between rooms in a building to transcontinental transmission.

As mentioned earlier, many ICs use serial I/O for interfacing with a microcontroller. An analog-to-digital converter has inputs to read analog data and sends the digital equivalent to a microcontroller using the serial interface. A microcontroller may send display data to an LED driver using the SPI. The SPI subsystem may also be used for processor-to-processor communications.

Multiprocessor systems are quite common. For example, an automobile typically uses several microcontrollers. Each controls a different system, such as antiskid braking, driver instrument panel, and engine control. The complete automobile control system also uses many sensors and actuators for mirrors, window motors, lights, door locks, and so on. Without the use of serial I/O, the system may require over 50 wires. To reduce wiring, the microcontrollers and peripheral I/O chips can communicate with one another using a serial network.

One serial standard is the automotive serial controller area network (CAN) by Bosch. Both Motorola and Intel chips can follow this standard. Other standards are also vying for acceptance by automobile manufacturers. Some of the ICs that use a synchronous serial interface are

Phase-locked-loop (PLL) frequency synthesizers (e.g., MC145156)
Seven-segment display decoder/drivers (e.g., MC14499)
LCD display decoders/drivers (e.g., MC145453)
Vacuum fluorescent display drivers (e.g., SN75512B)
Analog-to-digital converters (e.g., MC145041)
Digital-to-analog converters (e.g., MC144110, DAC-8840)
Shift registers (e.g., 74HC589, 74HC595)
Real-time clocks (e.g., MC68HC68T1)
ISDN transceivers (e.g., MC145472, MC145474)

A short note about the last item: ISDN stands for *integrated services digital network.* It is a standard for combining digital data and voice signals for telecommunications (telephone) systems. A *transceiver* is a device that allows a digital device to transmit and receive data over a telecommunications channel.

■ 10.5 THE SERIAL PERIPHERAL INTERFACE (SPI)

The serial peripheral interface (SPI) subsystem in the 68HC11 can be configured to interface directly with numerous standard product peripherals supplied by various manufacturers. The subsystem can be configured as a master or a slave device (see Section 8.6.1). A configuration routine can choose the clocking polarity, one of four clocking rates (master only), and one of two fundamentally different clocking protocols. The maximum master bit frequency

TABLE 10.3   SERIAL PERIPHERAL INTERFACE (SPI) PINS

Port D pin	SPI signal	Master mode	Slave mode
PD2	MISO (master in, slave out)	Input	Output
PD3	MOSI (master out, slave in)	Output	Input
PD4	SCK (serial clock)	Output	Input
PD5	$\overline{SS}$ (slave select)	Programmable	Input

is 1.05 MHz and the slave has a maximum bit frequency of 2.1 MHz. The subsystem allows simultaneous transmission and reception of data using the same clock. This is useful for multiprocessor communications (one SPI master allowed at a time). The SPI has flexible I/O pin control, which allows the direction of each pin to be controlled by software.

Refer to the block diagram in Figure B.1 in the Quick Reference (Appendix B). The system uses port D pins. Recall that the asynchronous serial communications interface (SCI) used pin PD0 for receive (RxD) and PD1 for transmit (TxD). The remaining pins can be used for the SPI, if enabled. Again, these pins can also be used as general-purpose I/O if the SPI subsystem does not use them. We shall see that register DDRD provides some configuration options for the SPI subsystem. Table 10.3 shows the SPI pins. The registers are

SPCR   SPI control register
SPSR   SPI status register
SPDR   SPI data register

We can include DDRD, the data direction register for port D, since it also has an important role to play.

■ 10.6 SPI TOPOLOGIES AND APPLICATIONS

In digital communications systems a *topology* is the way in which communicating devices (known as *nodes*) are connected. There are several types of topologies that are described in most books or articles about data communications or computer networks. References 7 and 8 in Appendix C explain more about the subject. For a network of peripheral chips using the SPI system, only two of these configurations, the bus and cascade topologies, are in common use.[3]

10.6.1   Bus Topology

In bus topology, each device or node taps off a communications bus. In this case the bus is made up of the two data lines and the clock line. The master has control of the bus. It uses additional control lines to select with which slave it wishes to communicate. It is possible to have a *floating master,* which means that devices may take turns at becoming the master. This may complicate the connections, as shown in Figure 10.9. However, most peripheral I/O networks in a microcontroller application use a *fixed master* system, as shown in Figure

---

[3]The cascade topology is like the ring or loop topologies used by some computer networks. A computer network usually is known as a local area network (LAN) or a wide area network (WAN).

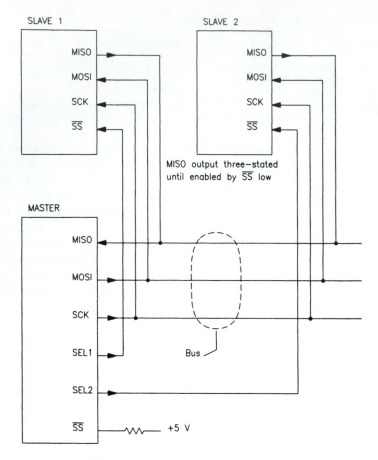

Figure 10.9   SPI System With Bus Topology

10.9. Whatever the system being used, only one master is allowed at any one time. In this system a 68HC11 master has its slave select pin $(\overline{SS})$ tied to logic high. It drives its select lines to select a slave. A slave is selected when its $\overline{SS}$ input is driven low. Any of the master's remaining output lines can be used for select logic.

Slaves can be peripheral ICs or other microcontrollers. The peripherals may have both serial input and outputs or may be input-only or output-only devices. The 68HC11 can be configured as master or slave by setting or clearing bit MSTR in control register SPCR. In Figure 10.9, when the master communicates with slave 1, it drives its SEL1 line low. To disable any other slave device, such as slave 2, it must drive the other select lines high. For example, it must drive SEL2 high (and others not shown). A high input to a slave's select $\overline{SS}$ input disables the external SPI interface for that device.

10.6.2 Microcontroller I/O With Bus Topology

Let's say that a system needs to sample 16 digital inputs. These could be switch inputs from a control panel for an industrial process controller. This system can use a parallel-input to serial-output shift register to latch in data from individual input pins as a binary number. The

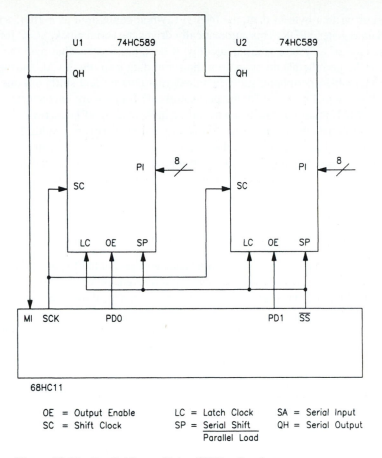

OE = Output Enable   LC = Latch Clock   SA = Serial Input
SC = Shift Clock   SP = Serial Shift   QH = Serial Output
                             Parallel Load

Figure 10.10   Parallel Inputs Using SPI Bus Topology

microcontroller can read this data by shifting in each data bit using the SPI subsystem. The 74HC589 is a parallel input IC with serial output. Each contains a data latch and a shift register. The data latch is used to store the parallel input data. The shift register is used to shift out this data. Figure 10.10 shows how two of them can be connected using bus topology. Using this configuration allows the microcontroller to use its other I/O pins for other functions.

The 68HC11 is the master device. It also uses slave select pin $\overline{SS}$ as an output. To read in data using the SPI, it drives $\overline{SS}$ low, then high again. A rising edge on the latch clock (LC) loads the parallel data on the parallel inputs into the peripheral's data latch. The 68HC11 drives $\overline{SS}$ low again. This, in turn, drives the SP inputs of peripheral chips U1 and U2 low. When SP is low, each peripheral chip transfers the data latch contents to its respective shift register. To read in data, the 68HC11 drives $\overline{SS}$ high again. Setting the peripheral pins SP and LC high is one of the conditions necessary for a peripheral chip to shift out the latched data. The MCU selects which peripheral to read by driving its output enable line low. When it has selected (addressed) one of the peripherals, the MCU drives the serial clock for eight cycles. This shifts in the peripheral's data bits into the master input (MI) of the MCU. We toggle the $\overline{SS}$ twice because it drives two different inputs, LC and SP. An alternative design could use separate control lines for each.

To read or write a byte of data, the 68HC11, when configured as master, writes a byte to SPI data register SPDR. This automatically drives the serial clock, SCK, for eight cycles. In the case of a system with inputs only, it does not matter what byte the MCU writes because the peripherals do not accept the serial data from the MCU. During each clock pulse, an enabled peripheral shifts in an external data bit and shifts out one of its data bits. In the case of Figure 10.10, the peripheral serial inputs are unconnected. After the transfer, the MCU reads the data that was shifted in by reading SPI data register SPDR.

Sending serial data out to an output "serial-to-parallel" peripheral would be similar. In this case, after the data is shifted into the peripheral, the MCU asserts a strobe pulse to transfer the peripheral's shift register to its latch register. For example, to update a seven-segment LED display with a serial interface, the SCI transfers bytes to it and then asserts a strobe pulse to update the display drivers.

10.6.3 Cascade Topology

In Figure 10.11 we see that for the master to send data to slave 2, it has to pass through slave 1 first. More slaves can be added to the cascade loop if desired. Other variations of the connections are possible. The slave select pins ($\overline{SS}$) can be connected together. In this case a ground connection enables the slaves.

In Section 9.3 we introduced LCD driver ICs. Some drivers have a serial input for data and are designed to be cascaded. Since an MCU has no need to read an LCD display, the return connection to the master's input (MI) is not necessary. For example, the MC145000 LCD Master Driver has 48 segments. It has serial input, serial output, and a *frame synchronization* output. The frame synchronization signal is a series of pulses to synchronize all LCD segment waveforms. Recall that LCDs require alternating waveforms.

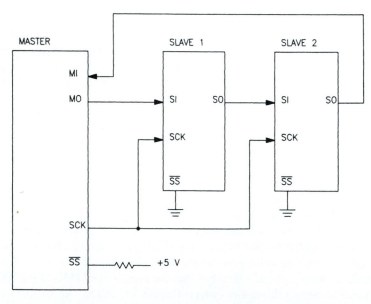

Figure 10.11   SPI System With Cascade or Loop Topology

Some LCD displays require more than 48 segments. To drive the extra segments, slave drivers can be cascaded from the master driver. The MC145001 is an LCD slave driver capable of driving 44 segments. The master driver's frame synchronization output drives the frame synchronization inputs of the slave drivers. This synchronizes all LCD segment waveforms.

Each slave driver also has a serial input and output. Thus to add more LCD segments, you can cascade additional slave drivers to the master driver. The MCU sends data bytes to the cascaded drivers. The first byte contains segment data for the last driver in the cascade chain. The number of bytes sent depends on the number of drivers. Unfortunately, this example illustrates some terminology confusion. The microcontroller sending the data to the LCD displays is the *serial system master.* The LCD master driver is a *serial slave device.* It is called a *master driver* because it supplies the frame synchronization signal necessary for the LCD segments.

10.6.4 Microcontroller I/O With Cascade Topology

Figure 10.12 illustrates an I/O interface with eight inputs and eight outputs. It turns out that the 68HC11 in this interface circuit sends out a parallel output and reads an input by doing the following: Initially, the slave select line ($\overline{\text{SS}}$) is high. The 68HC11 toggles $\overline{\text{SS}}$ twice.

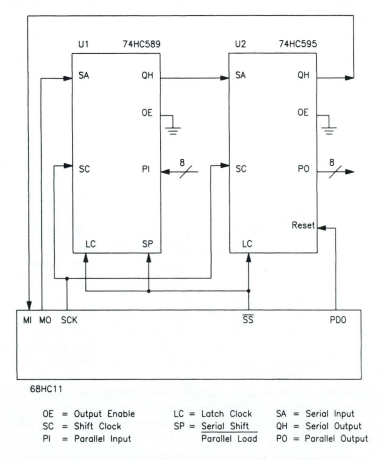

OE	= Output Enable	LC	= Latch Clock	SA	= Serial Input
SC	= Shift Clock	SP	= Serial Shift	QH	= Serial Output
PI	= Parallel Input		Parallel Load	PO	= Parallel Output

Figure 10.12 Parallel Input/Output Using SPI Cascade Topology

Then it writes the data byte to the SPI data register twice. It reads the parallel input byte by reading the SPI data register after this operation. To update the output, it toggles $\overline{SS}$ again. In the following paragraphs we will explain this process in more detail.

We have already discussed how the parallel-input, serial-output shift register (U1 in Figure 10.12) operated. The 74HC595 (U2 in Figure 10.12) is a serial-input, parallel-output shift register. It operates as follows: When the reset line is low, the shift register data is reset but the parallel outputs remain latched at their previous state (latch register unchanged). Data bits at its serial input (SA) are shifted into its shift register during every input clock pulse (SC). When a low-to-high transition occurs at its latch clock (LC), the shift register contents are transferred to its latch register. Because the output enable (OE) is tied low, the latch register always drives the parallel output pins (PO). During a serial transfer operation, both peripheral ICs also shift out a bit at their respective serial output (QH) as well as shifting in a bit at their respective serial input (SA).

The 68HC11 uses the interface as follows. Initially, $\overline{SS}$ is high. It drives $\overline{SS}$ low for U1 to latch the parallel input into its data latch as discussed in Section 10.6.2. The MCU drives $\overline{SS}$ high again to cause the outputs of U2 to be updated based on the current contents of its shift register. The 68HC11 toggles $\overline{SS}$ again in order to transfer the data latch contents of U1 to its shift register. The first write of the 68HC11 master causes the input data byte of U1 to be shifted into U2, and the output data byte desired is shifted into U1. The 68HC11 writes a second time. This causes the input data byte originally from U1 to be shifted into the microcontroller's master input line (MI). The desired output byte is now shifted into parallel output peripheral U2. The 68HC11 reads its SPI data register to get the input data and can update the parallel output pins by again toggling slave select line $\overline{SS}$.

10.6.5 A Summary of SPI Data Transfer Operations

In this section we will examine (and review) how the serial data transfer between a master and slave is performed. You may wish to refer back to Figure 8.3 when reading this section. Note that the master initiates data transfer and drives the serial clock to synchronize the transfer. Communication between master and slave occurs by shifting bits to each other's shift register. Eight data bits are shifted during each operation.

The master initiates transfer by writing data to its data register, SPDR. The data register contents are automatically transferred to the shift register. This drives the clock automatically for eight pulses. Remember that during this time, the shift register contents are shifted out to the master output (MOSI) pin and bits are shifted in from the master input (MISO) pin.

If the microcontroller is configured as a slave device, a different sequence occurs. It, too, initially writes to data register SPDR, which in turn is transferred automatically to the shift register. But unlike a microcontroller configured as a master, it must wait for a master to send clock signals before any bits are shifted in or out. For each clock pulse generated by a master device, the slave shifts out a bit to the slave output (MISO) pin and shifts in a bit from the slave input (MOSI) pins.

Upon completion of eight clock cycles, both master and slave do essentially the same thing automatically. The shift register data is transferred to the receive data register, SPDR. The SPI request status flag (SPIF) in status register SPSR sets. If enabled (if SPIE = 1 in SPCR), an interrupt is asserted. To clear SPIF, the 68HC11 must read or write to data register SPDR.

■ 10.7 SPI SOFTWARE

10.7.1 Configuration and Registers

The basic configuration process is to write to port D data direction register DDRD and then to SPI control register SPCR. To configure any subsystem, we need to know the structure of its registers. Figure 10.13 shows the SPI register structure.

Serial Peripheral Control Register (SPCR).   To enable the SPI subsystem, the configuration routine must set the SPI enable bit (SPE). If the subsystem is not enabled (SPE = 0), port D pins PD2 to PD5 are general-purpose parallel I/O pins. It is possible to make all port D outputs open-drain by setting bit DWOM. This also makes all SPI outputs open-drain. This is useful in communication systems with multiple masters. Such a

SPCR

	B7							B0
$1028	SPIE	SPE	DWOM	MSTR	CPOL	CPHA	SPR1	SPR0
RESET =	0	0	0	0	0	1	U	U

SPIE =   SPI Interrupt Enable
SPE =   SPI System Enable
DWOM =   Port D Wire-Or Mode (0 = normal, 1 = open-drain)
MSTR =   Master/Slave Select
              0 = Slave Mode
              1 = Master Mode
CPOL =   Clock Polarity, see Figure 10.14
              0 = active high clock, SCK idles low
              1 = active low clock, SCK idles high
CPHA =   Clock Phase Select, see Figure 10.14
SPR1,0 = SPI Clock (SCK) Rate Select Bits

SPR1	SPR0	E - clock Divided by
0	0	2
0	1	4
1	0	16
1	1	32

SPSR

	B7							B0
$1029	SPIF	WCOL		MODF				
RESET =	0	0	0	0	0	0	0	0

SPIF =   SPI Interrupt Request
WCOL =   Write Collision Status Flag
MODF =   SPI Mode Error Interrupt Status Flag

Figure 10.13   SPI Registers *(Continued)*

DDRD

	B7							B0
$1009			DDRD5	DDRD4	DDRD3	DDRD2	DDRD1	DDRD0
RESET =	0	0	0	0	0	0	0	0
Pin =			SS	SCK	MOSI	MISO	(TxD)	(RxD)

DDRD5 (Master)

    0 = SS input to detect mode fault
    1 = SS is general purpose output
    Slave = always input

DDRD4 (Master)

    0 = SCK output disabled
    1 = SCK output enabled
    Slave = always input

DDRD3 (Master)

    0 = MOSI output disabled
    1 = MOSI output enabled
    Slave = always input

DDRD2 (Slave)

    0 = MISO output disabled
    1 = MISO output enabled
    Master = always input

Figure 10.13 SPI Registers *(Continued)*

system may be prone to bus *contention* (fighting for control of the bus). This occurs if more than one master tries to drive the same line at the same time. When one chip tries to drive a logic high and another tries to drive a logic low, high currents result. This contention can be destructive to HCMOS chips because of the high currents. If the outputs are open-drain with pull-up resistors, contention does not result in damage since current is drawn through the resistors instead of through the gate outputs. Despite careful design to avoid a situation when two devices try to become a master at the same time, a possibility of error always exists.

A 68HC11 can be configured as an SPI master or slave. To make the MCU a master, set the master bit (MSTR) to 1. To make it a slave, clear bit MSTR. The next 2 bits, clock polarity (CPOL) and clock phase (CPHA), determine the clocking and data bit sampling relationship. Master and slaves should have the same setting. Many peripheral chips have a fixed relationship for their serial interface. The microcontroller configuration routine should initialize these bits so that it is using the same clocking relationship as those used by the peripherals.

The options that the SPI configuration routine can set up are as follows. When the clock polarity bit (CPOL) is clear, the master drives the clock output (SCK) low as a clock idle. For the slave, clearing bit CPOL causes it to recognize clock idle as a logic low. Setting bit CPOL causes the clock to idle high. Both devices shift out a bit at their respective outputs at the beginning of each serial clock cycle. At their respective inputs, each device

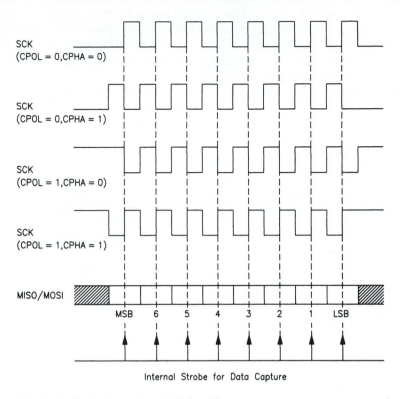

SS

SCK
(CPOL = 0,CPHA = 0)

SCK
(CPOL = 0,CPHA = 1)

SCK
(CPOL = 1,CPHA = 0)

SCK
(CPOL = 1,CPHA = 1)

MISO/MOSI

MSB   6   5   4   3   2   1   LSB

Internal Strobe for Data Capture

Figure 10.14   SPI Data Clock Timing Diagram

will latch in a bit at a clock edge. To latch bits on the first edge of each clock pulse (SCK), clear clock phase bit CPHA. To latch bits on the second edge of each SCK cycle, set bit CPHA. It turns out that to latch bits on a rising edge, the clock polarity and phase bits, CPOL and CPHA, should be identical. To latch on a falling edge, they should differ. Figure 10.14 illustrates this timing relationship.

The clock phase bit (CPHA) has another function for the 68HC11 configured as a slave. This does not apply to a master device. If CPHA is zero, a transfer will not occur until the slave select line ($\overline{SS}$) goes from a high to a low condition. The $\overline{SS}$ line must remain low for the entire 8-bit transfer. Another transfer will not occur until $\overline{SS}$ goes high and then low again. For this slave configuration, a master will have to control the slave's $\overline{SS}$ line. It cannot be left grounded. Note that the reset condition of bit CPHA is 1. If the configuration routine sets the SPI enable bit (SPE) and does nothing else, the MCU is configured as a slave. In this case, its select line ($\overline{SS}$) should be left grounded.

The final 2 bits are the SPI clock rate selects (SPR1 and SPR0). They have an effect only if the MCU is configured as a master. These bits set the serial transfer rate based on the internal system clock. Figure 10.13 shows their settings.

Port D Data Direction Register (DDRD). When the SPI subsystem is enabled, all SPI-defined inputs are configured to be inputs regardless of the data direction bits. However, all pins defined to be SPI outputs will be outputs only if their respective data direction bits are 1.

If you wish to use slave select ($\overline{SS}$) as a general-purpose output, the software should set bit DDRD5 in the data direction register and initialize control register SPCR. In this case the order is important. The configuration routine first writes to port D to make pin PD5 an output and then writes to register SPCR. If it initializes SPCR first, a mode fault may occur if the subsystem is initialized as a master. We discuss mode fault later.

If the SPI subsystem is configured as a master, bit DDRD4 must be set for a serial clock (SCK) output to be generated. If the SPI subsystem is configured and enabled as a slave, pin PD4 is a clock input regardless of the state of bit DDRD4.

Bit DDRD5 is an exception to the input rule. If the SPI subsystem is configured and enabled as a master, bit DDRD5 must be clear for the $\overline{SS}$ pin to act as an input. This means that the $\overline{SS}$ pin must be high for the device to remain a master. If a logic low is detected, a mode fault occurs (sets status bit MODF in status register SPSR), which immediately disables all SPI outputs (clears corresponding DDRD bits). The mode fault also disables the SPI system by clearing the enable bit (SPE) and clears the MSTR bit by forcing the device into slave mode. Due to the mode error detection mechanism of the SPI subsystem, some designers use the $\overline{SS}$ pin as an input to sense whether another device is trying to become a master. Detecting the condition prevents output contention. If bit DDRD5 is 1, the master's $\overline{SS}$ pin acts like a general-purpose output and mode error detection is disabled. This is the case for the systems shown in Figures 10.10 and 10.12. For a slave, bit DDRD5 has no effect since PD5 must be an input.

A data direction register bit can be changed at any time to affect SPI operation. There are situations when it is useful to enable and disable outputs. In some networks the master may disable its output if it only wants to read data. In others, the slave should disable its output when reading data.

Consider a network of devices using bus topology. Unlike the system shown in Figure 10.9, all slaves have their slave select pins ($\overline{SS}$) grounded. A protocol has to be established to enable the master to address a slave and read input from it only. Initially, all slaves disable their outputs. The master transmits address information. The slaves read in the address information. The addressed slave enables its output (MISO). The other slave devices do not enable their outputs. If they did enable their outputs, the serial clock would shift out their data as well, causing contention. The protocol also has to define how to end a data transaction between a master and slave. Again, this is a trade-off. The simplification of slave-selecting hardware is made at the expense of more complicated software. Again, open-drain outputs are generally desirable in systems where different devices take turns driving SPI pins.

SPI Status Register (SPSR). Recall the summary of SPI data transfer operations described in Section 10.6.5. The serial peripheral data transfer flag (SPIF) sets upon completion of data transfer between the MCU and an external device. If the interrupt is enabled (SPIE = 1), it also asserts an interrupt. This bit is cleared when the status register is read followed by a read or write to the data register (SPDR). The write collision bit (WCOL) is set when the MCU tries to write to the data register (SPDR) while a data transfer is taking place. The mode fault flag (MODF) indicates that another device may be attempting to be the master. We return to these two error flags later.

SPI Data I/O Register (SPDR). The MCU must write to this register in order to transmit and receive. A master writes to this register to initiate a data transfer. If a slave writes to the data register, it must wait for an external master to drive the serial clock to initiate a new transfer. The MCU has a buffer for reading data. After 8 bits have been transferred, the data is loaded into the buffer. This buffer must be read before another 8 bits are shifted in. Note that the transmit side of the SPI is not double buffered.

10.7.2 Master SPI Operation

After configuration, a master does the following for each data byte:

Writes a byte to data register SPDR (option, initially assert slave $\overline{SS}$ low).

Waits for the flag bit SPIF to set (option, deassert slave $\overline{SS}$ high).

Reads a byte from data register SPDR.

Listing 10.5 is a program example that illustrates a transfer of 1 byte. To transfer more bytes, repeat the same basic procedure for each byte. Similarly, the technique can be modified by using interrupts instead of polling. This can be done by replacing the polling loop with a WAI instruction. The interrupt service routine could be an RTI instruction. Of course, the initialization part of the program would also have to be modified to set the interrupt enable bit, SPIE.

```
*Listing 10.5
*Demonstrate SPI byte output and input with
*slave always enabled, its SS line wired low.
*This program uses polling.
*------------------------------------
DATA     EQU     0

         ORG     $100

         LDX     #$1000       ;point to registers
         LDAA    #$38         ;enable SPI outputs,
         STAA    DDRD,X       ;note that PD5 is general-purpose output
         LDAA    #$57         ;SPI as master, CPHA=1, CPOL=0
         STAA    SPDR,X       ;and transmit it
POLL
         TST     SPSR,X       ;wait for transfer complete
         BPL     POLL
         LDAA    SPDR,X       ;get data from slave
*                             ;this also clears flag, SPIF
DONE
         BRA     DONE         ;end of demo, now ACCA contains data
*                             ;received from slave.
```

C Code Example. C Listing 10.5 shows the equivalent in C language. In this case we are not using the library SPI routines read_spi which returns an unsigned character and write_spi which requires an unsigned character parameter.

```
/* C Listing 10.5 */

#define_SPI */Tell compiler to use SPI bit definitions in hc11.h */
#include <hc11.h>          /* Include register definitions */


/* Declarations */
unsigned char Data, SlaveData;

void main(void)
{
      /* Configure SPI */
      DDRD = 0x38;     /* enable SPI outputs */
      SPCR = 0x57;     /* SPI as master, CPHA=1, CPOL=0, /32 clock */
      SPDR = Data;     /* get data and transmit it */
      while ((SPSR & SPIF) == 0); /* wait for SPIF to set */
      SlaveData = SPDR; /* then get data from slave */
                                            /* also clears SPIF
                                            */


      while (1);       /* Loop forever */
}
```

10.7.3 Slave SPI Operation

Usually, the MCU is the master since it controls communication with peripheral I/O chips. However, it may be a slave if the SPI subsystem is used for multiprocessor communications. After configuration a slave does the following for each data byte:

Writes a byte to data register SPDR.

Waits for the flag bit SPIF to set.

Reads a byte from data register SPDR.

The difference from a master is that the slave does not know how long it will have to wait for the second step. This depends on the master device. The slave can use a program similar to that used for the master. This is shown as Listing 10.6.

```
*Listing 10.6
*Demonstrate SPI byte output and input with master.
*This slave is always enabled with SS line wired low.
*This program uses polling.
*------------------------------------
DATA   EQU       0

       ORG       $100

       LDX       #$1000   ;point to registers
       LDAA      #$04     ;enable MISO output,
```

```
          STAA      DDRD,X      ;note others forced as inputs
          LDAA      #$47        ;SPI as slave, CPHA=1, CPOL=0
          STAA      SPCR,X      ;clock rate is really don't care
          LDAA      DATA        ;get data from somewhere
          STAA      SPDR,X      ;and send it to data register
*                               ;wait for master clock to shift it out
*                               ;and to shift master's data in
POLL
          TST       SPSR,X      ;wait for transfer complete
          BPL       POLL        ;stays here until master
*                               ;completes transfer
          LDAA      SPDR,X      ;get data from master
*                               ;this also clears flag, SPIF
DONE
          BRA       DONE        ;end of demo, now ACCA contains data
*                               ;received from master.
```

Similarly to the master program of Listing 10.5, the slave program can be modified to use interrupts. Also, this program should begin execution before the master program so that it will wait for the master's data. If it starts later, it will miss the master's initial data, and similarly, the master will not get valid data from the slave initially.

C Code Example.   C Listing 10.6 shows the equivalent in C language.

```
/* C Listing 10.6 */

#define_SPI */ Tell compiler to use SPI bit definitions in hc11.h */
#include <hc11.h>          /* Include register definitions */


/* Declarations */
unsigned char Data, MasterData;

void main(void)
{
     /* Configure SPI */
     DDRD = 0x04;     /* enable MISO output, others forced as
                         inputs */
     SPCR = 0x47;     /* SPI as slave, CPHA=1, CPOL=0, x clock */
     SPDR = Data;     /* get data, send it to data register, wait
                         for master data */
     while ((SPSR & SPIF) == 0); /* wait for SPIF to set */
     MasterData = SPDR;         /* then get data from master */
                                      /* also clears SPIF */


     while (1);      /* Loop forever */
}
```

10.7.4 Sample Driver Routines

We now show sample software drivers to handle the peripheral interface shown in Figure 10.12. Subroutine IOSPI first ensures that the parallel input IC (U1) transfers the most recent input pin data into its shift register (Listing 10.7). Then it performs two serial byte transfers. At the end it updates the data pins of the parallel output IC (U2) with the data just sent there. You may wish to review the description of this interface in Section 10.6.4. Note that the MCU has to transfer data twice. To get the U1 input, it has to shift it through U2 first. To send data to U2, it has to shift it through U1 first.

```
*Listing 10.7
*Source code for two subroutines that handle
*I/O for the interface of Figure 10.12.
*------------------------------------

        ORG      $180


*Subroutine INITSPI
*Initializes the SPI interface for polling operation
* registers
*     IX = register block address
*No return registers except that CCR affected

INITSPI
      PSHA                        ;preserve registers
      BSET      DDRD,X $31        ;enable SCK, SS, PD0 as outputs
*                                 ;note mode fault disabled
      BCLR      PORTD,X $21       ;reset U2 output to zero
      BSET      PORTD,X $21       ;and drive SS high


      LDAA      #$5C              ;set MSTR, enable SPI,
*                                 ;clock on rising edge,
*                                 ;clock idles high


      STAA      SPCR,X
      LDAA      SPSR,X            ;clear SPIF if set
      LDAA      SPDR,X
      PULA                        ;restore registers
      RTS                         ;and return


*Subroutine IOSPI
*Sends data byte to parallel outputs (U2)
*and reads parallel inputs (U1)
*Calling Registers
*       IX = register block address
*       ACCA = output byte for U2
*Return Registers
*     ACCA = input byte from U1
*     CCR affected, rest unaffected
```

```
IOSPI
      PSHB                            ;preserve ACCB
*                                     ;following sequence to latch in U1 input

      BCLR    PORTD,X $20   ;toggle SS to transfer current
      BSET    PORTD,X $20   ;U1 inputs to latch
      BCLR    PORTD,X $20   ;toggle SS to transfer U1 latch
      BSET    PORTD,X $20   ;to shift register

      STAA    SPDR,X        ;transmit data for U2
POLL1
      BRCLR   SPSR,X $80 POLL1
      LDAB    SPDR,X        ;receive (don't care) data from U2
      STAB    SPDR,X        ;transmit (don't care) data for U1
POLL2
      BRCLR   SPSR,X $80 POLL2
      LDAA    SPDR,X        ;receive data from U1
*                           ;following sequence to load U2 outputs

      BCLR    PORTD,X $20   ;transfer U2 shift register
      BSET    PORTD,X $20   ;to latch and output pins

      PULB                  ;restore ACCB
      RTS                   ;and return
```

PRACTICE

Write software for the system of Figure 10.12. The software reads the U1 input data into ACCB and sends the byte $B5 to U2.

Solution

```
*Solution
      LDS     #STACK        ;initialize stack
      LDX     #REGBAS       ;point to registers
      JSR     INITSPI       ;configure SPI subsystem
      LDAA    #$B5          ;set up calling registers
      JSR     IOSPI         ;perform I/O transfer
      TAB                   ;store input in ACCB
DONE
      BRA     DONE          ;done for now
```

10.7.5 Transfer to Peripherals With Odd Word Lengths

The SPI subsystem handles 8-bit transfers automatically. Some peripherals, such as the MC144110 digital-to-analog converter (D/A), use 6-bit words. In this case the MCU software will have to reformat the block of 6-bit words for the converter chip into 8-bit words that it can transfer using the SPI.

The MC144110 D/A has six converter channels that convert a 6-bit word into analog voltage. It has six cascaded 6-bit shift registers whose contents can be transferred to the converters. Before sending data to it, a subroutine can convert the six 6-bit words (36 bits) into a block of five 8-bit words (40 bits). The upper nibble in the first byte of the new block is not used (logic don't care). The SPI subsystem can send the 5 bytes to the MC144110. The first 4 bits, which are logic don't care, end up being shifted out of the peripheral chip after the fifth byte is sent. The MCU can then drive a control line to latch the digital data to the converters.[4]

Another digital-to-analog converter with a serial interface is the DAC-8840 by Analog Devices. It has eight 8-bit D/As. But unlike the MC144110, it uses only one (12-bit) shift register. To send a byte to one of its channels, it is necessary to send 12 serial bits. The upper 4 bits specify the channel address. The rest of the 8 bits comprise the data byte for the channel. After the 12 bits are sent, the microcontroller asserts a load pulse to transfer the data byte to the channel D/A selected.

The 68HC11 SPI transfers data only as bytes. Its software would have to reformat each 12-bit word into 2 bytes. Again, the upper nibble of the first byte sent would be shifted out of the peripheral chip after the 2 bytes are sent. After each 2-byte transfer, it would pulse the chip's load pin to latch in the channel data.

■ 10.8 SPI ERROR HANDLING

The serial peripheral interface has two built-in error-handling mechanisms: one for mode fault (flagged by bit MODF), the other for write collision (flagged by bit WCOL).

Mode Fault. We have already mentioned this fault in earlier sections. It occurs when more than one device tries to be a master. The error is detected when another device, assuming the role as master, drives the slave select ($\overline{SS}$) line of the current master low. This protection exists only if the master's data direction bit for $\overline{SS}$ (DDRD5) is zero.

Assuming bits DDRD5 = 0 (in data direction register DDRD) and MSTR = 1 (in control register SPCR) when pin $\overline{SS}$ is driven low, the mechanism works as follows:

Enable bit SPE in control register SPCR clears.

Data output MOSI and clock output SCK tristate.

Mode bit MSTR in control register SPCR clears.

Mode fault flag MODF in status register SPSR sets.

If enable bit SPIE in control register SPCR is set, the SPI asserts an interrupt.

What happens after this depends on the application software you use. Typically, the software would wait until the other master has relinquished control. Then a designated device would attempt to reassume the role as master. To clear the mode fault flag, the software must read the status register (SPSR) and then write to the control register (SPCR).

Write Collision. This occurs when the microcontroller attempts to write to data register SPDR while a transfer is taking place. Normally, this could not happen to a master device.

---

[4]Section 8.7 of the *Reference Manual* covers this example in depth.

It should wait for the transfer complete flag (SPIF) to set before clearing it and writing to the data register again.

A write collision error is more likely to happen to a microcontroller configured as a slave. It does not know whether a master is already transferring data. If a master has initiated a data transfer and the slave writes to the data register during this transfer, the slave's shift register transfer continues undisturbed. However, the data written to the data register (SPDR) is not in fact transferred to the shift register. The write collision error flag (WCOL) sets to inform the device that this error occurred.

A device must poll the status register (SPSR) to determine if a write collision occurred. To clear this error flag, the MCU must read the status register and then read or write to the SPDR data register. Note that the program of Listing 10.6 will clear the error bit, although it does nothing if the bit sets.

■ 10.9 NETWORK COMMUNICATIONS

A *network* is a communications system for linking computers and associated devices. In today's information age, the interchange of information is vital in order for business, industry, and social services to function. Often, data from real-time controllers is required. Hence, microcontrollers are sometimes part of a network. For example, a weather service bureau sends commands to remote monitoring stations and reads sensor measurements. A distributed control system will use a network for communication with controllers located throughout a processing plant or factory. Naturally, there have to be standards covering how all of these devices, manufactured by various companies, can communicate with one another. We will look at some common protocols, such as the Internet through Ethernet.

The Open Systems Interconnection (OSI) Reference Model. The International Standards Organization (ISO) developed a logical model for network communications. It is a hierarchical model that breaks down a communications system into layers. Others develop well-defined interfaces for passing information between the layers. Thus, it is not necessary to have one protocol handle all aspects of communication. For example, if you write a letter to a friend, you follow the rules of using envelopes, stamps, and a mailing address. You do not need to know how the post office handles your mail (although you might worry about it). The protocol for labeling and stamping envelopes belongs to a different layer than that of the duties performed by post office workers.

The ISO communications model is called the Open Systems Interconnection (OSI) Reference Model. The OSI model layers are

Layer number	Layer name	Function
7	Application	Selection of communication services
6	Presentation	Data formatting and coding
5	Session	Establishment of connections
4	Transport	Control of message exchange
3	Network	Routing of message exchange
2	Data link	Error checking
1	Physical	Electrical and physical interface for each bit

For example, the RS-232 standard can specify the physical layer. The NRZ asynchronous serial character frame (Section 10.1.1) can be part of the data link layer. In this book we concern ourselves only with the lower layers since these are the layers implemented by semiconductor manufacturers. For information about the other layers, consult books or articles on the subject, such as reference 7 or 8 in Appendix C.

10.9.1 Physical and Data Link

Data Link Control Protocols. Two standard methods of sending synchronous data are High Level Data Link Control (HDLC) and Synchronous Data Link Control (SDLC). HDLC is part of an ISO standard for layer 2. SDLC is an earlier *de facto* standard developed by IBM. The synchronous protocols are similar. Both send data in blocks of bits called *frames*. Each frame is separated by a *flag* pattern. Figure 10.15(a) shows the frame structure for these protocols.

The flag byte is used to identify the frame boundaries, similar to the start and stop bits for the asynchronous frame. The flag byte pattern is a 0 bit, six 1 bits, and a 0 bit ($7E). This means that the other elements of a frame—address, control, information, and frame check sequence—cannot have six 1 bits in a row. If they did, the receiver could interpret the pattern as a flag. To avoid the occurrence of six 1 bits, the transmitter employs a technique called *bit stuffing*. The transmitter inserts an extra 0 bit after every sequence of five 1 bits. Similarly, the receiver removes a 0 bit after every sequence of five 1 bits. If the receiver sees six 1 bits in a row, it knows the sequence must belong to the flag.

| Flag | Address | Control | Information | | FCS | Flag |

FCS = Frame Check Sequence

(a)   Basic Format of a Data Link Frame
SDLC or HDLC

(b)   Manchester Encoding

Figure 10.15   Data Link Control Protocols

The frame check sequence (FCS) is used to verify whether data was corrupted during transmission. It is possible that noise causes bit errors. The errors usually affect a sequence of bits in one part of the frame. The FCS is usually a cyclic redundancy check (CRC). The transmitter uses a feedback mechanism to make a code word depend on the previous history of the message. Then the transmitter adds this CRC block to the end of the frame. The receiver also calculates the CRC based on the bits it receives. This includes the CRC character sent by the transmitter. If the receiver's CRC calculation results in zero, there is a very high probability that the message was received without error.

Synchronous data transfer may require that the clock signal be sent without using a physical clock connection. *Manchester encoding* (see Figure 10.15[b]) is a technique that includes the clock signal as part of the data signal. This technique uses the phase of a square wave to indicate a 1 bit or a 0 bit. A transition always occurs at a bit center to allow synchronization. A logic 0 is a falling edge and a logic 1 is a rising edge. When Manchester encoding is used, a *preamble* is used to begin a block of bytes. It is an alternating pattern of 1s and 0s used by receivers to synchronize their sampling rate. Some systems employ a variation of this technique.

Another encoding method used by some networks is frequency shift keying (FSK). We looked at FSK during our discussion of modems in Section 10.1.4.

10.9.2 Internetworking and Ethernet

Many embedded devices have become Internet appliances. They are connected to the Internet, often through a Local Area Network (LAN), or sometimes through a dial-up telephone connection or through a wireless connection. Using a Web browser, one could read data from or direct the actions of an embedded controller connected to the Internet.

Most LANs are computers, peripherals, and other devices that are physically connected using *Ethernet*. This is the IEEE 802.3 specification for cabling, signal characteristics, and data framing. Ethernet corresponds to layers 1 and 2 of the OSI model; Physical and Data Link. The standard incorporates Manchester encoding for defining clock and bit level signals, and the frame is a form of HDLC, both of which are shown in Figure 10.15. There is significantly more processing required for Ethernet than for asynchronous communications using SCI and synchronous communications using SPI. There is yet more processing required for handling Internet communications. This section gives a basic overview of the standards and one technique for using an 8-bit microcontroller interfaced to the Internet using Ethernet.

Packet Switching. The Internet is a packet switched network. This is unlike a circuit switched network such as a telephone system. In a telephone system, a telephone call proceeds as follows. The caller dials a number. A system of switching offices makes a connection between the calling telephone and called telephone. While the call is in progress, the offices maintain a physical connection between the two telephones. When one telephone is hung up, the offices remove the connection and can reuse circuit lines for connecting other telephones. In packet switching, there is no distinct circuit connecting two communicating devices. It is more like the postal system, where messages are put in packages and deposited at a mailbox. The postal system handles the delivery of the package to the address that the package specifies.

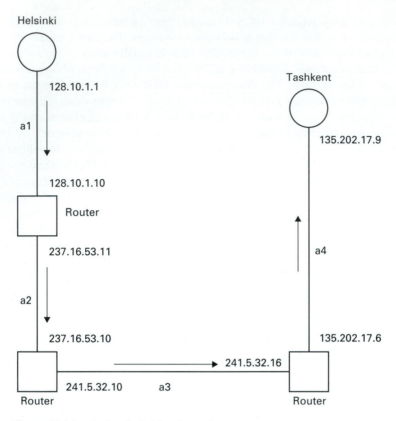

Figure 10.16   Packet Switching Scenario

Figure 10.16 shows an Internet packet switching scenario. A sender in Helsinki sends a message to Tashkent. The Internet consists of routers that receive messages and pass them on. A message may consist of one packet or several packets. A packet consists of control information that specifies how to deliver the message and the data itself, sometimes known as the payload. The packet(s) can travel using route a1, a2, a3, or a4.

Internet Addresses.   Every device connected to the Internet has to have a valid address. All addresses on the Internet are 32-bit binary numbers expressed in a decimal octet format.[5] For example the address of Tashkent is `135.202.17.9` which in binary is `10000111 11001010 00010000 00001001` (hex `87 CA 10 9`). Typically people are more familiar with word addresses such as Tashkent.microcontrol.net. The Internet has a system of domain name servers that map mnemonic addresses to numerical addresses.

TCP/IP.   The TCP/IP Internet Protocol Suite defines how networks are interconnected using the Internet. It essentially specifies how the Internet operates. TCP/IP refers to the two main standards Transmission Control Protocol (TCP) and Internet Protocol (IP). They are

---

[5]For the Internet standard IPv4. A newer standard, IPv6, uses 128-bit addresses expressed in hexadecimal format.

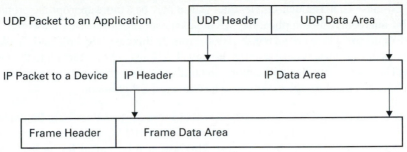

Ethernet frame packet at physical and data link layer. The data area contains the IP packet. The IP data area contains the UDP packet.

Figure 10.17   Packet Encapsulation

network technology independent. For example, a device can use the Internet using an Ethernet LAN connection or dial-up telephone connection. TCP defines protocols for maintaining reliable communications between two applications such as the Hypertext Transfer Protocol (HTTP) used by a Web browser to retrieve a Web page from a Web site server. TCP/IP standards correspond to the higher levels of the OSI model.

IP Datagram.   The IP part of the protocol suite specifies the fundamental data framing format. This is the IP datagram. It is a packet of information that contains the Internet destination and source addresses, some other routing information, and the actual payload data. In Figure 10.16, a datagram from Helsinki to Tashkent would have a source address of 128.10.1.1 and a destination address of 135.202.17.9. The other routing information gives basic directions about how to travel through the Internet to reach the destination. Figure 10.17 shows the format of an IP datagram. The addresses and routing information are part of the IP header.

Protocol Ports.   The Internet is about applications communicating. For example a computer sends e-mail to another computer, accesses a Web page, transfers a file, and handles a remote login session. The computer has one Internet address but several applications using the Internet in a different way. A protocol port is a 16-bit integer that is used to distinguish between different applications so that the software knows how to handle the data. For example, packets sent out for a Web page request come from a different port than for e-mail, and each application uses a different set of rules (protocol).

User Datagram Protocol (UDP).   The Internet Protocol (IP) specifies transferring of data between two devices on the Internet. The User Datagram Protocol (UDP) specifies transferring of data between two applications, each on different devices and connected by the Internet. The Transport Control Protocol (TCP) also specifies the same sort of thing, except that it also includes additional processes, such as a handshake from the destination application to make the connection more reliable. UDP is considerably more simple than TCP. An example of UDP is the ECHO service used to determine if a device is on the Internet. Some readers may be familiar with Ping command, which in fact sends a UDP ECHO datagram to a specified address. UDP is sometimes called "unreliable" because it contains no inherent checks to confirm whether a datagram has been successfully delivered. This doesn't stop applications from adding their own checking mechanism if using UDP.

Because UDP is much simpler than TCP, it may be a more appropriate protocol to use in 8-bit microcontroller applications. UDP is very useful in Ethernet networks. Typically 99.9% of the time there are no transmission errors. In this case, Helsinki and Tashkent could be names of two devices connected to the same LAN instead of referring to the geographic cities. The microcontroller application can build in its own error control mechanism on top of UDP that is much simpler than using a complete TCP protocol.

Layering of Protocols. The Internet protocols are universal and can be made to work with a variety of different physical standards. Figure 10.17 shows layering of protocols using a UDP, IP, and Ethernet scenario. This scenario is typical for an 8-bit embedded Internet device. A microcontroller application sends data to the Internet using UDP. Let's say Helsinki (see Figure 10.16) sends motor speed data to a terminal at Tashkent. The software in Helsinki makes up an appropriate UDP header and adds it to the data. This is now a UDP datagram. But this is not quite ready for transmission. The software has to form it into an IP datagram by making up an appropriate IP header and adding the complete UDP datagram as its IP data area. This is now an IP datagram that is logically ready for the Internet. However, it is not physically ready. In this case, an Ethernet is used for physical connectivity. Another layer of software has to make the data Ethernet ready. It does this by making the Ethernet Frame header and adds it to the complete IP datagram, which in this case is the Ethernet data payload or Frame Data Area. Figure 10.17 shows this encapsulation of datagrams or packets within each layer.

Because the Internet is standard, any Internet device can receive the microcontroller data. In this case, it could be a terminal at Tashkent (see Figure 10.16). At this end, the process is in reverse. The device receives the Ethernet packet, and removes the Frame header. This leaves the IP packet. The software removes the IP header, leaving the UDP packet. Then it can remove the UDP header, leaving it the data sent by the microcontroller. Thus Tashkent now has the motor speed in Helsinki. Many details are left out such as routing details. For microcontroller development, the key is to implement appropriate parts of the standards and let other parts of the infrastructure handle other details.

Ethernet MAC Addresses and Standard Features. Each Ethernet device has its own unique hardware address that is a 48-bit integer defined by the manufacturer. This is sometimes known as the Media Access (MAC) address. The TCP/IP suite includes various means for translating between Internet addresses and MAC addresses. The Ethernet standard also covers other physical details such as packet filtering, collision detection and recovery, media access, and checksum generation for error detection and correction. The collision detection and recovery part refers to the scenario of simultaneous access by multiple devices. Ethernet devices on a network segment share the same network bus. Because they may attempt to transmit data at the same time, this may cause collision. Ethernet devices have mechanisms to detect the situation to handle them.

10.9.3 68HC11 Ethernet Interface

We have examined the reasons for connecting an 8-bit microcontroller to the Ethernet, and at some background surrounding appropriate standards that permit open connectivity using the Internet. The 68HC11 does not have an inherent subsystem for handling the Ethernet. One common interface device is the Crystal CS8900A LAN ISA Ethernet Controller supplied by

Cirrus Logic. Many computers use it as a basis for Ethernet connectivity. For an 8-bit microcontroller, it is possible to use it in 8-bit mode. Cirrus Logic provides an Application Note called AN181 that describes how to use it with an 8-bit data bus. The reader can download this note from their Web site at www.cirrus.com. The 68HC11 will need to have sufficient memory and operate in expanded multiplexed mode so that it can use an external bus. An interface using the 68HC11F1 is described at www.wiresncode.com. Other implementations are possible provided that an external bus and sufficient memory for code is available.

In this interface example, the program arbitrarily sets up the MAC address because the target system is an internal LAN. Normally, the MAC address would be stored in an EEPROM in the CS8900A by a manufacturer of a Network Interface Card (NIC). The Wiresncode Web site includes details such as a circuit diagram and the source code of the software. In this section we will point out a few features of the software which is written in C.

The interface is a memory-mapped system. In 8-bit mode the CS8900A is accessed through 16-bit I/O ports as described in the application note AN181. We will focus mainly on frame transmission. The transmission steps are

Write a transmit command which is 00C0h to the memory-mapped address for the port TxCMD.

Write the length of the data (number of bytes) to port TxLENGTH.

Check the Rdy4Tx bit in register BusSt.

Write data, a byte at a time, to Receive/Transmit Data port 0.

If the CS8900A is set up for a default port block beginning at address 300h, the 16-bit TxCMD port is at addresses 304h and 305h for the low and high bytes, respectively. Thus, transmission would begin by writing the low byte C0h to address 304h. Then write the high byte 00h to address 305h. If the length is decimal 81, write 0051h to the TxLENGTH port at address 306h (51h to address 306h, 00h to address 307h).

A key feature of the CS8900A architecture is a 4-KB block of RAM known as PacketPage memory. This block contains internal registers among other things. Register access is done through a PacketPage Pointer port and two PacketPage Data ports. To access the BusSt register, write 0138h to the PacketPage Pointer port at address 30Ah (i.e., write 0138h to address 30Ah). This sets up PacketPage Data port 0 to contain the BusSt register contents. That is, read PacketPage Data port 0 at address 30Ch, and check bit 8, which corresponds to the Rdy4Tx flag. If this flag is set, one may begin transmission. Frame transmission proceeds by writing bytes to the Receive/Transmit Data port 0 at addresses 300h and 301h. This occurs as follows. Write the first byte to 300h, the second byte to 301h, byte 3 to 300h, byte 4 to 301h, and so on. Continue this alternating pattern until all frame bytes are written. The CS8900A stores the frame bytes in its internal RAM and automatically sends the frame after the last byte is written.

The following shows some code excerpts from the embedded Ethernet interface software from the wiresncode project published at the Web site www.wiresncode.com. The following is an example of sending a UDP packet.

```
udp_send(buffer, IPPORT_DISCARD, IPPORT_DISCARD, destIP, 0);
```

In this case, the statement sends a UDP datagram given a standard defined port for both destination and source, a destination IP address, and the payload data as specified by the argument buffer. The initial length of the payload data is specified as zero in this case. Not all UDP packets have data. This function in turn uses the following statement.

```
ip_send(aBuffer);
```

This statement sends the data that aBuffer points to (first array index of array holding packet data). It sends an IP packet where the protocol (UDP in this case), length, and destination address are already known. The packet is already in an Ethernet Frame format. It also takes care of specifying an appropriate IP header. The first argument of udp_send is buffer. It is passed to ip_send using its argument aBuffer. Function ip_send in turn calls cs8900_send. This function accesses Ethernet Frame data by using an array named txBuff containing elements that are the frame bytes. The function uses the argument aBuffer to initialize a pointer to the array. This specifies where to begin using the array. In this software, array txBuff defines the Ethernet Frame. In this case the Ethernet Frame Data Area encapsulates the IP packet which in turn encapsulates the UDP packet (see Figure 10.17). Here is the statement to initialize the transmit pointer named txPtr.

```
txPtr = txBuff[aBuffer];
```

The function cs8900_send also contains the following loop used to write the frame bytes alternatively to the Receive/Transmit Data port 0.

```
/* write to xmit buffer (convert length to words & round odd
bytes up) */
aLength = (aLength + 1) >> 1; /* Right shift multiplies by two */
for (i = 1; i <= aLength; i++)
{
        cpu_cs8900_reg_write(CS8900_DATA_PORT0_LSB, *txPtr++);
        cpu_cs8900_reg_write(CS8900_DATA_PORT0_MSB, *txPtr++);
}
```

Up to now, the software was generic for any 8-bit CPU. However the function cpu_cs8900_reg_write has to be written for a specific microcontroller. In this case, the software includes a file named hc11.c, which contains a function cpu_cs8900_reg_write that is defined as follows.

```
// write to 8 bit cs8900 register
void
cpu_cs8900_reg_write(const u_int8_t reg, const u_int8_t data)
{
    _io_ports[reg + CS8900_OFFSET] = data;
}
```

In this case, the function writes the byte specified by parameter data to the register specified by parameter reg. This is the same as writing the data to an address specified by reg + CS8900_OFFSET. The C compiler in this case will handle direct memory access using an array. `CS8900_OFFSET` would be the port block beginning address and reg would be the offset for the specific register in CS8900A. In this case the reg value would be 0 and 1 for `CS8900_DATA_PORT0_LSB` and `MSB` respectively. For the wiresncode project, the `CS8900_OFFSET` value is 60h, instead of the default value of 300h.

This section illustrated parts of a 68HC11 Internet and Ethernet interface. For more details, refer to the wiresncode Web site at www.wiresncode.com.

■10.10 SUMMARY

The 68HC11 has two serial I/O subsystems. The Serial Communications Interface (SCI) is asynchronous. It is intended primarily for data communications between remote devices. The Serial Peripheral Interface (SPI) is synchronous. It is designed for serial I/O with peripheral chips that support this kind of interface.

10.10.1 Asynchronous Communications

Since systems that use asynchronous communications do not use a common clock, each byte is framed by start and stop bits using the non-return-to-zero (NRZ) format. Sometimes a frame includes a parity bit. Baud rate represents the total number of bits (including framing bits) sent per second. Officially, it is the rate of change of the smallest signaling element.

Most asynchronous communications equipment uses the RS-232 interface standard. The signal levels are nominally ±12 V. The standard categorizes equipment as data terminal equipment (DTE) or data communications equipment (DCE). The standard specifies the signal directions for each type.

When devices communicate, they must follow a protocol for flow control. Usually, the software driver for the interface implements the flow control. One such protocol is acknowledge/not acknowledge (ACK/NAK). The device that receives a block of data sends an NAK signal to tell the sender to repeat the last transmission. It sends an ACK to tell the sender that it is okay to continue with the next block. Another flow control method is XON/XOFF. The receiver uses either ASCII character to tell the sender to continue or suspend transmission.

A communication channel may operate using simplex, half-duplex, or full-duplex operation. A modem (modulator-demodulator) is a device used for communications using a telephone line. Most modem standards are defined for full-duplex operation. The RS-232 interface includes modem control signals.

10.10.2 The Serial Communications Interface (SCI)

The SCI subsystem in the 68HC11 uses two port D pins, PD0 and PD1, for receive and transmit, respectively. It automatically handles framing requirements for 8 or 9 data bits (without parity) for the standard NRZ format. Communications may be interrupt driven, or status flags can be polled. Both the transmitter and receiver are double buffered.

It has three control registers, SCCR1, SCCR2, and BAUD. The latter is used to select the baud rate as an integral fraction of the system clock. The first (SCCR1) determines the wake-up and data bit selection (8 or 9). It also serves as the data register for the ninth data bit if used. The second (SCCR2) contains enabling bits for receiver, transmitter, and various interrupts. The status register (SCSR) has flags for receiver data register full (RDRF), transmit data register empty (TDRE), and receive error flags. The system uses one data register (SCDR) address to access the transmit and receive data registers. SCDR is used to shift out a data byte for transmitting or to shift in a data byte for receiving.

The basic transmit procedure is to check if the transmit empty register flag is set and then write to the data register. The basic receive operation is to wait for the receive register full flag to set and then read the data register. The transmit flag is cleared by reading SCSR and then writing to SCDR. The receive and error flags are cleared by reading SCSR and then reading SCDR.

10.10.3   Synchronous Serial Systems

Many semiconductor manufacturers build ICs that support data transfers using a synchronous serial interface. Display drivers and analog converters are typical. The serial interface uses an input line, output line, clock line, and control line. In such a system the master device controls the transfer. It selects a slave device, initiates the transfer, drives the clock, and ends the transfer. Network topologies can be bus or cascade. The cascade topology is more common for a fixed master system. Bits are simply shifted from one device to the next in the cascade loop.

10.10.4   The Serial Peripheral Interface (SPI)

The 68HC11 uses the port D pins as follows:

PD2 for MISO (master in, slave out)
PD3 for MOSI (master out, slave in)
PD4 for SCK (master clock output, slave clock input)
PD5 for $\overline{SS}$ (master input or output, slave select input)

The subsystem has a control register (SPCR), status register (SPSR), and data register (SPDR). The control register is used to enable the subsystem, enable its interrupt, select master or slave operation, and select the clocking relationship and the clocking rate (for master). The status register has a data transfer complete flag (SPIF) and two flags for error detection. The data transfer flag is cleared by reading SPSR and then accessing SPDR. The data register serves for both transmitting and receiving data.

The basic initialization (or configuration) procedure is to set up bits in the port D data direction register (DDRD), then initialize the control register (SPCR). The basic data transfer operation occurs when both devices write to the data register. The master and slave communicate by shifting bits to each other's registers. Upon completion of 8 bit shifts, the status flag (SPIF) sets and an interrupt is asserted if enabled. Then both devices read their respective data registers.

■ EXERCISES

1. Draw the CMOS and RS-232 signals for the asynchronous serial frame for data byte $A5. The frame specifications are 9600 baud, 8 data bits, no parity, and 1 stop bit.

2. Repeat Question 1 except that the data byte is character 'f' and the frame specifications are 1200 baud, 7 data bits, odd parity, and 2 stop bits.

3. Indicate the bit levels for each frame bit for character 'D'.
   (a) 7 bits + even parity + 2 stop bits
   (b) 7 bits + odd parity + 2 stop bits
   (c) 7 bits + even parity + 1 stop bit
   (d) 7 bits + odd parity + 1 stop bit
   (e) 8 bits + 2 stop bits
   (f) 8 bits + 1 stop bit
   (g) 8 bits + even parity + 1 stop bit
   (h) 8 bits + odd parity + 1 stop bit

4. Calculate the bits per second (BPS) and the character rate.

Frame Format	Baud
(a) 7 bits + even parity + 2 stop bits	2400
(b) 7 bits + odd parity + 2 stop bits	9600
(c) 7 bits + even parity + 1 stop bit	19.2K
(d) 7 bits + odd parity + 1 stop bit	300
(e) 8 bits + 2 stop bits	1200
(f) 8 bits + 1 stop bit	2400
(g) 8 bits + even parity + 1 stop bit	150
(h) 8 bits + odd parity + 1 stop bit	62.5K

5. What is the shortest time required to transmit a 2-kilobyte file using an NRZ format of 8 data bits, no parity, and 1 stop bit? The baud rate is 600.

6. Refer to Listing 10.1. Estimate the number of clock cycles that will occur from the time the program enters the TR loop to the time it exits. Estimate the same for the REC loop. Also estimate how many times each loop executes.

7. Change the program of Listing 10.1 so that the SCI is configured as follows:
   (a) 15.625 kilobaud, 9 bits, MSB = 0, wake up by idle line
   (b) 1200 baud, 8 bits, wake up by idle line, receiver asleep
   (c) 125.0 kilobaud, 9 bits, MSB = 1, wake up by address mark

8. Write a program similar to that of Listing 10.1. It transmits a block of 256 bytes, receives it, and stores it somewhere else in RAM. The SCI system uses interrupts instead of polling.

9. Two 68HC11 microcontrollers communicate with each other using the SCI. The transmitter uses an 8-MHz crystal. The receiver uses a 4-MHz crystal. The protocol specification is

> Baud rate of 15.625 kilobaud.
>
> 8-bit data or address character.
>
> A ninth bit is 1 for an address mark, 0 for a data mark.

Polling is used for all operations. When the transmitter has completed sending a block of data to the receiver addressed, it waits for the completion of transmission before sending a continuous break. Any receiver that is not addressed will go back to sleep when awakened. An ACK/NAK verification of data is used. The receiver responds by sending any byte with an MSB of 0 as an ACK signal. It responds by sending any byte with an MSB of 1 as an ACK signal. Write the transmitter instructions for the following operations.

(a) Set up the baud rate.

(b) Initialize the control registers.

(c) Wait for an empty transmit register.

(d) Send an address mark of $05.

(e) Send a data character of $9C.

(f) Send a continuous break after transmission is complete.

(g) Branch to location "RETRAN" to transmit if an NAK signal is received within 10 ms.

10. For the system described in Question 9, write the receiver instructions for the following operations.

(a) Set up the baud rate.

(b) Initialize the control registers.

(c) Wake up in response to an address mark.

(d) Wait to read the data character.

(e) Read any character.

(f) Go to sleep (receiver only!) when a break character is read.

(g) Send an ACK signal.

(h) Send an NAK signal.

11. Write a subroutine that checks for even parity in an 8-bit number received by the SCI. It returns CCR flag Z as logic 1 if no parity error has occurred. (From *Developing Systems with the MC68HC11 Lab Book,* published by Motorola, Inc.)

12. Your design team has a contract to develop a locomotive control system. The team has to decide how various controllers and I/O units should communicate with each other. The general choices are parallel, asynchronous serial, and synchronous serial. List some factors to consider when making the design decision.

13. Using only the Quick Reference (Appendix B) or the Data Book, identify the SPI pins and registers. Indicate port numbers and register addresses.

14. Draw diagrams showing the SPI system used in a network with cascade and bus topology.

15. To test the SPI and monitor its signals using an oscilloscope, you connect MISO to MOSI. Develop a program that continually sends out constant data and reads it back again. The clock should be an observable output.

16. Describe how you could modify the test procedure of Question 15 to use open-drain outputs.

17. The MC145000,01 LCD drivers have a serial interface. Data is clocked into the shift registers on the falling edge of the clock pulse. The drivers do not have an enable pin. Write the subroutine INITLCD to initialize the SPI for polling as the interface to the LCD drivers. It can use a data clock of up to 12.5 MHz (see Section 9.3).

18. Write the code for subroutine OUTLCD (see Chapter 9 Exercises, Question 9). Assume that the SPI system is configured as a master and uses polling.

19. Write the code to read four inputs using the SPI. The SPI is already configured as a master to use polling. Store the data in addresses 0 to 3.

20. Write I/O driver subroutines INITSPI and INSPI for the system shown in Figure 10.10.

```
*Subroutine INITSPI
*Initializes the SPI interface
*Calling registers
*         IX = register block address
*No return registers except that CCR affected

*Subroutine INSPI
*Reads parallel inputs from U1 and U2
*Calling registers
*         IX = register block address
*Return Registers
*         Address 00 = 16-bit data
*         CCR affected
```

21. Write I/O driver subroutines INITSPI (see Question 20) and IOSPI for the system shown in Figure 10.12 without referring to Listing 10.7.

```
*Subroutine IOSPI
*Sends data byte to parallel outputs (U2)
*and reads parallel input (U1)
*Calling Registers
*          IX = register block address
*          ACCA = output byte for U2
*Return Registers
*          ACCA = input byte from U1
*          CCR affected, rest unaffected
```

22. Write an instruction that branches to location FAULT if either SPI fault occurs.

23. At location FAULT of Question 22, write instructions that clear the fault flags.

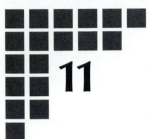

11 Programmable Timer Operations

OBJECTIVES

After completing this chapter, you should be able to

- Describe the operations, signals, and registers used in the programmable timer components: output compare, input capture, and pulse accumulator.
- Write software using output compare, input capture, and pulse accumulator.
- Design software for one-shot, square-wave, and pulse-width modulation (PWM) outputs.
- Design a stepper motor interface.
- Design software for pulse width and period measurement using input capture.
- Design software for event counting, and time accumulation using pulse accumulator.
- Design software using the real-time interrupt.

Many applications require that control decisions be made in *real time*. For these systems the programmable timer subsystem can be considered to be their bread-and-butter interface. In this chapter we will look at how the timer output compare functions generate timed control signals. The counterpart to the output compares are the input captures. We will examine how to use them for measuring pulse widths and related variables. Then we will examine two other features of the programmable timer subsystem: the pulse accumulator and the real-time interrupt.

■ 11.1 OVERVIEW

In Section 8.7 we noted the functions that can be performed by the programmable timer subsystem. In particular, Figure 8.5 illustrates the functions. They include time measurement

419

inputs and outputs that change state at preselected times. Sometimes the real-time interrupt (RTII) and the computer operating properly (COP) reset are considered to be part of this subsystem. This depends on which manual or data sheets you look at. In Section 3.10.2 we described RTII briefly. In this chapter we come back to this topic to examine how to set up RTII. In Section 3.9.2 we described the application and operation of the COP reset. The COP reset can be considered to be part of CPU monitoring and control. The subsystem has several modules (or sub-subsystems):

> Output compare
>
> Input capture
>
> Pulse accumulator

Registers.   This subsystem has more registers than the others. The programmable timer subsystem registers are

Control registers:

TCTL1	Timer control register 1
TCTL2	Timer control register 2
TMSK1	Main timer interrupt mask register 1
TMSK2	Miscellaneous timer interrupt mask register 2
PACTL	Pulse accumulator control register
OC1M	Action mask register
OC1D	Action data register

Data registers:

TCNT	Timer counter register
TIC1–TIC3	Timer input capture registers 1, 2, and 3
TOC1–TOC5	Timer output compare registers 1, 2, 3, 4, and 5
PACNT	Pulse accumulator count register

Status registers:

TFLG1	Main timer interrupt flag register 1
TFLG2	Miscellaneous timer flag register 2

Figure 11.1 shows the bit functions for the control, mask, and flag registers. In the following sections we discuss in detail what the bits do.

Pins.   Each of the port A pins has a corresponding timer subsystem function. Some have two possible functions. If the timer function is not required (not enabled), the corresponding port A pin can be used as a general input or output pin depending on the specific pin. The following lists the port A pin functions:

PA0	IC3	Input capture 3
PA1	IC2	Input capture 2
PA2	IC1	Input capture 1
PA3	OC5/OC1	Output compare 5 or 1
PA4	OC4/OC1	Output compare 4 or 1

TCTL1   Timer Control Register 1

	B7							B0
$1020	OM2	OL2	OM3	OL3	OM4	OL4	OM5	OL5
RESET =	0	0	0	0	0	0	0	0

OMx,OLx =   Output Compare Specification,
                 see Figure 11.4

TCTL2   Timer Control Register 2

	B7							B0
$1021			EDG1B	EDG1A	EDG2B	EDG2A	EDG3B	EDG3A
RESET =	0	0	0	0	0	0	0	0

EDGxB,EDGxA =   Input Capture Specification,
                     see Figure 11.5

TMSK1   Main Timer Interrupt Mask Register 1

	B7							B0
$1022	OC1I	OC2I	OC3I	OC4I	OC5I	IC1I	IC2I	IC3I
RESET =	0	0	0	0	0	0	0	0

OC1I - OC5I =   Output Compare Interrupt enable,
                     see Figure 11.4
IC1I - IC3I =   Input Capture Interrupt enable,
                     see Figure 11.5
                     0 = Interrupt inhibit
                     1 = Interrupt enable

TFLG1   Main Timer Interrupt Flag Register 1

	B7							B0
$1023	OC1F	OC2F	OC3F	OC4F	OC5F	IC1F	IC2F	IC3F
RESET =	0	0	0	0	0	0	0	0

OC1F - OC5F =   Output Compare Flag,
                     see Figure 11.4
IC1F - IC3F =   Input Capture Flag,
                     see Figure 11.5

* Write with bit(s) set to clear corresponding flag(s).

Figure 11.1   Timer Subsystem Control and Status Registers. For OC1M and OC1D, see Table B.1 in Appendix B. *(Continued)*

TMSK2   Miscellaneous Timer Interrupt Mask Register 2

B7							B0
TOI	RTII	PAOVI	PAII			PR1	PR0

$1024

RESET = 0   0   0   0   0   0   0   0

TOI =   Timer Overflow enable,
RTII =   Real Time Interrupt (RTII) enable
PAOVI =   Pulse Accumulator Overflow Interrupt enable,
             see Figure 11.9
PAII =   Pulse Accumulator Interrupt enable,
             see Figure 11.9
             0 = Interrupt inhibit
             1 = Interrupt enable
PR1, PR0 =   Timer Prescale select (Time protected),
                  see Figure 11.2

TFLG2   Miscellaneous Timer Interrupt Flag Register 2

B7							B0
TOF	RTIF	PAOVF	PAIF				

$1025

RESET = 0   0   0   0   0   0   0   0

TOF =   Timer Overflow Flag,
             see Figure 11.2
RTIF =   Real Time (periodic) Interrupt Flag
PAOVF =   Pulse Accumulator Overflow Flag
PAIF =   Pulse Accumulator Input edge Flag

* Write with bit(s) set to clear corresponding flag(s).

Figure 11.1 Timer Subsystem Control and Status Registers. For OC1M and OC1D, see Table B.1 in Appendix B. *(Continued)*

PA5	OC3/OC1	Output compare 3 or 1
PA6	OC2/OC1	Output compare 2 or 1
PA7	PAI/OC1	Pulse accumulator input or output compare 1

The Free-Running Counter (TCNT). The heart of the timer subsystem is the 16-bit timer counter register at address $100E,0F called TCNT. For every set of clock pulses from the system clock (E), the register increments. A program can read it any time to get timing information. The timer subsystem uses TCNT as a timing reference for most of its operations. Note that a program cannot write to the register. Only the system clock can drive (write to) TCNT. It is also known as a 16-bit *free-running counter.*

Figure 11.2(a) illustrates how the system clock drives this register. When the MCU is reset, register TCNT is also reset to zero. After this, it increments at a fixed rate. The prescale factor determines how many clock pulses are required to increment register TCNT each time. Another subsystem register, the miscellaneous timer mask register (TMSK2), has the timer prescaler select bits (PR1, PR0).

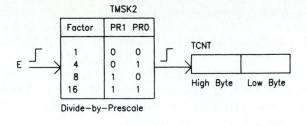

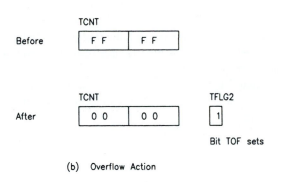

(a) Counter System

(b) Overflow Action

Figure 11.2 Free-Running Counter (TCNT)

The following segment of program code shows how to set the prescale factor to eight (IX = $1000).

```
BCLR      TMSK2,X $01      ;PR0 = 0
BSET      TMSK2,X $02      ;PR1 = 1
```

After the MCU executes these instructions, the free-running counter (TCNT) will increment by one every eight clock cycles. If the timer overflow interrupt enable bit (TOI) in register TMSK2 is set, an interrupt is also asserted. If it is not set, the MCU will have to poll flag register TFLG2 to determine if an overflow occurred. Since TCNT is a 2-byte (16-bit) register, it can increment up to 65,535 before it *overflows*. An overflow occurs every time TCNT counts past $FFFF to $0000. This causes the timer overflow flag (TOF) in register TFLG2 to set. This is shown in Figure 11.2(b).

Clearing Timer Flags. The flag bits in both registers TFLG1 and TFLG2 are cleared by writing a 1 to the bit to be cleared! For example, to clear timer overflow flag TOF, you can use the following instructions (IX = $1000):

```
LDAA      #$80            ;sequence to clear TOF
STAA      TFLG2,X         ;this is the most common sequence
```

Section 10.2.4 in the *Reference Manual* explains the procedure in more detail. It also explains why the bit set instruction (BSET) should not be used. We summarize the explanation here. If a BSET instruction is used, it will clear other flag bits as well. The reason is that the BSET instruction reads the operand (in this case the flag register) and then performs a logical OR. It writes this result back to the operand. If another flag bit had set, the result of the OR leaves the bit set. When the BSET instruction writes the result back to the flag register, it writes a byte with the bits set for other flag bits as well. Hence, executing a BSET with a flag register as the operand will clear any flag that set.

Consider the following situation. Let's say that bits TOF and PAOVF have set in register TFLG2 (TFLG2 has the value %1010 0000). Furthermore, you only want to clear flag bit TOF. Executing

```
BSET    TFLG2,X $80    ;this is not correct!
```

results in the logical OR of the mask byte %1000 0000 with the flag byte %1010 0000 resulting in the byte %1010 0000. The BSET instruction writes this result back to flag register TFLG2, resulting in bits 7 and 5 being cleared.

An alternative way to clear timer flag bits is to use the bit clear instruction (BCLR). In this case the BCLR mask bits are 1 for any bits that are not to be cleared. A 0 bit specifies the flag bit to be cleared. For example:

```
BCLR    TFLG2,X $7F    ;this also clears TOF
```

The reason this is acceptable is because the BCLR instruction ANDs the operand byte with the complement of the mask byte. In this case, operand byte %1010 0000 is ANDed with %01111111. This results in %1010 0000 ANDed with %1000 0000. Then the BCLR instruction writes the byte %1000 0000 to the flag register to clear bit 7.

Time-Protected Bits. Note that timer prescale bits PR1 and PR0 are *time-protected* bits (see Figure 11.1). These are bits that can only be modified once. Also, any modification must be done within the first 64 clock cycles after reset. The 68HC11 also uses time protection elsewhere. In Section 3.9.5 we explained that the IRQ interrupt can be made edge sensitive by setting bit IRQE in register OPTION. Some of the other bits in register OPTION are also time protected. Two of them (CR1, CR0) are bits that specify the COP timer rate. Another is the startup delay bit (DLY). It delays the MCU for approximately 4000 clock cycles after it exits from a STOP instruction. This is to allow the crystal oscillator to stabilize. In Section 6.8 we explained that the INIT register cannot be written to after the first 64 clock cycles. All INIT register bits are time protected.

Time protection is provided for those bits whose function is considered to be critical. If these bits were to be modified accidentally, catastrophic failure could result. In the case of real-time control applications typical for microcontrollers, this kind of protection can be necessary for safety. General-purpose computers used in an office do not usually have this protection. Their failures are not considered to be life threatening, although they may be thought to be financially threatening.

General Timer Software. The basic software to handle this subsystem is similar to that of the other subsystems. This is as follows:

Configure the control registers (and enable the subsystem).

Write to the data register if required.

Wait for a flag to set.

Clear the flag.

Read from or write to a data register as required.

Repeat if desired.

■ 11.2 OUTPUT COMPARE

The output compare module can output pulse waveforms such as square waves, variable-duty-cycle waves, and single pulses (see Figure 8.5[a] to [c]). It can also trigger an execution sequence to begin at a specified time without generating an external output.

Output compare functions can be used to generate time delays that do not depend on executing a program loop a specified number of times. For example, it can generate the 10-ms delay required for EEPROM programming and erasing (see Section 6.6). It may be used by an EPROM programmer to send the required pulses for its programming operations. It can trigger the MCU to sample signals at periodic intervals necessary for feedback control systems or digital filters (discussed in Sections 13.4 and 13.5).

An output compare function can generate repetitive signals such as a flashing light in an instrument or cockpit display panel. In Section 8.7 we list other applications, such as pulse-width modulation (PWM) for motor speed control. PWM can also be used to implement a simple digital-to-analog converter (D/A) using only a few extra components. The PWM output is sent to an analog low-pass filter, which in turn outputs a dc signal. We will return to the PWM applications later.

11.2.1 Sequence of Operations

Figures 11.3 and 11.4 illustrate the operation of the output compare functions. The illustrations are a simplification of the hardware actually used in the module. Refer to Figure 11.3 first. Recall that there are five output compare functions (and pins) labeled OC1 to OC5. Each has an associated 16-bit timer output compare register labeled TOC1 to TOC5, respectively. A software instruction writes a value to the register (TOCx). When timer counter register TCNT counts up (increments) and reaches the value stored in an output compare register (TOCx), the MCU automatically sets the associated output compare flag (OCxF) in timer flag register TFLG1. This event is called a *successful compare*. A successful compare will occur every time a *match* occurs between the output compare and counter register.

In Figure 11.3 the triangle represents the comparison part of the function. Its output is low as long as input A is not equal to input B. When input B increments to equal input A, the equality function outputs (latches) a logic high until it is cleared back to low. Even if the software writes a value to a register TOCx that is less than TCNT, the flag (OCxF) won't set until after TCNT overflows and counts up to the TOCx value.

To write a value to an output compare register, use a double-byte instruction. For example:

```
STD       TOC5,X
```

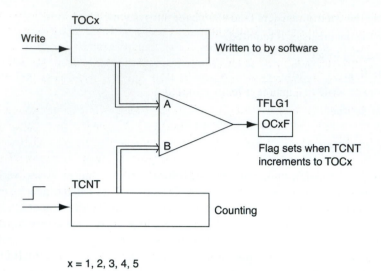

x = 1, 2, 3, 4, 5

Figure 11.3 Output Compare Action

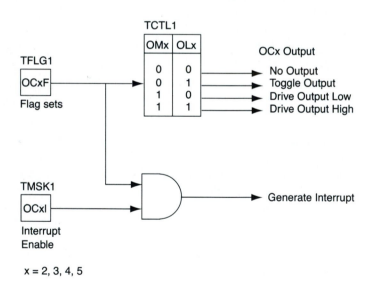

x = 2, 3, 4, 5

Figure 11.4 Results of a Successful Compare

Motorola recommends that your software use a double-byte instruction to write to an output compare register. If two separate instructions are used to write 2 bytes, there is the possibility that an erroneous output compare will occur. A double-byte instruction such as STD, STX, or STY writes to the operand high byte first and then to the low byte in the next clock cycle. An output compare function is inhibited for one bus clock cycle after a write to the high byte of its associated output compare register.[1]

For example, consider the following situation. Register TOC5 has the value $7F80. Suppose that the program will write the value $A000 to TOC5 at the very instant when counter register TCNT has the value $A080. Before the low byte is updated, registers TOC5 and TCNT will both have the value $A080. The one-cycle delay prevents an erroneous output compare from occurring because the low byte of TOC5 will be updated in the next cycle.

In the previous discussion we examined what causes the output compare flag (OCxF) to set. Figure 11.4 illustrates what actions can occur every time it sets. When an output compare flag (OCxF) sets, it causes an event to occur depending on the configuration settings in registers TCTL1 and TMSK1. The setting of bits OMx and OLx in control register TCTL1 determines how the output pin (OCx) is affected when its associated flag (OCxF) sets. The associated interrupt enable bit (OCxI) in mask register TMSK1 determines whether an interrupt will occur (as represented by the AND gate). Note that Figure 11.4 only describes the actions of outputs OC2, 3, 4, 5.

Exception for OC1. Each output compare (OCx) except for OC1 has one pin associated with it. Timer output function OC1 has two registers dedicated for its use. It can also control any combination of pins PA3 to PA7. The output compare 1 mask register (OC1M) determines which port A pins OC1 can control. A set bit connects an output to OC1 and a clear bit disconnects it. When a successful compare occurs between timer output compare 1 register (TOC1) and the counter (TCNT), the data in the output compare 1 data register (OC1D) is sent to the port A pins. Note that mask register OC1M determines which port A pins are affected.

For example, let's say that we want to drive PA3 and PA4 high and PA7 low for every successful TOC1 output compare. Then we can use the following instructions to set up output compare 1 as interrupt driven:

```
LDX     #REGBAS          ;point to registers
BSET    PACTL,X $80      ;make PA7 output
LDAA    #$98             ;set OC1M3,4,7
STAA    OC1M,X           ;to enable PA3,4,7
LDAA    #$18             ;set OC1D3,4 and clear OC1D7
STAA    OC1D,X           ;to drive PA3,4 high and PA7 low
BSET    TMSK1,X $80      ;enable interrupt
CLI
```

Note that two instructions are used to enable the interrupt. The program enables the local interrupt bit (bit OC1I in register TMSK1). It should also clear the interrupt mask (I) in the condition code register (CCR).

---

[1]If you have a copy of the *Data Book,* check out "Cycle-by-Cycle Operation" (Table 10.2) for instructions STD, STX, and STY.

OC1 is useful in situations when a single external device is driven by several outputs. It can also be used in combination with another output compare function to drive the same pin. For example, OC1 and OC2 can both drive pin PA6. Only one of them is needed to generate an interrupt, but both can generate successive edges. This situation is useful when narrow pulses are required. It is possible to generate a pulse as narrow as one clock cycle.

11.2.2 One-Shot Pulse Example

Consider the case of generating a high pulse that is 10 ms wide. Initially, the software sets an output line high using a parallel I/O operation. Then it configures an output compare to drive the output low. Finally, the software writes a value to the output compare register for the time-delay period.

To do the latter part, it reads the counter register (TCNT) and adds it to a time-delay value. Then the program writes this new value to the appropriate output compare register. The next thing is to determine the time-delay value. This is the number of TCNT cycles required. Let's take the case of a 2-MHz clock and a timer prescale of 1 (Listing 11.1). At 0.5 μs for every E cycle, 10 ms will then require 20,000 E cycles. With a prescale of 1, this means that the program will add 20,000 to the current counter value to trigger an event 10 ms later.

```
*Listing 11.1
*Drive one-shot high pulse for 10 ms
*with E = 2 MHz and prescale = 1

*REMINDER, see template listing in Appendix D, Cross Assembly
*----------------------------------

        ORG     $100

PWIDTH  EQU     20000

        LDD     TCNT,X          ;prevent premature
        STD     TOC2,X          ;OC2 compare
PULSE
*                               ;drive PA6/OC2 high
        BSET    PORTA,X $40
        LDAA    #$80            ;configure OC2 to clear
        STAA    TCTL1,X         ;and disconnect other OCx's
        LDAA    #$40            ;clear OC2F if set
        STAA    TFLG1,X
        LDD     TCNT,X          ;arm TOC2 for 10-ms trigger
        ADDD    #PWIDTH-17
        STD     TOC2,X
*wait for trigger
PULSE1
*                               ;by polling for OC2F high
        BRCLR   TFLG1,X $40 PULSE1
*                               ;now output OC2 low
        BCLR    PORTA,X $40     ;clear latch for PA6
```

```
        LDAA    #$40            ;then clear OC2F
        STAA    TFLG1,X         ;before
        BCLR    TCTL1,X $80     ;disconnecting OC2


        BRA     *               ;end for now
```

The program of Listing 11.1 is based on the registers initially being in their reset condition. It ensures that no accidental compares occur before it is ready by writing the latest TCNT value to TOC2. Admittedly, a premature successful compare is a low probability (1 in 65,536), but the possibility exists. By the time the most recently read TCNT value is written, counter TCNT would already have incremented past the value.

The program uses pin OC2 for the pulse output. Note that the program drives pin PA6 high as a general-purpose output before OC2 is enabled. Once a port A pin is configured for timer use, it cannot be used as a general-purpose output pin. The *Reference Manual* explains how to force a timer output without waiting for a successful compare. This can be done by writing to the timer compare force register (CFORC). To force a compare, write a 1 to the corresponding bit (FOCx). For example, to force a compare action for OC2, set bit FOC2 in register CFORC as follows:

```
    BSET    CFORC,X $40    ;note IX = $1000
```

Another feature to note in Listing 11.1 is the number added to the current count in register TCNT. Recall that a count of 20,000 provides a delay of 10 ms. The program adjusts this factor by subtracting 17. From the time the program drives pin PA6/OC2 high to the time it reads TCNT, the program has executed 17 clock cycles. Whether the time for program instructions to execute is important depends on the accuracy required. In this case omitting the adjustment would only increase the pulse width by 8.5 μs, an error of 0.085%.

After the successful compare automatically drives output pin OC2 low (and sets flag OC2F), the program clears PA6 (in an internal latch for port A) before disconnecting OC2. Data written to a port A output has no effect on the pin when the associated timer output function is connected. However, the internal latch for port A stores this value. When any port A output is disconnected from the timer subsystem, the output is driven according to the state of the port A internal latch. For this reason the program clears pin PA6 before clearing bit OM2 (OL2 = 0 already) to disconnect the PA6/OC2 pin from the output compare timer function.

C Code Example. C Listing 11.1 shows the one-shot pulse example using C. The program follows the same logic as the assembler version. One difference is that it uses a different correction factor for the adjustment used to account for the time required to read TCNT after driving PA6/OC2 high. In this case, we estimated the time to be a small multiple of the adjustment used for the assembly language version. Typical code created using C uses more equivalent assembly language instructions than those written directly in assembly. Also, different C compilers will generate different machine code (or different assembly language equivalents) for the same C source code.

```
* C Listing 11.1 */
/* Drive one-shot high pulse for 10 ms
with E = 2 MHz and prescale = 1 */
```

```
#include <hc11.h>          /* Include register definitions */

/* Declarations */
unsigned int PWidth;       /* Pulse width is E cycles */

void main(void)
{
        /* Initialize to 20000 E cycles for 10 ms */
        PWidth = 20000;
        /* Prevent premature OC2 compare */
        TOC2 = TCNT;
        /* Drive PA6/OC2 high */
        PORTA = PORTA | 0x40;
        /* Configure OC2 to clear */
        TCTL1 = 0x80;
        /* and disconnect other OCxs */
        TFLG1 = 0x40;
        /* Arm TOC2 for 10-ms trigger */
        /* 50 is estimate of E cycles from
        driving OC2 high to reading TCNT */
        TOC2 = TCNT + PWidth - 50;
        /* Wait for trigger of OC2F high */
        /* (Stay in while loop while OC2F low) */
        While ((TFLG1 & 0x40)==0);
        /* Then clear latch for PA6 */
        PORTA = PORTA & ~0x40;
        /* Then clear OC2F */
        TFLG1 = 0x40;
        /* Then disconnect OC2 */
        TCTL1 = TCTL1 & ~0x80;
        /* End now by looping forever */
        while (1);
}
```

Alternative Using Forced Compare.   Listing 11.2 shows an alternative way to generate one single high pulse. Note that forcing an output compare using register CFORC affects only the pin. It does not affect the associated flag and it does not assert an interrupt if it was enabled.

```
*Listing 11.2
*Drive one-shot high pulse for 10 ms
*with E-2 MHz and prescale = 1
*-------------------------------------

        ORG     $100

PWIDTH  EQU     20000

        LDD     TCNT,X          ;prevent premature
        STD     TOC2,X          ;OC2 compare
```

```
PULSE
*                                       ;PA6/OC2 high
        BCLR      PORTA,X $40           ;ensure that PA6 initially low
        LDAA      #$40                  ;configure OC2 to toggle
        STAA      TCTL1,X
        LDAA      #$40                  ;clear OC2F if set
        STAA      TFLG1,X
        STAA      CFORC,X               ;force first compare
*                                       ;to drive OC2 high
        LDD       TCNT,X                ;arm OC2 for 10 ms trigger
        ADDD      #PWIDTH-5             ;to drive it low
        STD       TOC2,X
*wait for trigger
PULSE1
        BRCLR     TFLG1,X $40 PULSE1
        BCLR      TFLG1,X $BF           ;clear OC2F (another way)
        BCLR      TCTL1,X $C0           ;disconnect OC2
        BRA       *                     ;end demo for now
```

Another way to clear the output compare flag (OC2F) is to use the following sequence:

```
        LDAA      #$40
        STAA      TFLG1,X
```

11.2.3   Square-Wave and Pulse-Width Modulation (PWM) Outputs

A square wave can be generated quite easily by configuring an output compare function to toggle every half-period. Listing 11.3 shows service routines to do this for pin OC3. Once the main program has called subroutine INITOC3 and has set up the variable HPERIOD, it can do other tasks. The interrupt service routine handles the OC3 output.

```
*Listing 11.3
*Shows routines for generating square waves
*using output OC3.
*A main application program varies frequency
*by modifying 16-bit data in addresses HPERIOD.
*Frequency = 1/(2 * HPERIOD)
*User responsibility to set up vector addresses
*and other initialization as required by application.
*------------------------------------

        ORG       $180
*Subroutine INITOC3
*Initializes timer output OC3 for
*square-wave output, interrupt driven.
*Calling registers
*       IX = register block address
*Return registers
*       None, except CCR affected
```

```
INITOC3
        PSHA                            ;preserve registers
        PSHB
        LDAA    #$10                    ;OM3:OL3=0:1
        STAA    TCTL1,X                 ;to toggle OC3
        LDAA    #$20                    ;clear OC3F if set
        STAA    TFLG1,X
        STAA    TMSK1,X                 ;set OC3l to enable
        CLI                             ;interrupt
        PULB                            ;restore registers
        PULA
        RTS                             ;return
*Service routine RTOC3
*Drives OC3 output for square wave by scheduling
*time delay for next edge to be toggled.
*The minimum permissible half-period is the number of clock
*cycles in this routine plus those required to enter
*and exit the routine.
*Static variable (2 bytes)
*Address HPERIOD = OC3 half-period time duration

*This routine executed after TOC3 == TCNT occurs
RTOC3
        LDX     #REGBAS                 ;point to registers
        LDD     HPERIOD                 ;update TOC3
        ADDD    TOC3,X                  ;by adding half period
        STD     TOC3,X                  ;to latest TOC3 value
*                                       ;to schedule the next interrupt
        BCLR    TFLG1,X $DF
*                                       ;clear flag OC3F
        RTI                             ;return from service
```

PRACTICE

Write a program that can run on the EVBU board. It outputs a square wave at a frequency of 10 kHz. The EVBU board uses a 2-MHz clock.

Solution

First calculate the required half-period setting.

$$\text{HPERIOD} = \frac{1}{2 \times \text{frequency}} = 50 \text{ ms} = 100T$$

where T is the E clock period of 500 ns. The program code follows.

```
*Solution
*square-wave output at pin PA5/OC3
*define registers and vectors by inserting HC11REG.ASM and
HC11VEC.ASM
```

```
HPERIOD  EQU       $0000          ;half-period in cycles

*can put code in RAM starting at $C000
         ORG       $0100

*EVBU monitor already sets up stack

         LDX       #REGBAS        ;point to registers
         LDAA      #$7E           ;set up OC3 interrupt jump table
         STAA      JTOC3
         LDD       #RTOC3
         STD       JTOC3+1
         JSR       INITOC3        ;initialize OC3 interrupt

         LDD       #100           ;specify square-wave frequency
                                  ; of 10 kHz
         STD       HPERIOD
         CLI                      ;enable interrupts
HERE
         BRA       HERE           ;OC3 interrupt drives program

;routine codes follow, see Listing 11.3.
```

Figure 8.5 shows two timer output lines with continuous output. They are examples of a pulse-width modulation (PWM) output. A PWM signal generates a pulse at fixed periodic intervals (fixed frequency). The *duty cycle* is related to the width of the pulse. It is the percentage of time the signal is high compared to the signal's period. The PWM signal shown in Figure 8.5(a) has a duty cycle of approximately 70%. A square wave has a duty cycle of 50%. Figure 8.5(b) shows a varying duty cycle.

We consider a PWM example. Listing 11.4 shows some routines that can be used by an application program. The main program would call the initialization subroutine early. Then the main program modifies the duty-cycle variables OC2HI and OC2LO as required. The interrupt routine would handle the output changes without further intervention by the rest of the software. A motor control program can increase motor speed by increasing variable OC2HI while decreasing variable OC2LO. To slow down the motor, it would do the opposite. Note that the sum of OC2HI and OC2LO does not change in order to keep the frequency constant.

```
*Listing 11.4
*Shows routines for handling PWM using output OC2.
*A main application program varies the duty cycle
*by modifying 16-bit data in addresses OC2HI and OC2LO.
*Duty cycle = 100% * OC2HI/(OC2HI + OC2LO)
*User responsibility to set up vector addresses
*and other initialization as required by application.
*-------------------------------------

         ORG       $180
```

```
*Subroutine INITOC2
*Initializes timer output OC2 for
*PWM output, interrupt driven.
*Calling registers
*       IX = register block address
*Return registers
*       None, except CCR affected
INITOC2
        PSHA                            ;preserve registers
        PSHB
        LDD     TCNT,X                  ;delay PWM generation
        STD     TOC2,X
        LDAA    #$C0                    ;OM2:OL2=1:1 to set
        STAA    TCTL1,X                 ;OC2 high first time
        LDAA    #$40                    ;clear OC2F if set
        STAA    TFLG1,X
        STAA    TMSK1,X                 ;set OC2I to enable
        CLI                             ;interrupt
        PULB                            ;restore registers
        PULA
        RTS                             ;return


*Service routine RTOC2
*Drives OC2 output for PWM by scheduling
*time delay for next edge. Also reconfigures
*next edge opposite to that of current edge.
*Note that routine will not work properly with
*duty cycles close to 0% or 100%.
*Static variables (2 bytes each)
*Address OC2HI = OC2 time duration for high pulse
*Address OC2LO = OC2 time duration for low pulse

*This routine executed after TOC2 == TCNT occurs

OC2HI   EQU     $0006
OC2LO   EQU     $0008


RTOC2
        LDX     #REGBAS                 ;point to registers
*                                       ;if low part of cycle
*                                       ;then load OC2LO
        BRCLR   TCTL1,X $40 GETOC2LO
        LDD     OC2HI                   ;else load OC2HI
        BRA     NEWTOC2
GETOC2LO
        LDD     OC2LO
*update TOC2
NEWTOC2
        ADDD    TOC2,X
        LDAA    TCTL1,X                 ;invert OL2 to toggle
```

```
        EORA      #%01000000          ;next OC2 edge
        STAA      TCTL1,X             ;by updating control reg.
        BCLR      TFLG1,X $BF         ;clear flag OC2F
        RTI                           ;return from service
```

The main feature of the interrupt service routine is to keep track of which part of the cycle is current. It schedules the next edge accordingly. It does this by checking the control register bits to determine the current type of edge (rising or falling) that is configured. It adds the value of OC2HI to TOC2 if the OC2 interrupt drove pin OC2 high. Otherwise, it adds OC2LO. Then the service routine toggles bit OL2 in the control register to reverse the process for the next interrupt. Of course, the service routine clears the flag bit before returning. If it did not clear the flag bit, an interrupt would assert itself every time the RTI instruction completed execution. Recall that executing RTI restores the previous condition code register value (meaning that bit I returned to zero).

The routines in Listing 11.4 can also generate a square wave by making OC2LO and OC2HI equal. It can increase the frequency by decreasing their values, and vice versa to decrease the frequency. But you may prefer to use a simpler technique, as shown in Listing 11.3.

PRACTICE

Use the EVBU board to generate a PWM signal with a duty cycle of 60% at a frequency of 5 kHz.

Solution

```
*Solution
*PWM output at pin PA6/OC2
*For brevity, will not show pseudo-ops to set up addresses

        LDX       #REGBAS             ;point to registers
        LDAA      #$7E                ;set up OC2 interrupt jump table
        STAA      JTOC2
        LDD       #RTOC2
        STD       JTOC2+1
        JSR       INITOC2             ;initialize OC2 interrupt

        LDD       #120                ;specify PWM frequency and duty
                                      cycle
        STD       OC2HI
        LDD       #80
        STD       OC2LO
        CLI                           ;enable interrupts
HERE
        BRA       HERE                ;OC2 interrupt drives program

*routine codes follow, see Listing 11.4.
```

C Code Example.   C Listing 11.4 shows a solution to the practice exercise using C. Some things to note are that INTR_OFF and INTR_ON are statements defined in the header file hc 11.h. They disable and enable the interrupt (bit I in CCR), respectively. The other is the use of the asm statement to embed assembly language statements in the C source code. In this case, we are setting up a jump vector for use in Buffalo. The final asm statement is a fudge of sorts. The compiler translates functions into subroutines. We are forcing an inclusion of the RTI instruction. The compiler will add an RTS instruction after it. The RTS instruction becomes redundant because it will never be executed when the program is run. In practice it may be appropriate to code interrupt service routines directly in assembly to make them more efficient.

```
/* C Listing 11.4 */

#include <hc11.h>                    /* Include register definitions */

/* Declarations */
unsigned int OC2hi, OC2lo;           /* Duty cycle data */

/* Prototypes */
void initoc2(void);
void rtoc2(void);

void main(void)
{
        INTR_OFF();
        /* set up jump vector for buffalo */
        asm(" ldaa #$7e");
        asm(" staa $dc");
        asm(" ldd #_rtoc2");
        asm(" std $dd");
        initoc2();
        while (1);                   /* Loop forever */
}

void initoc2(void)
/* Initialize OC2 */
{

        TOC2  =  TCNT;
        TCTL1  =  0xC0;
        TFLG1  =  0x40;
        TMSK1  =  0x40;
        INTR_ON();
}

void rtoc2(void)
/* Interrupt Service Routine for OC2 PWM */
{
        if ((TOC2 & 0x40) == 0)
                TOC2 = TOC2 + OC2lo;
```

```
            else
                    TOC2 = TOC2 + OC2hi;
            TCTL1 = TCTL1 ^ 0x40; /* invert OL2 to toggle next OC2 edge */
            TFLG1 = 0x40;      /* Clear flag OC2F */
            asm(" rti");       /* This is an interrupt service routine */
}
```

PWM Functions.   Although you may appreciate the flexibility of the output compare functions, you might be dismayed by the software overhead required to generate square waves and PWM outputs. Some of you may be aware that some microcontrollers and programmable timer chips provide automatic mechanisms for generating repetitive pulses. Remember that this book concentrates on one model of 68HC11 chip, the MC68HC11A8. Here we describe briefly the PWM function available in other models such as the MC68HC711K4. The 7 in the part number means that the chip has an internal EPROM instead of a ROM. It has many enhancements of the basic A8 part, but we will not detail them here.

The PWM module in the MC68HC711K4 uses four pins belonging to an extra port (port H). They provide four PWM channels. This MCU uses four control registers to select features such as clock source, channel polarity, channel enable, and clock scaling. Each channel also has a separate counter, period, and duty-cycle register. Once these registers are set up, the PWM module drives the outputs without further software intervention. If an application program needs to change the period or duty cycle, it can write to their respective registers. Other models in the G, K, and N series of the 68HC11 have PWM modules. The N series supports six PWM outputs.

11.2.4 Stepper Motor Outputs

One application for using output compare functions is the control of *stepper motors*. These are electric motors that rotate from one position to the next in discrete intervals (steps). You can say that they actually step from one position to the next. First, we will describe the stepper motor operations. Then we will examine how to control it using a microcontroller. A *step* refers to the fraction of a full 360° of the motor shaft (or rotor) rotation. Typical step angles are 1.8°, 3.6°, and 7.5°. With reduction gears, smaller step angles can be achieved to provide finer resolution.

The primary stepper motor application is position control. Some disk drives use stepper motors to position the read/write head. Some robots use them for positioning joints and tools. The dot matrix printer uses stepper motors to rotate the paper feed carriage (vertical paper position) and to position the print head horizontally relative to the paper. In Section 14.3 we describe the dot matrix printer in more detail.

A common type is the four-phase stepper motor (see Figure 11.5[a]). It has two sets of windings. The direction (polarity) of current through the windings determines which position the motor steps to next. These are the *rush* currents. A high level of current is required to cause shaft rotation. Each change of polarity at the terminal of a winding is called a *step of phase shift*. A sequence of logic pulses steps the motor from one position to the next (see Figure 11.5[b] and [c]). This causes the shaft (rotor) to rotate in precise angular increments per step. The phase sequence may begin anywhere but it must continue in order.

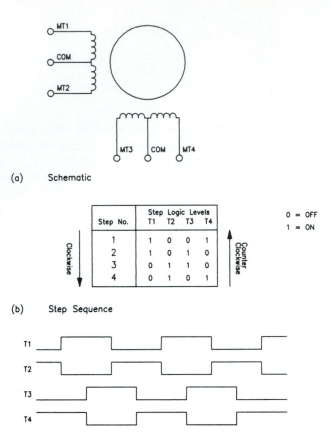

(a)   Schematic

Step No.	Step Logic Levels			
	T1	T2	T3	T4
1	1	0	0	1
2	1	0	1	0
3	0	1	1	0
4	0	1	0	1

0 = OFF
1 = ON

Clockwise

Counter Clockwise

(b)   Step Sequence

T1

T2

T3

T4

(c)   Sequence Waveforms   (Forward Direction)

Figure 11.5   Four-Phase Stepper Motor

Consider the following example (also refer to Figure 11.5[b]). The motor has a step size of 1.8°. The last signal to drive the motor was step 2. It is being presently held in position. We discuss how to hold the motor in position shortly. For now, consider how we can drive it 10.8° (six steps) clockwise. To do this, drive the motor through steps 3, 4, 1, 2, 3 to step 4. Let's say that we have to drive it 3.6° counterclockwise next. To do this, drive the motor through step 3 to step 2.

Typically, the stepper motor rotates a set number of steps as illustrated by the preceding example. Then it maintains that position until another position change is required. During the period when no rotation occurs, the motor windings must be supplied with a lower level of current called the *hold* current. Typically, the control system applies hold current by switching a current-limiting resistor between the power supply and the common (COM) terminals.

Since high currents are required to drive the stepper motor, a MOSFET latch with open-drain outputs is used to supply pulses to the windings. Figure 11.6 shows an interface circuit using the TPIC2406 Intelligent Power Quad MOSFET Latch by Texas Instruments. It can drive high loads (3 A of pulsed current) with built-in protection against inductive tran-

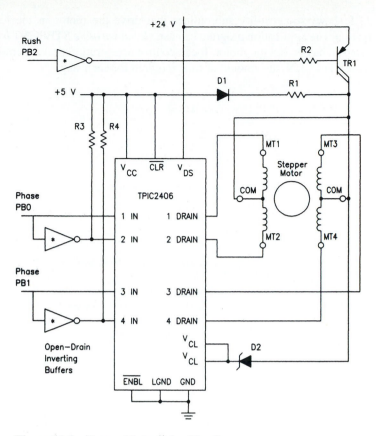

Figure 11.6 Stepper Motor Drive Circuit

sients. The port B pins of the 68HC11 supply the stepper motor signals. In this case the associated output compare function is configured to be disconnected. However, the software uses it to control port B.

Since the signals required for each terminal pair are complements, one output can control them by using an extra inverter (NOT gate) buffer. For example, in step 1, output PB0 is high. This supplies a high to 1 IN and a low to 2 IN of the MOSFET latch. This causes current to flow into the common terminal (COM), through the windings to MT2 and into output 2 DRAIN. Similarly with PB1 low, current flows into COM and out of MT3 into 3 DRAIN.

Output PB2 is used for rush/hold control. When PB2 is high, the inverter turns on transistor TR1 to connect 24 V to the stepper motor. This supplies the motor with the rush current. When PB2 is low, TR1 switches off and shunt current is supplied through diode D1 and resistor R1.

A stepper motor is complicated by the fact that a motor has inherent mechanical inertia. Hence, the motor should be accelerated to a constant speed by *ramping* up (increasing) the stepping rate. Similarly, it should be decelerated to a stop by ramping down (decreasing) the stepping rate.

Listing 11.5 shows some utility subroutines that drive the motor in the forward (clockwise) direction. The application program can use *global variable* STEPCNT to monitor how many steps it has driven the motor. It can control acceleration or deceleration by modifying variable STEPLEN, the amount of time required for each step.

```
*Listing 11.5
*Stepper Motor Control utility routines
*-----------------------------------

        ORG       $180

*Subroutine INITOC4
*Initializes OC4 to drive four-phase stepper motor
*Calling Register
*       IX = register block address
*No return registers except that CCR affected
INITOC4
        PSHA                     ;preserve registers
        PSHB
        BCLR      TCTL1,X $0C    ;OM4:OL4=0:0
        LDD       TCNT,X         ;avoid premature compare
        STD       TOC4,X
        LDAA      #$10           ;clear OC4F is set
        STAA      TFLG1,X
        STAA      TMSK1,X        ;OC4I=1 to enable interrupt
        CLI
        PULB                     ;restore registers
        PULA
        RTS                      ;return

*Service routine RTOC4
*Drives stepper motor to next step.
*It uses a look-up table to find the output pattern
*of the current step. Then it drives the outputs
*and moves the pointer to the next step. It also
*increments a step counter (STEPCNT) for use by an
*application program to keep track of number of
*steps driven.
*Driver Outputs:
*       MT1 <- PB0
*       MT2 <- INV PB0
*       MT3 <- PB1
*       MT4 <- INV PB1
*       RUSH <- PB2
*Global variables
*       STEPLEN = time duration of step
*       STEPPTR = pointer to step in sequence
*       STEPCNT = step counter

STEPLEN EQU       $0A
STEPPTR EQU       $0C
STEPCNT EQU       $0D
```

```
RTOC4
        LDX     #REGBAS         ;point to registers
        LDD     STEPLEN         ;schedule next step
        ADDD    TOC4,X
        STD     TOC4,X
        BCLR    TFLG1,X $EF     ;clears flag OC4F
        LDY     #STEPTAB        ;point to sequence for this step
        LDAB    STEPPTR
        ABY
        LDAA    PORTB,X         ;get previous drive outputs and
        ANDA    #$FC            ;modify by clearing last two bits
        ORAA    0,Y             ;then change them as per table
        STAA    PORTB,X         ;and update drive outputs
        INC     STEPPTR         ;stepptr ++
        LDAB    STEPPTR         ;if stepptr == 4
        CMPB    #4              ;(note stepper mtr has four phases)
        BNE     SKIP
        CLR     STEPPTR          ;then stepptr = 0
SKIP
        INC     STEPCNT         ;update stepper mtr count
        RTI                     ;return


*-----------------------------------


*Stepper motor sequence look-up table
*PB2 is high to supply rush current
STEPTAB
STEP1   FCB     %101
STEP2   FCB     %111
STEP3   FCB     %110
STEP4   FCB     %100
```

Basically, during each timeout period of function OC4, the service routine looks up the step outputs in the sequence and drives port B accordingly. Then it increments a pointer (STEPPTR) to point to the next step in the sequence.

Listing 11.6 shows an EVBU test program using the utility routines. For simplification, it ignores the ramping requirements to start and stop. Note that an application program can implement ramping if necessary by modifying the contents of STEPLEN.

```
*Listing 11.6
*Demonstration of utility routines to drive
*stepper motor for 16 steps using EVBU board.
*See Listing 11.5 for source code of utility routines.

*REMINDER, see template listing in Appendix D for
*necessary overhead, such as segment definition,
*stack setup, and interrupt vectors
*-----------------------------------
```

```
          ORG       $0A
STEPLEN   FCB       $01,00   ;period of time for each step
STEPPTR   RMB       1        ;points to step sequence in STEPTAB
STEPCNT   RMB       1        ;for use by application program to
*                            ;keep track of number of steps
*Remove above definitions from Listing 11.5

*-------------------------------------


          ORG       $100


*Stepper motor control part


          LDD       STEPLEN       ;init stepper motor step time
          CLR       STEPPTR       ;define first step
          LDX       #REGBAS       ;point to registers
          BSET      PORTB,X $04
*                                 ;supply rush current
          JSR       INITOC4       ;initialize OC4 for stepper
*                                 ;motor control
          CLR       STEPCNT       ;init step counter


*OC4 interrupt drives program here
LOOP
          LDAA      STEPCNT       ;for 16 steps
          CMPA      #16
          BNE       LOOP
          BCLR      TMSK1,X $10   ;then disable OC4 interrupt
          BCLR      PORTB,X $04   ;and supply hold current
          BRA       *             ;end of demo for now
```

■ 11.3 INPUT CAPTURE

An input capture function records the time (TCNT value) when an active transition occurred. The active transition can be rising, falling, or either. This makes it useful for applications that require measuring the time between edges. By using software, the MCU application program can calculate the period, frequency, or any other related quantity, such as speed. Typically, rotating mechanical equipment have sensors that generate a pulse for every fraction of rotation. With this input, the MCU can use the input capture function to measure speed, detect motion, and record the distance traveled.

By referring to the timer input line example in Figure 8.5(d), you should be able to see how input capture functions are able to measure frequency by recording (capturing) the times when falling (or rising) edges occur. The MCU software can then find the waveform period by subtracting two successive edge times to find their difference. It can then calculate frequency by dividing 1 by the period. Recall that frequency is the reciprocal of the period. Similarly, the input capture can measure pulse width (Figure 8.5[e]) by capturing the first edge and then the second edge. The pulse width is the difference between the TCNT counter values when the edges occurred.

Automobiles have many uses for timer input captures. The traveling speed of the automobile is measured by sensors attached to the wheels. This can be used by the MCU for cruise control (maintain constant speed). It can also warn drivers who tend to have a heavy foot to slow down. The ignition timing signals in the engine are referenced to an engine cylinder position called *top dead center* (TDC). The input capture detects this position by sensing crankshaft rotation. This is used to trigger an output compare function to generate the ignition signals. An antiskid braking system is another automobile example. The system senses wheel rotation to determine the braking pressure to apply.

Other applications are turbine flowmeters. They use spinning turbines to measure fluid flow. A similar device is a propeller airspeed sensor. The MCU measures flow rate by measuring the frequency of the sensor pulses. It can also keep track of total flow by counting the pulses since each pulse represents a unit of fluid.

The input capture functions can also serve as general-purpose edge-sensitive inputs. This is useful because they can be programmed for any type of edge. For example, two input captures can be used to sense the data available (DAV) input of an IEEE-488 interface (see Figure 9.21). One is configured to react to a rising edge, the other for a falling edge. In this case the captured time information is not required.

11.3.1 Sequence of Operations

The 68HC11A8 and other A-type versions have three input capture pins—IC1, IC2, and IC3—at port A pins 2, 1, and 0, respectively. Other versions have an optional fourth input capture (IC4). For example, the 68HC11E9 uses pin PA3 as either output OC5 or input IC4. This is a useful addition for applications such as antiskid braking systems since most cars use four wheels! Note that an MCU with only three input captures can control four wheels by using extra hardware.

We will limit ourselves to the functions provided in the 68HC11E9, Figure 11.7 illustrates the operation of the input capture sequence for an input *x*. As specified by timer control register 2 (TCTL2), an input capture can react to a rising, falling, or either type of edge. It can also be disabled altogether. When an edge is detected by the input capture pin, the value of the timer counter register (TCNT) is latched into the corresponding input capture register (TICx). The function also sets the input capture flag (ICxF) in timer flag register 1 (TFLG1). If the corresponding input capture interrupt enable bit (ICxI) in timer mask register 1 (TMSK1) is set, the detected edge also generates an interrupt. Figure 11.7 shows this as a logical AND operation.

Typically, the application program reads the value in the timer input capture register (TICx). Then it clears flag ICxF so as to rearm itself to capture the next edge. To detect whether a flag sets, the program can poll register TFLG1 or respond to an interrupt.

11.3.2 Pulse Width Example

We will consider the example of measuring the width (time duration) of a high pulse using pin IC1. First we will show the pseudocode to illustrate the basic algorithm (Listing 11.7).

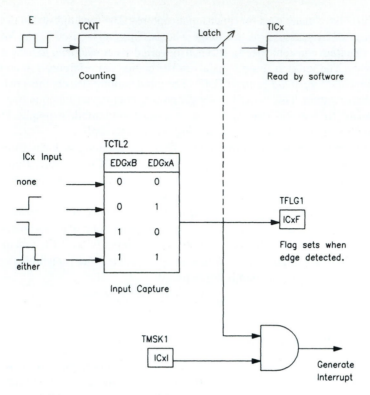

Figure 11.7 Input Capture Sequence

```
*Listing 11.7
*Pseudocode to measure pulse width of a high pulse

        Configure: capture on rising edge
        Clear capture flag
wait_for_rise
        If flag == 1,      then rise_time = capture time
                           else repeat wait_for_rise


        Configure: capture on falling edge
        Clear capture flag
wait_for_fall
        If flag == 1,      then fall_time = capture time
                           else repeat wait_for_fall
        pulse_width = fall_time - rise_time
```

There are limitations. If the pulse is too short, the second edge occurs before the software can capture it. One way to remedy this is to use two input capture functions. One is configured for a rising edge and the other is configured for a falling edge. If the pulse is longer than 65,536 TCNT cycles, overflow will give erroneous results. The program will have to count how many times the timer overflow bit (bit TOF in register TFLG2) has set.

We can write the pseudocode of Listing 11.7 in assembly language (Listing 11.8).

```
*Listing 11.8
*Measure time between a rising and a falling edge on IC1.


*-------------------------------------

RISETIME        EQU     $10
PULSEWIDTH      EQU     $12


                ORG     $100


                LDAA    #$10            ;config. to capture rising edge
                STAA    TCTL2,X
                LDAA    #$04            ;clear flag IC1F if set
                STAA    TFLG1,X
*wait for rising edge
POLLRISE
                BRCLR   TFLG1,X$04 POLLRISE
                LDD     TIC1,X          ;store the rise time
                STD     RISETIME
                LDAA    #$20            ;config. to capture falling edge
                STAA    TCTL2,X
                LDAA    #$04            ;clear flag IC1F
                STAA    TFLG1,X
*wait for falling edge
POLLFALL
                BRCLR   TFLG1,X$04 POLLFALL
                LDD     TIC1,X          ;read the fall time
                SUBD    RISETIME        ;width = fall - rise
                STD     PULSEWIDTH      ;store width
                BRA     *               ;stop here for now
```

Note that this program will work even if a timer overflow occurs (once only) during the pulse. For example, consider a rising edge occurring when TCNT = $000A. This is stored in variable RISETIME. If the pulse width is 65,528 E cycles (prescale of 1), register TIC1 will capture the value $0002 when the falling edge occurs. This is also an overflow situation. The 16-bit subtraction yields $0002 − $000A = $FFF8, the unsigned hex number for 65,528. The program stores this number in address PULSEWIDTH.

C Code Example.   C Listing 11.8 shows the equivalent in C.

```
/* C Listing 11.8 */
/* Measure time between rising and falling edge of IC1 */
#include <hc11.h>  /* Include register definitions */

/* Declarations */
unsigned int RiseTime, PulseWidth;
```

```
void main(void)
{
        TCTL2 = 0x10;     /* conf. to rising edge */
        TFLG1 = 0x04;     /* clear flag IC1F if set */
        /* wait for rising edge */
        while ((TFLG1 & 0x04) == 0);
        RiseTime = TIC1;
        TCTL2 = 0x02;     /* conf. to falling edge */
        TFLG1 = 0x04;     /* clear flag IC1F if set */
        /* wait for falling edge */
        while ((TFLG1 & 0x04) == 0);
        PulseWidth = TIC1 - RiseTime;     /* get pulsewidth */
        while (1);            /* Loop forever */
}
```

11.3.3 Period (Frequency, Speed) Measurement

The period of a repetitive signal includes both the high and low parts of the cycle. To measure period, a program needs to capture the time of two successive rising (or falling) edges. An application program can use the routines shown in Listing 11.9.

```
*Listing 11.9
*Service routines to measure period between two rising
*edges at the IC1 pin. Results valid in range of
*~27 E cycles to 65,535 TCNT cycles (overflows ignored)
*-----------------------------------

        ORG     0

IC1DUN  RMB     1         ;flag:0-not done, 1-pulse measured
IC1MOD  RMB     1         ;mode flag:FF-off,0-1st, 1-last edge
PER     RMB     2         ;period, 16 bits
*-----------------------------------

        ORG     $180

*Subroutine INITIC1
*Configures input capture IC1 to measure high pulse using
*interrupts.
*Calling register
*       IX = register block address
*No return registers except that CCR affected
*Return variables
*       IC1DUN = 0
*       IC1MOD = $FF, means measurement has not started

INITIC1
        PSHA                      ;preserve registers
        LDAA    #$10              ;EDG1B:EDG1A = 0:1 for rising edge
```

```
        STAA    TCTL2,X
        LDAA    #$FF            ;mode flag off
        STAA    IC1MOD
        CLR     IC1DUN          ;signal period not done

        BCLR    TFLG1,X $FB     ;clear IC1F if set
        BSET    TMSK1,X $04     ;enable IC1 interrupt
        CLI                     ;enable interrupts
        PULA                    ;restore registers
        RTS                     ;and return

*Service routine RTIC1
*Handles input capture IC1 interrupts.
*Return variables on first edge
*       IC1MOD = 0
*       PER = first capture time
*On second rising edge, calculates period and disables
*further rising edge interrupts.
*Return variables on second edge
*       IC1MOD = 1
*       PER = period
*       IC1DUN = 1
RTIC1
        LDX     #REGBAS         ;point to registers
        INC     IC1MOD          ;$FF -> 0 at first edge
*                               ;0 -> 1 at second edge
        BNE     EDGE2           ;if not 0 then this is second edge
*process first edge
EDGE1
        LDD     TIC1,X          ;capture first edge time
        STD     PER             ;and save it
        LDAA    #$04            ;clear flag IC1F
        STAA    TFLG1,X
        RTI                     ;and return
*process second edge
EDGE2
        LDD     TIC1,X          ;capture second edge time
        SUBD    PER             ;second time - first time
        STD     PER             ;is low 2 bytes of result
        BCLR    TCTL2,X $30
        LDAA    #1              ;and signal period measured
        STAA    IC1DUN
*clear flag IC1F
EXITRIC1
        BCLR    TFLG1,X $FB
        RTI                     ;and return from service routine
```

The application program first executes subroutine INITIC1 to start the process. Later it checks RAM variable IC1DUN periodically to determine if the measurement was completed. If it has (the IC1DUN value is 1), it can read the result in RAM variable PER. To make a new measurement, it should execute subroutine INITIC1 again. If a frequency value is required, the software will have to scale the period result into seconds or fractions of seconds and then take the reciprocal of the result.

PRACTICE

What is the pulse width if the software reads a value of $D15C in RAM variable PER? The 68HC11 uses an 8-MHz crystal. Assume that the timer prescale is eight.

Solution

The E clock is 2 MHz. Then the period is 500 ns and TCNT increments every $(8 \times 500 \text{ ns})$ = 4 μs.

$$\$D15C \times 4 \text{ μs} = 53{,}596 \times 4 \text{ μs} = 214.4 \text{ ms}$$

PRACTICE

Use the EVBU board to measure the width of a high pulse.

Solution

```
*Solution

*For brevity, will not show pseudo-ops to set up addresses
*pulse input at pin PA2/IC1

            LDX     #REGBAS             ;point to registers
            LDAA    #$7E                ;set up IC1 interrupt jump table
            STAA    JTIC1
            LDD     #RTIC1
            STD     JTIC1+1
            JSR     INITIC1             ;initialize IC1 interrupt
            CLI                         ;enable interrupts

REPEAT
            LDAA    IC1DUN              ;repeat while IC1DUN ! = 1
            CMPA    #1
            BNE     REPEAT
            LDD     PER                 ;get pulse width
HERE
            BRA     HERE                ;IC1 interrupt drives program

*routine codes follow, see Listing 11.9.
```

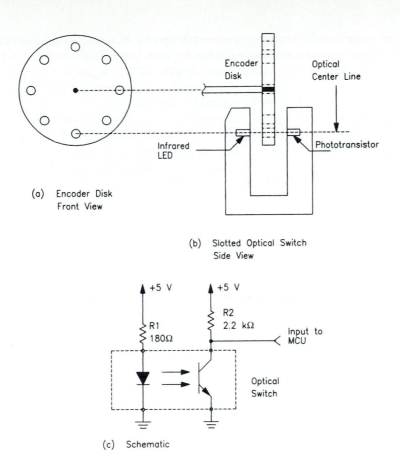

(a) Encoder Disk
Front View

(b) Slotted Optical Switch
Side View

(c) Schematic

Figure 11.8 Optoelectronic Position/Motion Detection

Speed Control.   In many cases the MCU uses input capture to monitor and/or control speed. A common type of speed sensor is an encoder disk mounted on a shaft. The disk is located between a slotted optical switch as shown in Figure 11.8. When the encoder disk rotates, the switch outputs pulses. The following explains its operation. An LED emits infrared light. Note that the encoder disk has holes that allow light to pass through. The phototransistor switches on only when the disk rotates to a position such that a disk hole is aligned between the infrared LED and phototransistor. The switched-on phototransistor pulls resistor R2 low. When the disk blocks light, the phototransistor is switched off and resistor R2 is pulled high.

In summary, the signal is low whenever an encoder hole is aligned between the LED and phototransistor. Whenever the encoder disk rotates, the input capture pin will see pulses. For the disk shown in Figure 11.8, eight pulses correspond to one disk revolution. The pulse frequency is related directly to the speed of revolution.

Let's say that a control system wants to maintain an electric motor at a constant speed. This desired speed has a corresponding period. It is the period that would be generated by the speed sensor if the motor ran at the desired speed. We can call this period the *set point*. The system varies the output signal to the motor until the measured period from the speed sensor equals that of the set-point period.

It does not matter whether the system measures period by using rising edges or falling edges. If a system needs to control speed, it can use something like the routines of Listing 11.9 to determine the period. The output signal could be a PWM signal that is generated using routines like those in Listing 11.4. Algorithms that can determine the PWM signal parameters is another topic, which we leave until Chapter 13.

To measure forward or reverse (bidirectional) speed, it is necessary to use an optical switch with two LED phototransistor pairs. This is known as a dual-channel slotted optical switch. The encoder disk will have two rings of holes. During rotation, two outputs generate pulses that are 90° out of phase. The rotational direction is determined by checking the pattern of the two outputs. In Chapter 13 we will look at a software technique for determining how many steps forward or backward the disk rotates.

Note that zero speed means that the next pulse will never arrive! In practice, the application requirements will determine what is considered to be zero speed. One simple way is to use a timer overflow to detect when the time between pulses is greater than 65,536 counts by the timer counter register (TCNT). In Section 11.3.4 we discuss ways of handling low speeds (long periods).

Another alternative to measuring speed is to count the number of pulses that occur during a set time period. This method is more appropriate when the pulse frequency is very high. For high frequencies, the measurement of time between the pulses becomes imprecise. For this technique an input capture interrupt routine can increment a software counter. An output compare interrupt will assert itself each time the set time period has expired. Then the output compare service routine reads the counter and then resets it. This counter value is proportional to speed. Instead of using the input capture to detect a pulse, the MCU could use the pulse accumulator function (Section 11.4).

The choice of method will depend on the speed range to measure and the speed sensor characteristics for that range.

Motion Detection. Other situations require that only motion is detected. This is the case for an antiskid braking system. To prevent skidding it is necessary to prevent any of the wheels from locking up. If any of the wheels stop turning while the brake pedal is pressed, the system adjusts the braking pressure to compensate. A wheel is considered to have stopped rotating when a time limit expires before the next edge is detected. Motion detection can also be used to prevent wheel slippage such as in a locomotive. The rotary speed of the wheels should match that required for the locomotive's true speed.

11.3.4 Measuring Long Periods

It is possible to measure longer periods. One way is to change the prescale for the free-running counter. Another way is to count the number of times a timer overflow occurs. This can be done by using a software counter. Listing 11.10 shows routines that can be used for this purpose. The global RAM variable OVCNT1 is the software counter that totals the number of overflows.

Counting the number of overflows is complicated because there may be situations when an input capture and timer overflow occur at or almost at the same time. The software should only count an overflow that occurs between the leading edge and trailing edge. Also note that a timer input capture interrupt has a higher priority than a timer overflow interrupt (see Table B.2 in the Quick Reference [Appendix B]).

If the first edge occurred with a negative value, an overflow will be detected later. If the value was positive, routine RIC1 checks if the TOF flag was set. If it was, that overflow occurred before the edge and its effect is canceled by decrementing counter OVCNT1. Note that only the overflow routine clears flag TOF, so that other routines can use it if necessary.

In the case of the second edge, routine RIC1 ignores an overflow (if any) if the edge occurred at a negative time. This means that the edge occurred before overflow. The routine also adjusts the overflow count if a borrow was required when subtracting the first edge capture value from the second. If longer periods need to be measured, the routines can be modified.

```
*Listing 11.10
*Service routines to measure period between two rising
*edges at the IC1 pin. Will count up to 255 overflows.
*Range is ~70 E cycles to 16,777,215 TCNT cycles.
*-----------------------------------

        ORG     0

IC1DUN  RMB     1               ;flag:0-not done, 1-pulse measured
IC1MOD  RMB     1               ;mode flag:FF-off,0-first,1-last edge
LPER    RMB     2               ;lower 16 bits of 24-bit period
OVCNT1  RMB     1               ;timer overflow count
*                               ;this is also upper 8 bits of
*                               ;24-bit period

*-----------------------------------


        ORG     $180
*Subroutine INITIC1
*Configures input capture IC1 to measure high pulse using
*interrupts.
*Calling register
*       IX = register block address
*No return registers except that CCR affected
*Return variables
*       IC1DUN = 0
*       IC1MOD = $FF, means measurement has not started
*       OVCNT1 = 0
INITIC1
        PSHA                    ;preserve registers
        LDAA    #$10            ;EDG1B:EDG1A = 0:1 for rising edge
        STAA    TCTL2,X
        LDAA    #$FF            ;mode flag off
        STAA    IC1MOD
        CLR     IC1DUN          ;signal period not done
        CLR     OVCNT1          ;reset overflow count
        BCLR    TFLG1,X $FB     ;clear IC1F if set
        BCLR    TFLG2,X $7F     ;clear TOF if set
        BSET    TMSK1,X $04     ;enable IC1 interrupt
        BSET    TMSK2,X $80     ;enable TOF interrupt
        CLI                     ;enable interrupts
        PULA                    ;restore registers
        RTS                     ;and return
```

```
*----------------------------------------
*Service routine RTOF
*Handles timer overflow interrupt.
*Increments OVCNT1 if first edge at IC1 already detected.
RTOF
         TST     IC1MOD           ;if IC1MOD negative then
*                                 ;period measurement has not started
         BMI     EXITTOF
         INC     OVCNT1           ;increment overflow count
EXITTOF
         LDAA    #$80             ;clear overflow flag
         STAA    REGBAS+TFLG2
         RTI                      ;and exit


*----------------------------------------
*Service routine RIC1
*Handles input capture IC1 interrupts.
*On first rising edge, saves input capture and adjusts
*overflow count if necessary.
*Return variables on first edge
*        IC1MOD = 0
*        LPER = first capture time
*On second rising edge, calculates period and disables
*further rising edge interrupts.
*Return variables on second edge
*        IC1MOD = 1
*        OVCNT1, LPER = period
*        IC1DUN = 1
RIC1
         LDX     #REGBAS          ;point to registers
         INC     IC1MOD           ;$FF-> 0 at first edge
*                                  ;0 -> 1 at second edge
         BNE     EDGE2            ;if not 0 then this is second edge
*process first edge
EDGE1
         CLR     OVCNT1           ;reset overflow count
         LDD     TIC1,X           ;capture first edge time
         STD     LPER             ;and save it
         BMI     EXITRIC1         ;done if captured before TOF
         LDAA    TFLG2,X          ;otherwise cancel effect of TOF
         BPL     EXITRIC1         ;if it occurred right after capture
         DEC     OVCNT1           ;note if ovcnt = −1, RTOF will
*                                 ;increment it back to 0
         BRA     EXITRIC1         ;finished with first edge
*process second edge
EDGE2
         LDD     TIC1,X           ;capture second edge time
         BMI     LRESULT          ;if MSB == 1, then skip TOF check
         TST     TFLG2,X          ;if TOF before second edge
         INC     OVCNT1           ;then count it
```

452   Chapter 11

```
LRESULT
        SUBD    LPER            ;second time - first time
        STD     LPER            ;is low 2 bytes of result
        BCC     LRESOK          ;if borrow req'd
        DEC     OVCNT1          ;then fix overflow count
*                               ;disable IC1 interrupt
LRESOK
        BCLR    TCTL2,X $30
        LDAA    #1              ;and signal period measured
        STAA    IC1DUN
*                               ;clear flag IC1F
EXITRIC1
        BCLR    TFLG1,X $FB
        RTI                     ;and return from service routine
```

If the prescale (PR1,0 in TMSK2) for TCNT is set for 1, the maximum period is about 8.38 s. This was calculated as follows.

255 overflow cycles multiplied by 65,536 cycles per overflow equals 16,711,680 cycles.

An additional 65,535 cycles can occur before the next overflow occurs.

This makes the total number of cycles equal to 16,777,215.

Multiply this result by 0.5 µs per clock cycle to get 8.38 s.

■ 11.4 PULSE ACCUMULATOR

The pulse accumulator is an 8-bit counter that can count input edges or measure pulse width depending on the operating modes used. Some typical pulse counting applications are monitoring tape position in a VCR or automobile odometer, and monitoring total volume of fluid flowing through a flowmeter. Recall that some flowmeters, such as turbine meters, generate pulses.

It is not as accurate as the input capture functions for pulse-width measurement. However, the pulse accumulator can identify a wide pulse from a narrow pulse much more easily. This makes it useful for decoding signals that use pulse width as part of their codes.

11.4.1 Function Description

The pulse accumulator has one input pin (PAI) at port A pin PA7. This is a bidirectional pin whose direction is configured by the data direction for PA7 bit (DDRA7) in the pulse accumulator control register (PACTL). Normally, bit DDRA7 is zero when using the pulse accumulator. However, pin PA7 still drives the pulse accumulator if bit DDRA7 is set.

Control register PACTL has a pulse accumulator system enable bit (PAEN) which must be logic 1 in order to use the function. Two other bits, pulse accumulator mode

(PAMOD) and pulse accumulator edge control (PEDGE), establish the function's operating modes. We discuss the two modes shortly. The function's control register PACTL also has the rate select bits for the real-time interrupt (RTII). These are bits RTR1 and RTR0.

The pulse accumulator count register (PACNT) can be read and written to. This means that the pulse accumulator can start counting from a preset count instead of zero. As we shall see, this provides for a specific number of events to be detected without having to compare two numbers. Unlike the other programmable timer data registers, it is an 8-bit register.

The function also makes use of registers TFLG2 and TMSK2 (see Figure 11.1). The pulse accumulator edge interrupt activates during a pulse edge. This pulse edge can be configured to be rising or falling by bit PEDGE in control register PACTL. The interrupt flag bit is the pulse accumulator input edge flag (PAIF) in flag register TFLG2. The enable bit (PAII) for this interrupt is mask register TMSK2. The pulse accumulator overflow interrupt activates whenever counter PACNT counts past $FF to $00. The pulse accumulator overflow flag bit (PAOVF) is in register TFLG2 and the enable bit (PAOVI) is in register TMSK2.

11.4.2 Event Counting

First we dissect the function in detail. Then we illustrate an application for factory automation. Figure 11.9 shows the sequence when the pulse accumulator is in the event counting mode. Setting the pulse accumulator enable bit (PAEN) in the pulse accumulator control register (PACTL) enables the pulse accumulator system. Clearing the pulse mode bit (PAMOD) puts the function in event counting mode. The edge control bit (PEDGE) selects the active edge for counting. For example, if PEDGE is logic zero, a falling edge increments counter PACNT. Every time an edge is detected at pin PAI, the pulse accumulator edge flag (PAIF) sets. The flag can be polled by software, or alternatively, it can generate an interrupt. To assert an interrupt, its associated enable bit (PAII) should be set. Like other timer flags, the software should write a 1 to flag PAIF to clear it.

When pulse accumulator register PACNT overflows (counts past $FF to $00), the function sets the overflow flag (PAOVF). Similarly, its interrupt is enabled by setting the enable bit (PAOVI); an overflow condition asserts an interrupt. Clearing flag PAOVF is also accomplished by writing a 1 to it.

Consider a factory assembly line with a section to paint cars. A certain number of cars are to be painted one color. Then another group of cars are to be painted a different color. The microcontroller energizes the appropriate paint spray guns for each group of cars. It uses the pulse accumulator function to keep track of how many cars were painted for each color. Once the preset number of cars have been painted, the microcontroller turns off the spray guns and begins the sequence for the next group of cars. The spray paint system uses the pulse accumulator overflow interrupt to tell the application program that all cars in a group have been painted. A sensor located at the end of the paint line generates a high pulse whenever a car passes it. This sensor is connected to the pulse accumulator input (PAI). The sensor can be a photoswitch that operates similarly to the optical switch shown in Figure 11.8.

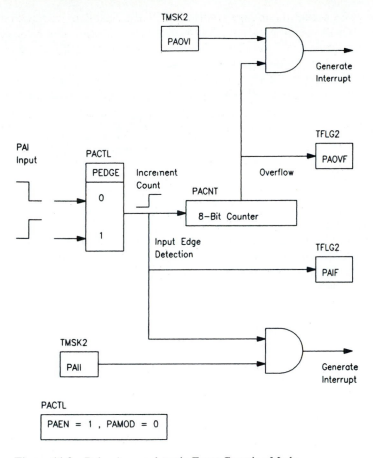

Figure 11.9 Pulse Accumulator in Event Counting Mode

Short Counts. The pulse accumulator can be set up to produce an interrupt after a preset number of edges have been detected. This is quite simple if the preset is less than 256 (an 8-bit number). To do this, write the two's complement of the preset to counter register PACNT. When the preset number of events have been detected, PACNT would have counted past $FF to cause an overflow condition. For example, consider a preset of 24 ($18). The two's complement is $E8. The counter will start at $E8. When the twenty-third car has passed, PACNT will have incremented to $FF. The next car (the twenty-fourth) will increment the counter past $FF to $00, which is an overflow condition.

Listing 11.11 shows some pulse accumulator routines that can be used for this type of application. Note that subroutine INITPA uses bit manipulation instructions so that other control register bits are not affected.

```
*Listing 11.11
*Pulse Accumulator routines for spray paint line
*to paint a preset number of cars
```

```
        ORG      $180

*Subroutine INITPA
*Initializes PA for counting rising edges but it
*does not enable interrupts
*No calling or return registers

        PSHX                      ;preserve IX
        LDX      #REGBAS          ;point to register block
*                                 ;both PAEN, PEDGE = 1
        BSET     PACTL,X $50
        PULX                      ;restore IX
        RTS                       ;return

*Subroutine CNTSPA
*Counter (short) for pulse accumulator
*Loads PA function with preset and enables it
*for overflow (PAOV) interrupt
*Calling Registers
*       IX = register block address
*       ACCA = preset, < 256
*No return registers except that CCR affected
CNTSPA
        PSHA                      ;preserve registers
*                                 ;clear PAOVF
        BCLR     TFLG2,X $DF
*                                 ;PAOVI = 1
        BSET     TMSK2,X $20
        NEGA                      ;two's complement
*                                 ;and store to pulse accum counter
        STAA     PACNT,X
        CLI                       ;enable interrupts
        PULA                      ;restore registers
        RTS                       ;and return

*Interrupt Service Routine RPAOV
*Handles pulse accumulator overflow
RPAOV
        LDX      #REGBAS          ;point to registers
*                                 ;PAOVI = 0
        BCLR     TMSK2,X $20
*                                 ;clear PAOVF
        BCLR     TFLG2,X $DF
*                                 ;stop spray guns
        JSR      STOP_PAINT
        RTI                       ;return

*Note that source code for subroutine STOP_PAINT not shown
```

PRACTICE

Write a program to paint 30 cars red using subroutine START_PAINT. The subroutine uses ACCA as a calling register that holds the color code. It has no return registers. After returning from START_PAINT, spraying operation continues until subroutine STOP_PAINT is called (by RPAOV). Assume that vectors are defined in ROM.

Solution

```
*Solution
*paint 30 cars red
*pin PA7/PAI is input from car detection sensor
*use routines and macro from Listing 11.11

        LDX     #REGBAS         ;point to registers
        JSR     INITPA          ;initialize PA
        LDAA    #30             ;preset count for 30
        JSR     CNTSPA
        CLI                     ;enable interrupts
        LDAA    #RED_CODE       ;paint cars red
        JSR     START_PAINT
WAIT
        WAI
        BRA     WAIT            ;until 30 done

*note that WAIT loop responds to any interrupt
```

Longer Counts. More than 256 events can be counted by using software to keep track of the number of pulse accumulator overflows. If the counter starts at zero, 256 events will cause an overflow. Hence, the number of overflows required is the preset divided by 256.

Double accumulator D proves itself to be very useful. The preset is a 16-bit number that can be loaded into this register. The accumulator A part has a number that represents multiples of 256, so it holds the required number of overflows, and accumulator B holds the remainder. Counter PACNT can be loaded with the two's complement of this remainder value. If the remainder is nonzero, an extra overflow count will be required. Consider the following: Let's say that the preset is 770 ($302). The software loads PACNT with the complement of $02, which is $FE. When two events have been detected, an overflow occurs. Then it should count three more overflows (768 events).

Listing 11.12 shows routines that can be used to count up to 65,535 events. An application program will need to check the RAM variable OVCNT for zero to determine if the preset number of events have occurred.

```
*Listing 11.12
*Routines for counting large number of events
```

```
*For INITPA, see Listing 11.11
*-------------------------------------

OVCNT   EQU     $14

        ORG     $180

*Subroutine CNTLPA
*Counter (long) for pulse accumulator
*Loads PA function and overflow counter with preset
*and enables it for overflow (PAOV) interrupt
*Calling Registers
*       IX = register block address
*       ACCD = preset, max is 65,535
*No return registers except that CCR affected
*Return variables
*       OVCNT = overflow count
CNTLPA
        PSHA                    ;preserve registers
        PSHB
*                               ;clear PAOVF
        BCLR    TFLG2,X $DF
*                               ;PAOVI = 1
        BSET    TMSK2,X $20
        TSTB                    ;test for remainder count
        BEQ     NOINC           ;if none, skip INCA
        INCA
NOINC
        NEGB                    ;two's complement remainder
*                               ;and store to pulse accum counter
        STAB    PACNT,X
        STAA    OVCNT           ;store overflow count
        CLI                     ;enable interrupts
        PULB                    ;restore registers
        PULA
        RTS                     ;and return

* Routine RPAOV
* Handles pulse accumulator overflow by decrementing count
* Main application program checks if preset count achieved
* by checking RAM variable OVCNT for zero
RPAOV
        LDX     #REGBAS         ;point to registers
*                               ;clear PAOVF
        BCLR    TFLG2,X $DF
        DEC     OVCNT           ;decrement overflow count
        RTI                     ;return
```

11.4.3 Gated Time Accumulation

In this mode the pulse accumulator counts up once every 64 clock cycles when its PAI input is at the active level. It does not count edges. Figure 11.10 illustrates this mode of operation. Setting the pulse accumulator mode bit (PAMOD) to 1 puts the function in gated time accumulation mode. The edge bit (PEDGE) specifies the input that will *inhibit* time accumulation. For example, if PEDGE is 0, the counter will not increment as long as input PAI is low. When PAI is high, it increments once every 64 clock (E) cycles. If PAI changes level, it continues counting from where it left off once the input returns to the enabling level again.

The flag and interrupt bits work identically as in the event counting mode. For example, flag bit PAIF sets every time input PAI goes from a high to a low if control bit PEDGE is 0. Thus, you can think of flag PAIF setting when the trailing edge of an active pulse is detected. Bit PEDGE specifies the trailing transition. For example, if PEDGE equals 0, the trailing edge is falling.

Pulse-Width Discriminator.   A common use for this mode is to *discriminate* a wide pulse from a narrow pulse. This is done by writing a value to counter PACNT that is halfway between the values of the narrow and wide pulse widths. When a narrow pulse arrives, the

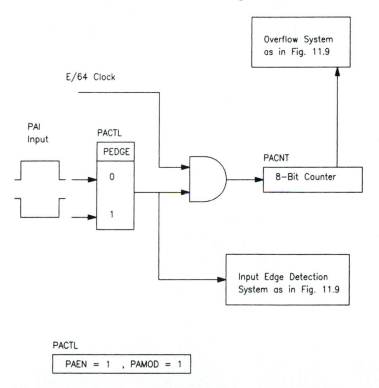

Figure 11.10   Pulse Accumulator in Gated Time Accumulation Mode

counter accumulates and the trailing edge is detected before an overflow occurs. When a wide pulse arrives, the counter accumulates and overflows before the trailing edge is detected. Thus, if the edge flag (PAIF) sets first, the pulse was narrow. If the overflow flag (PAOVF) sets first, the pulse was wide.

We will consider an audiocassette interface. Admittedly, this is an old-fashioned way to store data. However, it does illustrate the use of pulse width to encode data. Each bit is represented by a low pulse followed by a high pulse. A condition of no data is always high. For a logic 1, each pulse width is 0.25 ms (narrow). For a logic 0, each pulse width is 0.5 ms (wide). Halfway between the two ranges is 0.375 ms (or 375 μs). The counter PACNT increments every 64 cycles. This is 32 μs for a 2-MHz clock. Divide 375 by 32 to get 11.719. This is approximately 12 ($0C). The two's complement is $F4. Listing 11.13 shows routines that use pulse-width discrimination to read in cassette data. This system uses a low pulse for each bit because an idle line is high. Also, the routines do not account for possible errors.

Subroutine INPA uses two interrupts. If an edge interrupt occurs first, the data bit is assumed to be a logic 1. Routine RPAIE also reinitializes the counter. If an overflow occurs first, the data bit is assumed to be a logic 0. However, routine RPAO waits for the edge before it restores the preset to the counter and exits.

```
*Listing 11.13
*Routines that read bytes from an audiocassette interface
*using the pulse accumulator to demonstrate a gated time
*accumulation mode application
*1 kHz = 0, 2 kHz = 1, idle is high
*-----------------------------------

          ORG      $180

*Subroutine INPA
*Reads a set number of data bytes from cassette
*interface using pulse accumulator.
*Another routine has started cassette motors.
*Data stored in RAM block starting at location STBLK.
*Exits when all bytes read.
*No calling or return registers except that CCR affected.
*RAM variables
*         STBLK = start address of block where data stored
*         ENDBLK = last address of block

STBLK   EQU      $20
ENDBLK  EQU      $2F


INPA
          PSHA                       ;preserve registers
          PSHB
          PSHX
          PSHY

          LDY      #STBLK            ;initialize pointers to data buffer
          LDX      #REGBAS           ;register block
```

```
*                               ;config. pulse acc., gated mode
        LDAA    #%01110000
        STAA    PACTL,X         ;low level enables accumulation
        LDAA    #$30            ;enable local interrupts
        STAA    TMSK2,X
        STAA    TFLG2,X         ;and clear flags if set
        LDAA    #$F4            ;preset pulse accumulator
        STAA    PACNT,X
NEXTBYTE
        LDAB    #8              ;for (bit count=8; bc >= 1; bc -)
        CLR     0,Y             ;clear data byte
NEXTBIT
        CLI                     ;enable interrupt and wait for
        WAI                     ;either interrupt to shift in a bit
        DECB                    ;bit count--
        BNE     NEXTBIT         ;until 8 bits shifted in
        INY                     ;byte ptr++
        CPY     #ENDBLK         ;until last byte read
        BNE     NEXTBYTE
        SEI                     ;mask interrupts
        PULY                    ;restore registers
        PULX
        PULB
        PULA
        RTS                     ;and return


*Service routine RPAO
*Pulse Accumulator Overflow Service Routine
*If overflow occurred first then bit is a 0 (1 kHz)
RPAO
        LDAA    0,Y             ;get old data byte
        CLC                     ;shift in a zero bit
        ROLA
        STAA    0,Y             ;and store it
*inhibit rising edge interrupt
WAIT1
        BRCLR   TFLG2,X $10WAIT1
*                               ;clear flags PAOVF, PAIF
        BCLR    TFLG2,X $CF
        LDAA    #$F4            ;restore acc. preset
        STAA    PACNT,X
        RTI                     ;and return


*Service routine RPAIE
*Pulse Accumulator Edge Detection Service Routine
*If rising edge occurred first, then bit is a 1 (2 kHz)
RPAIE
        LDAA    0,Y             ;get old data byte
        SEC                     ;shift in a one bit
        ROLA
```

```
            STAA        0,Y                 ;and store it
*                                           ;clear flag PAOVF, PAIF
            BCLR        TFLG2,X $CF
            LDAA        #$F4                ;restore acc. preset
            STAA        PACNT,X
            RTI                             ;and return
```

■ 11.5 REAL-TIME INTERRUPT

Back in Chapter 3 we briefly mentioned the real-time interrupt (RTII). This can also be considered to be one of the programmable timer functions. It asserts itself at periodic intervals. This is useful for timing to schedule other events. Typically, it makes the application program do something at specified intervals. This might be to take measurements for recording. It may drive software counters to keep track of real time. For example, the application software can reference the counters to determine when an alarm condition occurred.

The RTII interrupt rate bits (RTR1, RTR0) determine when the interrupt asserts itself. This is shown in Table 11.1. The reset condition for both bits is zero. This means that 2^{13} ($2^{13} = 8192 = \$2000$) clock cycles are acquired to assert the interrupt. When each clock cycle is 0.5 μs (E = 2 MHz), the interrupt rate is

$$2^{13} \times 0.5 \text{ μs} = 4096 \text{ s} \cong 4.10 \text{ ms}$$

To enable the real-time interrupt, set its enable bit (also called RTII) in mask register TMSK2. When the time limit has elapsed, the flag bit (RTIF) in flag register TFLG2 sets. The service routine can clear the flag by writing a 1 to it.

Listing 11.14 shows routines that implement a crude 24-hour clock. This is a *foreground* timer. An application program can read the elapsed hours, minutes, and seconds at any time. Initially, the application program sets up the start time in the RAM variables and calls subroutine INITRTII. Then routine RRTII handles the rest. There is some inaccuracy since the RTII interrupt rate is not an integer fraction of 1 s. The end-of-chapter exercises explore these inaccuracies further. The value for FCNT is an approximation. In Section 13.7 we explain what is meant by "foreground."

```
*Listing 11.14
*Utility routines for 24-hour clock


            ORG         0                   ;start address of data used by program
HR          RMB         1                   ;Hour count in BCD
MIN         RMB         1                   ;Minute count in BCD
```

TABLE 11.1   REAL-TIME INTERRUPT RATES

PACTL bits			
RTR1	RTR0	Divide E by:	Rate (ms) for E = 2.0 MHz
0	0	2^{13}	4.10
0	1	2^{14}	8.19
1	0	2^{15}	16.38
1	1	2^{16}	32.77

```
SEC        RMB      1              ;Second count in BCD
FSEC       RMB      2              ;Fraction of second count
*                                  ;for RTII routine in hex

FCNT       EQU      244            ;number of RTII interrupts to
*                                  ;make 1 second
*------------------------------------

           ORG      $180

*Subroutine INITRTII
*Initializes real-time interrupt foreground timer
*No return registers except that CCR affected
*Application program should initialize
*following RAM variables:
*        HR = Hour count in BCD
*        MIN = Minute count in BCD
*        SEC = Second count in BCD
*Calling registers
*        IX = register block address
INITRTII
           BSET     TMSK2,X $40 ;set RTII flag

           CLI
           CLR      FSEC           ;initialize FSEC
           CLR      FSEC+1
*FSEC is fraction of second count for RTII routine in hex
           RTS                     ;return

*Service routine RRTII
*Handles real-time interrupt
*Uses 16-bit software counter FSEC to count fractions
*of seconds. Updates RAM variables SECONDS, MINUTES, and
*HOURS as BCD values for use by any other routine
RRTII
           LDX      #REGBAS
           LDAA     #$40           ;clear RTIF
           STAA     TFLG2,X
           LDY      FSEC           ;count # of RTII
           INY                     ;occurrences
           STY      FSEC
           CPY      #FCNT          ;add 1 to SEC if
           BEQ      SECONDS        ;enough occurred
           RTI
SECONDS
           CLR      FSEC           ;reset fraction sec count
           CLR      FSEC+1         ;to zero
           LDAA     SEC
           ADDA     #1             ;add 1 to seconds count
           DAA                     ;adjust for decimal arithmetic
           CMPA     #$60           ;check for seconds == 60
```

```
            BEQ     MINUTES         ;adjust minutes if == 60
            STAA    SEC
            RTI
MINUTES
            CLR     SEC             ;reset seconds to 0
            LDAA    MIN
            ADDA    #1              ;add 1 to minutes count
            DAA                     ;adjust for decimal arithmetic
            CMPA    #$60            ;check for minutes == 60
            BEQ     HOURS           ;and adjust hours if == 60
            STAA    MIN
            RTI
HOURS
            CLR     MIN             ;reset minutes count
            LDAA    HR
            ADDA    #1              ;add 1 to hours count
            DAA
            STAA    HR
            CMPA    #$24            ;check for hours == 24
            BNE     RETURN          ;and reset to 0 if == 24
            CLR     HR
RETURN
            RTI
```

These routines can be expanded to update counters for days and months, although the accuracy will not be good enough without adjustments. If accuracy is required, it is better to use a real-time clock chip such as the 68HC68T1.

PRACTICE

Assume that the code for Listing 11.14 and all vectors are loaded in memory. Write an instruction sequence to sound an end-of-shift alarm at 4:30 PM by turning on pin PB6 for 1 min.

Solution

```
*Solution
*turn on alarm at 4:30 PM (16:30 hours)
*note that application code repeats continuously while
*RTII updates 24-hour clock

*some extra RAM variables to specify alarm time of 16:30
HRSET   FCB     16
MINSET  FCB     30

*       ;applicable initialization code here
*       ;following code repeats continuously
REPEAT

*       ;other application code here
```

```
*                   ;the following instruction sequence handles the alarm
          LDX       #REGBAS        ;do this if IX != REGBAS
          LDAA      HR             ;if hr != hrset then alarm = off
          CMPA      HRSET
          BNE       OFF_ALARM
          LDAA      MIN            ;else if min != minset then
alarm = off
          CMPA      MINSET
          BNE       OFF_ALARM
*                                  ;else alarm = on
          BSET      PORTB,X $40
          BRA       SKIP_OFF       ;and continue past
*                                  ;instruction to turn off alarm
OFF_ALARM
          BCLR      PORTB,X $40
SKIP_OFF
*         ;more application code here

          BRA       REPEAT
```

Real-Time Executive.   In Section 13.7 we explain what a real-time executive does. For now, we'll just say that a real-time executive for the 68HC11 can use the RTII as the time-keeper to control when it performs its duties.

■ 11.6 SUMMARY

The major functions of the programmable timer subsystem are output compare, input capture, and pulse accumulator. The heart of the subsystem is the free-running counter (TCNT), which increments every integral number of clock cycles depending on configuration. An output compare function continually compares the stored value in an output compare register with TCNT. When they are equal, it asserts a change in output depending on configuration. An input capture waits for an edge (depending on how it is configured) and records the time when the edge occurs (value of TCNT) in an input capture register.

The basic software to handle these functions is to enable an external pin by setting the appropriate bits in a control register (TCTL1 or TCTL2). Then enable (or disable) the local interrupt by setting the corresponding bits in a mask register (TMSK1 or TMSK2). When the function is active, a corresponding bit in a flag register (TFLG1 or TFLG2) sets. A program detects this by polling or by responding to an interrupt. The handler routine clears the flag by writing a 1 to it. After the flag is cleared, the function is ready for the next event.

To use the output compare function, write a value to an output compare register. Then wait for TCNT to equal its value (wait for a successful compare). Application software can use output compare to generate single pulses, square waves, or PWM signals. To use the input capture function, wait for an active edge and read the input capture register. The program can use this captured value to determine pulse width or signal period. A program can use this to calculate signal frequency, speed of a device, or any other quantity related to a periodic signal.

The pulse accumulator has two modes of operation. In one of them, event counting mode, the pulse accumulator (PACNT) increments once for every input edge, depending on configuration. This is often used to detect when a set number of events (edges detected) have occurred. In the other mode, gated time accumulation mode, register PACNT increments once every 64 clock cycles when a gate signal is active. This is often used for pulse-width discrimination.

The real-time interrupt (RTII) asserts itself every interval depending on configuration. This interval is determined by the setting of the RTII rate select bits (RTR1,0 in PACTL). The RTII interrupt is used mainly for timekeeping.

■ EXERCISES

1. Identify the pins for output compare (OCx), input capture (ICx), and the pulse accumulator (PAI).

2. Identify the data registers associated with each pin asked about in Question 1.

3. Describe how to change the divide-by-prescale of the free-running counter.

4. How can you change the prescale in the EVBU board? Note that the EVBU board has a jumper called J2. When the EVBU board is reset, it starts execution at the Buffalo monitor address of $E000 to check the status of J2. If J2 connects pins 2 and 3, the Buffalo monitor continues execution past address $E00A. If J2 connects pins 1 and 2, Buffalo jumps to the start of EEPROM (address $B600).

5. List seven possible instruction sequences to clear flag bit TOF.

6. Write a subroutine called DLY10. It is a 10-ms time delay that can be used for EEPROM programming. Use function OC3 and a prescale factor of 8. Also use polling.

7. Modify subroutine DLY10 of Question 6 to drive pin PA5 high for the duration of a 10-ms delay.

8. For the program of Question 6 modified to have a prescale factor of 1, calculate the actual time delay from the time the calling program calls the subroutine to the time it returns. Assume that no interrupts occurred.

9. Modify the program of Question 8 (if necessary) to have an exact delay of 10 ms assuming no interrupts. Note that this precision is not required for EEPROM programming.

10. Describe how you can modify the program of Question 9 to make the time delay independent of interrupts (more or less). Explain why it is not possible to guarantee an exact time delay.

11. Write a program using interrupts to generate a square wave at output pin PA6 using OC2.

12. Write a program to produce two PWM signals at pins PA5 and PA6. Function OC1 controls the two pins in conjunction with OC2 and OC3. OC1 drives the period and the scheduling of OC2 and OC3.

13. A conventional telephone uses pulse dialing. (The modern type uses dual-tone multifrequency [DTMF] dialing.) The number of breaks (logic low) in a pulse train represents the digit dialed with the exception of the zero digit. It uses 10 breaks. Each digit pulse is low for 60 ms and high for 40 ms. Each digit is separated by a high signal interval that is nominally 700 ms long. A 700-ms high connect signal should be generated before dial pulses are sent. Write a subroutine that generates dial pulses for a telephone number stored in RAM as 7 BCD bytes, one for each digit.

14. Determine the shortest pulse width that can be measured by the program of Listing 11.8.

15. Rewrite the program of Listing 11.8 to measure pulse width using IC2.

16. Develop a system to measure the width of a high pulse. Assume that the pulse occurs 10 microseconds or longer after the program starts. The width of the pulse ranges from 1 to 65,535 E cycles.

17. Write a program to generate an interrupt after 100 pulses are detected.

18. Write a program to generate an interrupt after a preset number of objects have been counted. The preset can be any 16-bit unsigned integer.

19. Without looking at Listing 11.13, write a program to read data from an audiocassette interface and store it in a block of memory. A 0 is a 1-kHz pulse and a 1 is a 2-kHz pulse. An idle is high.

20. Write a program that starts up a process heater in a chemical plant by running the heater through a purge cycle. The purge cycle operates as follows. The control system turns on a purge valve (PB1) for 10 s. This forces air into the heater box to force out any explosive vapors. After the purge, the ignitor (PB0) is turned on for 0.5 s. The control system reads a flame detector 10 s later. If no flame is detected, the heater system shuts down and the entire sequence must be repeated to attempt another start. A flame is sensed by calling subroutine ADC4, which returns a value in accumulator A. A return value less than a threshold value (#TH) means that no flame exists.

21. What is the inaccuracy of the 24-hour clock example of Listing 11.14?

22. Rewrite the 24-hour clock program of Listing 11.14 so that it is more accurate. It is possible to adjust the fractional second counts such that the program will increment the seconds count most of the time after 244 interrupts but will occasionally adjust by using 245 interrupts to slow down.

12 The Analog Converter Subsystem

OBJECTIVES

After completing this chapter, you should be able to

- Use concepts and terminology to represent analog information using binary numbers.
- Design fundamental circuitry using transducers, signal condition, and actuators.
- Describe the operation of the R-2R ladder for digital-to-analog conversion.
- Design the hardware and software for a digital-to-analog interface.
- Identify different techniques for analog-to-digital conversion and describe the successive approximation technique.
- Describe the operation of the 68HC11 built-in analog-to-digital converter (A/D).
- Use the 68HC11 A/D for designing software to handle analog inputs.

Analog refers to physical quantities that vary continuously instead of discretely (in steps). For example, room temperature can be hot, cold, or anywhere in between. It would not be possible for temperature to change from 20° to 30° without passing through all the other values in between. Although they do not change in steps, analog values such as voltage, current, pressure, speed, and other variables can be represented by binary or hexadecimal (digital) numbers. This chapter covers binary codes used to represent analog information. It also covers some common transducers, a method to convert digital data to analog data, and a method to convert analog data to digital data. The rest of the chapter covers the analog-to-digital converter subsystem used in the 68HC11.

For review, refer to Figure 8.6. Very briefly, a digital-to-analog converter (D/A) has a digital input that specifies an output whose value changes in steps. These step changes are in volts or amperes. A D/A is also known as a DAC. The analog-to-digital converter (A/D) has an input that can vary from a minimum to a maximum value of volts or amperes. The output is a digital number that represents the input value. An A/D is also known as an ADC.

Clearly, there are many uses for A/Ds in a control system. Many variables, such as temperature, pressure, flow rate, and more, are analog values that a control system needs to monitor. Other control systems use D/As for outputs to servomotors. These are dc motors that rotate to a fixed position based on an analog signal. Some disk drive heads are positioned by sending an analog signal to a voice coil actuator. The voice coil uses electromagnetic force to move a read/write head. A change in the analog signal changes the electromagnetic force. Signal processing systems use both types of converters.

The 68HC11 has a built-in A/D subsystem, but it does not have a D/A, although it can easily be interfaced to one. More accurately, the 68HC11 does not have a built-in D/A that can output an external analog signal. It turns out that it uses an A/D, which in turn uses a D/A to generate internal signals as part of the process to convert analog to digital. This is common to many types of A/Ds.

■ 12.1 CONCEPTS AND TERMINOLOGY

For the sake of brevity, we use standard short forms for units of measurement. Note that Appendix D lists units of measurement.

Analog signals have a *ratiometric* range. These range from a low value to a high value. Some examples are

0 V to 5 V
−2.5 V to +12.5 V
4 mA to 20 mA

A binary code can represent the percentage of any analog range. For example, $00 (0% of $100) represents 0 V and $FF (almost 100% of $100) almost represents 5 V for an 8-bit binary code representing the analog range 0 to 5 V. Generally, a binary numbering system that represents the analog range is unsigned. A number with all 0s represents the lowest analog value. A binary number with all 1s (almost) represents the maximum analog value. Later we will see why the maximum analog value can only be approximated digitally. A binary number in the middle (MSB logic 1, remainder logic 0s) represents the middle of the analog range. Referring to our earlier example, $80 (50% of $100) represents 2.5 V. In the section, "Fractions and Normalizing," in Appendix F we also describe scaling of analog ranges.

A binary/digital code has a fixed number of bits. Common bit sizes are 6, 8, and 12. If the analog range starts from a nonzero minimum value, it is usually represented by an *offset binary code*. A binary zero represents the minimum analog value. The last two sample analog ranges have a corresponding offset binary code. Binary zero represents −2.5 V and 4 mA for each range, respectively.

Definitions. To talk about analog scales, we need to use standard definitions.

Offset	Minimum analog value.
Span	Maximum to minimum analog value (difference).
Weight	The analog change corresponding to a change in a bit in the digital number.

Step size Span/2^n, where n is the number of bits in the corresponding digital code. *Note:*

$n - 1 =$ bit position of MSB
$n = 0$ for a 1-bit number

Resolution Same as step size. It is the smallest analog change resulting from changing one bit in the digital number. It is also the weight of the LSB. Sometimes resolution is specified as the number of bits (n).

$$\% \text{ Resolution} = \frac{\text{step size}}{\text{span}} \times 100\%$$

$$= \frac{1}{2^n} \times 100\%$$

Equations. To convert analog and digital values to one another in the upcoming examples, we use the following equations:

$$\text{analog number} = (\text{digital number} \times \text{step size}) + \text{offset} \qquad (12.1)$$

$$\text{digital number} = \frac{\text{analog number} - \text{offset}}{\text{step size}} \qquad (12.2)$$

$$= \frac{(\text{analog number} - \text{offset}) \times 2^n}{\text{span}}$$

These equations hold true as long as the numbers are within the range. This is similar to normalizing as shown by equations (F.5) to (F.7) in Appendix F.

Examples. The following illustrates the use of the definitions and equations. Consider an analog range of -5 to $+5$ V. The digital range is 8 bits. We can also say that we are using 8-bit resolution ($n = 8$).

$$\text{offset} = -5 \text{ V}$$
$$\text{span} = 5 \text{ V} - (-5 \text{ V}) = 10 \text{ V}$$
$$\text{step size} = \frac{10 \text{ V}}{2^8} = \frac{10 \text{ V}}{256} = 39.1 \text{ mV} = \text{resolution}$$
$$\% \text{ resolution} = \frac{1}{256} \times 100\% = 0.391\%$$

Table 12.1 illustrates the weights for this range. Each bit is weighted according to its position within the binary number.

Next, we show the digital number corresponding to some of the analog values in Table 12.2. We did not show all 256 numbers. Hopefully, you will not be disappointed. As you can see, the maximum digital number can only approximate the maximum analog value in the range. This is because the numbering starts from digital zero. An extra bit is required to code the maximum analog value. Also note that we have chosen to show values that increment by 6 bits (weight for bit position 5 is 1.25 V).

TABLE 12.1 BIT WEIGHTS FOR A −5 V TO +5 V
ANALOG RANGE TO 8-BIT DIGITAL CONVERSION

Digital bit	Bit weight (V)
7	$10 \times 2^{-1} = 10/2 = 5$
6	$10 \times 2^{-2} = 10/4 = 2.5$
5	$10 \times 2^{-3} = 10/8 = 1.25$
4	$10 \times 2^{-4} = 10/16 = 0.625$
3	$10 \times 2^{-5} = 10/32 = 0.313$
2	$10 \times 2^{-6} = 10/64 = 0.157$
1	$10 \times 2^{-7} = 10/128 = 0.078$
0	$10 \times 2^{-8} = 10/256 = 0.039$

TABLE 12.2 −5 V TO +5 V ANALOG RANGE TO 8-BIT
DIGITAL CONVERSION TABLE

Analog (V)	Digital (hex)
−5	00
−3.75	20
−2.5	40
−1.25	60
0	80
1.25	A0
2.5	C0
3.75	E0
5−0.0391 = 4.961	FF

Equation (12.1) is a shortcut for calculating the analog value corresponding to the digital number. We show another form of the equation that is more representative of how bit weights are added to make the analog value:

$$\text{analog number} = (b_{n-1} \times 2^{-1} + b_{n-2} \times 2^{-2} + \ldots + b_1 \times 2^{-(n-1)}$$
$$+ b_0 \times 2^{-n}) \times \text{span} + \text{offset} \tag{12.3}$$

where $n - 1$ is the bit position of the MSB and b_i is the logic 1 or 0 value for bit position i. Equation (12.3) more accurately represents the method used by D/A and A/D hardware to perform conversions.

Another Example. A 6-bit D/A has an analog output range of −2.5 to +5.0 V. Calculate the analog output when the input is %010101 (decimal 21).

We can work this out as follows:

$$\text{offset} = -2.5 \text{ V}$$
$$\text{span} = 5.0 \text{ V} - (-2.5 \text{ V}) = 7.5 \text{ V}$$
$$\text{step size} = \frac{7.5 \text{ V}}{2^6} = 0.1172 \text{ V}$$

Using equation (12.3), the analog value is

$$\text{analog number} = \left(0 \times \frac{1}{2} + 1 \times \frac{1}{4} + 0 \times \frac{1}{8} + 1 \times \frac{1}{16}\right.$$
$$\left. + 0 \times \frac{1}{32} + 1 \times \frac{1}{64}\right) \times 7.5\text{ V} + (-2.5\text{ V})$$
$$= 1.875\text{ V} + 0.469\text{ V} + 0.117\text{ V} - 2.5\text{ V}$$
$$= -0.039\text{ V}$$

Using equation (12.1), the analog value is

$$\text{analog number} = 21 \times 0.1172\text{ V} - 2.5\text{ V}$$
$$= -0.039\text{ V}$$

Outside the Range. If the input of an A/D is below the minimum or above the maximum of the range, the corresponding digital value will be minimum or maximum, respectively. For example, if an 8-bit A/D with a range of 0 to 5 V has an input of -0.5 V, the digital output will be $00. If the input is 5.5 V, the output will be $FF. In practice, extreme input swings beyond the specified range may damage the device.

Variable Step Sizes. We mention this briefly to inform you that not all systems use uniform step sizes. Pulse-code modulation (PCM) is used in telecommunications systems. It is a method of transmitting data in which analog signals are sampled by an A/D and are then converted into digital pulses. Since noise is more of a problem when the signal is weak, the step size is reduced for smaller signal amplitudes and is increased for larger amplitudes. The step size is approximately proportional to the signal amplitude.

Companding (Compressing–expanding) is another technique used to reduce noise in telecommunications signals. Most companding systems follow the *mu* law. It defines how different analog levels are represented. Basically, it divides the signal range into fifteen 4-bit segments: seven negative, one centering at zero, and seven positive. Each segment has 16 equal steps. The step sizes increase by a factor of 2 when moving from a lower magnitude segment to a higher-magnitude segment.

Other systems may use a *nonlinear* range. Examples are *pH, decibels* (dB), and square root. Acidity is measured using pH. It is the negative logarithm of hydronium ion concentration (pH $= -\log[H_3O^+]$). Decibels are commonly used to measure audio, video, amplifier, and filter signals. It is based on the logarithm of the ratio of output power to input power:

$$\text{dB} = 10 \log (P_{out}/P_{in})$$

Square-root scales are used in some flow measurement sensors. You may wish to refer to equation (2.1) in Section 2.5.10. The sensor outputs a signal linearly proportional to the differential pressure between two points. The fluid flow rate is proportional to the square root of the differential pressure.

For simplicity, we will refer mainly to linear scales since most sensors output an analog signal that varies linearly with the physical quantity being measured. If this is not the case, a look-up table can be used.

■ 12.2 TRANSDUCERS

The "real world" is analog in nature and microcontroller I/O must interact with this world. When cruising down the highway in your sports car, the microcontroller may output a signal to change fuel valve position based on readings from an input indicating speed. *Transducers* are used to interface the I/O with the process, in this case an automobile.

A transducer is a device that converts a process variable (e.g., car speed) into an electrical signal, or vice versa.[1] For example, one type of transducer converts temperature into a current signal in the range 4 to 20 mA. One type of output transducer converts a signal in the range 0 to 5 V to a corresponding shaft position. Other terms are *sensors* and *actuators*. They refer to input and output transducers, respectively.

In this section we provide an overview of the hardware used for interfacing with the analog world. This will give you an idea of the possibilities. Since this is only a brief description of the subject, you may wish to refer to other literature for detailed treatment. Some sources are references 10 to 12 in Appendix C. The following sections cover common sensors, signal conditioning circuits, and methods to drive actuators.

12.2.1 Sensors

The following are common sensors used to measure basic physical quantities.

Position: Potentiometer
                    Linear variable displacement transformer (LVDT)
Force: Strain gauge
              Piezoelectric device

Temperature: Thermistor
                        Thermocouple
Light Intensity: Photoconductive cell
                          Phototransistor
Current: Current transformer
                SENSEFET

Often, other physical quantities are measured using variations of the sensors listed above. For example, flow can be measured using force or temperature. Pressure is the amount of force per unit area. Flow rate can be determined by measuring the pressure drop across an element (also discussed in Section 12.1). Another way to determine flow rate is to measure the amount of heat removed from a heated element by the fluid. This is an indirect use of temperature since this type of flowmeter actually measures the current needed to maintain a heat probe at a constant temperature. Hence, it must also measure temperature.

---

[1]A dictionary may state that a transducer is "a device that converts variations of one quantity into those of another" (*Oxford Dictionary*). Some control systems use pneumatic or hydraulic signals instead of electrical signals.

The potentiometer (Figure 12.1[a]) outputs the angular position of the shaft. When the wiper moves the voltage changes because of the change in resistance. It is the simple application of *Ohm's law:* Voltage equals current multiplied by resistance. Note that the current through the potentiometer is constant.

The LVDT (Figure 12.1[b]) works on the principle of a moving iron core changing the property of the transformer. A transformer is a device with a primary winding and secondary winding. The voltage at the secondary winding varies with the primary voltage at a fixed ratio. Also, the transformer operates only with ac electricity. The position of the iron core changes the primary/secondary voltage ratio. The LVDT has two secondary windings, so the two secondary voltages differ from one another, depending on the iron core position.

When force is applied to a solid object, the object stretches, although the stretch may be microscopic. A strain gauge pad (Figure 12.1[c]) is a resistor whose resistance varies with the amount of stretching (strain). Typically, a strain gauge is mounted at the place where force is exerted. The change in resistance is detected by measuring the voltage change in a Wheatstone bridge circuit (Figure 12.1[d]). Wheatstone bridges or their variations are often used in instrumentation systems because constant-current sources are not required.

Thermocouples output a voltage that is related to the temperature at the tip of the probe (Figure 12.1[e]). They work on the principle of the *Seebeck effect.* If two dissimilar metal conductors are joined together at each end, current will flow if the two ends are at different temperatures. Thus, a difference of temperature at the two junctions generates a voltage. This voltage is very minute (in millivolts). The relationship between voltage and temperature is nonlinear. Hence, a microcontroller will have to use a look-up table to find the temperature corresponding to a millivolt signal.

A phototransistor (Figure 12.1[f]) acts like a variable-current source. The amount of current varies with the light intensity. This is detected as a voltage change at the external resistor. It can sense position because an object blocks a varying amount of light, depending on its position. You may recall the phototransistor as being part of an optical switch (Figure 11.8). In Chapter 11 it was used to generate pulses. In this case the phototransistor detected only the presence or absence of light.

A thermistor is a device whose resistance varies with temperature. A photoconductive cell is a device that conducts more current when more light strikes it. Hence, it too, acts like a variable resistor. For both devices resistance change is detected by sensing voltage using a potentiometer circuit or Wheatstone bridge circuit.

In systems that control electrical machinery, it is particularly important to monitor current used by the machine or machines. The traditional way is to use resistors with a low resistance value but high power rating. The controller measures the voltage drop across the resistor to calculate current using Ohm's law. A variation is to use a current transformer to induce a proportional amount of smaller current through a resistor.

A more recent method is to use a current-sensing power MOSFET (a type of transistor) known as a SENSEFET. It splits the total machine current into power and sense components. The sense current may be one-thousandth of the value of the power current. A control system can measure the sense current using a resistor to give an indication of the power current. Reference 13 in Appendix C provides more information.

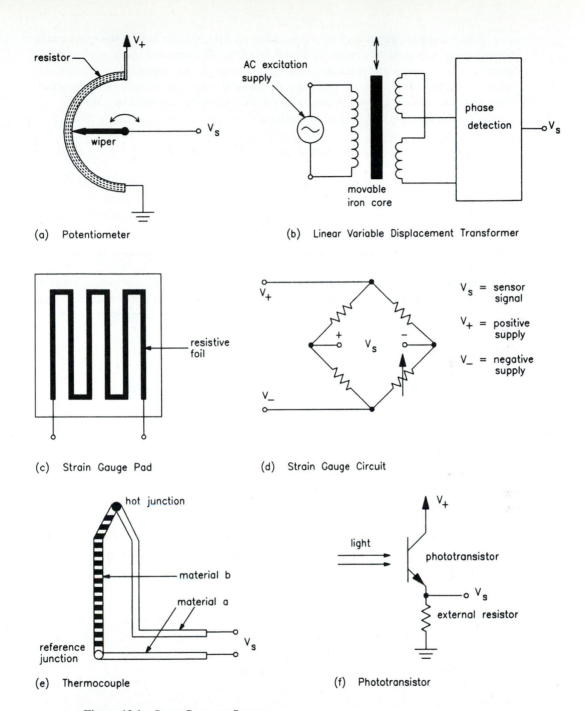

(a) Potentiometer

(b) Linear Variable Displacement Transformer

V_s = sensor signal

V_+ = positive supply

V_- = negative supply

(c) Strain Gauge Pad

(d) Strain Gauge Circuit

(e) Thermocouple

(f) Phototransistor

Figure 12.1 Some Common Sensors

Since a sensor output is often connected to an A/D, we can look at the relationship between physical measurements and digital numbers. The sensor's measurement range specifies the minimum and maximum limits it can measure. Its output range is the analog electrical signal it outputs to the A/D. Thus, we can also relate specifications such as offset, span, step size, weight, and resolution to the physical quantity measured by a sensor.

The general form of a sensor output is

$$\text{analog number} = m \times \text{measurement} + k \tag{12.4}$$

where m is a slope constant and k is an offset constant. To find the analog output for any measurement within the range, you will have to find m and k. To do this, you can solve two simultaneous equations. In general, these will be the minimum and maximum of the range.

PRACTICE

A temperature sensor has a measurement range of -10 to $+140°C$. The output range is -2.5 to $+5.0$ V. The sensor is connected to an 8-bit A/D. Indicate the offset, span, step size, and resolution. Also, what is the digital output of the A/D if the temperature is $+10°C$?

Solution

Offset: -2.5 V, $-10°C$

Span: 5.0 V $- (-2.5$ V$) = 7.5$ V, $140°C - (-10°C) = 150$ C$°$

Step size: $\dfrac{7.5 \text{ V}}{256} = 29$ mV, $\dfrac{150 \text{ C}°}{256} = 0.59$ C$°$

Resolution: 29 mV at 8-bit, 0.59 C$°$ at 8-bit

Find the analog output of the sensor, then find the digital output of the A/D. To find the analog output of sensor, solve two simultaneous equations.

$$140°C\, m + k = 5 \text{ V} \tag{a}$$
$$-10°C\, m + k = -2.5 \text{ V} \tag{b}$$

Subtract (b) from (a) to get

$$150°C\, m = 7.5 \text{ V}$$

Then

$$m = \frac{7.5 \text{ V}}{150°C} = 0.05 \text{ V/C}°$$

Then solve for k:

$$140°C \times (0.05 \text{ V/C}°) + k = 5 \text{ V}$$
$$7 \text{ V} + k = 5 \text{ V}$$
$$k = 5 \text{ V} - 7 \text{ V} = -2 \text{ V}$$

The analog output is

$$(0.05 \text{ V/C}°) \times 10°\text{C} - 2 \text{ V} = -1.5 \text{ V}$$

Digital output from equation (12.2) is

$$\frac{-1.5 \text{ V} - (-2.5 \text{ V})}{0.029 \text{ V}} = 34.8$$

$$\cong 34 \text{ (rounded to an integer)}$$
$$= \$22 \text{ (the hex number)}$$

12.2.2 Signal Conditioning

This is sometimes necessary because raw sensor outputs are not always suitable for analog-to-digital conversion. Signal conditioning circuits typically amplify the raw signal from a sensor. For example, an A/D will not be sensitive enough to convert thermocouple signals accurately. Signal-conditioning circuits also provide buffering so that the sensor signal is not affected by anything else connected to the circuit.[2]

Most signal-conditioning circuits employ operational amplifiers (op amps). These are shown schematically as trianglar symbols, as illustrated in Figure 12.2. The inverting and noninverting amplifier circuits are useful for boosting weak sensor signals so that an A/D can convert them into digital values.

The differential amplifier outputs a signal proportional to the difference of two voltage signals. This is useful for converting 4- to 20-mA current signals to voltage. Also note that many standard process control instruments use a 4- to 20-mA standard. Conversion to voltage is accomplished by connecting a resistor across the two terminals. The current flowing through this resistor causes a voltage drop that is measured by the differential amplifier. Note that none or very little of this current flows into the amplifier circuit itself. Other examples are differential voltage signals, which appear at thermocouple leads or electrocardiogram electrodes. Both signals are relatively weak and require significant amplification.

Differential amplifiers should have minimal common-mode gain. This refers to the output amplification with zero differential input. For example, if both inputs are zero, the output should be zero. If both inputs are 2 volts, hence zero differential input, the output should also be zero. In practice, differential amplifiers do have outputs that also depend on the absolute input values as well as the difference. Ideally, they should have no common-mode gain at all. The differential amplifier shown in Figure 12.2(c) has a potentiometer with a resistance value close to that of the feedback resistor (R_f). It is adjusted to reduce common-mode gain. One of the specifications of amplifiers is their common-mode rejection ratio (CMRR), which is the differential gain divided by the common-mode gain. This should be made as high as possible.

---

[2]More precisely, the conditioning circuit minimizes circuit loading to the sensor because of the high-impedance input of the conditioning circuit. It also provides a low-impedance output.

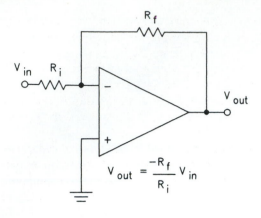

$$V_{out} = \frac{-R_f}{R_i} V_{in}$$

(a) Inverting Amplifier

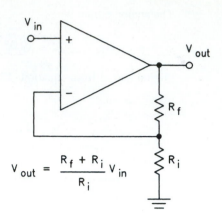

$$V_{out} = \frac{R_f + R_i}{R_i} V_{in}$$

(b) Noninverting Amplifier

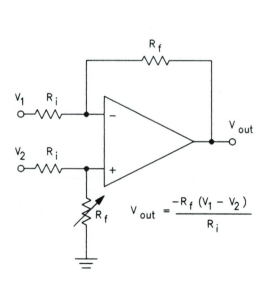

$$V_{out} = \frac{-R_f (V_1 - V_2)}{R_i}$$

(c) Differential Amplifier

$$V_{out} = -R_f \left(\frac{V_1}{R_1} + \frac{V_2}{R_2} + \frac{V_3}{R_3} \right)$$

(d) Summing Amplifier

Figure 12.2 Amplifier Circuits for Signal Conditioning

The summing amplifier is useful for introducing offsets. A two-input (no V_3) summing amplifier can have a fixed offset as one of its inputs and the other as the variable signal. This is useful for A/Ds that have limited input ranges. For example, the A/Ds used by the 68HC11 cannot input negative voltages. The summing amplifier connected to an inverting amplifier can scale a range such as -5 to $+5$ V to the range 0 to 5 V. Both input resistors (R1 and R2) would have a value of twice the R_f value. One input would be fixed at $+5$ V to add the required offset.

478   Chapter 12

Another classification of signal-conditioning circuits comprises the filters. Many signals tend to be noisy. If the signal is known to be restricted to certain frequencies, the unwanted signals at other frequencies can be removed by using filters. Low-pass filters are most commonly used since many analog signals are at lower frequencies. The low-pass filter will allow the low-frequency component to pass through. It will remove the higher-frequency noise components. For example, a temperature signal changes slowly over time. There should not be any sudden (higher-frequency) changes. The filter can remove noise introduced by electromagnetic interference.

Passive filters are made using passive components such as resistors and capacitors. Active filters use operational amplifiers and passive components. Filter design is a subject of its own. The performance of filters varies with the number of components and their relative values. References 5 and 10 in Appendix C describe filters in more detail. It is also possible to filter signals digitally using software known as digital filters. In Section 13.5 we will describe digital filters.

12.2.3 Actuators

Some common actuators are solenoids, relays, *Darlington* transistors, triacs, and silicon-controlled rectifiers (SCRs) or thyristors. In Figure 12.3(a) an increasing analog signal at the gate of the MOSFET increases the amount of current drawn through the dc motor shunt field. This type of control using the field current is called *field control*. As field current is increased, the motor speed decreases. A shunt dc motor has the following speed equation:

$$\text{speed} = \frac{V_a - I_a R_a}{\text{flux}} \tag{12.5}$$

where V_a is the armature voltage, I_a is the armature current, and R_a is the armature resistance of winding. Flux is a magnetic field property analogous to current for electricity. It is proportional to the field current over a limited range. If the field current is lost, the dc motor will speed up into a dangerous condition known as *runaway*. A practical field control circuit must include some protection against field current loss.

As the voltage at the MOSFET gate (see the lower transistor in Figure 6.2) is lowered, the resistance between the drain to source increases. For limited conditions, the resistance of a MOSFET is proportional to the voltage between drain to gate. This means that a MOSFET can also be used as a potentiometer to act as a voltage divider. Hence, a D/A can output a signal to the MOSFET gate to control the output level of amplifier circuits: for example, the volume control of an audio amplifier.

The MOSFET transistor can also be controlled digitally (on–off) as a switch using pulse-width modulation (PWM). A low gate signal turns off the switch and a high signal turns it on. Changing the duty cycle when using PWM effectively changes the average amount of field current. Recall our discussion of PWM in Section 11.2. Also, in Section 6.1.2 we discussed the use of transistors as digital switches. In fact, many actuators can be controlled using PWM since they are digital (on or off) by nature. Instead of using a physical D/A, the PWM output is used to effectively change the dc output value. In the next section we explain some of the theory behind this.

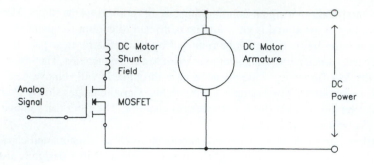

(a)   Power MOSFET Transistor Controlling a DC Motor

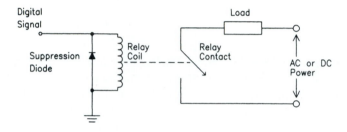

(b)   Relay Circuit for On/Off Control

Figure 12.3   Two Common Actuators

A relay is one such device. It is an electromagnetic switch with a coil and one or more contacts. Applying voltage to the coil (energizing the relay) will cause any *normally open* contacts to close and any *normally closed* contacts to open. Again, using PWM changes the duty cycle of contact closure, which will effectively change the dc current value drawn through the load (see Figure 12.3[b]). When its contacts must carry large amounts of current, the relay is known as a *contactor.*

The suppression diode is used to eliminate inductive kickback. When a coil is deenergized, the current rushes out and causes a high-voltage spike. Such a spike at an IC pin can damage it. Any inductive load should have some form of suppression such as a diode mounted as close as possible to the coil. An inductive load is any electromagnetic coil used in devices such as relays and motors.

A solenoid is like a relay. It has a coil but, instead of moving electrical contacts, it moves a mechanical object such as a cylinder, plunger, or valve stem. A Darlington transistor is a combination of two bipolar transistors designed to switch large amounts of current. Triacs and SCRs are semiconductor devices used to switch ac currents.

■ 12.3 DIGITAL-TO-ANALOG CONVERTERS

Many digital-to-analog converters (D/As) use an R–$2R$ ladder network. It is called that because the resistors are connected in a pattern that resembles a ladder. The resistors have one resistance value or twice that value. Figure 12.4 shows an example. It is a type known as an inverted ladder circuit. Note that it is similar to the summing amplifier shown in Figure 12.2(d). A D/A IC has the ladder circuit, switches, and operational amplifier built in. The switches are analog switches controlled by digital signals. Data sheets for different D/As may show this circuit or variations of it. Figure 12.4 serves to illustrate the conversion method.

The output voltage (V_O) is proportional to the binary input. Each branch of the ladder network contributes current whose value is proportional to the bit weight of that branch. The amplifier circuit sums the current components to produce a voltage proportional to the binary input.

If you do not have an electronics background, you may wish to skip the next three paragraphs. They present a more detailed explanation of the circuit.

Circuit Explanation. The resistors are precision resistors of value R or twice the value of R ($2R$). The value of R typically ranges from 100 Ω to 1 k Ω. Each bit (b_i) controls a switch. If the bit is a 0, the switch connects a $2R$ resistor to ground. If a bit is a 1, the corresponding $2R$ resistor connects the branch to the input of the operational amplifier. When an operational amplifier operates in the linear range, the negative input is *virtually* equal to the positive input. Hence, a switch connects a branch to ground when its bit is 0. It connects a branch to virtual ground and to the amplifier input when the bit is 1.

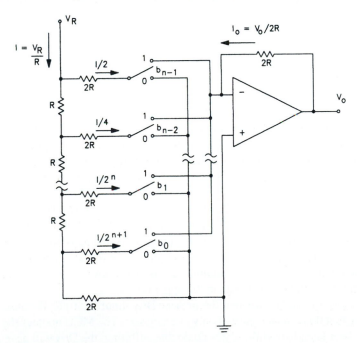

Figure 12.4 Inverted R–$2R$ Ladder Circuit

The operational amplifier circuit with the feedback resistor (2R) acts like a current amplifier. With all negative (branch) inputs at virtual or real ground, the output voltage is the total current multiplied by the 2R value. The equivalent resistance of the entire ladder network is 1R. You should be able to verify this using the basic series and parallel resistor connection equations. Also note that the input impedance of an operational amplifier is very high. With V_R as the reference (input) voltage, the total current is V_R/R using Ohm's law.

The ladder network is also equivalent to two 2R resistors in parallel. Hence, half the total current flows into the branch corresponding to b_{n-1} for the MSB. The other half flows into the rest of the network. Similarly, the next significant branch takes half of this current. This halving process repeats for each branch for successively lower bit weights.

A Parallel D/A Interface. Digital-to-analog converter ICs are available with different bit resolutions. We will consider interfacing a 12-bit D/A to an 8-bit MCU. This will serve to show how to mate different bit resolutions with different-size data buses. Also, it will illustrate *double buffering.*

Figure 12.5 shows an interface for an MCU operating in expanded multiplexed mode. A D/A has two sets of pins. One set is used for the digital connections. The other is used for the analog signal. To drive the internal R–$2R$ network and amplifier, most D/As require a negative voltage (V_{EE}) for current sinking. The analog range is selected by connecting the SPAN and OFFSET inputs to voltage references such as potentiometers and zener diodes. For ranges beyond that which can be handled by the D/A, additional signal-conditioning circuitry (see Section 12.2.2) can scale the output to produce another range. If the ANALOG OUT signal is also proportional to the supply voltage (V_{CC}), the D/A is known as a multiplying D/A.

The D/A shown in Figure 12.5 has an internal latch. When its latch line ($\overline{L}$) is driven low by a control signal from the MCU, the D/A latches in the current digital input and converts it to analog. The interface circuit shown in Figure 12.5 employs an additional latch. It serves two purposes. One is to convert a sequence of 2 bytes into one 12-bit word. The other is to provide double buffering.

To output a 12-bit word, the MCU writes the data in two steps. First, it puts the upper 4 bits on the lower nibble of the data bus. Then it selects the latch address. The address decoder (not shown) decodes the address and drives the latch ($\overline{CS}$) inputs low. Also, the MCU drives address line A0 low (an even address) so that only the 4-bit latch accepts data from the data bus. Then the MCU repeats the sequence for the lower 8 bits except that it drives A0 high (an odd address).

The other purpose of the latches is to provide the double buffering to which we alluded previously. This means that the D/A input updates during an additional step instead of when data is written to the latches. In this case the updated latch data is not transferred to the D/A until the MCU drives the control line to latch in data to the buffer in the D/A. Remember that the D/A does not change its output until $\overline{L}$ is driven low. As we shall demonstrate, this is important when there is a change in the upper nibble.

Consider the situation when the present D/A input is $0FF. The next input will be $100. Note that this is an increase of only one step size. The MCU updates the input by writing to the 4-bit latch first. If there was no double buffering, the D/A will have $1 in the 4-bit latch with $FF in the 8-bit latch before the update is complete. The result is a spike in the D/A output. With double buffering the spike can be avoided by enabling the second latch after the first latch has been updated.

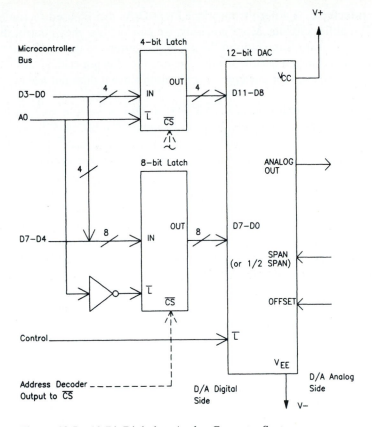

Figure 12.5 12-Bit Digital-to-Analog Converter System

Most 8-bit MCUs have an instruction to transfer 2 bytes using one instruction. For the 68HC11 the store double accumulator (STD) instruction can be used. Accumulator D holds the 12-bit word. The instruction is used with an even address operand (A0 = 0), which is the D/A address. Consider the following instructions:

```
        STD     DA_ADDR         ;send data to D/A #1 latch
*                               ;DA_ADDR is an even address
        LDAA    PORTB,X         ;(maintain old port B data)
        STAA    PORTB,X         ;pulse STB line which is connected
*                               ;to L of D/A
```

When the STD instruction puts the accumulator A contents on the data bus, the address bus has the value DA_ADDR to drive A0 low. The next cycle puts the accumulator B contents on the data bus and the address bus has the value of DA_ADDR + 1 to drive A0 high.

If the D/A latches in data only when its latch input sees a rising edge, the design can be economized by connecting the latch input to the chip selects of the first latches. The address decoder output has a rising edge only after the last cycle of the STD instruction.

A Serial D/A Interface. Earlier we mentioned that D/As can be used to control the output level of amplifier circuits. More precisely, we can say that they control the amplifier *gain,* where the gain is the ratio of the output magnitude to the input magnitude. This type of output is changed once in a while; hence, speed is not critical. A D/A with a serial interface can be used for this type of application since it does not need to be updated very often.

You may recall our discussion of the MC144110 D/A in Section 10.7.5 when describing SPI transfers to peripherals with odd word lengths. The MC144110 D/A has six converter channels that convert a 6-bit word into analog voltage. If desired, two channels can be combined to make one 12-bit channel. The channel output representing the upper 6 bits can be connected to the input of a summing amplifier with resistance 1R. The other output is connected to a summing input with resistance 64R. The summing amplifier output may be connected to the input of an inverting amplifier to restore signal polarity.

The circuit designer may choose other methods to build a 12-bit D/A interface because the method just described uses several components. It must also use precision resistors for the summing amplifier. This is to ensure that each upper word bit has a weight 64 times that of the corresponding lower word bit.

Waveform Generation. With a D/A a processor can, theoretically, generate any kind of waveform. Each new output can be calculated or be looked up using a table of values stored in memory. Since the outputs are in discrete steps, low-pass filters at the output can smooth the waveform. However, there is a signal frequency limitation. It takes time for a controller to output each new output value.

Even with a look-up table, it will take the 68HC11 twenty or more instruction cycles to update each new output. Assuming that the system uses 256 steps to represent the swing from minimum to maximum signal value, it will take 2.56 ms ($256 \times 20E \times 0.5$ μs/E) for each waveform cycle. The maximum signal frequency is then roughly 390 Hz. This is fastest if the 68HC11 does nothing but generate a wave. This is a waste of resources, although it is useful for experimentation. Faster frequencies can be obtained by using fewer steps; hence, the MCU updates the output less often per cycle. The output is then sent to a low-pass filter for smoothing.

Another solution is to use a processor with a faster clock, but this does not help much if very high signal frequencies are required. Some useful waveforms can be generated by using the timer subsystem to generate periodic pulses, which can then be conditioned using analog circuits. Another option is to use special processors that are designed to transfer data quickly to D/A outputs. These are digital signal processors (DSP). We will look at DSPs briefly in Chapter 13.

PRACTICE

Write the pseudocode to generate a sine wave.

Solution

Use an output compare interrupt service routine and a look-up table.

```
timer interrupt entry point
        set up next sample time
        D/A = *tableptr
        tableptr++
        if tableptr >= maxptr
                then tableptr = minptr

        exit
```

PRACTICE

Develop the software to output a 60-Hz sine wave. Assume that vectors, symbols, and tables have been set up. Also, the required initialization has been executed.

Solution

Use a table with 64 samples of one sine-wave cycle. A 60-Hz wave requires 16,667 E cycles (2-MHz E clock). Hence, interrupt should occur every 260 E cycles.

```
*Solution

*some static variables in Page0
        ORG     0

HDLY    FDB     260             ;sample interval
SINPTR  RMB     2               ;stores pointer to sine look-
                                up table
SINTAB  EQU     $B600           ;sine look-up table
*using OC2 interrupt
RTOC2
*executes every 260 E cycles
*() parentheses indicate E cycles per instruction
*(15) overhead to enter interrupt routine

        LDD     HDLY            ;(4), schedule next sample
                                time
        ADDD    TOC2+REGBAS     ;(6)
        STD     TOC2+REGBAS     ;(5)
        LDAA    #$40            ;(2), clear flag OC2F
        STAA    TFLG2+REGBAS    ;(4)
        LDX     SINPTR          ;(4), get sine table pointer
        LDAA    0,X             ;(4), look up sine value
        STAA    PORTB+REGBAS    ;(4), sent it to D/A
        INX                     ;(3), point to next cycle
                                sample
        CPX     #SINTAB+64      ;(4), begin new cycle if
        BLT     RTOC2A          ;(3), current cycle complete
```

```
            LDX        #SINTAB              ;(3)
RTOC2A
            STX        SINPTR               ;(4), update pointer
            RTI                             ;(12), return
```

Maximum time for RTOC2 to execute is 77 E cycles, which gives the MCU some time to do other things before another update is required. Note that each cycle requires 16,640 E clocks (64 × 260) and a 60-Hz cycle requires 16,667 E clocks. Hence, the generated wave is slow by 0.16%. The local power company typically supplies 60 Hz at higher accuracy than this.

PWM Outputs and Fourier Series.   In our discussion of output compare functions in Section 11.2, we explained that pulse-width modulation (PWM) can be used as a D/A. Sending a PWM signal to a low-pass filter will produce an analog signal proportional to the duty cycle at the output of the filter. The reason this works is based on the mathematical principle of the *Fourier series.*

Any periodic waveform can be expressed as an addition of different sinusoidal waveforms. This includes the sine wave with a frequency of zero. A zero-frequency sine wave (or any other wave) is a signal that does not change with time. In other words, a periodic signal has a dc component. The other frequency components are integer multiples of the waveform's *fundamental frequency.*

Consider a square wave with a frequency of 1 kHz. Its fundamental frequency is 1 kHz. The other frequency components will be 2, 3, 4, and 5 kHz, and so on, until infinity. These are also known as harmonics.[3] It turns out that half-wave symmetric waveforms such as a square wave have zero magnitude for the even harmonics. Each higher odd harmonic has a successively lower magnitude. In theory, you could produce a square wave using only a dc signal, sine waves at the harmonic frequencies, and a summing amplifier. We say "in theory" because the summing amplifier would need an infinite number of inputs.

We have just said that a square wave (and PWM) has a dc component with sinusoidal harmonics. A low-pass filter can remove the harmonic components, leaving only the dc. This is the analog output. The dc component is proportional to the duty cycle. Since a low-pass filter is used, it is not possible to change the analog output faster than a fraction of the original signal's period. This is fine for slowly changing analog output requirements: for example, volume control of an audio amplifier using a MOSFET (see Section 12.2.3). Some microcontroller data sheets also label their PWM output pins as D/A outputs because the duty cycle is effectively related to the analog value.

On-Chip D/A.   Some versions of the 68HC11 have a built-in D/A subsystem that converts 8-bit digital data to a corresponding analog voltage. The MC68HC11 N-series can use port G pins PG4 and PG5 as D/A output channels. To enable a channel, set the corresponding enable bit in the D/A converter control register (DACON). Then write a byte to the D/A data register for channels 1 or 2 (DA1 or DA2). The subsystem outputs an

---

[3]It is the different magnitudes of harmonics that make the same musical note from one instrument sound different from that of another. For example, middle C from a piano certainly differs from a guitar's middle C. Cheap electronic music sounds artificial because of the lack of harmonics.

analog voltage between supply voltages V_{SS} and V_{DD} based on the byte in the data register. It works on the principle of the $R-2R$ ladder circuit described earlier. The maximum output current is 1.0 A at 5 V. If higher currents are required, you will have to connect the D/A output channel to the input of an amplifier circuit for signal conditioning (see Figure 12.2).

■ 12.4 ANALOG-TO-DIGITAL CONVERTERS

To read information about the outside world, a microcontroller inputs voltage signals from sensors.

Different Techniques.   There are several techniques used by analog-to-digital converters (A/Ds) to perform the conversion. The following summarizes the common methods with their characteristics. Some other methods are a hybrid of the techniques listed. References 9 to 11 in Appendix C give more information about the following and other techniques.

Successive Approximation.   This method is common in computer systems. It has a medium conversion speed (nominally 20 s). This is the technique we will concentrate on shortly.

Integration.   The integration method is used in digital meters. It has good accuracy and noise immunity, although it is relatively slow.

Flash or Simultaneous.   Since a flash A/D has a fast conversion rate, it is used for signal processing. However, a flash A/D is expensive because it uses a lot of hardware.

Sigma-Delta.   This is a more complex technique with very good resolution. Essentially, the A/D samples the analog signal at a very high rate (oversampling). Then it converts each sample to a 1 or 0 bit based on how the sample changed relative to the previous sample. Next, the converter uses a digital decimation filter to convert each group of bit samples to a high-resolution multibit sample. The result is a slower rate sequence of high-resolution samples. We describe digital filters in Section 13.5.

Successive Approximation.   This is the most common technique used in general-purpose microcontroller applications. The 68HC11 internal A/D uses a variation of this technique. It is important to have some understanding of the technique since its limitations will have an effect on the accuracy of the data obtained.

Figure 12.6 illustrates the technique. The A/D has a built-in D/A. A sample-and-hold circuit stores an analog input. The A/D logic steps through a sequence of trial-and-error guessing to find the digital equivalent of the input. It begins the sequence by sending a digital signal that is at midrange to the D/A. The analog output (V_{OUT}) from the D/A is compared to the analog input (V_{IN}) from the sample-and-hold circuit. Thus the A/D determines (by using C_{OUT}) whether the analog input is above, at, or below half scale. If the input is in the upper half, the A/D will send a signal to the D/A that is halfway between the upper half (three-fourths of range). If the input is in the lower half, the A/D will send a signal to the D/A that is halfway between the lower half (one-fourth of range). It continues determining to which half of the next range selection the analog input belongs. The A/D continues this sequence until the least significant bit is determined. Thus, it successively narrows the possibilities until it finds the closest binary match.

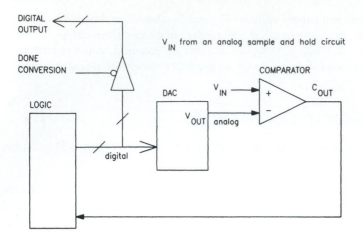

(a)   Block Diagram of Successive–Approximation
       System

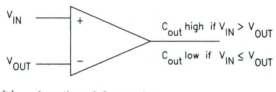

(b)   Operation of Comparator

Figure 12.6   Analog-to-Digital Conversion Using Successive Approximation

When the approximation sequence is complete, the A/D enables the digital signal to the digital output port. In Figure 12.6(a) this is accomplished by driving the DONE CONVERSION signal low to enable the output. The number of steps required to perform the conversion is equal to the bit resolution. For example, an 8-bit A/D requires eight steps for the conversion process.

Figure 12.7 serves to illustrate an example of a conversion sequence. To keep the example simple, we use a 6-bit A/D with the range 0 to 5 V. The principles will also apply to A/Ds with different bit resolutions and ranges. In Figure 12.7 the first step tells the A/D that the analog input is in the upper half; hence, the MSB must be high. The sequence continues until it obtains the closest possible result of %110101. Hence, the A/D outputs a digital number that is equivalent to an analog signal of 4.14 V, which differs from the actual input of 4.16 V. This difference is smaller than the step size of 0.078 V.

In fact, the system has an inherent inaccuracy of plus one step size. For example, if the input had been 4.218 V, it would still output %110101 instead of %110110. Note that %110110 corresponds to an analog value of 4.21875 V. This system can be modified to provide an accuracy of plus or minus half a step size by applying an offset of minus half a step size to the D/A output.

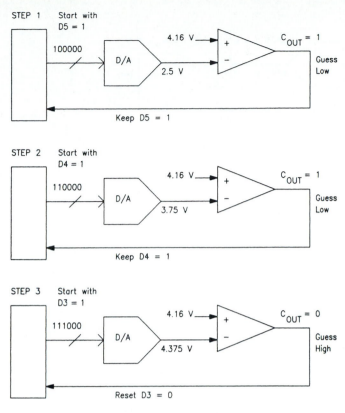

Figure 12.7 Successive-Approximation Example *(Continued)*

Since the conversion process is a sequence of steps, the conversion time depends on the time required for each step and the number of steps. A successive approximation A/D IC requires a clock input to drive it through the steps. Often, an A/D IC uses an external resistor (R) and capacitor (C) to drive an internal oscillator to provide the clock pulses. This is known as an *RC clock*. An A/D IC will have an input pin as a start conversion signal and an output pin to indicate when conversion is done. Often, these pins can be tied together so that the chip converts continuously.

Charge Redistribution A/D. The A/D in the 68HC11 uses a conversion technique similar to the successive-approximation technique just described. Its A/D uses a network of capacitors connected to a comparator. The capacitors are switched into different configurations. For each conversion, the system switches into a sample mode, hold mode, and approximation mode. During the sample mode, the capacitors are charged. This total charge is proportional to the input analog signal. For the duration of the conversion process, this total charge remains conserved. The A/D switches to the hold mode and then to the approximation mode. The approximation mode is a sequence of successive-approximation

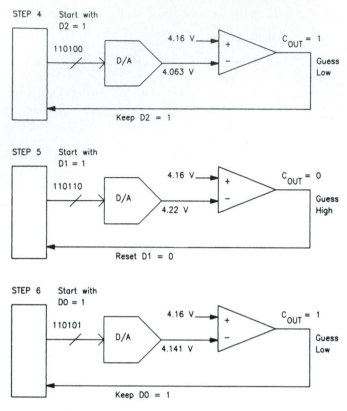

Figure 12.7 Successive-Approximation Example *(Continued)*

steps that switches different capacitors to modify the input voltage to the comparator. The result of each step sets or resets a corresponding bit in a successive-approximation register (SAR). Section 12.1 in the *Reference Manual* explains this procedure in great detail. It also explains how the system can convert with an accuracy of plus or minus half a step size.

■ 12.5 THE BUILT-IN A/D SUBSYSTEM

Analog converter software is relatively easy compared to that required for other I/O subsystems. It has eight inputs at port E called *channels*. Depending on configuration of subsystem registers, the software can read any of the eight possible inputs.

Charge Pump and Oscillator. The software initialization procedure for the A/D includes enabling the charge pump and defining the conversion clock source. To charge the capacitors for the successive-approximation circuit, the 68HC11 uses a charge pump. A charge pump is also a system of switching capacitors to redistribute a stored charge. You may recall that the EEPROM programming subsystem (Section 6.6.1) and the MC145407 Driver/ Receiver IC (Section 10.1.1, RS-232 Interface Standard) also use charge pumps. The charge pump for the A/D can develop 7 or 8 V so that analog signals as high as 6 V can be converted.

To use the A/D, the software must enable the charge pump. To enable the charge pump, it sets the A/D power-up bit (ADPU) in the system configuration options register (OPTION). Then the MCU should wait at least 100 μs. It is only necessary to enable it once after a power-on reset ($\overline{\text{RESET}}$). Some 68HC11 systems disable the charge pump to conserve power when analog data is not required.

The successive-approximation circuit can use the system clock (E) if it is greater than 750 kHz. This low-frequency limit is necessary to minimize charge leakage. The software selects E as the clock source by clearing the clock select bit (CSEL) in register OPTION. To select an internal *RC* oscillator (approximately 2 MHz), it sets bit CSEL. Bit CSEL also selects the clock source for the EEPROM programming subsystem (Section 6.6.2). Motorola recommends that CSEL be set for A/D conversions if the E clock is slower than 750 kHz.

Remember, if you intend to use the A/D, your application program must specify the clock source and enable the charge pump. You can use an initialization subroutine as shown in Listing 12.1. Because μ(mu) is not an ASCII character, we use us to represent μs (microseconds).

```
*Listing 12.1
*Initialization subroutine for A/D
*-----------------------------------

        ORG       $180


*Subroutine STARTAD
*Powers up A/D. Must be run before using it
*after reset or a power-down sequence.
*In this case, uses E clock as source.
*No calling or return registers except CCR affected

STARTAD
        PSHX                       ;preserve IX
        LDX       #REGBAS          ;point to registers
        BSET      OPTION,X $80     ;ADPU = 1 to power up
        BCLR      OPTION,X $40     ;CSEL = 0 to select E clock
        JSR       DLY100           ;wait 100 us
        PULX                       ;restore IX
        RTS                        ;return
```

Reference Voltages and Channels.   Refer to the pins assignment diagrams in the Quick Reference (Appendix B). The voltage at the analog reference high pin (V_{RH}) sets the analog value for digital data $FF. In other words, it sets the upper value of the analog range. Similarly, the voltage at the analog reference low pin (V_{RL}) sets the low limit of the range, the analog value of $00. The analog reference limits are

$$V_{RH} \text{ maximum } V_{DD} + 0.1 \text{ V}$$
$$V_{RL} \text{ minimum } V_{SS} - 0.1 \text{ V}$$

The suggested minimum span is

$$V_{RH} - V_{RL} = 3.0 \text{ V}$$

TABLE 12.3 REGISTER ADCTL SETTINGS FOR A/D SINGLE CHANNEL SELECTION (MULT = 0)

CC	CB	CA	Channel
0	0	0	AN0
0	0	1	AN1
0	1	0	AN2
0	1	1	AN3
1	0	0	AN4
1	0	1	AN5
1	1	0	AN6
1	1	1	AN7

The A/D subsystem can use any of the eight port E inputs to read analog signals. Each pin is known as a channel. Pin PE0 is channel zero (AN0), and so on. Channels AN4 to AN7 are not available for the DIP package of the 68HC11 because pins PE4 to PE7 are not included.

The subsystem can operate in a single-channel or multichannel mode as selected by the multiple-channel control bit (MULT) in register ADCTL. For either method the subsystem performs four conversions requiring 32 cycles each. At the end of each conversion it puts the digital data in a result register. The result registers are updated in sequential order (ADR1, ADR2, ADR3, ADR4). The setting of the bits in the control register (ADCTL) determines how this is done. After conversion is done, the MCU sets the conversion complete flag (CCF).

Single-Channel Operation.   If the software clears bit MULT, the subsystem operates in single-channel mode. Only one channel is used and its result is placed in all four result registers when conversion is complete. Table 12.3 shows the channel select bit settings that determine which channel has its input converted to digital.

Actually, the subsystem performs four consecutive conversions and places their results sequentially in each result register. After the four conversions are complete, the A/D sets the conversion complete flag (CCF) in control register ADCTL. Hence, the A/D control register is also a status register. The software can then read any result register or read all of them to obtain an average. The subsystem can perform the four conversions once or perform them continuously. This is determined by the continuous scan control bit (SCAN) in control register ADCTL. If SCAN is 0, the A/D performs four conversions and then stops. With SCAN set to 1, the A/D performs conversions continuously.

To start conversion, the application program writes to control register ADCTL. To read the result, the program must poll flag bit CCF (no interrupt available) to determine when the conversion is done. Once the flag has set, the program reads any result register. For continuous scan conversion, the application program can obtain a new reading simply by reading a result register at any time. If continuous scan was not selected, the software must repeat the procedure to start conversion and wait for the flag to set. Flag CCF is cleared automatically every time (any) data is written to register ADCTL.

Listing 12.2 shows a subroutine that can be used for single-channel operation.

```
*Listing 12.2
*----------------------------------

        ORG     $180
```

```
*Subroutine INAD
*I/O handler routine for analog input
*using continuous scan. Calling program
*must execute this subroutine after subroutine
*STARTAD (Listing 12.1) and before reading
*analog result.
*Calling registers
*        IX = register block address
*        ACCA = channel number from 0 to 7
*Return registers
*        ACCA = analog result
*        CCR affected

INAD
        ANDA    #$07            ;ensure that bits 3 to 7 clear
*                               ;note that others select channel
        ORAA    #$20            ;SCAN=1, MULT=0
        STAA    ADCTL,X         ;start conversion
*                               ;also clears CCF
*wait for first conversion done
INAD1
        BRCLR   ADCTL,X $80 INAD1
*                               ;CCF set at this point
        LDAA    ADR4,X          ;read any result register
*                               ;in this case, the most recent
        RTS                     ;return
```

After the application program has called the subroutine once, it can read any result register at any time to obtain an analog result. In the worse case, the result will be 128 cycles old because it takes 32 cycles to make each of the four conversions. To read a different channel the application program will have to set up the calling register and call subroutine INAD again. For non-continuous conversion, a similar subroutine can be used. The subroutine will not have the ORAA instruction. The application program will have to call it each time it wants an analog input.

PRACTICE

Interrupt service routine RTOC3 calls subroutine AN2SAM to get the analog data at channel AN2 every 20,000 clock cycles. Write the code for subroutine AN2SAM. Assume that the initialization part of the main program has called subroutine STARTAD (see Listing 12.1).

Solution

```
*Subroutine AN2SAM
*read analog input at channel AN2
*No calling or return registers

        LDX     #REGBAS
        LDAA    #$02            ;configure for single channel (AN2)
*                               ;single scan
```

```
            STAA      ADCTL,X      ;start conversion
*                                  ;also clears CCF
*wait for conversion done
AN2SAM1
            BRCLR     ADCTL,X $80 AN2SAM1
*                                  ;CCF set at this point
            RTS                    ;return, most recent result in
                                   ADR4
```

PRACTICE

Write an instruction sequence to store 512 AN2 samples in RAM. Assume that these are the only instructions that set flag CCF. The system could be an EVBU board modified to use external RAM.

Solution

```
*instruction sequence to store 512 analog samples
            ORG       $D000           ;location of external RAM
BLOCK                 RMB             512
            ORG       $100
            LDX       #REGBAS
            LDY       #BLOCK
REPEAT
*                     ;wait until RTOC2 completes A/D conversion
            LDAA      ADCTL,X         ;if CCF == 1, conversion done
            BITA      #$80
            BEQ       REPEAT
            STAA      ADCTL,X         ;then clear CCF
            LDAA      ADR4,X          ;and store result
            STAA      0,Y             ;and point to next location
            INY
            CPY       #BLOCK+512      ;until block full
            BLT       REPEAT
DONE
            BRA       DONE            ;stop here for now
```

Note that if other routines could cause CCF to set, each routine should set a unique flag in RAM to identify which one completed its task.

Multichannel Operation. If the software sets bit MULT, the subsystem operates in multichannel mode. In this mode either the first four channels (AN0 to AN3) or the last four channels (AN4 to AN7) are converted. The subsystem sets the conversion complete flag after the four result registers have been updated. Table 12.4 shows which channel inputs are loaded into which result register, depending on the setting of bits CC and CD.

TABLE 12.4 REGISTER ADCTL SETTINGS FOR A/D MULTICHANNEL SELECTION (MULT = 1)

CD	CC	Channel	Result register
0	0	AN0	ADR1
		AN1	ADR2
		AN2	ADR3
		AN3	ADR4
0	1	AN4	ADR1
		AN5	ADR2
		AN6	ADR3
		AN7	ADR4

Like single-channel operation, there is the option of using continuous conversion. If this is not selected by clearing bit SCAN, the four channels selected are converted once every time the software writes to control register ADCTL. With continuous scan (SCAN = 1), the software only needs to write to ADCTL once and wait for flag CCF to set. After this, it can read a channel by reading the corresponding result register. Listing 12.3 shows a subroutine that can be used for multiple-channel operation. After executing this subroutine once, the calling application program can read the desired A/D result register at any time. Again, it will not be necessary to call the subroutine again.

```
*Listing 12.3
*------------------------------------


        ORG     $180


*Subroutine INMAD
*I/O handler routine for analog input
*using continuous scan and multiple-channel
*operation. Calling program must execute
*this subroutine after subroutine
*STARTAD (Listing 12.1) and before reading
*any analog result. Returns result in RAM variables.
*Calling registers
*       IX = register block address
*       ACCA = $00 to select channels AN0 to AN3
*       ACCA = $04 to select channels AN4 to AN7
*       C = 1 if error, otherwise 0
*Return analog results (ACCA=$00)
*       ADR1 = AN0, ADR2 = AN1
*       ADR3 = AN2, ADR4 = AN3
*Return analog results (ACCA=$04)
*       ADR1 = AN4, ADR2 = AN5
*       ADR3 = AN6, ADR4 = AN7


INMAD
        PSHA                ;preserve ACCA if no error
        CMPA    #$00        ;check if ACCA is legal
```

```
        BEQ     INMAD1
        CMPA    #$04
        BNE     INMAD3  ;exit to return error code
*                       ;if not legal
INMAD1
        ORAA    #$30    ;SCAN=1, MULT=1
        STAA    ADCTL,X ;start conversion
*                       ;also clears CCF
*wait for first conversion done
INMAD2
        BRCLR   ADCTL,X $80 INMAD2
*                       ;CCF set at this point
*return section if no error
        PULA            ;restore ACCA
        CLC             ;C=0 for no error
        RTS             ;return results
*return section if error
INMAD3
        PULA            ;restore ACCA
        SEC             ;C=1 for error
        RTS             ;and return
```

Subroutine INMAD can be modified for noncontinuous conversion by changing the ORAA instruction to

```
ORAA    #$10    ;SCAN=0, MULT=1
```

In this case, the calling application program will have to call INMAD each time it wants to read an analog input. The option of using single- or multiple-channel modes allows the designer to use some port E pins as general-purpose digital inputs and others as analog inputs. In all of the examples, bit CD in control register ADCTL was 0. It is set for certain factory test conditions.

■ 12.6 SUMMARY

Digital numbers can represent analog values, usually voltage or current. Simply put, a D/A converts a digital number into an equivalent analog value. An A/D does the reverse.

Each step in an analog range is represented by a corresponding digital number. A range's offset is its minimum value. Its digital representation is 0 (all bits 0). The span is the difference between the maximum and minimum analog values. The maximum value is represented digitally by a number with all bits set to logic 1. The resolution of a converter is determined by the maximum number of bits that a digital value can have. Each bit has a corresponding analog weight that represents a fraction of the range (negative power of 2).

For linear ranges, the step size is the weight of the LSB. This value is the span divided by 2 to the power n, where n is the number of bits. For nonlinear ranges the step size will vary throughout the range. Software can compensate for this by using a lookup table.

Transducers are signal converters. Output transducers (actuators) modify electrical signals to affect the real-world process. Input transducers (sensors) convert physical properties (example temperature) into electrical signals. Most sensors output a varying voltage signal to represent a physical quantity. It may be necessary to process a sensor output further using signal-conditioning circuits so that a D/A can convert it accurately. Conditioning circuits modify a range by multiplying the span by a factor and adding an offset (positive or negative). Some systems also employ filters to reduce noise. Many actuators are only switched on or off. These can be made effectively analog by using PWM. Power transistors can be used as true analog outputs by varying their voltage–current characteristics (biasing).

The most common type of D/A uses an *R–2R* ladder network. This type of network adds current components, where each component represents a corresponding bit weight in the digital input. If the corresponding bit is 0, no current component is added. Computers often use an A/D based on successive approximation. In this technique the converter compares an internally generated analog signal (digital value known) with the analog input (digital value unknown). Based on the result of the comparison, it varies the analog signal until a close match is obtained. To do this, the converter steps through a sequence from MSB to LSB to more closely approximate the final result. The digital equivalent of the approximated result is then sent as the digital output.

The 68HC11 has an A/D based on successive approximation using a capacitor charge redistribution network. It has eight analog inputs called channels at port E. The subsystem can convert data from one channel four times or from a group of four once each. To use the A/D, the software must enable the charge pump for the capacitor network and specify the clock source. It does this by setting up bits ADPU and CSEL in register OPTION. Control register ADCTL specifies other options, such as single- or multiple-channel (bit MULT) operations and single or continuous conversions (bit SCAN). The channel select bits (bits CA, CB, CC, CD) determine which analog channels can be read.

Basically, the software starts conversion by writing to control register ADCTL (which also clears flag bit CCF) and then waits for a conversion complete signal. It does this by polling ADCTL to check when flag CCF sets. There are no interrupts in this subsystem. After conversion is complete, the software can read the result from the appropriate result register (one of ADR1, 2, 3, 4). If continuous mode was selected, the software can read the most recent result any time after the first conversion is done.

■ EXERCISES

Questions 2, 10, 13, and 14 require some circuit theory or electronics knowledge not explained in this book.

1. Describe how the 68HC11 with a 5-V power supply can output a signal at 2.5 V.

2. Describe how the 68HC11 can measure electrical power given a voltage and current sensor.

3. Calculate the span, offset, step size, resolution, and percentage resolution for an analog range of 0 to 5 V for 8-bit digital conversion. Also calculate the bit weights.

4. For the system in Question 3, calculate the analog output of a D/A with the following digital inputs: $15, $4D, $7E, $E7.

5. For the same analog and digital values stated in Question 3, calculate the hex output of an A/D with the following inputs in volts. The A/D accuracy is ±1/2 LSB.

6. Repeat Question 3 for an analog range of 4 to 20 mA.

7. Repeat Question 4 for an analog range of 4 to 20 mA.

8. For an analog range of 4 to 20 mA and 8-bit resolution, calculate the hex output of an A/D with the following inputs (in mA): 3, 5, 7.5, 12, 13.9, 21.

9. A process heater uses a flame detector input (see Question 20 in Chapter 11). The flame detector sensor outputs an analog signal with a range of 4 to 20 mA. The flame detection threshold is 9 mA. What is the hex value for the threshold (variable TH)?

10. The input leakage at an A/D input pin for the 68HC11 can be as high as 400 nA. Would you connect a 1- or a 100-kΩ potentiometer to it? Explain why.

11. Describe ways for a microcontroller to find the position of a robot's arm. A robot's gripper (hand or end effector) may have to pick up a light, fragile object such as an egg. Describe how a microcontroller can control the position of the gripper fingers.

12. A dc motor's speed is controlled by controlling field current using a MOSFET power transistor. A tachometer measures its speed. It operates like a generator to increase its voltage output as speed increases. To maintain constant speed, should an MCU increase or decrease its D/A output when an increasing tachometer signal is detected? Why?

13. Prove that the current in each branch in an R–2R network in a D/A is half that in the preceding branch.

14. Derive the equation for the current-to-voltage converter amplifier circuit shown in Figure 12.4.

15. Write a program to generate a sawtooth wave for an 8-bit D/A connected to port B. The D/A has a parallel interface and is always enabled (no internal latch).

16. A 10-bit A/D uses successive approximation. It is driven by an 800-kHz clock. Each step requires four clock cycles. Estimate the conversion time.

17. The 68HC11 has V_{RL} connected to 1 V and V_{RH} connected to 4.5 V. What are the span, offset, and step size?

18. Design an interface using the 68HC11 internal A/D for the analog range -2.5 to $+7.5$ V.

19. Repeat Question 18 for the range 4 to 20 mA.

20. For the 68HC11 internal A/D subsystem, write a program to read inputs from channels 1, 3, and 7 with the system clock running at 100 kHz (to conserve power).

21. Write pseudocode to display the analog input at pin PE2 as a hex number. The source code will use a previously developed subroutine for displaying a hex digit to a seven-segment display.

22. Refer to Question 20 in Chapter 11. Write the code for subroutine ADC4 that reads an analog value from the flame detector using channel AD4.

23. Write the code for subroutine STARTADC. It is called by the reset routine for the system described in Question 22.

13 Application Control Software

OBJECTIVES

After completing this chapter, you should be able to

- Use Boolean logic and sequencers for typical control applications.
- Design and use sequential machines.
- Design software for feedback control using a proportional integral derivative (PID) technique.
- Design some basic digital signal processing software.
- Use more C language features.
- Describe principles of how C function calls use the stack.
- Describe principles about how microcontrollers can handle multitasking.
- Use fuzzy logic in applications.

Microcontrollers did not get their name for nothing. Obviously, the word *micro* means that they are small. The controllers part refers to their use. In previous chapters we have concentrated on I/O and the instruction set. We have seen how an MCU can input data to measure the outside world. Also, we have seen how it can output data to change the outside world. In some chapters we examined how I/O is used for communications between an MCU and other devices or people. In this chapter we look at what happens between input and output. In other words, we look at the software to determine what outputs to send based (sometimes) on the past input history. By *application* we mean something that the MCU controls, such as an automobile engine or washing machine.

Sections 13.1 to 13.3 cover discrete event systems. Section 13.4 covers a common control technique used for continuous (or analog) systems. In Section 13.5 we give a brief presentation of digital signal processors (DSP). Experts predict that they will have as much impact on the control industry as that of the microcontroller. Section 13.6 brings us back to the C language to make use of the material covered since we introduced this language in Section 2.3. Section 13.7 is a conceptual look at multitasking. Typical microcontroller systems perform several tasks. Section 13.8 covers fuzzy logic. This is a technique that permits an MCU to control processes that were previously difficult to handle.

■ 13.1 BOOLEAN LOGIC CONTROL

Boolean logic control is a common technique used for applications whose outputs depend only on the present inputs. It is possible to use Boolean logic for systems whose outputs also depend on previous input. However, other techniques are more suitable for these applications when the logic becomes very complex. *Programmable logic controllers* (PLCs) are industrial real-time control computers designed to work in a factory environment.[1] They typically use Boolean logic. A common PLC application is to turn motors on and off depending on process measurements.

PRACTICE

An automobile instrument panel (or dashboard) has a warning light that illuminates when any of the doors are not properly closed. Write the pseudocode to do this for a four-door passenger car.

Solution

$$\text{door warning} = \overline{\text{door1}} \cdot \overline{\text{door2}} \cdot \overline{\text{door3}} \cdot \overline{\text{door4}}$$

```
*a logical OR of inverted inputs
*input high when door is closed
*warning light on when output high
```

An alternative solution is

$$\text{door warning} = \overline{\text{door1} \cdot \text{door2} \cdot \text{door3} \cdot \text{door4}}$$

```
*a logical NAND
```

If you are familiar with digital logic, you may recognize the two solutions as one of De-Morgan's theorems.

Three-Wire Motor Starting. A simple and very common control application is the three-wire start-stop system for electric motor control. *Three-wire* refers to the traditional way an

---

[1]Some in the control industry use the term *programmable controller* (PC). We use PLC instead of PC to distinguish it from the popular *personal computer* (also PC).

electromechanical switch (contactor) was connected. When MCU control is used, an output is connected to a transistor or analog switch, which in turn would drive a contactor. Recall our discussion of relays and contactors in Section 12.2.3. The MCU sends a signal to open or close the contactor switch, which turns the motor on or off.

The motor start-stop is a system with two push buttons, for start and stop, respectively. This sounds simple so far, but it gets more complicated. The push buttons are momentary contact switches. A momentary contact switch bounces back to its original position after being pressed. This differs from a mechanically held switch such as a typical household light switch, which stays in a locked position. Basically, the system works as follows: Pressing the start button causes the motor to run as long as the stop button is not pressed at the same time. Releasing the start button latches in the motor (keeps it running). If the stop button is pressed, the motor is unlatched (stops running). It overrides the latching action of the start button.

In three-wire control a human operator must press the start button to restart the motor if the motor should stop for any reason. This is important for industrial safety reasons. It is important that the motor does not start up automatically after a power failure is corrected. Note that with a mechanically held switch in the closed position, the motor would start (perhaps unpredictably) when power is restored.

If you have a digital logic background, you may recognize this to be similar to a flip-flop circuit. An MCU can use software to do the same thing. Note that it would be economical to use an MCU only if it did things other than starting and stopping one motor in response to push-button commands. Listing 13.1 shows pseudocode that illustrates Boolean logic control for this application. For safety reasons, a real stop input is active (low) when there is no power. This ensures that loss of power to push-button wiring will not prevent a stop action.

```
*Listing 13.1
*Pseudocode for motor start-stop control
*Stop is active low, higher priority.
*Start is active high.

        motor = 0                   ;motor off initially
loop
*               Repeat loop continuously, note that
*               the motor does not change state if
*               no inputs are active.
        if stop == 0            ;stop motor
                then motor = 0
        else if start == 1      ;start motor
                then motor = 1
        repeat loop

*       Also note that stop overrides start.
```

This example is relatively simple but a system gets more complicated with additional I/O and time requirements.

A sump pump control system uses logic similar to that of three-wire motor control. A sump well collects water runoff. The sump pump pumps out the well whenever it fills up. When the level rises to a *high limit*, the pump turns on. It stays on until the level drops to a *low limit*. Then the pump shuts off. It does not turn on again until the well fills up to the *high limit*. (See also Question 20 in Chapter 2.)

■ 13.2 SEQUENCERS

Many applications, such as vending machines, execute a specific sequence when activated. Another example is an automatic washing machine. Note that both machines existed before integrated circuits were invented. Their controllers were built using relays and electric-motor-driven timers. Now the MCU provides cheaper and more versatile control.

Consider the washing machine, an essential part of civilization. A washing machine goes through a sequence of filling the tub with water (detergent added), agitating the water-detergent mixture by driving a rotor, and draining the mixture by pumping it out. Then it does a spin sequence to force out water from clothing using centrifugal force. The entire cycle repeats to rinse the detergent from the clothing.

Some inputs are switches for the operator or consumer to set options and start or stop the wash cycle. Other inputs are process measurements such as switches to sense water level and whether the tub lid is closed. Outputs are solenoids to open and close the hot and cold water valves. Contactors control spin and agitate by switching the pump and rotor motor(s) as required.

For illustration of sequential control, Table 13.1 shows the sequence of operation for a simplified machine that has only one water temperature. In this case outputs for steps 1 to 4 are identical to those for steps 5 to 8. Generally, this would not be true for all wash sequences. Table 13.1 does not specify the control logic completely. The machine should stop its sequence whenever the consumer turns off the start-stop switch or opens the lid. The wash cycles resume when the switches are restored to their operating positions. Additional flexibility can be obtained by allowing user-selectable time delays and water levels. The software is a program using a look-up table. The program can run a different sequence simply by using a different table.

TABLE 13.1 SIMPLIFIED WASHING MACHINE SEQUENCE

Process	Step	Valve PB3	Pump PB2	Spin PB1	Agitate PB0	Transition condition
Wash fill	1	1	0	0	0	Until level high
Wash	2	0	0	0	1	10 min
Drain	3	0	1	0	0	Until tub empty
Spin	4	0	1	1	0	1 min
Rinse fill	5	1	0	0	0	Until level high
Rinse	6	0	0	0	1	10 min
Drain	7	0	1	0	0	Until tub empty
Spin	8	0	1	1	0	1 min
Stop	9	0	0	0	0	Until restart

Note: Active high logic. 0 = off, 1 = on.

We illustrate software for a 68HC11-based system with the following connections. The outputs are connected to pins PB0 to PB3 as shown in Table 13.1. To illustrate the software with fewer complications, we use a level switch with built-in hysteresis. That is, when the water level reaches a high point, the switch closes but does not open again until the tub is empty. The level switch input is PC1.

The start-stop switch and lid switch are wired in series to input PA7. Hence, if either of the switches open, a falling edge occurs. The MCU system can use a pulse accumulator edge interrupt routine to suspend the sequence until both switches are closed again. It can also use the pulse accumulator gated mode to run the software timers only when both switches are closed. Listing 13.2 shows the major parts of the washing machine software. The code required for initialization is not shown. Subroutines DLY10 and DLY1 implement the time delays. Subroutine STOPIT waits until the start—stop switch and lid switches are both closed (wired in series) before returning to the main program.

```
*Listing 13.2
*Major part of washing machine sequence control
*-----------------------------------

        ORG     $100

MAIN
        JSR     STOPIT
        LDAB    #2          ;in this case,
*                           ;wash and rinse sequences identical
SECOND
        LDY     #SEQTAB  ;point to sequence table
        LDAA    0,Y      ;fill tub
        STAA    PORTB,X
*                           ;until full
FULL
        BRCLR   PORTC,X $02 FULL
        INY
        LDAA    0,Y         ;agitate for 10 min
        STAA    PORTB,X
        JSR     DLY10
        INY
        LDAA    0,Y         ;drain tub
        STAA    PORTB,X
*                           ;until empty
EMPTY
        BRSET   PORTC,X $02 EMPTY
        INY
        LDAA    0,Y         ;spin for 1 min
        STAA    PORTB,X
        JSR     DLY1
        DECB                ;repeat for rinse sequence
        BNE     SECOND
        INY
        LDAA    0,Y         ;else stop
```

```
            STAA      PORTB,X
            BRA       MAIN


*------------------------------------
*look-up output sequence table
SEQTAB
            FCB       %00001000       ;valve
            FCB       %00000001       ;agitate
            FCB       %00000100       ;pump
            FCB       %00000110       ;pump and spin
            FCB       %00000000       ;stop
```

In Listing 13.2 note the importance of having subroutines that preserve registers. The program could be rewritten using bit manipulation instructions BSET and BCLR instead of using table look-up. With a look-up table program, the washing machine manufacturer can incorporate changes for different machine models more easily by using the same instructions. Only the table contents will differ. The EEPROM is a good candidate for holding the table.

PRACTICE

Write a sequencer program to control traffic lights. During the first minute, a red light and the opposite green light turn on. During the next 15 seconds, the red light stays on while the opposite yellow light turns on. During the following minute, the green light turns on while the opposite red light turns on. Fifteen seconds later, the yellow light turns on while the opposite red light stays on. The entire cycle then repeats.

Solution

```
            ORG $100


*Sequencer program to control traffic lights
*Outputs are
*PB0 = N-S RED
*PB1 = N-S YEL
*PB2 = N-S GRN
*PB3 = E-W RED
*PB4 = E-W YEL
*PB5 = E-W GRN


AGAIN
            LDX       #TABLE          ;start cycle
NEXT
            LDAA      0,X             ;get output from sequence table
            STAA      PORTB+REGBAS    ;and drive lights
            JSR       DLY15           ;delay 15 s
            INX                       ;point to next step
            CPX       #TABLE+10       ;until last step
            BLT       NEXT
            BRA       AGAIN           ;when cycle complete, repeat it
```

```
*sequence table
TABLE
*               binary pattern in 15-s increments

*               ;N-S RED, E-W GRN ON for 1 min
        FCB     %00100001
        FCB     %00100001
        FCB     %00100001
        FCB     %00100001


*               ;N-S RED, E-W YEL ON for 15 s
        FCB     %00010001


*               ;N-S GRN, E-W RED ON for 1 min
        FCB     %00001100
        FCB     %00001100
        FCB     %00001100
        FCB     %00001100


*               ;N-S YEL, E-W RED ON for 15 s
        FCB     %00001010
```

■ 13.3 SEQUENTIAL MACHINES

A sequential machine is similar to a sequencer except that it is more formalized. Also, the sequence is not necessarily linear. It can go backward at times, and branch or loop at other times. The sequential machine uses an algorithm based on the concept of *state transitions*. Theoreticians have developed theorems and definitions to describe digital systems whose outputs depend on present and past inputs. Such a system changes when it makes transitions to conditions called *states*. A state represents the past history of a system. Each state determines the next state the system will go to depending on the present inputs.

For example, a combination lock will open only if a unique sequence of numbers is entered. You can model this as a system that has to enter some specific states before it can enter the final state that allows it to open. Another example is elevator control. An elevator controller will remember that it has passengers to pick up while it is transporting others. If it is traveling from the fourth floor to the ground floor, it will return to pick up a passenger on the fifth floor who has requested service.

A system that behaves as described is also known as a *finite-state machine*, meaning that it has a finite number of states. There are two types of finite-state machines; these are named after the theoreticians Mealy and Moore. We will work with a *Mealy machine*, which has the property that its *outputs* (not next state) can be determined by its present state and present inputs. In contrast, a Moore machine's outputs depend only on its present state—not the inputs. The transitions between states can be shown in two ways, the state-transition diagram and the state table.

Elevator Control Application. Figure 13.1 illustrates the two methods using a much simplified elevator control system that uses only two floors. Each floor represents a state.

STATES   INPUTS   OUTPUTS

A = First Floor   u = up button $0 = stop, door open
                      d = down button $1 = motor up, door closed
B = Second Floor $2 = motor down, door closed

    (a)   Definitions

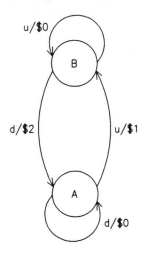

    (b)   State–Transition Diagram

Present State	Inputs u	d	Inputs None	Both
A	B, $1	A, $0	A, $0	A, $0
B	B, $0	A, $2	B, $0	B, $0

    (c)   State Table

Figure 13.1   Sequential Machine Example: Simple Elevator Control

The inputs that cause a transition from one state to another are the up and down buttons. The outputs are the reversible motor contractors and door solenoid. Limit switches indicate the current state. In Figure 13.1(b) each circle represents a state. The arrows represent the transitions. Each transition is labeled to indicate the input causing it and the output it produces. The state table shown in Figure 13.1(c) is another way to show the actions of a sequential machine.

Although possible, it is awkward to control an elevator using Boolean logic as described in Section 13.1. State-transition techniques are more appropriate. For the elevator example, we make further simplifications for the sake of clarity. The elevator always starts in state A. A mechanical interlock shuts off the motor when a floor is reached. Software will start the motor again when necessary.

For an elevator system with three or more floors, software can be used to shut off the motor. This software can be in the form of a more complicated state-transition algorithm that includes limit switches as part of the inputs. However, we stick with the two-floor example to keep it simple.

The software can use a table that contains the information shown in the state table. The program of Listing 13.3 illustrates this. Subroutine INPUT uses accumulator A as a return register. It is a code specifying the input. Subroutine OUTPUT updates the appropriate port lines to drive the motor and door as specified by a code in accumulator A, its calling register.

```
*Listing 13.3
*Sequential machine to control elevator
*See Figure 13.1 for illustration of transitions.
*------------------------------------
*Arbitrarily chosen values
u          EQU     $01          ;input is up
d          EQU     $00          ;input is down
ZERO       EQU     %01100110    ;drive some outputs
ONE        EQU     %00110011    ;drive other outputs
TWO        EQU     %11001100    ;etc.


           ORG     $100


           JSR     INITIALIZE   ;initialize as required


           LDX     #A           ;point to initial state A
           LDAB    #3           ;set up table incrementer
*MAIN PROGRAM LOOP*****
GETINPUT
           JSR     INPUT        ;read input state
           CMPA    #u           ;if input == u then transit
           BEQ     TRANSIT
           ABX                  ;else point to next potential state
           CMPA    #d           ;if input == d then transit
           BEQ     TRANSIT
           ABX                  ;else assume no or both inputs


TRANSIT
           LDAA    2, X         ;read output state
           JSR     OUTPUT       ;and drive output
           LDX     0, X         ;get address of next state
           BRA     GETINPUT     ;repeat sequence
*END OF MAIN PROGRAM LOOP*****
*STATE TABLE - the heart of the sequential machine


*State A
A
```

```
        FDB     A               ;next state if input d
        FCB     ZERO            ;and transition output
        FDB     B               ;next state if input u
        FCB     ONE             ;and transition output
        FDB     A               ;next state if none or both
        FCB     ZERO            ;and transition output
*State B
B
        FDB     A               ;next state if input d
        FCB     TWO             ;and transition output
        FDB     B               ;next state if input u
        FCB     ZERO            ;and transition output
        FDB     B               ;next state if none or both
        FCB     ZERO            ;and transition output
*END OF STATE TABLE
```

In the case of Listing 13.3, each state is defined by a 9-byte block. Each subgroup of 3 bytes specifies the destination state and the corresponding transition output for an input condition (note the table incrementer in Listing 13.3). In computer science terminology, this is an example of a *linked list*. This is a collection of blocks with the same internal structure. Each block has one or more pointers to other blocks. In our case each state is a block.

Consider the example when the elevator is on the first floor (index register X points to address #A) and a passenger presses the up button. The program detects this and loads the code for the output (third byte in 3-byte subgroup) and calls subroutine OUTPUT to update the output. Then it loads index register X with the address of the destination state (B in this case). The cycle repeats except that the loop starts with a different value in index register X to indicate a different state. For this program to work, subroutine OUTPUT returns control back to the main program after the elevator has reached a floor.

High Alarm Annunciator.   Annunciation systems can also be implemented as a sequential machine. Control rooms in chemical processing plants often use annunciators to warn plant operators of a damaging or dangerous condition such as an overheated reactor. Typically, they flash very prominent lights and sound loud horns.

The annunciation sequence depends on the application. Here we will consider a high-alarm situation. The process variable (PV) should not exceed a high value, called the alarm set point (ASP). In our case, the ASP is the highest safe temperature and PV is the measured temperature. During normal conditions, the annunciator displays the PV value. When the PV climbs up to the ASP value, the annunciator gets the operator's attention by flashing the PV display and turning on an alarm light. Other systems may also sound a horn.

The operator responds by pressing the acknowledge button (ACK). This turns off the alarm light, but the display still flashes the PV value. When the PV is restored to a value below the ASP, the display is steady again. However, this sequence changes if the operator does not acknowledge the alarm. If the PV drops below the ASP before the operator presses the ACK button, the display stops flashing but the alarm light remains on. This warns the operator to check for possible damage even though the temperature returned to normal. Pressing the ACK button at this point will turn off the alarm light.

Next we show a state-transition diagram analysis. This sequential machine has four states, two inputs and two outputs. The two inputs are the PV (high temperature) switch and the ACK button. The outputs are the PV display and alarm light (ALARM). Again, we have simplified the situation. A horn is not included. The four states can be defined as

A   Process normal
B   Alarm condition, not acknowledged
C   Alarm condition, acknowledged
D   Return to normal without acknowledging alarm

The state-transition inputs are

a   PV <=ASP, ACK off
b   PV <=ASP, ACK on
c   PV > ASP, ACK off
d   PV > ASP, ACK on

The state-transition outputs are

$0   Display PV steady, ALARM off
$1   Display PV steady, ALARM on
$2   Display PV flashing, ALARM off
$3   Display PV flashing, ALARM on

Figure 13.2 shows the state-transition diagram for this system. Listing 13.4 shows an excerpt of the annunciator program. Like the elevator control program of Listing 13.3, it uses a main loop to determine which state to go to next from the current state and inputs. Again, we do not show necessary initialization, subroutine, and interrupt service routine codes.

```
*Listing 13.4
*Annunciator program code to control state transitions.
*See Figure 13.2.
*-------------------------------------
*Arbitrarily chosen values
aa      EQU     $01           ; state-transistion input a
bb      EQU     $00           ; state-transition input b
cc      EQU     $03           ; etc.
dd      EQU     $04
ZERO    EQU     %01100110     ;drive some outputs
ONE     EQU     %00110011     ;drive other outputs
TWO     EQU     %11001100     ;etc.
THREE   EQU     %00011110

        ORG     $100
        LDX     #A            ;point to initial state A
        LDAB    #3            ;set up table incrementer
*MAIN PROGRAM LOOP *****
GETINPUT
        JSR     INPUT         ;read input state
        CMPA    #aa           ;if input == aa then transit
        BEQ     TRANSIT
```

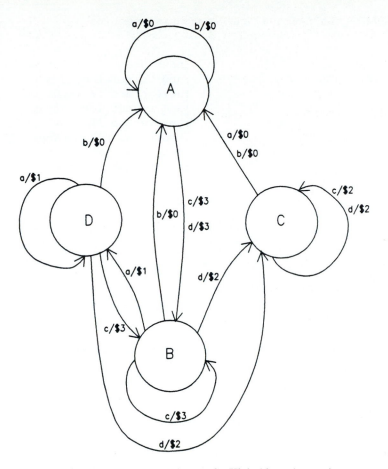

Figure 13.2 State-Transition Diagram for High-Alarm Annunciator

```
        ABX                     ;else point to next potential state
        CMPA    #bb             ;if input == bb then transit
        BEQ     TRANSIT
        ABX                     ;else point to next potential state
        CMPA    #cc             ;if input == cc then transit
        BEQ     TRANSIT
        ABX                     ;else point to next potential state
        CMPA    #dd             ;if input == dd then transit
        BEQ     TRANSIT
        JMP     ERROR           ;else have error,
*                               ;exit this main loop
TRANSIT
        LDAA    2,X             ;read output state
        JSR     OUTPUT          ;and drive output
        LDX     0,X             ;get address of next state
        BRA     GETINPUT        ;repeat sequence
*END OF MAIN PROGRAM LOOP *****
```

```
*STATE TABLE - the heart of the sequential machine
*State A
A
          FDB       A                         ;next state if input aa
          FCB       ZERO                      ;and transition output
          FDB       A                         ;next state if input bb
          FCB       ZERO                      ;and transition output
          FDB       B                         ;next state if input cc
          FCB       THREE                     ;and transition output
          FDB       B                         ;next state if input dd
          FCB       THREE                     ;and transition output
*ditto for rest of table
B
          FDB       D
          FCB       ONE
          FDB       A
          FCB       ZERO

*etc., etc. until end of table

*END OF STATE TABLE
```

We leave the rest of the table as an exercise for the reader to complete (see Exercises, Question 5).

PRACTICE

Write the state table for a dual-channel optical switch. The purpose of the state table is to increment the position counter each time the disk rotates to the next position in the forward direction. Conversely, the position counter should decrement if the disk rotates in the reverse direction. Note that Figure 11.8 shows a single-channel optical switch. It cannot indicate direction of rotation.

The dual-channel switch uses an encoding disk with two rings of holes. It has two outputs, P1 and P2. When the disk rotates, the outputs are pulse trains that are 90° out of phase. They resemble T1 and T3 (or T2 and T4) of Figure 11.5. When the disk rotates in the forward direction, the (P1 P2) sequence is 10, 11, 01, 00, 10, and so on. When the disk rotates in the reverse direction, the (P1 P2) sequence is 10, 00, 01, 11, 10, and so on.

Solution

```
                    ORG       $130    ; arbitrary start address

*arbitrary assignments for outputs
NO_CHANGE           EQU       $00
DEC_COUNT           EQU       $01
INC_COUNT           EQU       $02
NOT_POSSIBLE        EQU       $03
```

```
*STATE TABLE for dual-channel slotted optical switch

A       FDB     A                       ;next state if input 00
        FCB     NO_CHANGE
        FDB     D                       ;next state if input 01
        FCB     DEC_COUNT
        FDB     B                       ;next state if input 10
        FCB     INC_COUNT
        FDB     NULL                    ;next state if input 11
        FCB     NOT_POSSIBLE
B       FDB     A                       ;next state if input 00
        FCB     DEC_COUNT
        FDB     NULL                    ;next state if input 01
        FCB     NOT_POSSIBLE
        FDB     B                       ;next state if input 10
        FCB     NO_CHANGE
        FDB     C                       ;next state if input 11
        FCB     INC_COUNT

C       FDB     NULL                    ;next state if input 00
        FCB     NOT_POSSIBLE
        FDB     C                       ;next state if input 01
        FCB     INC_COUNT
        FDB     B                       ;next state if input 10
        FCB     DEC_COUNT
        FDB     C                       ;next state if input 11
        FCB     NO_CHANGE

D       FDB     A                       ;next state if input 00
        FCB     INC_COUNT
        FDB     D                       ;next state if input 01
        FCB     NO_CHANGE
        FDB     NULL                    ;next state if input 10
        FCB     NOT_POSSIBLE
        FDB     C                       ;next state if input 11
        FCB     DEC_COUNT

*Nothing in null
NULL
*end of state table
```

■ 13.4 FEEDBACK CONTROL OF THE ANALOG WORLD

Figure 13.3 shows a block diagram of the basic feedback control system. For simplification we ignore systems with multiple inputs and outputs. An example of feedback control of an analog system is automobile cruise (speed) control. In this system the driver determines the desired cruise speed and the cruise controller adjusts the engine throttle position

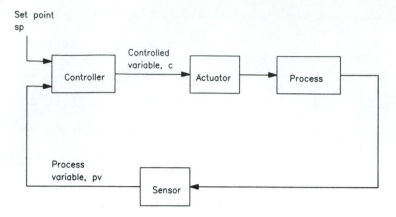

Figure 13.3 Feedback Control

to maintain the speed as set by the driver. Note that the cruise controller can be part of a microcontroller system that also controls other engine functions.

Generally, feedback control works as follows: The controller compares the desired response to the measured response. If they differ, the controller adjusts the output until they match or almost match within a specified tolerance. In control terminology the desired response is known as the set point (*sp*) and the measured variable is known as the process variable (*pv*). The output to an actuator is the controlled or manipulated variable.[2] When the automobile climbs a hill, the speed will drop and the controller should increase the throttle signal to bring the speed back up. When the automobile reaches the top of the hill and goes back down, the increased speed causes the controller to cut back on the throttle.

Feedback control is also known as *closed-loop control*. This distinguishes it from *open-loop control*. In open loop, there is no feedback signal to the controller from the sensor. Examples of open-loop systems are basic traffic lights, vending machines, and talking toys.

Proportional-Integral-Derivative (PID) Control. One traditional method of control is the proportional-integral-derivative (PID) controller. It outputs a manipulated variable based on the following equation. The variables are functions of time.

$$c = K_p e + K_i \int e\, dt - K_d \frac{de}{dt} \qquad (13.1)$$

$$e = sp - pv \qquad (13.2)$$

where c = controlled variable

K_p = proportional gain

K_i = integral gain

---

[2]The sensor converts the process variable to a measured variable. We consider the process variable to be the same as the measured variable since the controller reads only the measured variable to get information about the process.

K_d = derivative gain

e = error = difference between set point and process variable

sp = set point

pv = process variable

A simplified controller is the proportional controller:

$$c = K_p + c_0 \tag{13.3}$$

where c_0 is the offset. Consider the cruise control example. From equations (13.1) and (13.2) note that the output (c) increases when the process variable (pv) decreases because this increases the error (e).

If you happen to refer to other sources about control theory (such as reference 11 in Appendix C), you may note some differences between Figure 13.3 and control loops shown in these other sources. Most show a separate unit to subtract pv from sp. This results in the error as the input to the controller block. We show the controller as a device that includes the subtraction unit. This is the case for a microcontroller that senses the measured variable and compares it to a set point.

Historically, PID controllers became popular because they were easy to build using analog electronics or pneumatic bellows. Using an MCU for PID control requires some changes since it cannot process continuous changes in sensor signals. Instead, it samples the sensor signal using an A/D at discrete time intervals. Similarly, it cannot continuously change outputs. Hence, the MCU updates an analog output using a D/A at discrete time intervals. To keep the analysis simple, all time intervals are equal.

Figure 13.4 shows the effect of sampling an analog signal at discrete intervals. The sampling rate must be high (fast) enough. Figure 13.4(c) shows the effects of sampling too slowly. We return to the question of sampling rate later in this section. For now, we examine how to implement a digital version of a PID controller. This is shown as equation (13.4). We have spared the details of its derivation from the analog version of equation (13.1).

$$c_n = c_{n-1} + A_0 e_n - A_1 e_{n-1} + A_2 e_{n-2} \tag{13.4}$$

where n = integer referring to the nth sample

$$A_0 = K_p + K_i \Delta t + \frac{K_d}{\Delta t}$$

$$A_1 = K_p + 2\frac{K_d}{\Delta t}$$

$$A_2 = \frac{K_d}{\Delta t}$$

Δt = sampling time interval

Note that

$$e_n = sp - pv_n \tag{13.5}$$

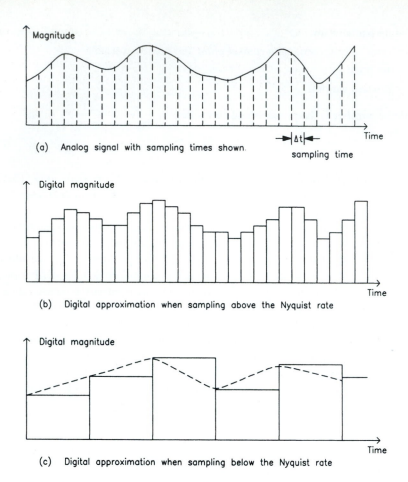

(a) Analog signal with sampling times shown.

sampling time

(b) Digital approximation when sampling above the Nyquist rate

(c) Digital approximation when sampling below the Nyquist rate

Figure 13.4 Sampling Analog Signals

These are the digital PID equations.

It is not necessary to save the PID controller outputs and errors (or inputs) for every sample (n). To do so would soon fill up the available RAM space. Instead, the software can update variables after each sample time. Then equation (13.4) can be rewritten as

$$C_0 = C_1 + A_0\,E_0 - A_1\,E_1 + A_2\,E_2 \qquad (13.6)$$

where C_0 = present output (c_n)

C_1 = previous output (c_{n-1})

E_0 = present error (e_n)

E_1 = previous error (e_{n-1})

E_2 = error preceding previous error (e_{n-2})

We can call this the PID sample equation. The variables are stored in scratchpad RAM memory.

To start updating the control output, the equation must use defined initial values. Initial error values of zero can be chosen. The choice of initial control output values depends on knowledge of the process. Usually, it can be the process variable multiplied by a constant. Eventually, repeated execution of the PID equation during each sample time *should* bring the error to zero. Soon, we'll examine factors that determine whether zero error is achieved, but now we present Listing 13.5, a PID algorithm in the form of pseudocode.

```
*Listing 13.5
*Pseudocode for PID controller algorithm.
*Note that outputs are not clamped at limits.
*SP = set point, PV = process variable

        define constants A0, A1, A2
        define SP
        initialize E1, E2 = 0
        input PV from A/D subsystem
        C1 = factor pv
        sample-time delay
loop
        output C1 to D/A or PWM subsystem
        input PV from A/D subsystem
             ;calculate PID sample output
        E0 = SP - PV ;sample error
        C0 = C1 + A0 E0 - A1 E1 + A2 E2
*            ;now reassign variables in scratchpad RAM
        C1 = C0
        E2 = E1
        E1 = E0
        sample-time delay
        repeat loop
```

Note that the time delay can be triggered using an output-compare interrupt or real-time interrupt. The service routine can include the rest of the loop. We return to this algorithm again when we discuss C programming in Section 13.6. The *pseudocode* of Listing 13.5 does not take into account the bit size limit of the numbers. The controller output must fall between maximum and minimum limits to avoid overflow and underflow resulting from the calculations. In other words, the outputs are *clamped.* This is analogous to the analog PID (no pun intended) controller, whose outputs are naturally clamped due to the high and low limits of the power supply. We will show later that the C code will include clamping.

Equation (13.4) is also known as the incremental form of the PID algorithm because it uses the previous output to calculate the next one. Another form, known as the *absolute form,* has the disadvantage of requiring integral limits as well as output limits.

For further information, you may wish to consult Motorola's application note (AN1215/D), which describes assembly source code and C source code to implement PID for the 68HC11K4 and 68HC11N4 microcontrollers. One would have to modify the code significantly because these microcontrollers have features not available for the 68HC11E9 type that we normally discuss.

What to Know About Control Theory in 500 Words or Less. Control theory is used to determine whether and how a process can be controlled. The goal is to minimize the error. Basically, three factors determine if a zero (or constant) error occurs:

1. Values of the gain constants
2. Sampling time
3. Process dynamics

The combination of these factors determines whether or not a control system is *stable*. It is our intention to present you with an intuitive overview of control theory. Control theory is a specialized discipline. For those who are not shy of higher mathematics, you may wish to consult literature specializing in this subject, such as reference 14 in Appendix C.

A stable system is a system whose output eventually reaches a constant value when the set point is constant. There are degrees of stability. Also, there is a desirability of obtaining *optimal* response. Optimal control is the technique of minimizing *cost*. The cost may be a combination of factors, such as process error, amount of fuel, and amount of contaminants or pollutants. It is not necessarily equivalent to the amount of money. Control engineers and technologists use mathematical models of the process and control theory techniques to determine gain constant and sampling-time values for stable and optimal responses.

Traditional control-theory techniques work if a system has an output proportional to input (linear) and it does not change its properties over time (time invariant). In physical reality this is rarely true but this does not prevent the use of control theory. For control purposes, a machine's behavior can be approximated as linear time invariant within a specified range of input limits. A cruise controller will probably not take into account changes in the weather or in the car's mass as fuel is consumed.

The relationship between input and output of a process is described mathematically by a *transfer function*. For analog systems the output (as a function of time) can be predicted from the input and transfer function using a mathematical trick known as the *Laplace transform* and its inverse transform. The magic involves changing differential equations to algebraic equations and then back again. Discrete-time systems are those that use digital samples of analog signals. They use a technique called the *z-transform* and its inverse. Alternatively, one may use computer modeling programs that simulate linear and nonlinear systems.

Consider automatic position control of an airplane. It is important that the gain constants are not too high. A slight change in position may cause an output change that overshoots the correct position for an increased error in the opposite direction. The result is another overreaction to this error and the errors rapidly increase. This is instability. Conversely, if the constants are too low, the airplane responds too slowly (stable but poor dynamic response).

Sampling Time. Now we will concern ourselves with sampling time since the gain constants (A_0, A_1, A_2) are affected by it. Intuitively we can note that a faster sampling rate is desirable (see Figure 13.4). A landing aircraft may travel at 50 m per second. If the controller took samples of aircraft position every second, the aircraft could have strayed off course by 50 m before correction can be applied.

In Figure 13.4 we have shown an analog signal being sampled at regular intervals. The *Nyquist sampling theorem* states that all analog information can be recovered if the sampling rate is equal to or more than twice the frequency of the highest-frequency component of the original signal. The Nyquist sampling rate is defined as twice the frequency of the maximum frequency in a signal. Recall (see Section 12.3 about PWM outputs) that adding sine waves at different discrete frequencies, magnitudes, and phase shifts can produce a periodic signal of any shape. Similarly, a nonperiodic signal can be made up by adding sine (and cosine) waves at all possible frequencies from zero to the maximum (which may be infinite). Mathematically, this is done by using the *Fourier transform,* which is similar to the Laplace transform mentioned earlier.

Since a signal may have very high frequency components, it is not always possible to recover the entire signal. The inaccuracy (error) in losing analog information at frequencies at greater than half the sampling rate is known as an *aliasing* error. This is quite visible in Figure 13.4(c). In practice, an MCU system samples at several times the Nyquist rate (Figure 13.4[a] and [b]).

In real signals the higher-frequency components are relatively small. The *bandwidth* (BW) is defined to be the frequency of the signal component that has half the power (70.7% of magnitude) of the zero-frequency (or dc) component. The bandwidth can be approximated by the following equation:

$$f_{BW} \cong \frac{0.35}{t_r} \tag{13.7}$$

where f_{BW} is the bandwidth frequency and t_r is the rise time of the signal. The rise time is defined as the amount of time required for the signal to change in response to a sudden (step) change in input. It is the amount of time required for the output to rise from 10% to 90% of its final value. A rule of thumb is to sample at a rate of at least five times the band-width. This allows the control system to obtain at least five samples during a rapid rise or fall in signal level.

A raw analog input can be bandlimited by passing it through a low-pass filter. This also removes unwanted high-frequency noise components. The A/D then samples the output of the low-pass filter.

PRACTICE

A low-pass filter is connected to an A/D input. It limits the signal bandwidth to 1 kHz. At what rate should the control system sample the signal?

Solution

At a rate of 5 kHz or faster.

Other Control Techniques.   So far we have kept the set point constant. This is not necessarily the case since the behavior of the process needs to change. For example, we do not drive a car at the same speed all the time. Another possibility is cascade control. The output of one PID control loop is the set point of another. A robotic position controller may use cascade. The position control loop output is the speed set point for the motor speed control loop. Other systems use expanded versions of the PID equation that modify an output based on many previous errors and outputs. Also, many systems have multiple inputs and outputs. One common method is to use *state equations.* They involve adding the results of many multiplications. We'll say no more about state equations. In Section 13.8 we describe another control technique called *fuzzy logic.*

■ 13.5 DIGITAL SIGNAL PROCESSORS

A digital signal processor (DSP) is a specialized processor. It is designed to execute arithmetic instructions faster than can a general-purpose processor. A DSP application is more math intensive. DSP programs use a higher proportion of arithmetic instructions compared to data moving and I/O instructions. We provide a brief introduction to DSPs since more control systems are using them. This will give you an idea as to what they are and what they can do. Also, it will show you a limitation of general-purpose microcontrollers such as the 68HC11.

Typically, a DSP samples analog inputs after being passed through a low-pass filter. Then it calculates new outputs, which in turn may be sent to D/A converters. It must do this at high speeds. For this reason DSPs are designed to execute arithmetic instructions quickly. The main arithmetic operation is the multiply and accumulate. Soon we shall see why this is important.

DSP applications include machine vision used for parts identification in a factory, position control of robot arms, and detection of telephone tone signals. Common DSP calculations are PID control loops, digital filters, and fast Fourier transforms (FFTs). We have already examined PID control. We will examine the rest now.

Digital Filters. Recall our discussion of the PID controller in Section 13.4 (see equation [13.4]). In general, an MCU can output a signal that is the weighted sum of previous inputs (or errors) and outputs. This is also known as a *digital filter*. For simplicity, we consider a simpler filter first. It generates an output based on past inputs only. This is the *finite impulse response* (FIR) *filter*. Its output is expressed as

$$y(n) = b_0\, x(n) + b_1\, x(n-1) + b_2\, x(n-2) + \ldots + b_M\, x(n-M) \tag{13.8}$$

where M = maximum number of samples

$y(n)$ = output of sample time n

$x(k)$ = input at sample time k

$b(k)$ = FIR coefficient for sample time k

Alternatively, this equation can be written as

$$y(n) = \sum_{k=0}^{M} b_k x(n-k) \tag{13.9}$$

Note: The symbol Σ represents summation.

Figure 13.5 shows how memory resources can be used to implement an FIR.

The difficulty in using an MCU such as the 68HC11 increases with the *length* of the filter (M+1). You should recognize the RAM space for inputs as a first in, first out (FIFO) queue. With an architecture like the 68HC11, the MCU must invoke software such as Listing 13.6 after each sample input is obtained.

```
*Listing 13.6
*Pseudocode for FIR filter
        read current input and store in top of FIFO
```

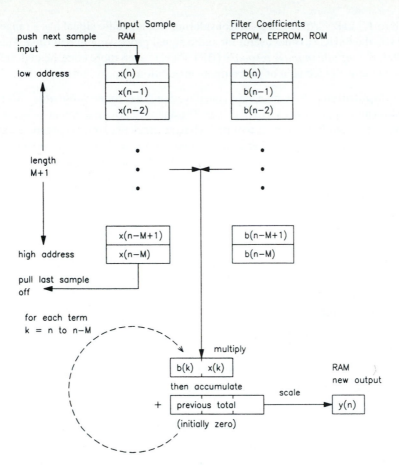

Figure 13.5 Digital Filter Calculations for Finite Impulse Response (FIR) Filter

```
initialize accumulator to zero
multiply an input sample with corresponding coefficient
add this product to the accumulator
        ;this is the multiply and accumulate operation
repeat the multiply and accumulate operation
        until all samples handled
scale to appropriate bit accuracy for output
shift down samples in FIFO
```

Each of these steps will take several instructions. Shifting the FIFO alone will take at least five cycles per sample assuming direct mode load and store instructions. This still ignores necessary software overhead. It also assumes that 8-bit arithmetic is used. Do not forget that an MCU must also do other things.

The following illustrates the bandwidth limitation using the 68HC11. A filter with a length of 10 will require at least 300 cycles to calculate each new output. This means that it can sample every 150 μs when the clock is 2 MHz. This limits the acceptable signal bandwidth to 1.3 kHz [$1/(5 \times 150 \,\mu s)$]. Considering that an audio signal has a range of 20 to 20,000 Hz, the 68HC11 is not useful for audio signal processing in real time. The other type of filter, the *infinite impulse response* (IIR) *filter*, is even more complicated. Its equation includes the weighted sum of past outputs in addition to past inputs.

Matrix Multiplication. So far we have considered only *vector multiplication*. This is the situation shown in equations (13.8) and (13.9). The answer is a one-element value. In some applications *matrix multiplication* is required. Here an array of elements organized as rows and columns is multiplied by another array to produce a new array. The number of arithmetic operations increases dramatically. Equation (13.10) shows the calculations required for each matrix C element when matrices A and B are multiplied. Equation (13.11) shows an alternative expression.

$$c_{ij} = a_{i1} b_{1j} + a_{i2} b_{2j} + \ldots + a_{in} b_{nj} \tag{13.10}$$

$$c_{ij} = \sum_{k=1}^{n} a_{ik} b_{kj} \tag{13.11}$$

where i = row index

j = column index

n = number of columns in matrix A

  = number of rows in matrix B

a_{ik} = element of matrix A, the multiplier

b_{kj} = element of matrix B, the multiplicand

c_{ij} = element of matrix C, an accumulated sum of products

The state equations mentioned at the end of Section 13.4 use matrix multiplication. Graphic processors manipulate the picture elements (pixels) using matrix multiplication. Typical operations for graphic objects on a screen are scaling (expanding and shrinking), rotating, moving, and overlapping with other objects.

Multiply and Accumulate Operations. Equation (13.10) illustrates the fundamental DSP operation, multiply and accumulate. Elements a_{ik} and b_{kj} are multiplied and added to form element c_{ij}. Listing 13.7 shows how the 68HC11 can execute a simple multiply and accumulate.

```
*Listing 13.7
*-----------------------------------

        ORG     0
SUM     RMB     2               ;accumulated sum

        ORG     $180

*Subroutine MAC
*Multiply and accumulate two unsigned 8-bit integers
*It adds (scaled 8-bit) product to 16-bit sum
*Calling variables
*       Address SUM = current accumulated 16-bit sum
*               note that it is a page 0 address
*       ACCA = multiplier
```

```
*          ACCB = multiplicand
*Return variables
*          ACCD = new SUM = old SUM + ACCA × ACCB
MAC
*                         ;6 cycles overhead for JSR extended
          MUL             ;10 cycles
          ADCA    #0      ;2 cycles
          TAB             ;2 cycles
          CLRA            ;2 cycles
          ADDD    SUM     ;5 cycles for direct mode
          STD     SUM     ;4 cycles for direct mode
          RTS             ;5 cycles
*                         ;Total of 36 cycles
```

From Listing 13.7 it is apparent that it would take the MCU a long time to perform many multiply and accumulate operations. For a signed multiply and accumulate, it will take even more instructions. In practice the 68HC11 could only do real-time DSP operations for very low-frequency signals. Thus it makes more sense to use a DSP chip for signal processing.

To illustrate a chip with a built-in DSP, we use the 68HC16. This is a Motorola microcontroller with more processing capability. It can execute 68HC11 programs as well as make use of its expanded instruction set. In other words, the 68HC16 is source code upwardly compatible with the 68HC11. It has instructions and associated internal hardware for DSP capability. We use the 68HC16 as an example since it is closely related to the 68HC11. In Chapter 15 we will describe other features.

PRACTICE

Using subroutine MAC, write an instruction sequence for a simple digital filter (FIR type) of length 3.

Solution

```
*Solution, simple FIR of length 3
*arbitrary equates
INFIFO  EQU     $140

COEFF   EQU     $B600

        ORG     $100

        LDX     #INFIFO
        LDY     #COEFF
        CLR     SUM
        CLR     SUM+1
        LDAA    0,X
        LDAB    0,Y
        JSR     MAC
        LDAA    1,X
        LDAB    1,Y
        JSR     MAC
```

```
         LDAA    2,X
         LDAB    2,Y
         JSR     MAC
*now can do something with result in SUM
```

Multiply and Accumulate Using the 68HC16.   The 68HC16 has extra registers, some of which are dedicated to multiply and accumulate operations. The H register is a 16-bit multiplier register. The I register is a 16-bit multiplicand register. Register AM (the MAC accumulator) is a 36-bit multiply and accumulate register. The 68HC16 has several extra instructions, including MAC (multiply and accumulate) and RMAC (repeat MAC). If you wish, take a preview look at Figure 15.3.

The MAC instruction multiplies the contents of registers H and I and adds the product to register AM. After this, it loads registers H and I with the next elements to be multiplied. The RMAC instruction repeats MAC the number of times specified by the contents of register E, another new 16-bit register. The 68HC16 can run with a maximum clock of 16.78 MHz. Like other DSPs, its main speed improvement is the reduction of clock cycles required to execute a multiply and accumulate. It executes MAC in 12 cycles. RMAC requires six cycles plus 12 cycles for every time it executes a MAC. Even so, this is still modest performance by DSP standards. The DSP56000 can do 24-bit multiply, 56-bit accumulate, and load data for the next operation during one clock cycle at 33 MHz.

The Fast Fourier Transform (FFT).   Often, it is easier to process signals using frequency instead of using time samples. A fast Fourier transform (FFT) is an algorithm to convert time information about a signal into frequency. As mentioned in our discussion of sampling time (Section 13.4), a signal contains frequency components.

One FFT application is to detect telephone tone signals. A touch-tone phone transmits dual-tone multifrequency (DTMF) signals when someone makes a call. Each key is transmitted as a combination of two sine waves at different frequencies. At the other end, a DSP samples the DTMF signal and performs an FFT to find the frequency of the two sinusoidal components. This identifies a telephone key.

Another application is signal filtering. A DSP samples the input and performs an FFT to get its frequency response. Then the DSP multiplies the input frequency response with the filter frequency response. For example, to remove a frequency component, it multiplies by zero. After the multiplication, the DSP performs an inverse FFT and sends the result to a D/A as a filtered output.

Now we will say more about the FFT operation itself. It is not our intent to give a complete explanation, since this can be obtained from other sources (references 19 and 20 in Appendix C). We will summarize the concepts to give you an idea of the computational requirements. The FFT is an algorithm using a divide-and-conquer strategy. It breaks the calculations down into simpler computational blocks known as *butterflies*. Basically, the algorithm uses the butterfly module for pairs of input samples. This is pass 1 (or stage I). Then it makes a second pass to produce new results from the results of pass 1. The process continues similarly for n passes where the number of samples is 2^n. Finally, the algorithm sorts the results into the proper sequence using a bit-reversing procedure. For example, result index 25 (%000011001) of 512 points is bit-reversed to become a frequency index of

304 (%100110000). The end result is the magnitude of N frequency components for frequencies between zero and the sampling frequency. Equation (13.12) shows the value of each frequency component.

$$f_k = \frac{f_s}{N}k \quad 0 \le k < N \tag{13.12}$$

where f_k = frequency point indexed by k

f_s = sampling frequency

N = number of samples

k = integer index for FFT points

Telephone Dial Tone Decoding Example.   Although specialized chips can be used for DTMF decoding, some designers may choose to use a DSP in a telephone receiver. To illustrate this, consider a 256-point FFT using a sampling frequency of 6700 Hz. Using equation (13.12) to find f_k, we find that

$$f_k = 26.2k$$

This means that the FFT result shows frequency magnitudes in increments of 26.2 Hz. With a sampling frequency of 6700 Hz, it will take 38.2 ms ($256 \times [1/6700 \text{ Hz}]$) to collect the samples. During DTMF tone generation, each digit is sent for at least 40 ms followed by a minimum interdigit interval of 40 ms. Hence, the DSP can collect the samples during the interval of digit transmission. Figure 13.6 shows the FFT indices used to detect a telephone key. For example, if the FFT shows peak magnitudes for indices 33 and 51, the number eight key was pressed (852 Hz and 1336 Hz).

The Butterfly.   As mentioned earlier, a butterfly is a computational block. It has two inputs and calculates two outputs. Both inputs and outputs are complex numbers. If you are familiar with complex number theory, you will realize that numbers have both real and imaginary parts. If you are not, you will have to accept this information at face value. The butterfly equations are the following:

$$\begin{aligned}
C_r &= A_r + B_r \cos \theta + B_i \sin \theta \\
C_i &= A_i + B_i \cos \theta - B_r \sin \theta \\
D_r &= 2A_r - C_r \\
D_i &= 2A_i - C_i \\
\theta &= \frac{2\pi}{N}k
\end{aligned} \tag{13.13}$$

where A, B are inputs

C, D are outputs

r refers to the real component

i refers to the imaginary component

N = number of samples

k = integer whose value depends on location of butterfly

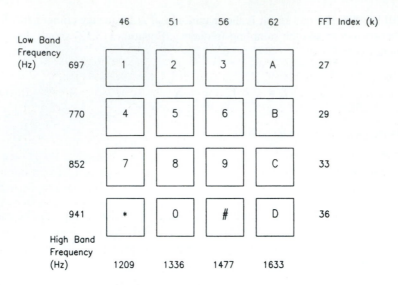

Example based on a 256-point FFT with a sampling frequency of 6.7 kHz.

Figure 13.6   Touch-Tone Decoding Using an FFT

The sine and cosine values are constants that can be referenced using a table. Many details are missing, such as necessary scaling and protection against overflow. Also, we have not indicated how the *k* values are obtained. It is explained well enough in literature specializing in DSP and numerical techniques. If you look at the available literature about FFTs, you will note variations of the algorithm.

■ 13.6 USING THE C LANGUAGE

In Section 2.3.2 we mentioned using C for programming microcontrollers. In particular, the program of Listing 2.3 illustrates a typical C program. As well, we have included C code examples throughout various chapters to illustrate data processing and I/O features. Recall from Chapter 2 that a compiler is a translation program. It translates source code from a high-level language into object code. It is also possible to tell the compiler to produce an equivalent assembly language listing for documentation purposes. The examples we illustrate are based on ImageCraft's ICC11, a DOS-based C cross compiler written by Richard Man. In particular, we will use pre-general-release version 0.48 since it is available from ftp sites on the Internet (see Appendix H). Appendix C lists other suppliers of C compilers in the sources section. Note that ImageCraft does offer a commercial version of ICC11 at a low price. Unless indicated otherwise, any mention of ICC11 refers to the pre-general-release version. You may find that it has more than enough features.

13.6.1 Why C?

Many use C to program microcontrollers because of its efficiency and ease of use relative to assembly.[3] Efficient program code is one that does the most things possible using the least amount of memory (code and data) and time. Another reason is increased productivity. Typically, a software developer can write more code to do more things using C than when using assembly. This is important since at least half the cost of developing an MCU application is in paying people to write the software.

Also, there are more people who know how to program in C than in assembly. This is because C source code is standardized (Kernighan and Ritchie or ANSI) and portable. In 1990 the American National Standards Institute (ANSI) published its C language standard as X3.159-1989. You can write your C source code anywhere and then compile it for the target processor of your choice. This assumes that you have a C cross compiler available for your chosen processor. Normally, one uses separate files for code that can run on any processor and code that is specific to a type of processor. Then one produces the executable code by linking the general object files with the appropriate processor-specific object files.

As we shall see (no pun intended), writing microcontroller software often requires knowledge of the bits and registers, even if we are programming in C. However, this bit and register information can be disguised if someone else has written appropriate *functions*. A function in C is the equivalent of the subroutine in assembly language (more or less). Also, a good knowledge of memory resources is required since the typical 8-bit microcontroller does not have the luxury of the large RAM space available to computers.

C is considered to be a good programming language for real-time control applications because it has some of the compactness and speed features of assembly. Yet it has high-level language features of portability. Also, flow control is more flexible and easier to use (see Listing 2.3). Recall how I/O can be handled using the ICC11 compiler.

```
PORTB = 0x5c;
```

In C you indicate a hex number by prefixing it with 0x. To read port E and store it at a location referenced by the label in 1, the instruction might be

```
in1 = PORTE;
```

It is possible to manipulate I/O or memory at the C source level using ICC11. However, one has to use some fairly sophisticated definitions or an appropriate header file. To review basic I/O port handling consider Listing 13.8.

```
/* Listing 13.8 */
/*Test register use by ICC11 */
#include<hc11.h>

/* Main program */
unsigned char in1;      /* a variable to store input */
```

---

[3]The Ada language is also highly efficient.

```
void main(void)
{
        PORTB = 0x5c;    /* write number $5C to PORTB */
        in1 = PORTE;     /* assign variable in 1 with the contents
                    of PORTE */
}/* End of Program Listing */
```

After heaping praises on the virtues of C, is there still a place for assembly? Yes, there is. Many applications require small (and cheap) microcontrollers in order to keep the price reasonable. Examples of such applications are microwave ovens, videocassette recorders, and coffeemakers. These applications use MCUs with limited memory resources. C programs use the stack very often and the 256-byte RAM in a single-chip 68HC11 does not leave a lot of room for excess data storage. Also, if fast I/O response is required, assembly language code executes faster.

To get the best of both worlds, programmers will use both. They use assembly language where speed is critical. C has a provision to allow assembly language code to be embedded within it. Also, some companies have developed optimized compilers that can sometimes produce machine code only a few bytes larger than one produced from an assembly language program doing the same thing. Alternatively, one can write sections of application code in C and time-critical sections in assembly. Then the object codes are linked to produce the final machine code.

13.6.2 Control Examples

Now let's take a look at another example (Listing 13.9). The program calls functions. This is basically like calling subroutines in assembly language.

Calling the Functions in a Motor Control Example

```
/* Listing 13.9
C source code to demonstrate three-wire motor control.
It also reviews how a main program can call
functions. */

#define TRUE 1
#include <hc11.h>

/* Library functions have the following prototypes
to tell the compiler how to call the functions. */

void motoron(void); /* Prefix void means no result returned. */
void motoroff(void); /* Void in parentheses means no argument re-
quired.*/
int stop(void); /*Prefix int means result returned as 16-bit inte-
ger*/
int start(void);

/* Main program follows */
```

```
void main(void) /* One 'main()' statement is necessary for
every C program (which may include several files).
Also every group of statements including the
instructions in main are enclosed in braces {}.
*/
{
        motoroff(); /* call function to initialize motor off */
        while (TRUE) /* repeat forever */
        {
                if(stop()) motoroff();
                else if (start()) motoron();
        }/* end while loop */
}/* end main */


/* Function definitions */

void motoron(void)
{
        /* motor on is PB0 high */
        PORTB = PORTB | 0x01;
}
void motoroff(void)
{
        /* motor off is PB0 low */
        PORTB = PORTB & ~0x01;
}
int stop(void)
{
        int Result= 0; /* initially assume false */
        /* return true (1) if stop (PE7) low */
        if ((PORTE | 0x7f) == 0x7f)
                Result = 1; /* if low return true */
        return(Result);
}
int start(void)
{
        int Result = 0; /* initially assume false */
        /* return true (1) if start (PE6) high */
        if ((PORTE | 0xbf) == 0xff)
                Result = 1; /* if high return true */
        return(Result);
}
```

Compare this code with the pseudocode of Listing 13.1. It is reasonably intuitive to guess what the C program is doing.

C Expressions and Statements. The first thing the main program does is to turn off the motor by calling a function to do so. After returning from the function, it enters a control loop as indicated by the *while* statement. A while statement evaluates the *expression* enclosed within its parentheses ( ). Then it executes any *statement* following it until the condition is false.

A C expression is defined as a combination of operations and operands that evaluates to a single value. In this case the expression is a constant label. A statement can be single or compound. A single statement consists of instructions, operations, and operands that end in a semicolon (;). A compound statement is a group of single statements enclosed in braces {}. Any expression that is not zero is considered to be true. Hence, the *while* statement always executes because its expression is the label TRUE with a constant value of 1.

The compound statement has if-else flow control statements. The *if* statement also evaluates the expression enclosed inside its parentheses. In this case it calls the function stop. If this function returns with a true value (nonzero), the *if* statement proceeds to execute the rest of the statement, in this case to call function `motoroff`. If the return value of stop is false (zero), execution continues with the next statement, in this case the *else if* statement. The else if statement operates similarly.

Functions. Recall that a function is the equivalent of a subroutine. After the program calls either `start` or `stop`, the function returns an integer value to the main (calling) program to indicate true or false. These functions are defined as *int*, meaning they return an integer value. Hence, the main program can evaluate the returned values in the *if* and *else if* statements. The functions `motoroff` and `motoron` do not return values to the calling program (defined as *void*). Instead, they simply perform an action. None of the functions in Listing 13.9 required *arguments*. Arguments are the actual values enclosed within the parentheses of the function. If used, they indicate input information required for the function to perform its task.

Bit Manipulation and Checking. So far, there is not much to indicate how this program controls I/O. For this we can initially examine another source code. Consider first setting or clearing bit flags as shown in Listing 13.10.

```
/*Listing 13.10
  C source code to demonstrate setting and clearing
  bit flags. */

/* Use directives to define bit positions */
#define BIT0     01
#define BIT1     02

/*Global variables */
unsigned char flag1;   /* A declaration statement*/

/* Main program follows*/
void main(void)
{
        /* Initialize flag 1 */
        flag 1 = 0x5c; /* Instruction 1*/
        /* Set the BIT0 and BIT1 flags in flag 1*/
        flag 1 |= (BIT0 | BIT1); /*Instruction 2*/
        /* Clear the BIT0 and BIT1 flags in flag 1 */
        flag 1 & = ~(BIT0 | BIT1); /* Instruction 3*/
}/* end main */
```

When the program executes, it initializes the variable `flag1` to the value of $5C. The prefix "0x" is C's method to indicate a hex value. Then the program sets bits 0 and 1. Then it clears them. The logical bit operator symbols are the following:

	OR
&	AND
~	NOT

The logical bit reassignment operators are

|=	left variable = left variable OR right expression
&=	left variable = left variable AND right expression

In the case of Instruction 2, $5C (contents of `flag 1`) is ORed with ($01 OR $02) to reassign `flag 1` with the value $5F ($5F = $5C OR $03). In Instruction 3, $5F (now the content of `flag 1`) is ANDed with NOT ($01 OR $02); hence, $5C = $5F AND NOT($01 OR $02) = $5F AND $FC.

A simpler way to do all this is to use constant mask values to define all bits, such as

```
flag1 |= 03;      /* flag1 = flag1 OR $03 */
flag1 &= 0xfc;    /* flag1 = flag1 AND $FC */
```

Some programmers may choose to use each one individually since most flag bits have a specific purpose. There are several ways to implement bit manipulation in assembly language. Some compilers can be forced to use an efficient method.

Declaration.   When variables are used, it is necessary to *declare* them. A declaration statement tells the compiler how to handle any reference to variable names. This is done by using reserved words to specify the *type* of variable. C allows for the use of different types of variables, such as signed or unsigned 8- or 16-bit integers and other types we mention later. The reserved words unsigned char tell the compiler that the variable is an unsigned 8-bit integer.

In this case the compiler will set up variable `flag1` as a fixed address in RAM. This is known as a global variable because any function in the program can reference it. Declarations outside the body of the main program cause variables to become global. We chose a global declaration to make the resulting assembly code simpler. This tells the compiler that it is not necessary to save stack space for variables used by a function. In C, main is also a function. After it is called during execution, the program returns to the code of the startup file. Local variables (declared within the body of main or functions) are set up as nonfixed addresses in the stack, which are referenced using a pointer. We will say more about local variables later.

Assembly Source File and Library File.   Listing 13.11 shows the resulting assembly language code created when the program of Listing 13.10 is compiled. We added some comments to cross-reference assembly source code with the C source code. Note that flag1 is referenced using extended addressing (a word address).

```
*Listing 13.11
_main::
        ldab #92;           flag1 = 0x5c (= decimal 92)
        stab _flag1
        ldy #_flag1
        bset 0,y,#3         ;flag1|= (BIT0 | BIT1)
        ldy #_flag1
        bclr 0,y,~#-4       ;flag1 &= ~(BIT0 | BIT1)
L1:
        rts
```

Note that main is compiled as a subroutine. We show assembly language instructions in lowercase letters because the compiler does the same when told to produce an assembly language listing. The underscore (_) prefix means that a label is global, meaning that the same function or variable can be used in any source code file. The resultant hex file includes the startup (and library if applicable) code as well as the hex code of the compiled C source file. Note that ~#−4 is the same as #3.

Bit Checking. Next we will examine how to control program flow (branching) based on bit flag values (Listing 13.12).

```
/* Listing 13.12
   Demonstrate program flow control based on
   bit flags. */

/* Use directive to define bit positions. */

#define BIT0 01
#define BIT1 02
/* declare two global variables */
unsigned char flag1, out1;

void main(void)
{
        /* If bits 0 and 1 in flag1 are set, out1 = 1 */
        if ((flag1 & (BIT0 | BIT1)) == (BIT0 | BIT1))
                out1 = 1;
        /* If bits 0 and 1 in flag1 are clear, out1 = 2 */
        if ((flag1 & (BIT0 | BIT1)) == 0)
                out1 = 2;
} /* end of program */
```

The symbol == is the *equality relational operator*. If the left side equals the right side, the expression is evaluated as true (nonzero). If they are not equal, the expression is evaluated as false (zero). In fact, this is a good time to refer to "C Conventions" in Appendix D. The equivalent assembly language program follows in Listing 13.13.

```
* Listing 13.13
_main::
        ldab _flag1     ;if ((flag1 & (BIT0|BIT1)) == (BIT0|BIT1))
```

```
              andb    #3
              cmpb    #3
              bne     L2          ;(else evaluate next if statement)
              ldab    #1          ;out1 = 1
              stab    _out1
      L2:
              ldab    _flag1      ;if ((flag1 & (BIT0|BIT1))==0)
              andb    #3
              tstb
              bne     L4          ;(else no change to out1)
              ldab    #2          ;out1 = 2
              stab    _out1
      L4:
      L1:
              rts
```

PRACTICE

Write C statements to set bits 1 and 3 in variable var1 and clear bits 1 and 3 in variable var2.

Solution

```
var1 |= 0xa;
var2 &= ~0xa;
```

13.6.3 The PID Controller Using C

In Section 13.4 we discussed PID control (see Listing 13.5). Listing 13.14 shows one way to implement it using C. It also serves to illustrate software issues of scaling, particularly when integer arithmetic is used.

```
/* Listing 13.14
   C source code to calculate PID output.
    C0 = C1 + A1*E0 - A1*E1 + A2*E2 */

#include <hc11.h>
/* function prototypes */
void startad(void);       /* start up the a/d */
unsigned char analog(unsigned char); /* analog input from speci-
fied channel */

/* PID controller gains */
#define         A2      17
#define         A1      21
#define         A0      75

/* Reserve RAM area for variables */
```

```c
static unsigned char SP, PV;
static int C1, C0;
/* Will scale C1, C0 to 0 - 16383 range */
static int E2, E1, E0;

void main(void)
{
    /* Initialize RAM variables */
    SP = 64; /* scaled halfway between 0 and 127 */
    E2 = 0; E1 = 0;
    /* Read first input, PV. */
    startad(); /* get first analog input */
    PV = analog(0);
    /* 1st output is half of PV,
    caled between 0 - 16383 */
    C1 = PV << 6;

    while(1) /* Repeat loop continuously. */
    {
            /* Scale output to byte size and send it to D/A connected
            to port B. */
            PORTB = C1 >> 6;

            PV = analog(0); /* Get next input. */
            PV = PV >> 1;

            E0 = SP - PV;    /* Calculate next error */
            /* Now calculate next output. */
            C0 = C1 + A0*E0 - A1*E1 + A2*E2;

            /* Clamp output at limits in case of over/underflow
            */
            if (C0 > 16383) C0 = 16383;
            if (C0 < 0) C0 = 0;

            /* Then reassign variables. */
            C1 = C0;
            E2 = E1;
            E1 = E0;

            /* will require a time delay here,
    use embedded assembly to call BUFFALO 10-ms time delay */
            asm(" jsr $e2d9");
            /* Actual address depends on BUFFALO version */
    } /* end while loop */

} /* end main */

void startad(void)
{
```

```
                    /* start up the a/d */
                    OPTION = OPTION | 0x80;    /* ADPU = 1 to power up */
                    OPTION = OPTION & ~0x40; /* CSEL = 0 tjo select clock */
                    asm(" jsr $e2d9"); /* time delay */
        }

        unsigned char analog(unsigned char Port)
        {
                    /* analog input from specified channel */
                    INTR_OFF();    /* disable interrupts */
                    ADCTL = Port;
                    while ((ADCTL & 0x80) == 0); /* wait for conversion complete
                                                   */
                    INTR_ON();    /* enable interrupts */
                    return(ADR1); /* return a/d result */
        }
```

The source code declares the variable as 16-bit signed integers. The function startad will initiate A/D conversion. The function analog returns a value from the specified analog channel, in this case channel 0.

The initial value of C1 is the analog input shifted left by six. The left-shift operator is << and the right-shift operator is >>. The program does this because the loop portion must scale the calculated 16-bit output to fit into 8 bits. Then it is possible to send the 8-bit output to the D/A at port B. It does this by shifting the calculated output result right six times (C1 >> 6). Recall that left and right shifts are equivalent to multiplying and dividing by powers of 2, respectively. Then it calls the function out to send the data to port B. In the *while* loop, the process variable (PV) is scaled by a right shift to ensure that no overflow occurs when calculating the error (E0).

After this, the program calculates the controller output (C0). The asterisk (*) is the multiplication operator. The equivalent assembly language code for this statement handles signed 16-bit addition and multiplication. However, it will not detect or correct overflow.

The two *if* statements ensure that the software does not try to output a controller output value beyond the range chosen. Without this protection, problems could occur. For example, the input sensor may be damaged. In this case, the controller may never see a change in the process variable; hence, it will always find the same error. The controller will continue to increase (or decrease) the output. Without clamping, an overflow (or underflow) will occur. This results in a sudden swing of the 8-bit number sent to the D/A in the opposite direction.

Scaling.   For example, a previous output of $FF (100%) could overflow to become something like $7C (48%). Let's see how this could have happened. Assume that the previous output was $3FFE (16,382). This means that $FF was the last value sent to the D/A. Remember to scale by a 6-bit right shift. Now, let's say that the error is always $70 because of the damaged sensor. Then the PID output becomes $5F0F (E0, E1, E2 = $70). After shifting right by six, the value sent to the D/A becomes $7C. A sudden swing from 100% to 48% is quite drastic.

The PID gains were chosen arbitrarily with the following condition:

$$A_0 + A_1 + A_2 < \frac{16,383}{127} \tag{13.14}$$

This ensures that no 16-bit signed overflow will occur when the gains are multiplied by the errors. Otherwise, the software will have to check for overflow/underflow during calculations.

Floating-Point Math.   The problem with integer arithmetic is the limited range of values available. Using floating-point arithmetic solves this problem but introduces another. When floating-point math is used, the resulting compiled code is longer and will take longer to execute. This is a problem for 8-bit microcontrollers that have to update results quickly for real-time control applications. A floating-point calculation may take up to 100 times longer than an integer calculation. For a review of floating-point numbers, refer to Section 2.7.7. Also note that Section 3.4.2 described the floating-point library for assembly language. ICC11 also supports floating-point math.

Embedded Assembly.   Most C compilers support embedded assembly. In our case we used embedded assembly to use the BUFFALO time-delay subroutine. An alternative would be to write a function or to modify the library to include a time-delay function.

13.6.4   C Function Library

Using Functions.   C compilers are supplied with functions to handle some of the I/O subsystems for the target microcontroller. Each of the library functions have different conventions as to the type of variables or expressions it will accept as a calling parameter.

The manuals or help system supplied with the C compiler will specify any conventions and conditions for using the library functions. This includes required header files and variable types of *parameters*. A parameter specifies a variable that a calling function passes to the called function and/or a variable that the called function passes to the calling function. The manuals will also state configuration requirements. Sometimes, they include examples.

ICC11 contains most standard ANSI C functions. One example is abs, which returns the absolute value of its argument. To use it, you must include the header file `stdlib.h`, which prototypes the function for the compiler. Another is the `printf` function, which writes information to the standard output device, typically a monitor screen. Header file `stdio.h` prototypes this function. Consider Listing 13.15, which demonstrates two standard functions with the commercial version of ICC11.

```
/* Listing 13.15
Program demonstrates using two library functions */.

/* Necessary header file to use function abs*/
#include <stdlib.h>

/* Necessary header file to use function printf*/
#include <stdio.h>

/* Necessary code to start main program */
void main(void)
{

        /* declare integer variable to use as a parameter */
        int par;
```

```
    /* Declare another integer variable to store the result.
    It is the value returned by the function */
    int retval;

    /*Assign value to the parameter to make it an argument */
    par = -5;

    /*Call function abs with parameter par (argument of -5) */
    retval = abs(par);

    /* Now variable retvar has the value of 5 because |-5| = 5
    Next, call function printf to display result to the
    monitor */
    printf("Answer is %d\n", retval);

} /*End program*/
```

To use the standard functions, the program must include the appropriate header files. The compiler may have to be configured to include the library itself. It is not necessary to define the functions since they have already been coded in a *library*. The user program simply uses them as shown in Listing 13.15.

More About Standard I/O.   Standard I/O is normally used in a general computer system. It uses a keyboard as the standard input device and a monitor screen as the standard output device. Hence, *standard* I/O is not very meaningful in a typical microcontroller system which generally uses customized I/O, depending on the application. However, ICC11 does support a rudimentary form of the printf function. It relies on a BUFFALO subroutine to implement a function called putchar which in turn is used by printf. Function putchar outputs the contents of accumulator A to the monitor screen. If your system does not use BUFFALO, you will have to provide your own version of putchar.

Run-Time Libraries.   In general, a run-time library contains code for common functions and the essential startup code required for a particular computing environment. Normally, it is in object code form (already compiled). The programmer writes code that references the library functions and then compiles it to produce object code. The linker combines the library and programmer-written object code to produce the final executable code. Hence, each executable program carries its own copy of library code. This leads to duplication of code when running several programs that all use the same library. Some operating systems, such as Windows, permit executable code to call upon one single copy of the library as required. In the case of ICC11, the startup code is in the object file crt11.0 and it includes various library files such as one for floating-point math.

13.6.5 User-Defined Functions

When the user writes his or her own functions in the C source code, he or she must prototype it and then define it. Listing 13.16 demonstrates a function.

```
/* Listing 13.16
Program demonstrates user-defined functions */

/* Function prototype */
void callme(unsigned int arg1, unsigned int arg2);

/* Main program */
void main(void)
{
        unsigned int mvar1, mvar2; /* local variables to main */
        mvar1 = 5; /* assign initial values to local variables */
        mvar2 = 15;
        callme(mvar1, mvar2); /* call the function */
        mvar1 = mvar2;
}


/* Function definition */
void callme(unsigned int arg1, unsigned int arg2)
{
        unsigned int var1, var2; /* local variables to callme */
        var1 = 2*arg1;
        var2 = 2*arg2;
        arg2 = var1 + var2;
}
```

We defined function callme to be *void* in order to simplify the future explanation of stack use. The calling program (main) passes variables mvar1 and mvar2 to the function as its arguments. It modifies the second argument but will not return it back to main. To cause function callme to return a value, one could define (and similarly prototype) it as follows:

```
unsigned int callme(unsigned int arg1, unsigned int arg2)
{
        return 2*arg1 + 2*arg2;
}
```

Then in main, use the statement : mvar2 = callme(mvar1, mvar2);

Local Variables.   Listing 13.16 also illustrates local variables. Variables var1 and var2 are accessible only while callme is executing. Variables mvar1 and mvar2 are available only while main is executing. Normally, it is good programming practice to use local instead of global variables. This prevents unexpected variable modifications by other functions. Functions using only local variables are reentrant. If an interrupt calls the same function that it was interrupting, each instance of the function will use its own variables. In Section 13.6.6 we will look behind the scenes to see how the program uses the stack to work with local variables.

Global Variables.   A global variable is one that is available to any part of the program. To specify a global variable, you declare it ahead of main. The following illustrates a global declaration.

```
/* Simple example to declare a global variable */
int globvar; /* declare it outside main */
void main(void)
{
        globvar = 5; /* main and any other function can use it*/
}
```

After compiling the program, the equivalent assembly code will reference global variables as absolute addresses in memory. Sometimes it is necessary to work with global variables. A typical embedded control system runs several tasks to deal with different I/O and computation activity. They usually have to communicate with each other. Any part of the final program can use the variable to keep track of how many milliseconds have expired since it was reset. Because global variables are fixed locations, any function that uses them cannot be reentrant.

13.6.6 Behind the Scenes: Using the Stack

Stack Use.   Another noteworthy point is C's use of the stack. Remember that the result of compilation is an equivalent assembly language program that does the same thing as the original C source. The resulting program uses the stack to store function variables. Both the calling program (also a subroutine) and called subroutine (nested inside calling subroutine) use a portion of the stack as a *frame* for accessing variables. In general, a called routine sets up its own frame by using the stack pointer to set up a frame pointer. Typically, it uses an index register as a frame pointer to address the variables stored in the frame. The exact implementation of this depends on the C compiler being used. Normally, we do not worry about it since the compiler does the job for us. However, if you want to code your own library functions using assembly language, you must understand how the stack is used. The following provides more details using a program compiled by an earlier version of ICC11 (pre-general-release version 0.48) as an example, since it uses a simpler method. The main point is that C programs require significant stack space.

Listing 13.17 shows the result of compiling the program of Listing 13.16. Figures 13.7 and 13.8 illustrate stack use for the program.

```
*Listing 13.17
*Shows assembly code produced by compiling the C source code
*of Listing 13.16 with additional comments manually inserted
*by the author. Note that FP = frame pointer and each type int
*variable uses two bytes of space.

*Assembly mnemonics are lowercase because that's the
*way ICC11 creates them.
        sect 0
_main:
* mvar2 -> 4,x
* mvar1 -> 6,x
*                       ;4 pshx's to create main frame
        pshx            ;set up space for parameter passing
        pshx
        pshx            ;unsigned mvar2 - space for mvar2
```

```
                pshx                 ;unsigned mvar1 - space for mvar1
                tsx                  ;SP+1 -> IX, assign FP for main
                ldd #5               ;mvar1 = 5
                std 6,x
                ldd #15              ;mvar2 =15
                std 4,x
                ldd 6,x              ;callme(mvar1, mvar2)
                std 0,x              ; - store arguments to
                ldd 4,x              ; pass parameters to callme
                std 2,x
                jsr _callme          ; - push return address onto stack
        *                            ; and invoke function
                tsx                  ;SP+ 1 ->IX, reassign FP for main
                ldd 4,x              ;mvar1 = mvar2
                std 6,x
                pulx                 ;4 pulx's to remove main frame
                pulx
                pulx
                pulx
                rts                  ;and return to call by monitor
L1.callex:
_callme:
*    var2 -> 0,x        - local variables
*    var1 -> 2,x
*    arg2 -> 8,x        - arguments
*    arg1 -> 6,x
*                            ;2 pshx's to create local variable space
*                            ; in callme frame
                pshx                 ;unsigned int var1
                pshx                 ;unsigned int var2
                tsx                  ;SP+1 ->IX, assign FP for callme
                ldd 6, x             ;var1 = 2* arg1
                lsld
                std 2,x
                ldd 8,x              ;var2 = 2* arg2
                lsld
                std 0,x
                ldd 2,x
                addd 0,x             ;arg2 = var1 + var2
                std 8,x
                pulx                 ;2 pulx's to remove local variables
                pulx
                rts                  ;and return to calling routine
L2.callex:
```

Refer to Figure 13.7. A calling function stores input parameters at the top of its frame (step 1). The subroutine call pushes the return address onto the stack (step 2). The called function may set up additional stack space for its own local variables (step 3). Then it loads an index register with the current stack pointer (SP) value to use as its frame pointer (step 4)

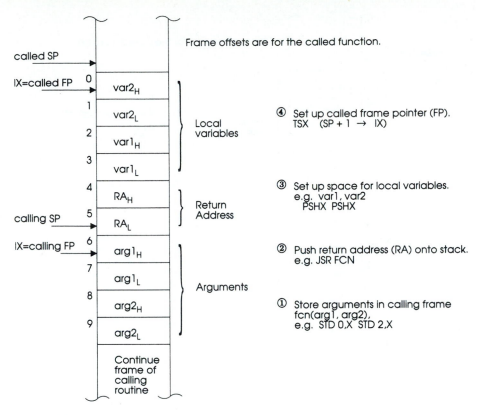

Figure 13.7 Setting Up the Stack Frame during a Function Call

Actually, the TSX instruction increments the stack pointer prior to transfer since the stack pointer always points *above* the stack. The function may write the results to memory referenced by the frame pointer.

The function (subroutine) can now use the stack frame for local variables as shown in the code for subroutine _callme in Listing 13.17. Each variable uses 2 bytes of space. The subroutine uses an index addressing offset to identify a local variable or argument. The frame of the routine called actually overlaps the frame of the calling routine. The parameters (arguments) passed between the two is in the overlap area. Other C compilers may produce non-overlapping frames.

Refer to Figure 13.8. Before returning, the called function removes (deallocates) the local variable space (step 1) by pulling them off the stack. Then it executes a return by pulling the return address off the stack (step 2). The calling function (main in this case) restores the old frame pointer (step 3) by using the TSX instruction. Then it reads the arguments from the frame. Some of these arguments may be results that the calling function needs.

Sometimes the function itself returns a value. Function analog is one example. The following statement causes variable in 1 to be assigned the input value in analog channel 0.

```
in1 = analog(0);
```

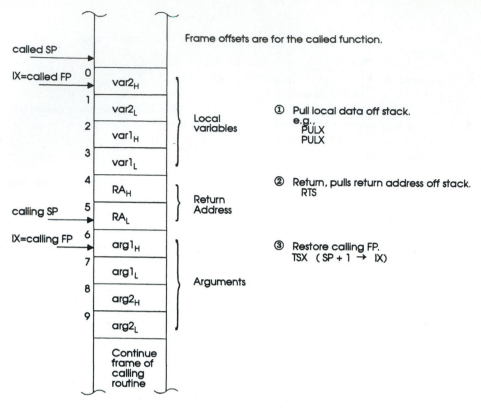

Figure 13.8   Returning From a Function Call

When a function returns a value, the equivalent assembly language code could use accumulator D to return a value of data type integer (16 bits). Stack use causes a lot of software overhead. This can be reduced by declaring local variables as *static*. This tells the compiler to assign them to fixed locations in RAM, but it does cause some problems. This leaves less space available for other uses. Note that a microcontroller may have limited RAM space. Also, any functions using static variables cannot be reentrant; that is, an interrupt cannot reuse the function.

13.6.7 Pointers

Pointer Basics   A program stores each variable's data in memory. The address of each variable is the address of its first storage location. A *pointer* variable is one that stores the *address* of another variable. Hence it *points* to another variable. Pointers are typically used to access individual elements in a block of variables. This is one of the most important features of C.

Consider using two variables.

```
char cvar;        /* declare variable of type character */
int ivar; /* declare variable of type integer */
```

```
cvar = 5;          /* assign value */
ivar = 258;        /* assign value */
```

In this discussion we use C syntax to indicate hex numbers. For example, 0×d000 means the same thing as $D000. Perhaps address 0×d000 is used to store the byte value of cvar. This would be 0×5. Also suppose that addresses 0×d001 and 0×d002 are used to store the 2-byte value of ivar (0×102). This would be 0×1 and 0×2, respectively. Now consider their pointers. The character pointer would have a value of 0×d000 to indicate the address of cvar. The integer pointer would have a value of 0×d001 to indicate the first address of ivar. This sounds a lot like indexed addressing. In fact, the processor uses index registers to implement pointers.

Address and Dereferencing Operators   To use pointers in C, there are two unary operators, the address operator and the dereferencing operator. The operator symbols are ampersand (&) and asterisk (*), respectively. Some people use the term *indirection* operator instead of *dereferencing* operator.

When you apply the address operator & to a variable, the result is the *address of the variable*. For example,

```
cptr = &cvar;   /* cptr is assigned the "address of" cvar */
iptr = &ivar;   /* iptr is assigned the "address of" ivar */
```

Hence, cptr and iptr would have values of 0×d000 and 0×d001, respectively. Both of these variables are pointers.

When you apply the dereferencing operator * to a pointer variable, the result is the *value at the address* assigned to the pointer variable. For example:

```
cget = *cptr;    /*    cget assigned contents of address
                       specifed by pointer */
iget = *iptr;    /*    iget assigned contents of address(es)
                       specified by pointer */
```

Hence, cget is assigned the value of 0×5 because this is what was stored in address 0×d000. Remember that pointer variable cptr has a value of 0×d000. Similarly, iget is assigned the value of 0×102 because this was stored in address 0×d001 as a 2-byte number. Note that iptr is an integer-type pointer with a value of 0×d001.

Pointer-Type Declaration.   Like any other variable, you have to declare pointers before you can use them. To indicate a pointer variable, the variable name must be prefixed by an asterisk (*). For example:

```
char *cptr;     /* declare pointer to type character */
int *iptr; /* declare pointer to type integer */
```

Note that the declaration allocates space for the named pointer, but not what it points to. Remember that a pointer contains an address. Consider the following:

```
iptr = ivar;    /* assign value of 0x102 to iptr */
iget = *iptr;   /* assign contents of addresses 0x102
                    and 0x103 to iget */
```

Perhaps this is not what you intended to do. Instead, you may have intended something like

```
iptr = &ivar;  /* assign address 0xd001 to iptr */
iget = *iptr;            /* assign contents of addresses 0xd001
                            and 0xd002 to iget */
```

You may ask: Why not do something simpler, such as

```
iget = ivar;     /* assign iget with value of ivar */
```

This is true in this case. However, there are situations when one must pass many parameters to a function. This will cause a lot of overhead. To reduce overhead, pass a pointer instead to reference a block of parameters. Also remember that the 68HC11 uses memory-mapped I/O, meaning that it accesses I/O ports as if they were memory locations. To access I/O registers at the C source code level, you must use pointers.

Using Pointers to Access Registers.   In Section 13.6.1 we stated that a C program could directly access registers with something like

```
PORTB = 0x5c;  /* drive output port */
in 1 = PORTE;            /* read input port */
```

As it stands, the compiler will not understand the names PORTB and PORTE. To define them, use the #define directive to assign a constant value to a symbolic name. With ICC11, you can use something like the following:

```
#define PORTB *(unsigned char *) (0x1004)
#define PORTE *(unsigned char *) (0x100a)
```

Recall that the addresses of port B and port E are 0×1004 and 0×100a, respectively. This assumes that you are using a default register start address of 0×1000. The directives define each name as the *contents* of a constant *pointer* to a single byte (unsigned char *).

Header File HC11.H.   It would be inconvenient to define registers explicitly for every program. ICC11 provides a header file to define the memory-mapped registers in the 68HC11. It uses the pointer method described previously.

Start-Up File CRT11.O.   When compiling to executable, the compiler links the object code from your source with the startup object file crt11.o. It may also link the object code of appropriate library files. The final executable begins with the startup code. Your reset vector should also be the start address of this executable. This startup code will initialize the stack, initialize variables, and set up the heap (a pool of memory locations reserved for dynamic data). Then it calls the main function, and waits for main to return where it loops forever as illustrated in the following assembly source code segment.

```
;call user main routine
        jsr _main
_exit::
        bra     _exit
```

It is possible to modify the source `crt 11.s`, rename it, and modify it to perform other functions such as returning to Buffalo (if used) and setting up time-protected bits within the first 64 clock cycles. Then assemble the source using ICC11 to create a new object file. With the compiler, set up the compile options to specify your startup object file.

13.6.8 Interrupt Programming

The three features for interrupt programming are

Define the interrupt service routine(s),

Set up the vector(s),

Disable and enable the interrupt(s).

The following will show how to create interrupt service routines and define vectors directly in C. Let's start with a variation of the pulse width modulation example covered earlier as C Listing 11.4 in Section 11.2.3.

```
/* C Listing 13.18, a variation of C Listing 11.4 */

#include <hc 11.h>        /* Include register definitions */

/* Declarations */
unsigned int OC2hi, OC2lo;              /* Duty cycle data */

/* Prototypes for subroutine and service routine */
void initoc2(void);
void rtoc2(void);

void main(void)
{
        INTR_OFF();
        initoc2();
        while (1);       /* Loop forever */
}

void initoc2(void)
/* Initialize OC2, a subroutine */
{
        TOC2 = TCNT;
        TCTL1 = 0xC0;
        TFLG1 = 0x40;
        TMSK1 = 0x40;
        INTR_ON();
}
```

```
#pragma interrupt_handler rtoc2
void rtoc2(void)
/* Interrupt Service Routine for OC2 PWM */
{
    if ((TOC2 & 0x40) == 0)
            TOC2 = TOC2 + OC2lo;
    else
            TOC2 = TOC2 + OC2hi;
    TCTL1 = TCTL1 ^ 0x40; /* invert OL2 to toggle next OC2 edge */
    TFLG1 = 0x40; /* Clear flag OC2F */
}
/* Define vectors by including a definition of a vector table */
#include <vectors.c>
```

One difference is that C Listing 11.4 used embedded assembly to set up jump vectors because the target system would be an EVBU board. It also fudged the definition of the service routine by embedding an assembly RTI instruction. Let's examine how to define a service routine, and set up vectors.

Defining an Interrupt Service Routine.   Defining a function causes the compiler to generate an assembly RTS instruction because normal function behavior is to return control back to the calling routine. It is necessary to use the preprocessor directive #pragma for declaring interrupt service routines. In general, the meaning of #pragma (if used) depends on the compiler. The general form for declaring a service routine in ICC11 is

```
#pragma interrupt_handler <name of service routine>
<'function' definition of service routine>
```

For example:

```
#pragma interrupt_handler rirq
void rirq(void)
{
/* code for rirq here */
}
```

Interrupt Vectors.   ICC11 has an example file named vectors.c to illustrate how to set up the vectors. Copy it to the INCLUDE directory and modify it as necessary to define the vectors for your application. At the end of your source, use the include statement to include the file. C Listing 13.19 shows the file that was included in the interrupt example for C Listing 13.18. Note that unlike Listing 11.4, we are defining vectors and are not using Buffalo jump vectors. The code illustrates another use of pragma, in this case to force loading of a vector table at a specific address.

```
/* C Listing 13.19
Rename this file as vectors.c and save it in the INCLUDE directory
for ICC11. This is a modified vectors.c file to include in source
code.
```

In this case, define a vector for rtoc2 by entering rtoc2. */

```c
/* As is, all interrupts except reset jumps to 0xffff, which is most
 * likely not going to be useful. To replace an entry, declare your
function,
 * and then change the corresponding entry in the table. For example,
 * if you have a TOC2 handler (which must be defined with
 * #pragma interrupt_handler . . .) then in this file:
 * In the TOC5 entry, change:
 *       DUMMY_ENTRY,
 * to
 *   the name of the TOC2 routine,
 */

extern void _start(void); /* entry point in crt??.s */

#define DUMMY_ENTRY (void (*)(void))0xFFFF

#pragma abs_address:0xffd6
/* change the above address if your vector starts elsewhere
 */
void (*interrupt_vectors[])(void) =
        {
        /* to cast a constant, say 0xb600, use
           (void (*)())0xb600
        */
        DUMMY_ENTRY,                /* SCI */
        DUMMY_ENTRY,                /* SPI */
        DUMMY_ENTRY,                /* PAIE */
        DUMMY_ENTRY,                /* PAO */
        DUMMY_ENTRY,                /* TOF */
        DUMMY_ENTRY,                /* TOC5 */      /* HC12 TC7 */
        DUMMY_ENTRY,                /* TOC4 */      /* TC6 */
        DUMMY_ENTRY,                /* TOC3 */      /* TC5 */
        rtoc2, /* TOC2 */           /* TC4 */
        DUMMY_ENTRY,                /* TOC1 */      /* TC3 */
        DUMMY_ENTRY,                /* TIC3 */      /* TC2 */
        DUMMY_ENTRY,                /* TIC2 */      /* TC1 */
        DUMMY_ENTRY,                /* TIC1 */      /* TC0 */
        DUMMY_ENTRY,                /* RTI */
        DUMMY_ENTRY,                /* IRQ */
        DUMMY_ENTRY,                /* XIRQ */
        DUMMY_ENTRY,                /* SWI */
        DUMMY_ENTRY,                /* ILLOP */
        DUMMY_ENTRY,                /* COP */
        DUMMY_ENTRY,                /* CLM */
        _start /* RESET */
        };
#pragma end_abs_address
```

Enabling and Disabling Interrupts.   To do this, use the macros in the header file hc11.h. The following shows a segment from the file. Note that the macro definitions use embedded assembly.

```
#define INTR_ON()      asm(" cli")
#define INTR_OFF()     asm(" sei")
```

Vector Jump Table.   If you are using Buffalo's jump vectors, it is possible to write C source code to do this, as illustrated by C Listing 13.20.

```
/* C Listing 13.20, another variation of Listing 11.4 */

#include <hc11.h>          /*Include register definitions */

/*Declarations */
unsigned int OC2hi, OC2lo;        /* Duty cycle data */

/* Prototypes */
void initoc2(void);
void rtoc2(void);

void main(void)
{
    INTR_OFF();
    /* Set up TOC2 jump vector for Buffalo */
    *(unsigned char *) 0xdc = 0x7E; /* 7E is "jmp" */
    *(void (**)()) 0xdd = rtoc2;
    initoc2();        /* Initialize the TOC2 interrupt */
    while (1);        /* Loop forever */
}

void initoc2(void)
/* Initialize OC2 */
{
    TOC2 = TCNT;
    TCTL1 = 0xC0;
    TFLG1 = 0x40;
    TMSK1 = 0x40;
    INTR_ON();
}

#pragma interrupt_handler rtoc2
void rtoc2(void)
/* Interrupt Service Routine for OC2 PWM */
{
    if ((TOC2 & 0x40) == 0)
            TOC2 = TOC2 + OC2lo;
```

```
        else
                TOC2 = TOC2 + OC2hi;
        TCTL1 = TCTL1 ^ 0x40; /* invert OL2 to toggle next OC2 edge */
        TFLG1 = 0x40; /* Clear flag OC2F */
}
```

Note how the interrupt jump table is set up.

```
        /* Set up TOC2 jump vector for Buffalo */
        *(unsigned char *) 0xdc = 0x7E; /* 7E is "jmp" */
        *(void (**)())0xdd = rtoc2;
```

This translates to the following assembly code.

```
;       /* Set up TOC2 jump vector for Buffalo */
;       *(unsigned char *)0xdc = 0x7E; /* 7E is "jmp" */
        ldab #126
        stab 0xdc
;       *(void (**)())0xdd = rtoc2;
        ldd #_rtoc2
        std 0xdd
```

13.6.9 Other Variable Types and Features

An ANSI C program can use many other variable types and features which we will not discuss in detail. If in doubt, consult your compiler manual. The features are

Floating point: See Section 2.7.7.

Array: A block of variables (elements).

String: A special array with ASCII characters.

Structure: A feature that allows grouping of variables of different types.

Union: A feature that allows different types of data to be stored at the same memory location at different times. This is important for 8-bit microcontrollers because they have limited RAM space.

Embedded Control Features. C compilers for microcontrollers must also provide some additional features. Typically, these are header files and library functions to handle processor-specific resources. They normally run an optimizing compiler to reduce the number of machine instructions in the code compiled. Other features are interrupt handling and vector generation. ICC11 also uses the preprocessor directive called #pragma for declaring interrupt service routines. In general, the meaning of #pragma (if used) depends on the compiler. The vectors.c file is an example included to show how to define vectors.

Last Words.   Again, this is a very brief presentation of C. If you need to find out more, refer to any book about C.

■ 13.7 MULTITASKING

So far we have concentrated on distinct tasks. Typically, an MCU performs many tasks within an application. The hardy camper uses a Swiss army knife to do different things. One situation may call for a screwdriver and another may require a corkscrew. One would not buy this expensive knife to use solely as a can opener in the kitchen. Similarly, it does not make sense to use something as powerful as a 68HC11 just to turn a light on and off.

Different Tasks in a Control Application. Consider controlling an elevator. A simple breakdown of elevator control tasks is

Speed and position control of the elevator itself

Position control of elevator doors

Interaction with people (keyboard and displays)

A *task* is a piece of code. Since a computer or microcontroller is essentially a serial device (remember the fetch/execute cycles), it can execute only one task at a time. However, it is possible to make it *appear* as if the microcontroller is running several tasks in parallel. Generally, one can classify tasks as *foreground* (high priority) or *background* (low priority) tasks. Foreground tasks are interactive and background tasks are noninteractive. A typical background task is a control sequence to direct the behavior of a system. A typical foreground task manages I/O. Interrupts will cause foreground tasks to execute. A multitasking system schedules execution of background tasks subject to interruption by foreground tasks.

We can use Listings 11.5 and 11.6 to illustrate a simple multitasking system. In this case the background is the LOOP in Listing 11.6, which ensures that the motor has advanced 16 steps. Service routine RTOC4 (Listing 11.5) is the foreground to drive the outputs for the steps.

Another example of a foreground routine is the PWM driver routine (RTOC2) of Listing 11.4. The background task could be a PID controller subroutine that modifies the pulse-time durations (OC2HI, OC2LO). This program will differ slightly from that shown in Listing 13.14. It does not send an output to a D/A at port B. Instead, the output is used by the PWM routine to define duty cycle. In this case the PWM frequency should be faster than the background frequency of changing duty cycle.

Now, let's return to the elevator control system. We can break it down into individual tasks. For example:

Background:

Calculate motor speed to determine PWM duty cycle (for ramping and constant speed).

Schedule sequence of next floors in which to travel.

Determine door solenoid outputs.

Determine state of indicating lights.

Foreground:

Handle pulse inputs that indicate actual motor speed.

Update PWM output to motor.

Shut off door solenoids (door limit control).

Detect floor position (and interrupt to stop elevator if applicable).

Remember that a microcontroller (MCU) can only do one thing at a time.[4] In a multitasking system, different tasks execute during different time slots. With frequent time slots, it may appear that the MCU is executing all tasks simultaneously, although more slowly. The MCU does not necessarily repeat execution of all tasks in a cycle.

Real-Time Executive. The different tasks often operate asynchronously. There must be a "master" program that can schedule the execution of the foreground and background tasks. This software is known as a *real-time executive* (RTE). Other names for a real-time executive are *scheduler* or *control program*. An RTE is similar to the *operating system* used in general-purpose computers.

An RTE has to ensure that no conflicts occur between tasks. For example, a background PID controller may be in the process of updating an output variable. The update may require several instructions. What would happen if a PWM service routine asserts itself in the middle of these instructions? The RTE environment has to ensure that the PWM routine uses the latest valid output variable. It should not use the one that is in the process of being changed! To prevent this error, the RTE can use *mutual exclusion*. Basically, this means that the software disables interrupts before executing critical sections of code such as updating an output. Afterward, it will enable interrupts.

An RTE provides an environment whereby different tasks execute and share resources. At various times these tasks will compete for system resources, such as memory, I/O, and execution time. In a typical real-time environment, tasks are assigned priorities. Tasks of lower priority can be interrupted by those with higher priority. Note that the 68HC11 provides an interrupt priority scheme. Software can modify the interrupt priorities to some extent by writing to the highest-priority interrupt register (HPRIO). Higher-priority tasks can interrupt those with lower priority. Otherwise, the RTE may allocate execution time to background tasks, perhaps in a rotating sequence. The lowest-priority task could be a do nothing loop. It simply waits for another task to start or continue from where it left off.

Task Communication. Tasks must communicate with one another using semaphores or flags. A *semaphore* is a variable used for signaling between tasks. For example, the PID background sets a status flag saying that it will update an output. After the update is complete, it clears the status flag. A PWM foreground routine reads the flag. If it is set, the routine will use the previous duty cycle instead of the controller output variable.

Another communication method is message transmission. One task transmits a message to another by sending data to the other's *mailbox*. Instead of using a flag, the PWM routine always reads its next duty-cycle information from a memory location (its mailbox). The PID background writes the duty-cycle data to this mailbox address after calculating it.

A third method is to use a FIFO *queue*. A queue is a list of data waiting to be processed. This is analogous to people lining up in front of a hot dog stand to eat their hot dogs. Queues are normally used for blocks of data such as characters for serial transmission.

---

[4]Any processor based on the *von Neumann* architecture is essentially a serial device. It executes one instruction at a time. Multiprocessor systems or processors based on a parallel architecture can truly execute several instructions at a time.

Time Scheduling. An RTE often allots time slots for the different tasks to execute. For example, a person enters new set point data using a keyboard. The RTE must then assign some time for the keyboard data input task to execute while assigning execution time for other tasks, such as a PID controller.

For a 68HC11 system, an RTE can use the real-time interrupt (RTII) to generate clock ticks (or tocks). The RTE can then assign a task a certain number of clock ticks before calling on the next task to execute. A computer operating properly (COP) reset can ensure that the MCU is not spending all of its time executing one task.

Motorola Real-Time Executive. If you want to use an RTE for the 68HC11, you can download the source code for the MicroController eXecutive (MCX11) from the Internet. When assembled, the code takes up only about 500 bytes. It supports tasks, semaphores, timers, queues, messages, and more. The executive includes documentation (manual) file and a sample application file. Some ftp sites include freeware versions of a C interface and SCI driver to MCX11. Reference 16 in Appendix C provides thorough coverage of multitasking based on the C language.

Other Real-Time Executives. One system is Interactive C from the Massachusetts Institute of Technology (MIT). Typically, one uses it with the Mini Board. It can control small mobile robots, powered by Lego motors, such as the *Rug Warrior* (see Reference 27 in Appendix C). ImageCraft supplies another, known as REXIS.

Petri Nets. This is a design methodology for representing finite-state machines (see Section 13.3) that takes into account multiple interrupt sources. This section is a brief and informal introduction to using Petri nets. It is a graph with two types of nodes, known as *places* and *transitions*. Program execution is a sequence of transitions to different places. For a transition to occur, an input must have one or more *tokens*. A transition fires by removing one token from each input place and adding one token to each output place. (Reference 25 in Appendix C gives a more detailed description of Petri nets.) It is a first step prior to writing code for a multitasking system. After developing a Petri net, the systems developer can more easily write program code using the Petri net as a guide.

The Petri net focuses on a *data flow* approach. The data itself determines when certain subroutines and interrupt service routines execute. Since an event (or signal) can trigger an interrupt at any time, the controller must be able to process the events when they occur. Flag, interrupt enable, and other control or status bits determine the state of the processor. These bits are the tokens. The main program is considered to be the *foreground*, which acts like a task dispatcher. Each task executes when certain token conditions occur. They modify certain tokens when the task is complete. A token can be defined in software or it can be hardware specific, such as the real-time interrupt flag (RTII).

The Petri net shows states, tokens, and actions. There are two types of tokens. *Enabling tokens* are required to execute an action. The action passes *output tokens* to the next state. Figures 13.9 and 13.10 show a partial Petri net for a simplified motor control scenario. Circles show states. Blocks show actions (formal Petri nets use bars). Each arrow from a state to a transition shows enabling tokens. Each arrow from an action to a state shows output tokens. Hence, a state passes enabling tokens on to an action, which passes output tokens on to the next state. In our example, the system usually returns to the same state. Our simplified

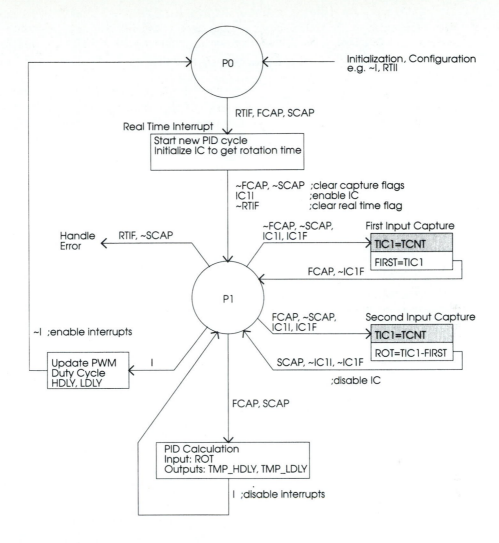

See Figure 13.10 for PWM Generation.

Shaded boxes show hardware action.

Figure 13.9   Partial Petri Net for Motor Speed Control

motor control system does not interact with a user to modify the speed setpoint or display actual motor speed. The system measures motor rotation using an encoding disk. It drives the motor using PWM.

Main Routine: Perform PID calculation to update output duty cycle.

Real-Time Interrupt Service Routine: Schedule next input sample.

Input Capture: Find actual rotation time (reciprocal of speed) of motor.

Output Compare: Generate PWM signal to drive motor.

Application Control Software   553

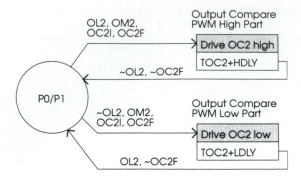

Figure 13.10   Petri Net for PWM Generation in Motor Speed Control System

This Petri net shows states P0 and P1. Table 13.2 lists the abbreviations to assist with interpreting the Petri net. Note that the approximate character (~) means logic 0 (*Not* 1). It is also used as the one's-complement operator when programming in C. Examination of Figures 13.9 and 13.10 reveals that the controller clears appropriate interrupt flags when returning from an interrupt service routine.

First the system performs necessary initialization and configuration (not shown) such as enabling interrupts. In state P0 it waits for a real-time interrupt. When the interrupt occurs, the controller clears two capture flags (FCAP, SCAP) to indicate that a new input capture sample will be taken. It also arms the input capture subsystem by setting the enable bit (IC1I). Now the controller is in the dispatcher state (P1) waiting for input edges to be detected. The capture routine records the first edge capture time. When the second edge occurs, it calculates the time difference between the two edges. Note that if another real-time interrupt triggers prior to capturing both edges, the controller enters an error condition. This could occur if the speed is too slow or the sensor fails. While this is going on, the output compare interrupt is generating PWM signals to the motor (see Figure 13.10) using duty-cycle (high and low) times stored in memory. After the controller has calculated the actual rotation time segment (time difference between two encoder disk holes), it calculates the new duty cycle. However, it does not give these new values to the output compare routine right away. It disables interrupts to eliminate the possibility of a successful compare event occurring while it is modifying the duty-cycle values. In other words, it locks out the output compare when modifying the variables it uses. When the duty-cycle update is complete, the controller enables the interrupt. The output compare interrupt can resume operation using the updated duty-cycle values.

The systems designer still has to consider interrupt latency, the time required to service an interrupt. What could happen if an input capture routine does not return in time for the output compare routine to do its job? Refer to Figure 13.11(a). This might occur if the high delay time is too short. One solution would be to limit the duty cycle as follows:

```
HDLY = TMP_HDLY  -  BUF
LDLY = TMP_LDLY  +  BUF
```

TABLE 13.2 PETRI NET TOKENS, REGISTERS, AND VARIABLES

Abbreviation	Meaning (hardware register or bit unless indicated otherwise)
Tokens	
I	Interrupt request (IRQ) mask
RTII	Real-time interrupt enable
RTIF	Real-time interrupt flag
FCAP	First input capture flag (in software)
SCAP	Second input capture flag (in software)
IC1I	Input capture 1 interrupt enable
IC1F	Input capture 1 interrupt flag
OL2	Output compare 2 pin control—set if high, clear if low
OM2	Output compare 2 pin control—set to enable output control
OC2I	Output compare 2 interrupt enable
OC2F	Output compare 2 interrupt flag
Registers/variables	
TIC1	Input capture 1 register
TCNT	Timer counter register—free running
FIRST	First capture time variable (in software)
ROT	Rotation time segment variable (in software)
TMP_HDLY	Temporary high delay time variable for PWM (in software)
TMP_LDLY	Temporary low delay time variable for PWM (in software)
HDLY	High delay time used in PWM generation (in software)
LDLY	Low delay time used in PWM generation (in software)
TOC2	Output compare 2 register
Others	
PID	Proportional-integral-derivative control
IC1	Input capture 1
OC2	Output compare 1 output pin

BUF is the minimum number of clock cycles allowed between two output compare events. The designer would set this value to the longest possible latency time of the other interrupts. If the value of BUF is too small, the time required to service another interrupt (latency) may be longer than the time between two output compare events. Another problem could be a service routine which takes more time to execute than the time between two input captures. Figure 13.11(b) illustrates this possibility. The designer should keep the time to service an interrupt as short as possible.

Petri Table. A Petri net is a useful design tool but its graphical form becomes awkward when many states, tokens, and interrupts are involved. Reference 25 in Appendix C describes an extension to the Petri net method called the Petri table. It presents information in a tabular form.

■ 13.8 FUZZY LOGIC

13.8.1 Why Use It and What Is It?

A washing machine will determine how dirty a load of clothing is to choose the appropriate wash cycle. A camcorder will record steady video even if the person using it is shaking.

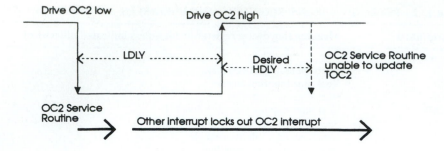

Dashed line shows where OC2 is supposed to be driven low.

Note that hardware automatically drives OC2 even though
interrupt is disabled (Timer connected to output).

(a) Other interrupt overlaps Output Compare

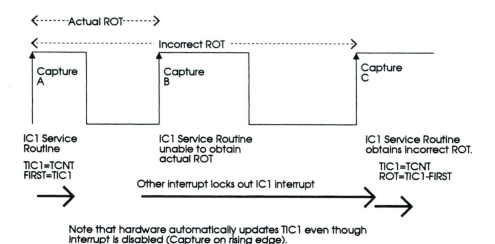

Note that hardware automatically updates TIC1 even though
interrupt is disabled (Capture on rising edge).

(b) Other interrupt overlaps Input Capture

Figure 13.11   Scenario: Interrupt Lock Out Caused by the Latency of Another Interrupt Source

Air-conditioning systems operate more effectively. A marketing company will run software
to predict the success of a new product. An optical character recognition (OCR) device will
identify the text printed on a shipping label. All of these represent fuzzy logic applications.
For example, cement manufacturing uses high-temperature kilns to convert limestone to lime.
Today, most kilns use fuzzy logic control. Lofti Zadeh invented fuzzy logic back in the 1960s.
The Japanese were the first to use it in commercial products in the 1970s. The rest of the world
followed. Business forecasters predict a market in billions of dollars.

Many fuzzy logic controllers use the 68HC11. This section covers enough fuzzy logic so that you can use Motorola's free software tools, as well as those provided by Fuzzy Technologies. It will omit generalizations and theory necessary for a complete coverage of fuzzy logic. Refer to references 22 to 24 in Appendix C for details.

In Section 13.4 we looked at controlling an analog output using a proportional integral-derivative (PID) control algorithm. A PID feedback controller reads a precise input and calculates a precise output based on a mathematical algorithm (see equations [13.4] to [13.6]). A fuzzy controller mimics human reasoning by interpreting inputs and outputs. When taking a shower, you open the cold-water valve more if the water is hot. If the water is very hot, you open the valve much more. Binary logic uses two distinct states, such as hot or cold. Fuzzy logic uses continuously varying degrees of states (membership functions). Temperature can be very hot, hot, warm, lukewarm, cool, cold, and very cold. The controller assigns *membership grades* to variables. It may interpret an analog temperature of 35°C to be 80% warm and 35% hot. The controller can vary the valve position to be large open, medium open, and small open, depending on a set of rules. For example, if the water is warm, then the cold-water valve is medium open. If the water is hot, then the cold-water valve is large open. Both rules will contribute to governing valve opening. Hence, the final valve position will be a percentage of large open and medium open. In a sense, each rule *votes* to determine the final outcome. However, some votes carry more weight than others. In the rest of this section we cover terminology such as membership function, membership grade, and rules, in addition to other terms not yet mentioned.

13.8.2 Overview

Figure 13.12 illustrates the process using Motorola's free MS-DOS software. You can obtain this software from the Internet. To start the process, you run the *knowledge-base generator* (KBG). The executable file KBG.EXE is a preprocessor that creates assembly code in the form of look-up tables. The tables define information about the system to be controlled. This information or knowledge base defines membership functions and rules (to be explained shortly). After KBG has created the assembly source code, you add (paste) the code to the beginning of another assembly source code, called FUZZ11B3.ASM. This other assembly code is the fuzzy inference engine that executes fuzzy logic. The fuzzy inference engine looks up the information in the knowledge base tables to perform the steps of fuzzification, rule evaluation, and defuzzification (also to be explained shortly). It does not handle I/O directly. Instead, it reads raw input data from RAM, calculates outputs, and writes this information to RAM.

A user-written program is responsible for I/O. It reads the calculated output values from RAM to drive actuators such as motors and valves. The user program is also responsible for obtaining sensor data and writing them to RAM for the inference engine to read. Then it branches or jumps to the start of the fuzzy inference engine to repeat the process all over again. You also have to add appropriate initialization software and whatever accessories your application needs. Assemble, link, and load all the code files—and presto, you have a fuzzy controller. Motorola also supplies Windows-based software for implementing fuzzy logic. It is called the FUzzy Design GEnerator (FUDGE).

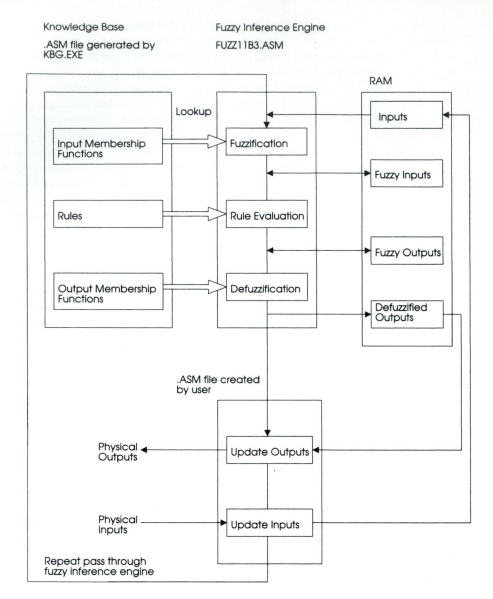

Knowledge Base
.ASM file generated by
KBG.EXE

Fuzzy Inference Engine
FUZZ11B3.ASM

RAM

Lookup

Input Membership Functions → Fuzzification

Rules → Rule Evaluation

Output Membership Functions → Defuzzification

Inputs

Fuzzy Inputs

Fuzzy Outputs

Defuzzified Outputs

.ASM file created by user

Physical Outputs ← Update Outputs

Physical Inputs → Update Inputs

Repeat pass through fuzzy inference engine

Figure 13.12 68HC11 Fuzzy Logic Overview. Assemble, link, and load source code (.ASM) program(s) to execute.

13.8.3 Create a Knowledge Base

Motorola supplies a tutorial to illustrate how to create a knowledge base. It uses the *classic* inverted pendulum problem as an example. Figure 13.13 illustrates the situation of balancing a rod by moving the balance point left or right. Here, we use a modified version of this example to explain things such as membership functions and rules. We will not explain the

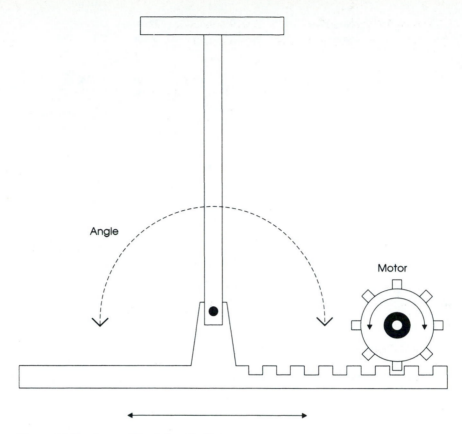

Figure 13.13 Inverted Pendulum Problem

keyboard commands necessary to use the knowledge-base generator (KBG). The menus in KBG are easy enough to use once you understand what a knowledge base is.

The system inputs are Angle and Delta_Angle. Angle is the pendulum's vertical orientation in degrees, as shown in Figure 13.14. Delta_Angle is the pendulum's angular speed in degrees per second. A sensor attached to the pendulum's pivot point could measure the inputs. A potentiometer, linear variable displacement transformer (Section 12.2.1), or optical encoder disk (Section 11.3.3) could serve as a sensor to measure and calculate both inputs. The output is the motor current required to move the balance point right or left. Gravitational force causes torque that causes the rod to fall.[5] Hence, the system should manipulate motor torque to counteract the torque caused by gravity. Motor torque is directly proportional to the motor's armature current with a constant field flux. Hence, the system uses a current amplifier instead of a voltage amplifier to manipulate the motor. This differs from motor speed control as described in Section 12.2.3.

---

[5]Torque is the twisting force. It is the product of the magnitude of gravitational force, the pendulum length, and the sine of the angle between the force direction (straight down in this case) and pendulum rod.

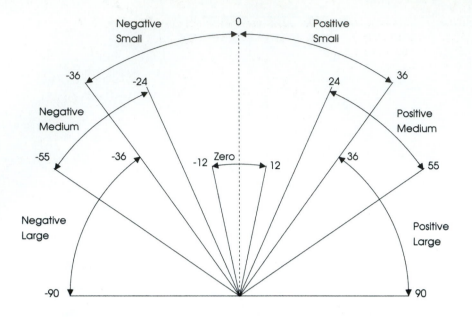

All angles in degrees.

Figure 13.14   Pendulum Angle as Fuzzy Input

Membership Functions.   A membership function (or fuzzy set) provides a numerical definition of each of the fuzzy logic states. In the temperature control example, the states were very hot, hot, warm, and so on. A membership function defines the range of analog values that define a fuzzy logic state. It also defines the degree of membership of each analog value within the range. With the inverted pendulum, the two inputs and one output each have seven membership functions defining a range of levels. Their labels are

NL	Negative_Large
NM	Negative_Medium
NS	Negative_Small
ZE	Zero
PS	Positive_Small
PM	Positive_Medium
PL	Positive_Large

An analog value may belong to two states with varying degrees of membership. Figure 13.14 shows that an angle of 10° is a certain percentage Zero and a certain percentage Positive_Small. Figure 13.15 illustrates the membership functions in graphical form. Figure 13.16 shows one sample function in more detail. KBG will create a table for each variable. Each table uses only 4 bytes to define each function.

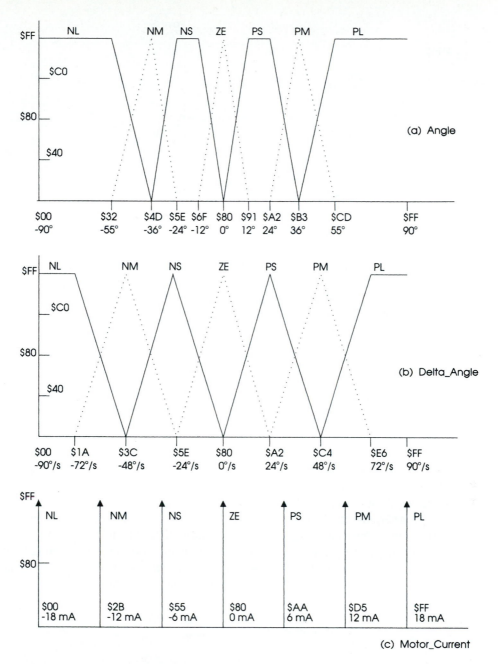

Figure 13.15 Membership Functions

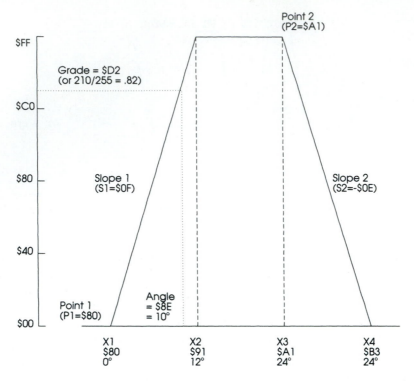

Figure 13.16 Membership Function for Angle Positive_Small, and Corresponding Grade for Angle of 10°

For KBG to create the table, the user must enter data in the form of *x*-axis points X1, X2, X3, X4 for each function (see Figure 13.16). KBG converts this to a normalized point-slope representation (to be explained shortly). This is so that the inference engine will be able to determine the variable's membership *grade* (or degree of membership) within a function given the variable's analog value. Although KBG does this process for you, we will explain how so that you will understand how the entire system works. First, KBG normalizes the analog numbers the user enters. If necessary, refer to "Fractions and Normalizing" in Appendix F. To normalize the *x*-axis points, use the following equation:

$$X_{normalized} = \frac{(2^n - 1)(X_{analog} - offset)}{span} \tag{13.15}$$

where $span$ = maximum analog value-minimum analog value

$offset$ = minimum analog value

n = number of bits

X_{analog} = analog value of *x*-axis point

$X_{normalized}$ = normalized number representing point

Remember that the 68HC11 is an 8-bit microcontroller. Hence for variables Angle and Delta_Angle, the normalization equations are

$$X_{normalized} = \frac{255(X_{analog} + 90)}{180}$$
(13.16)

For Motor_Current, we have

$$X_{normalized} = \frac{255(X_{analog} + 18)}{36}$$
(13.17)

Refer to Figure 13.16. An input membership function looks like a trapezoid. We can define a trapezoid using two points (P1 and P2) and two slopes (S1 and S2) knowing that the height is always 100% full scale (normalized 1). In Figure 13.15 we also see many input membership functions that look like triangles. We can also represent these functions using two points and two slopes. They are special cases where point X2 equals X3. In general, output membership functions could also be trapezoids or triangles. However, KBG defines them as *singletons*. A singleton has only one value for each membership function. Hence, the knowledge base needs only 1 byte to store each singleton position. This will simplify the work to be done by the inference engine when it calculates the final output. Now we show how KBG calculates the normalized point-slope representation (see Figure 13.16) for each input membership function. Points are easy to calculate.

$$Pl = X1_{normalized}, P2 = X3_{normalized}$$
(13.18)

The slope equations are

$$S1 = \frac{255}{X2_{normalized} - X1_{normalized}}$$
(13.19)

$$S2 = \frac{255}{X4_{normalized} - X3_{normalized}}$$
(13.20)

In theory, a vertical slope ($X1 = X2$ or $X3 = X4$) is infinity. KBG shows this as an overflow from the maximum of 255 to 0. From Figure 13.16, the careful reader will note that KBG actually calculates the negative of a slope for slope 2 (S2). This is done in order to use positive numbers. Tables 13.3 to 13.5 show all the points and slopes for our inverted pendulum application. Hence, KBG can define each input membership function using 4 bytes (two points and two slopes).

Rules.   The knowledge base also requires rules for the inference engine to evaluate. Often the rules are intuitive. Refer again to Figure 13.13. If the pendulum falls slowly a bit to the left, then move the balance point slowly a bit to the left. If the pendulum falls quickly to the left, then move the balance point quickly to the left. Rules are IF-THEN statements. For two inputs and two outputs, the general form of a rule is

IF Input1 IS Membership_Function_i AND Input2 IS Membership_Function_j

TABLE 13.3 INPUT MEMBERSHIP FUNCTIONS FOR ANGLE

Fcn	Analog trapezoid points (−90° to +90°)				Normalized decimal (0 to 255)				Normalized hex ($00 to $FF)			
	X1	**X2**	**X3**	**X4**	**X1**	**X2**	**X3**	**X4**	**X1**	**X2**	**X3**	**X4**
NL	−90	−90	−55	−36	0	0	50	77	$00	$00	$32	$4D
NM	−55	−36	−36	−24	50	77	77	94	$32	$4D	$4D	$5E
NS	−36	−24	−12	0	77	93	111	128	$4D	$5D	$6F	$80
ZE	−12	0	0	12	111	128	128	145	$6F	$80	$80	$91
PS	0	12	24	36	128	145	162	179	$80	$91	$A2	$B3
PM	24	36	36	55	162	179	179	205	$A2	$B3	$B3	$CD
PL	36	55	90	90	179	205	255	255	$B3	$CD	$FF	$FF

Fcn	Normalized point/slope (decimal 0 to 255)				Normalized point/slope (hex $00 to $FF)			
	P1	**S1**	**P2**	**S2**	**P1**	**S1**	**P2**	**S2**
NL	0	0	50	9	$00	$00	$32	$09
NM	50	9	77	15	$32	$09	$4D	$0F
NS	77	15	111	15	$4D	$0F	$6F	$0F
ZE	111	15	128	15	$6F	$0F	$80	$0F
PS	128	15	162	15	$80	$0F	$A2	$0F
PM	162	15	179	9	$A2	$0F	$B3	$09
PL	179	10	255	0	$B3	$0A	$FF	$00

Note: Boxes indicate an 8-bit vertical slope (256 = $100) shown as zero. Set carry bit assumed. Table represents slope 2 (S2) with the *negative* of the actual slope to maintain positive numbers.

THEN Output1 IS Membership_Function_k AND Output2 IS Membership_Function_1

Another feature is the AND operator, which is similar to the Boolean AND operator. You can extend the rule for more inputs and outputs by using extra AND operators. The keyword IF begins the rule. It precedes the *antecedents*. These are the conditions to check. Each antecedent begins with the input label followed by the keyword IS, which is followed by the membership function label. The keyword THEN precedes the *consequents*. These specify what action to take if the antecedents are true. KBG stores each rule in a rules look-up table as follows:

One byte per antecedent where some bits (a field) identify an input and others identify the membership function

One byte per rule consequent where some bits identify the output variable and others identify the membership function (singleton position)

One byte to mark the end of the rules

Thus, the inference engine looks up each rule from the rules table to evaluate it. The current version of KBG.EXE will handle up to 8 inputs, 4 outputs, 8 input membership functions per input, and 8 singletons per output. It can also handle up to 1024 rules.

TABLE 13.4   INPUT MEMBERSHIP FUNCTIONS FOR DELTA_ANGLE

Fcn	Analog trapezoid points (−90° to +90° per second)				Normalized decimal (0 to 255)				Normalized hex ($00 to $FF)			
	X1	X2	X3	X4	X1	X2	X3	X4	X1	X2	X3	X4
NL	−90	−90	−72	−49	0	0	26	59	$00	$00	$1A	$3B
NM	−72	−49	−48	−25	26	59	60	94	$1A	$3B	$3C	$5E
NS	−48	−25	−24	−1	60	92	94	127	$3C	$5C	$5E	$A2
ZE	−24	−1	0	23	94	127	128	161	$5E	$7F	$80	$A1
PS	0	23	24	47	128	161	162	195	$80	$A1	$A2	$C3
PM	24	47	48	72	162	195	196	230	$A2	$C3	$C4	$E6
PL	48	71	90	90	196	230	255	255	$C4	$E5	$FF	$FF

Fcn	Normalized point/slope (decimal 0 to 255)				Normalized point/slope (hex $00 to $FF)			
	P1	S1	P2	S2	P1	S1	P2	S2
NL	0	0	26	8	$00	$00	$1A	$08
NM	26	8	60	8	$1A	$08	$3C	$08
NS	60	8	94	8	$3C	$08	$5E	$08
ZE	94	8	128	8	$5E	$08	$80	$08
PS	128	8	162	8	$80	$08	$A2	$08
PM	162	8	196	8	$A2	$08	$C4	$08
PL	196	8	255	0	$C4	$08	$FF	$00

Note: See note in Table 13.3

TABLE 13.5   OUTPUT MEMBERSHIP FUNCTIONS FOR MOTOR_CURRENT

Fcn	Singleton (x-axis) (−18 mA to +18 ma)	Normalized decimal (0 to 255)	Normalized hex ($00 to $FF)
	X	X	X
NL	−18	0	$00
NM	−12	43	$2B
NS	−6	85	$55
ZE	0	128	$80
PS	6	170	$AA
PM	12	213	$D5
PL	18	255	$FF

An example of a rule is

IF Angle IS Negative_Small AND Delta_Angle IS Zero
THEN Motor_Current IS Positive_Small

The KBG tutorial mentioned previously lists 15 rules for the inverted pendulum problem. Since we are using only two inputs, Table 13.6 shows the rules in the form of a matrix with rows and columns. From now on, we refer to rules using their row, column coordinate. Our

TABLE 13.6 MATRIX OF FUZZY RULES[a]

Delta_ Angle	Delta_Angle						
	NL	NM	NS	ZE	PS	PM	PL
Angle							
NL				PL			
NM				PM			
NS				PS	PS		
ZE	PL	PM	PS	ZE	NS	NM	NL
PS			NS	NS			
PM				NM			
PS				NL			

[a]Each output action is associated with a pair of inputs.

example rule is rule 3,4. Mechanically, the user enters rules by selecting items from appropriate menu windows. KBG will translate this information into the rules table.

Final Compilation.   After the user has entered all the inputs, outputs, membership functions, and rules, the user selects the menu item to create assembly source code. The code consists of pseudo-ops (see Chapter 4) to define the tables.

13.8.4 Fuzzy Logic Inference

The microcontroller executes fuzzy logic decision making. If the pendulum angle is 10° and it is moving −10° per second (to the left), then set the motor current to −2.52 mA. We will show how the controller makes this decision.

The inference process produces output values through fuzzy rules given input values. It uses the knowledge base described in the preceding section. Refer back to Figure 13.12. Remember that the user pastes the assembly code created by KBG into the start section of the inference engine code (FUZZ11B3.ASM). The inference process involves three steps: fuzzification, rule evaluation, and defuzzification.

Fuzzification.   Fuzzification converts analog inputs into grades of membership. To do this it uses the point/slope representation of the membership functions. A grade of membership expresses the fraction of or the percentage of belonging to a particular function. One example is a rainbow (color spectrum). Where does red end and orange begin? Consider analog inputs of 10° and −10° per second. Refer to Figure 3.15. An angle of 10° would belong to the membership functions Zero Angle and Positive_Small Angle. An angular rate of −10° per second would belong to functions Negative_Small Delta_Angle and Zero Delta_Angle. During fuzzification, the inference engine will determine "how much" the variables belong.

Consider Figure 13.15. An angle of −15° belongs to the membership function Negative_Small Angle by 100%. Hence, its grade is 1. An angle of −40° has a grade of 0 in Negative_Small Angle. An angle of −30° has a grade somewhere between 0 and 1. Figure 13.16 illustrates the Grade for Angle Positive_Small. The inference engine calculates membership grades given the point-slope representation stored in the knowledge base. It uses the following equations (see Figure 13.16):

$$\text{Input} < \text{Point 1: Grade} = 0 \qquad (13.21a)$$

$$\text{Point 1} \leq \text{Input} < \text{Point 2: Grade} = (\text{Input} - \text{Point 1}) (\text{Slope 1})$$
$$\text{Limit maximum to } \$FF \qquad (13.21b)$$

$$\text{Input} \geq \text{Point 2: Grade} = \$FF - ((\text{Input} - \text{Point 2}) (\text{Slope 2}))$$
$$\text{Limit minimum to } \$00 \qquad (13.21c)$$

With normalized input values of

Angle = $8E (or 142/255 = 0.56) and Delta_Angle = $71 (or 113/255 = 0.44)

Using equations (13.21), the membership grades are

Grade in Zero Angle = $2D (or 45/255 = 0.18) using equation (13.21b)
Grade in Positive_Small Angle = $D2 (or 210/255 = 0.82) using equation (13.21c)
Grade in Negative_Small Delta_Angle = $67 (or 103/255 = 0.40) using equation (13.21b)
Grade in Zero Delta_Angle = $98 (or 152/255 = 0.60) using equation (13.21c)

The inference engine FUZZ11B3.ASM calculates grades using a subroutine called GET_GRADE. The calling registers are

```
ACCA   normalized input
IX     pointer to membership function in table (created by KBG)
```

The return registers are

```
ACCB   normalized grade
IX     IX + 4, points to next membership function
```

Now we are ready for the rule evaluation step.

Rule Evaluation.   The rule evaluation step computes the grades of each output membership function. Recall that rules are IF-THEN statements with antecedents and consequents (see "Rules" in Section 13.8.3). The fuzzy AND operator means to take the minimum value of the grades being evaluated. This is identical to the Boolean AND gate since its output is the minimum of the inputs. Refer to the rules in Table 13.6. Consider rule 4,3.

IF Angle IS Zero AND Delta_Angle IS Negative_Small
THEN Motor_Current IS Positive_Small

becomes

$2D AND $67 = Minimum ($2D, $67) = $2D

Hence, the output of function Positive_Small (PS) is (has grade of) $2D (or 45/255 = 0.18). Similarly (see Table 13.6),

Output function Zero = $2D AND $98 = $2D, using rule 4,4.
Output function Negative_Small = $D2 AND $67 = $67, using rule 5,3.
Output function Negative_Small = $D2 AND $98 = $98, using rule 5,4.

Two or more rules may have the same consequent function. Rules 5,3 and 5,4 both evaluate to Negative_Small. In this case use the fuzzy OR operator to evaluate the grade of the function. The fuzzy OR means to take the maximum value of the grades. This is identical to a Boolean OR gate since its output is the maximum of the inputs. The two rules evaluate to Negative_Small grades of $67 and $98. Hence, ORing the two evaluates to $98, the maximum of $67 and $98. This showed how the inference engine calculates fuzzy outputs for each output membership function. The next step is defuzzification.

Defuzzification.   The defuzzification step converts the fuzzy outputs into a *crisp* (analog) number. This will be the final decision as to what the physical outputs should be. Defuzzification involves taking the weighted average of all fuzzy outputs associated with a final output. Since the inference engine defines output membership functions as singletons, it uses the following equation:

$$\text{Output} = \frac{\sum_{i=1}^{n} F_i S_i}{\sum_{i=1}^{n} F_i} \tag{13.22}$$

where F_i = fuzzy output (i.e., the grade of membership)
S_i = singleton position
n = number of singleton functions

This is like finding the center of gravity (or centroid) of the outputs. Consider balancing a narrow beam with several piles of pennies. The beam is 255 cm long and the position of each pile corresponds to a normalized singleton position of S_i. The number of pennies in each pile (maximum of 255) corresponds to the grade of F_i. Where can you balance the beam with one finger? Equation (13.22) gives the answer. In our example, we have grades of $98, $2D, and $2D for output singletons Negative_Small, Zero, and Positive_Small, respectively. Refer to Table 13.5 and evaluate equation (13.22).

$$\text{Output} = \frac{(\$98 \times \$55) + (\$2D \times \$80) + (\$2D \times \$AA)}{\$98 + \$2D + \$2D} = \$6D(or\ 109/255 = 0.43)$$

The inference engine stores $6D in RAM as the final output. To find the equivalent analog output, rearrange equation (13.17) to solve for X_{analog}. Hence, the analog output is

$$(0.43 \times 36\ \text{mA}) - 18\ \text{mA} = -2.52\ \text{mA}$$

This will move the balance point slightly to the right.

The fuzzy inference engine will support up to eight singletons per output variable. Using equation (13.22) it must be able to sum eight products of 8-bit numbers and divide this by a sum of eight 8-bit numbers. Thus, the inference engine will have to perform a 19-bit by 11-bit division for defuzzification. The curious reader could investigate the software trick to do this. Examine the source code of FUZZ11B3.ASM (look for label DEFUZ).

Remember that we are describing the use of Motorola's freeware software for the 68HC11. More sophisticated fuzzy systems use trapezoidal or triangular output membership functions. The controller truncates each function to the grade evaluated by the corresponding rule. Then it combines (fuzzy OR) the functions to produce a *fuzzy set* for each output variable. To defuzzify, the controller must find the center of gravity (COG) of each set.

PRACTICE

If the pendulum angle is $-8°$ and it is moving $4°$ per second (to the right), find the motor current.

Solution

Normalize inputs. Use equation (13.16). Angle = $74 (or 116/255 = 0.45) and Delta_Angle = $86 (or 134/255 = 0.53).

Fuzzification. Use equations (13.21).

        Grade in Negative_Small Angle = $B4 (or 180/255 = 0.71)
        Grade in Zero Angle = $4B (or 75/255 = 0.29)
        Grade in Zero Delta_Angle = $CF (or 207/255 = 0.81)
        Grade in Positive_Small Delta_Angle = $30 (or 48/255 = 0.19)

Rule evaluation. See Table 13.6.

        Output function Positive_Small = $B4 AND $CF = $B4, using rule 3,4.
        OR output function Positive_Small = $B4 AND $30 = $30, using rule 3,5.
        Hence, output function Positive_Small = $B4.
        Output function Zero = $4B AND $CF = $4B, using rule 4,4.
        Output function Negative_Small = $4B AND $30 = $30, using rule 4,5.

Defuzzification. Use equation (13.22).

$$\text{Output} = \frac{(\$30 \times \$55) + (\$4B \times \$80) + (\$B4 \times \$AA)}{\$30 + \$4B + \$B4} = \$92 (\text{or } 146/255 = 0.57)$$

Analog output. Modify equation (13.17) to find analog number.

$$(0.57 \times 36 \text{ mA}) - 18 \text{ mA} = 2.52 \text{ mA}$$

13.8.5 Development and Further Information

Freeware Summary. Refer back to Figure 13.12. Use the knowledge-base generator (KBG.EXE or FUDGE.EXE) to create input membership functions, rules, and output membership functions. Compile the knowledge base to create assembly code defining the knowledge-base tables. Paste the code into the beginning of the fuzzy inference engine (FUZZ11B3.ASM). Write source code to handle I/O. This could be attached to the end of the inference engine. Assemble and link. When the inference engine executes, it fuzzifies (assigns grades of membership) the inputs, evaluates the rules to define fuzzy output singletons, and defuzzifies the singletons using center of gravity to create a crisp number for each output variable. You have created a fuzzy controller but it is only as good as the knowledge base. How do you know whether the knowledge base is any good? Note that you may simulate the inference engine outputs using KBG before compiling the knowledge base. In the next section we address this design issue.

System Development. Like any software design process, it is important to follow an established design methodology. Defining rules for a coffeemaker is *easy*, but finding rules for active suspension in an automobile is not. Also, different references use different terms to say the same thing. We continue to use the terms used in the freeware documentation.

The following is a quick overview to give you a flavor of one process. Reference 22 in Appendix C shows the process in more detail. Remember that you are trying to control a physical system. Step 1 is to define a model of how the physical system behaves. This includes identifying variables. One nice thing about fuzzy logic is that you do not need a detailed mathematical model. Step 2 is to define the membership functions. This is also known as *defining the control surface*. There are some rules of thumb. The number of functions per variable should be five, seven, or nine. Note that these are odd numbers. Membership functions should overlap between 10 and 50% (see Figure 13.15). For any analog (crisp) input, the sum of the overlapping membership functions should not be greater than 1. Step 3 is to write the rules. For most situations you write a rule for each possible combination of input membership functions. For example, two inputs with seven functions each makes a total of 49 possible rules. It is possible to use fewer rules (see Table 13.6) in the inverted pendulum problem. In other cases it can be risky to use fewer rules. Step 4 is to select a method of defuzzification as described in references 22, 23, or 24 in Appendix C. The singleton method described earlier is easy to implement in an 8-bit microcontroller.

FuzzyTech Demo. The disk includes a demo version of FuzzyTech which is a comprehensive windows-based development tool from Fuzzy Technologies. It provides more options such as viewing control surfaces as well as using different types of membership functions. The full commercial version can generate code for various target processors and in C. The demo version will generate code for the 68HC11 that is restricted to 2 inputs, 1 output, and 1 rule block. It has a very good tutorial about fuzzy logic in its help system.

Optical Character Associator (OCA). Motorola application note AN1220/D describes how to use the MC68HC11E9 for an optical character recognition (OCR) application. Pattern matching on a bit-by-bit basis is not very effective. Parts of the same character may appear in two sensor samples, character height and width may vary, and characters may be skewed if a document is misfed. Fuzzy logic is an effective solution. The system encodes input data using a Texas Instruments TSL21 charge-coupled device (CCD) sensor. Then it applies a fuzzy rule set to generate an output corresponding to the character being read. There are six fuzzy inputs based on the ratio of dark areas to light areas. The controller applies 14 rules to determine 14 characters. They are numeric characters zero through nine and four special characters. For a copy of the application note, contact Motorola as described in Appendix C.

13.8.6 Fuzzy Logic Support for the MC68HC12

In 1996, Motorola introduced the MC68HC12 microcontroller, a compatible upgrade to the MC68HC11. We will say more about the 68HC12 in Section 15.2.5. Of particular interest here is that the 68HC12 has additional instructions and hardware to support fuzzy logic. Instead of using the software engine required for the 68HC11, it can use its built-in functionality. Typically, a fuzzy logic program will be one-fifth the size and fifteen times faster than the equivalent 68HC11 code. The 68HC12 includes four fuzzy logic instructions and some other instructions that are useful for fuzzy logic as shown in Tables 13.7 and 13.8.

A Code Sample. Listing 13.21 shows a sample use of fuzzy logic. It illustrates how short the code can be. Later we will examine the instructions in more detail.

```
Listing 13.21
*A Sample Fuzzy Inference Engine for the 68HC12
*Based on 2 analog inputs, 7 membership functions, 1 output
*as in the inverted pendulum system

*First fuzzification occurs
```

TABLE 13.7   68HC12 FUZZY LOGIC INSTRUCTIONS

Instruction	Description
MEM	Determine grade of membership for fuzzification.
REV	Perform unweighted rule evaluation.
REVW	Perform weighted rule evaluation.
WAV	Perform weighted average defuzzification.

TABLE 13.8   ADDITIONAL 68HC12 INSTRUCTIONS USEFUL FOR FUZZY LOGIC

Instruction	Description
EMINM	Find minimum of two 16-bit unsigned integers.
MAXM	Find maximum of two 8-bit unsigned integers.
EMAXM	Find maximum of two 16-bit unsigned integers.
TBL	Table Look-up and Interpolate.
ETBL	Extended Table Look-up and Interpolate.
EMACS	Extended Multiply and Accumulate (Signed).

```
FUZZIFY
        LDAA    ANGLE           ;get angle analog input
        LDX     #INMEMFCN       ;point to input membership functions
        LDY     #FUZINPUT       ;point to fuzzified input table
        LDAB    #7              ;initialize membership function count
GRDANGLE
        MEM                     ;evaluate grade of membership
        DECB                    ;for all angle functions
        BNE     GRDANGLE
        LDAA    DELTA           ;get delta_angle analog input
        LDAB    #7              ;initialize membership function
count
GRDDELTA
        MEM                     ;evaluate grade of membership
        DECB                    ;for all delta_angle functions
        BNE     GRDDELTA
*Then have Rule Evaluation Section
RULEEVAL
        LDAB    #7              ;clear previous fuzzy outputs to zero
CLROUT
        CLR     1,Y
        INY
        DECB
        BNE     CLROUT
*Set up to evaluate rules, then evaluate using REV
        LDX     #RULESTART      ;point to rule list
        LDY     #FUZINPUT       ;point to fuzzy inputs
        LDAA    #$FF            ;necessary overhead for REV instruction
        REV                     ;now do it! Evaluate the rules to
                                  generate fuzzy outputs
*Then the defuzzification phase to generate analog output
DEFUZZYFY
        LDY     #FUZOUTPUT      ;point to list of fuzzy outputs
        LDX     #SGLTNPOSN      ;point to singleton positions
        LDAB    #7              ;initialize singleton position count
        WAV                     ;weighted average calculation
*now 24-bit sum of products in Y:D and 16-bit sum of weights in X
        EDIV                    ;a 68HC12 extended division
                                  instruction for Y:D/X
*now Y contains 8-bit analog output as 16-bit value
        XGDY
        STAB    ANOUT           ;store analog output
*Code to handle I/O follows
```

Fuzzification.   The MEM instruction determines the grade of membership for fuzzification. As with the 68HC11 fuzzy inference engine, the 68HC12 stores input membership functions as 4-byte values in a table. To use this instruction:

> load accumulator A with the analog input value
>
> load index register X with the start address of the applicable membership function.
>
> load index register Y with the address of where the fuzzy input is to be stored.

Note that the membership function (pointed at by IX) is the set of the P1, S1, P2, S2 points for the trapezoidal function as illustrated in Figure 13.16 and Tables 13.3 and 13.4. For complete fuzzification, execute several MEM instructions in a loop. The MEM instruction updates index registers X and Y automatically.

Rule Evaluation.   The REV and REVW instructions perform fuzzy logic rule evaluation for unweighted and weighted evaluation, respectively. They generate the fuzzy outputs. In an unweighted system, each rule is considered equally important. The rules in the knowledge base (see Table 13.6) are preset as a table of pointers. It consists of 8-bit antecedent offsets followed by 8-bit consequent offsets. The special value $FE separates antecedents, consequents, and successive rules. The value $FF marks the end of the rule list. To use the REV instruction:

   clear the fuzzy outputs to zero

   load index register X with the first element in the rule list

   load index register Y with the base address of fuzzy inputs

   load accumulator A with the value $FF

   clear bit V in the condition code register

Note that the fuzzy output addresses must immediately follow the fuzzy input addresses. Also note that the instruction LDAA #$FF places the correct value in accumulator A and clears bit V. During execution of REV, complex processing occurs. For more details, refer to the CPU12 Reference Manual for the 68HC12 (document CPU12RM/AD).

Defuzzification.   With the 68HC12, it is possible to use the WAV (weighted average) instruction to calculate the defuzzified output. We can use WAV and the new extended division instruction EDIV to calculate the weighted average as shown in equation (13.22). To use the WAV instruction:

   load index register X with start address of operand list for sum of products and weights

   load index register Y with start address of operand list for sum of weights

   load accumulator B with number of elements

Execution of WAV stores the sum of products in index register Y and accumulator D. This is the numerator part of equation (13.22). Execution of WAV also stores the sum of weights in index register X. This is the denominator part of equation (13.22). To perform the division, use the new extended division instruction EDIV. The EDIV instruction divides a 32-bit unsigned integer by a 16-bit unsigned integer. To use the EDIV instruction:

   load index register Y with the upper part of the 32-bit divisor

   load double accumulator D with the lower part of the 32-bit divisor

   load index register X with the 16-bit dividend

Execution of EDIV places the quotient in index register Y and the remainder in double accumulator D.

More Details?   Chapter 9 in the CPU12 Reference Manual describes fuzzy logic support in detail. You can order this document from Motorola or a Motorola representative. The document number is CPU12RM/AD.

■ 13.9 SUMMARY

We know that a microcontroller must handle I/O for a control application. It must also manipulate data to compute the outputs it sends out. Since an MCU is a digital device, it has a natural ability to control discrete events. We have seen how it can use Boolean logic functions such as AND, OR, and NOT to make decisions about changing outputs. The software typically uses an if-else control structure.

The motor control example is a simple one. A more complex situation occurs in a system of conveyor lines. A motor drives a section in a conveyor line. The control logic ensures that the motors are interlocked. A motor can run only if the motors farther down the line are also running. This ensures that the material does not pile when a section shuts down.

An extension of Boolean logic is the sequencer. The software is a loop that periodically updates the outputs using a look-up table. Again, we can consider a system of conveyor lines. To move material from one end of a factory to the other requires switching the appropriate conveyor motors on and off in a sequence.

Next, we looked at the sequential machine, which is based on using state transitions. A system is described as a collection of states. Controlling such a system involves movement (transition) from one state to another. The transition from one state to the other is specified by the inputs. The software is based on using a state table. The table indicates the outputs to drive. The table also contains an address that specifies where the table pointer should point to next (the next state). Formally, this type of table is known as a *linked-list structure*.

For the analog world, we looked at a traditional equation, PID. The aim of PID (and most other control methods) is to reduce the difference between the actual process response (variable) and the desired response (set point). The controller's effectiveness for any application is determined by the gain constants and sampling times. It is important to sample the input at a rate above the Nyquist frequency to reduce aliasing error. Typically, a digital controller samples analog data at a rate five times the bandwidth.

Numerically intensive control applications use digital signal processors (DSPs). Such applications typically use a multiply and accumulate operation. Any general-purpose microcontroller can perform DSP calculations using software. However, real-time implementation restricts the bandwidth; hence, it is practical to use specialized DSP chips instead. These include the TMS320C40 and others of the TMS320 series by Texas Instruments. Motorola builds the DSP56000, DSP56001, and DSP96002.

Assembly language is good for fast and efficient code to handle I/O but does have a disadvantage. It may not be easy to use for making complex control decisions and calculations. To beat the competition, a company wants to introduce their new product as soon as possible. Most of the time required to develop a product is in writing the software. Using C reduces this time. Yet the resulting code is usually small enough to fit in memory.

In this chapter we gave an insight into how C can be used for the 68HC11. Both the motor control and PID controller examples show how C can handle I/O. The motor control program used C to handle individual bits. The PID example demonstrated some of its computational power for calculations. C code tends to be more readable than assembly. Still, its syntax (grammar) may take some getting used to.

Most application programs execute several tasks. An MCU system can use a real-time executive (RTE) to integrate and coordinate execution of different tasks.

Fuzzy logic control has gained prominence in many applications ranging from coffeemakers to automobile suspension control. Until hardware implementation becomes common, existing microcontrollers offer the lowest-cost implementation. Motorola supplies free software for use on the 68HC11. It has two main parts. The knowledge-base generator (KBG.EXE) defines the input membership functions using point-slope representation of trapezoids and output membership functions using singletons. It also defines IF-THEN rules for the inference engine to use. The inference engine (FUZZ11B3.ASM) performs the operation of fuzzification (determine input membership grades based on crisp inputs). Then it evaluates rules using fuzzy AND and fuzzy OR operators. The result of a fuzzy AND is the minimum of its operands. The result of a fuzzy OR is the maximum of its operands. Finally, it defuzzifies output singleton grades to crisp outputs.

■ EXERCISES

1. What kind of control technique would you use for each of the following systems?
   (a) A conveyor interlock system
   (b) A traffic light (see Chapter 2, Question 19)
   (c) The detection part of a security system (If an intruder is detected, turn on an alarm bell and send a message.)
   (d) The safe lock control part of a security system (The MCU energizes the open lock solenoid when the correct combination code is entered.)
   (e) A robot arm (see Chapter 2, Question 22)
   (f) Purge-cycle control of a process heater (see Chapter 11, Question 20)

2. For the security system of Question 1(c), write the pseudocode.

3. Write pseudocode to start up a conveyor line. Consider conveyor interlock requirements.

4. Write pseudocode for a subroutine that a main program calls periodically to shut down the necessary conveyors if any conveyor stops. How can this subroutine be called?

5. Complete the state table for the program of Listing 13.4.

6. Write a table to specify when the MCU energizes a lock solenoid (see Question 1[d]). Assume that the keyboard has only three digits. One sequence of three digits opens the lock. This can happen even if a person has previously entered wrong digits.

7. An MCU system controls flow of fluid through a pipe by varying the opening of a valve.
   (a) What control technique should be used to control flow?
   (b) Describe the MCU control system.
   (c) Write the pseudocode that can be used to control flow.
   (d) During a test of the system, the MCU sends a step signal to the valve. Measurements showed that it required $3\mu s$ for the flow rate to reach 90% of the new steady state from 10% of the new steady state. Recommend a minimum sampling rate.

8. A digital PID controller is used. Error is an 8-bit number. The gain constants are 64, 16, and 9, respectively. The 8-bit A/D has a range of $00 to $FF for the analog range.
   (a) How can the software ensure that no overflow occurs when calculating error?
   (b) What is the maximum calculated output (C0) in terms of the previous output (C1) if 16-bit arithmetic is used?
   (c) How can the software determine when the maximum and minimum output limits are reached?

9. A 68HC11 implements an FIR digital filter with a length of six. It reads the input from the A/D subsystem and sends the output to an 8-bit D/A connected to port B. Assume that it only needs to use unsigned arithmetic.
   (a) Write the program.
   (b) What is the bandwidth limit of the input signal?

10. A DSP uses a sampling frequency of 4000 Hz to decode DTMF signals for a telephone receiver.
    (a) Determine the number of samples required for an FFT.
    (b) Make a table similar to that of Figure 13.6.
    (c) What telephone number was dialed if the following FFT indices indicated peak magnitudes after every interdigit interval? The (low k, high k) sequence is

    (25,47) (25,43) (27,39) (27,47) (22,39) (30,43) (25,43)

11. Your company intends to design an automatic weather station controlled by an MCU. Convince the manager of the product design department that it will be a good investment to purchase a C compiler. In this case the design team still needs to decide which MCU to use.

12. Is it possible to monitor execution of a program in the form of C source code?

13. Explain some trade-offs when deciding to use integer or floating-point variables in a C program.

14. Explain some trade-offs when deciding to use static variables.

15. What features of C would be useful for the following purposes?
    (a) Save RAM space
    (b) Control a robot arm (see Question 1[e])
    (c) Multi-I/O control system (see end of Section 13.4)

The following exercises ask you to write C programs. Since we have not covered everything about C, it is not necessary to write complete C code with all the necessary overhead.

16. Write a C program that converts the least significant nibble of a byte to a hex character.

17. A motor is controlled by two push-button stations. Pressing either start button turns on the motor. Pressing either stop button shuts it down.
    (a) Write the pseudocode.
    (b) Write the main C code. Note that & and | are logical bit operators. && and | are logical AND and OR operators for true (nonzero) and false (zero) values. Otherwise, make your own assumptions about the functions to be called.

18. For the ICC11 compiler, write instructions to make PC0 to PC3 input and PC4 to PC7 output. Then send $7B to port C and read port C and store the result at location IN_VAR.

19. The motor control system of Question 17 has I/O as follows:

    Motor output is PB0.
    Start1 is PE6.
    Stop1 is PE7.
    Start2 is PD1.
    Stop2 is PD0.

    Write the C program as main without calling any functions.

20. Write a C program for the security system described in Question 1(c). A library includes the following functions with an integer parameter.

    read_bit(input1)   Returns true if input1 is on
    set_bit(output1)   Turns on output1, returns void
    clear_bit(output1) Turns off output1, returns void

    Use the serial port to send the message.

21. A microcontroller is balancing an inverted pendulum. Find the motor current if the pendulum angle is $-5°$ and it is moving $5°$ per second (to the right). Use fuzzy logic and the knowledge base as described in Section 13.8.

22. Repeat Question 21 for an angle of $2°$ and movement of $27°$ per second.

14 Applications

OBJECTIVES

After completing this chapter, you should be able to

- Describe how microcontrollers handle automotive-engine control.
- Describe how microcontrollers handle automatic-camera control.
- Describe how microcontrollers handle dot matrix printer control.
- Expand upon the illustrated applications to further develop them or other applications.

There are hundreds of applications, and in this chapter we will look at three of them. This will give you an idea of how microcontrollers are used to do something for us. Engine control is an example of an automobile electronics application. In fact, microcontrollers were built in the first place primarily because of automobiles. Then we examine a consumer electronics application, an automatic camera. Finally, we look at a computer peripheral, the dot matrix printer. We look at (or review) how each application's device works. Next we examine the I/O requirements and overall software. Then we show details for a part of the system. For reasons of brevity and confidentiality, we do not show all details.

■ 14.1 AUTOMOTIVE-ENGINE CONTROL

History involves interesting relationships. We often think of chip manufacturers as those who lead the way for the rest of us in technological development. To ensure sales, a semiconductor manufacturer has to predict what its customers need. This is a difficult task. As a result, if the designers of a large (potential customer) company say that they need a chip with such-and-such features, the manufacturer listens very carefully. In 1976 General Motors (GM) selected Motorola to develop a General Motors Custom Microcomputer (GMCM) chip, an upgrade of the 6800 microprocessor. Later, this became a standard off-the-shelf chip, the 6801. Again, GM defined and sponsored another upgrade. This became the 68HC11A8.

A microcontroller has many automotive uses. Typical ones are

Engine control module (ECM)

Entertainment systems

Antitheft systems

Climate control

Instrument panel control and lighting

Transmission

Power steering

Antilock braking

We will concentrate on the engine control module (ECM). It controls spark timing and the fuel and air going to the engine. It does this to minimize emissions and maximize fuel economy and performance. To understand what the ECM does, it is necessary to understand how an automobile engine operates.

The Four-Stroke Cycle.   Most automobile engines are of a type known as the four-stroke internal combustion engine. Figure 14.1 illustrates its operation for each cylinder. An engine has several cylinders. Each cylinder has a piston connected to a common crankshaft. The rotating crankshaft moves the pistons up and down. Turning on the ignition key starts an electric motor to rotate the crankshaft. Once this motion is established, the firing of the cylinders keeps the motor running.

Now, we concentrate on a single cylinder. In automotive terminology, the top of the cylinder is known as *top dead center* (TDC). Conversely, the bottom is known as *bottom dead center* (BDC). Each cycle has four strokes, beginning with the intake stroke (Figure 14.1[a]). As a result of the previous cycle, the piston moves down and the intake valve opens. This causes a vacuum that pulls the air/fuel mixture into the cylinder. When the piston moves up again, the valve is closed and the air/fuel mixture is compressed (Figure 14.1[b]).

The spark plug ignites this mixture shortly before (sometimes after) the cylinder reaches the top. The force of hot expanding gas drives the piston down (Figure 14.1[c]). This expansion powers the rest of the cycle and the automobile itself. When the piston reaches the bottom, the exhaust valve opens. The mixture is forced out through this valve when the piston moves up again (Figure 14.1[d]).

Remember that the engine has several pistons connected to a common crankshaft. Their power strokes occur at different times to deliver continuous power to the crankshaft. The continuous ignitions heat up the engine. Therefore, it is necessary to circulate coolant fluid to remove the heat.

Engine Control Module (ECM).   To control the power and speed of the engine, the ECM varies the ignition timing and the mixture of air and fuel. The ECM uses the spark control system for the former and an air/fuel controller for the latter.[1]

---

[1]Some automobiles use a closed-loop carburetor controller (CLCC) to control the air/fuel mixture.

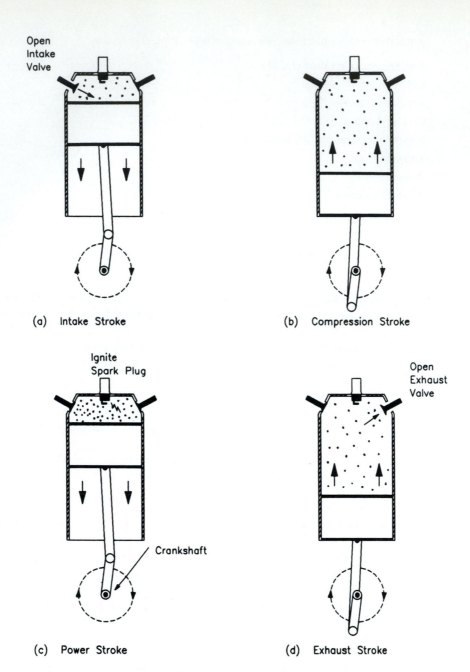

Figure 14.1 Four-Stroke Internal Combustion Engine Cycle

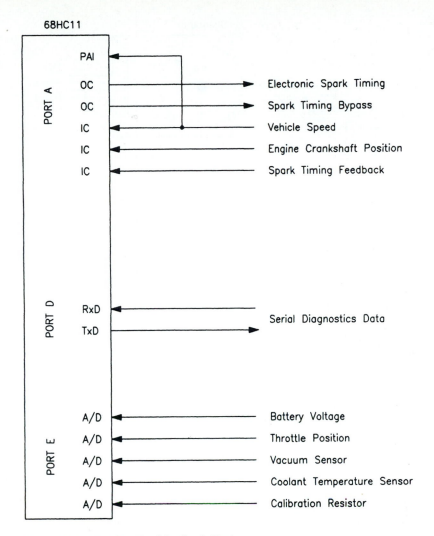

Figure 14.2 68HC11 Used for Spark Timing

Spark Control System. The ECM adjusts the spark timing to occur before the cylinder reaches top dead center, when it reaches it, or after the piston has started moving down. One setting provides the most torque, which is required to accelerate the vehicle quickly. Another setting provides the lowest fuel consumption. A third setting is better for reducing pollution. The ECM adjusts the timing to take all three into account. Changing the spark timing relative to piston position is also known as angle-based control.

Figure 14.2 illustrates the required sensors and actuators. The crankshaft position and spark timing feedback are used to tell the ECM when to trigger the next spark signal. The spark timing signals go to a unit that distributes it to the different cylinders.

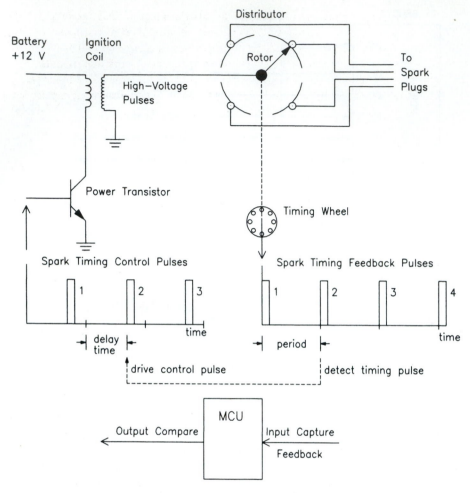

Figure 14.3 Spark Timing Control

Figure 14.3 shows more detail for part of the system. The crankshaft drives a distributor shaft. This drives a rotor that makes contact with each spark plug in sequence. The crankshaft/distributor system also drives a timing wheel. The timing wheel generates pulses to indicate when each successive piston is in the top dead center position during the power stroke. The sensor may be magnetic or an optical sensor of the type described in Section 11.3 (see Figure 11.8).

As shown in Figure 14.3, the power transistor switches the primary coil of the ignition coil (a transformer) on and off. The switching action induces high-voltage pulses on the secondary side. These pulses travel to the spark plugs as determined by the distributor rotor position.

The spark control system adjusts the spark timing pulses relative to the spark feedback pulses. In this case the MCU generates output pulses to occur before the cylinder reaches top dead center. Typically, the feedback pulses occur every 10 ms. The software can use a look-up table to calculate the delay time for the next spark control pulse. The look-up table stores the time delay value to correspond with each measured feedback period (related to engine speed). For the 68HC11, this can be done by using an input capture service routine. The pseudocode of Listing 14.1 shows the basic strategy.

```
*Listing 14.1
*Pseudocode for input capture service routine
*used for spark control timing.
*Feedback pulse edge asserts interrupt.
SVIC
        read input capture (IC) register
        period = IC register - previous capture value
        table pointer = (modified)period
*look up time delay
        time delay = *table pointer
*arm output compare (OC) pulse
        OC register = IC register + time delay
        previous capture value = IC register
        return from interrupt


*note that an output compare service routine
*will handle the spark control output pulse
```

Other parts of the software can modify the contents of the look-up table when operating conditions change. Listing 14.1 is a simple system. More complex systems use a multi-dimensional table to take into account changes in operating conditions. They use a pointer that depends on several factors. Note that Figure 14.3 shows a system using a distributor. Some car manufacturers build distributorless systems in which each spark plug has its own set of coil connections. An MCU sends pulses to each spark plug individually and in sequence.

Air/Fuel Controller.   Recall (Figure 14.1) that an air/fuel mixture enters the cylinder during the intake stroke. The ratio of air to fuel affects engine efficiency, power output, and makeup of exhaust emissions. Again, different ratios are optimum for torque, fuel economy, and pollution control. Also, the amount of mixture affects engine speed. The MCU can adjust the air/fuel ratio to optimize a combination of factors for different operating conditions. For example, a ratio of 14.7/1 is optimum for cruise conditions. Cold start and wide-open throttle positions require a higher ratio (richer mixture). Deceleration requires a lower ratio (leaner mixture).

The fuel injection system varies both the amount and ratio of air and fuel. The throttle valve position (varied by pressing the gas pedal) determines the amount of air flowing in as shown in Figure 14.4. A fuel pump supplies the fuel to the injection valve. The injection valve forces out droplets of fuel that evaporate to form a combustible vapor mixed with the air.

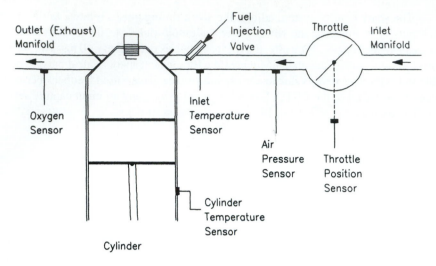

Figure 14.4 Simplified Fuel Injection System

A solenoid is the actuator for the injection valve. The MCU sends PWM signals to the solenoid to vary the duty cycle of the valve opening. This in turn varies the amount of fuel injected. The software calculates the amount of fuel to be injected by determining mass airflow. Many engines do not use sensors that measure mass airflow directly. Instead, the MCU calculates it from measured engine speed, manifold inlet air (vacuum) pressure, and air temperature. This is an example of open-loop control since it estimates (calculates) mass airflow instead of measuring it.[2]

The oxygen sensor provides an input for closed-loop feedback control. Little or no oxygen in the exhaust indicates a low air/fuel ratio. A feedback control program can then fine-tune the PWM signals sent to the fuel injection solenoid. Since the oxygen sensor operates when the exhaust air is very hot (900°C), the software should not use closed-loop control until the engine has warmed up. Figure 14.5 shows the sensors and actuators that can be used by a 68HC11 used for air/fuel control.

Summary.   For both the spark timing control system and air/fuel controller, other discrete I/O is used for instrument panel indicating lights and switches. Most I/O is connected using signal conditioning circuits for protection against electrical transients. In Chapter 10 we explained how the SPI subsystem can reduce wiring. For example, an MCU located near the instrument panel receives a turn indicator signal. It sends this request to the MCUs located near the front and back. They in turn drive the signal lights. Note that the same wiring carries all other data as well. This *drive-by-wire* design gives the automobile designer more options. The gas pedal does not have to be connected directly to the throttle. The EEPROM is used primarily for calibration and maintenance data. It can also store odometer readings.

---

[2]Some automobiles do use a mass airflow sensor; the LAMBDA sensor built by Bosch is one example.

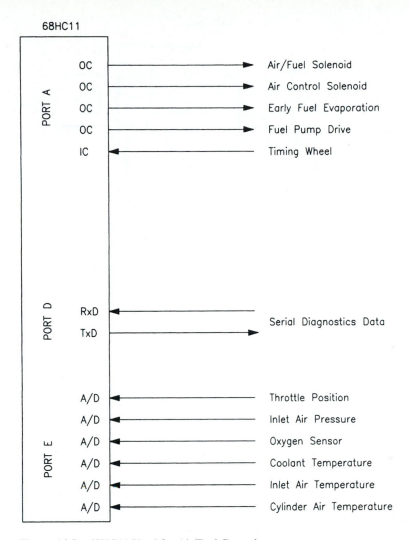

Figure 14.5   68HC11 Used for Air/Fuel Control

■ 14.2 CONSUMER ELECTRONICS: AUTOMATIC CAMERA

If you are a photography buff, you may already be familiar with the Canon EOS^{tm} model cameras. All EOS cameras (with the exception of the EOS 10) use a 68HC11 microcontroller. The cameras use interchangeable lenses which also have a microcontroller. Before examining what the MCUs do, we will review basic camera principles. Then we will look at some of the control activities performed by the MCUs.

Basic Camera Principles.   To take a picture, the photographer aims the camera at the subject. Either he (or she) or the camera control system focuses the image and adjusts the aperture and shutter speed for correct exposure. Then the photographer presses the shutter

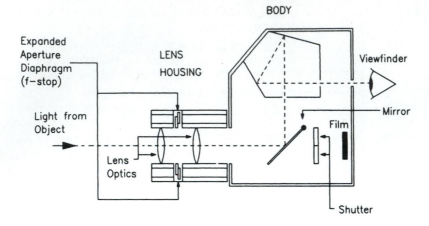

(a) Basic parts of a camera shown when aiming

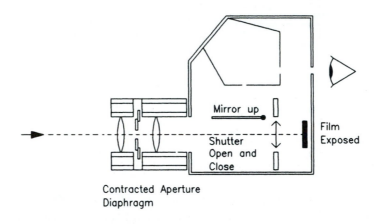

(b) Action of camera when taking a picture

Figure 14.6   Basic Principle of Camera Action

button and an image is formed on the film. To make another picture, the photographer must advance the film to the next frame. Depending on the camera, it is possible to perform variations of the procedure.

Figure 14.6 illustrates the general operation of the single-lens reflex (SLR) camera. During the aiming phase, the photographer uses the viewfinder to *frame* the picture. The distance from the lens to the film plane determines whether the image will be in focus or blurred. Light intensity is controlled by setting the shutter speed. If it is open too long, the film will be overexposed, and the image will be too bright. If the shutter is not open long enough, the film will be underexposed, and the image will be too dark. Varying the diameter of the aperture (f-stop setting) affects the *depth of field* and the required shutter speed to maintain proper light intensity. When a picture is taken, the aperture diaphragm contracts to its f-stop setting. For a narrow aperture diameter, the shutter should remain open for a longer time than for a wider aperture diameter.

To increase depth of field, the aperture diameter must be made smaller. The depth of field refers to how much of the image is in focus. With a small depth of field (wide aperture) only the main subject and objects close to it are in focus. With a large depth of field, objects farther away and closer than the subject will also be in focus.

Because different types of film have different sensitivities, the correct exposure setting depends on film type and actual lighting conditions. For a fast-moving subject, the photographer will have to sacrifice depth of field to use a fast shutter speed to obtain the correct exposure.

Many cameras have interchangeable lenses. This allows the photographer to switch lenses for different requirements. For example, a telephoto lens is used to magnify faraway objects such as birds in a tree. A macro lens is used to photograph objects at close range such as flowers and insects. For photography in places where there is not enough natural light, a flash unit is used to emit the required light.

Basic MCU Operations.   The electro-optical system (EOS) model cameras use a 68HC11 microcontroller in the camera body, which we will refer to as the *main MCU*. Each EF lens (meaning the complete unit with the aperture, lens optics, and other parts) has its own microcontroller. We call this the *lens MCU*. Figures 14.7 and 14.8 show block diagrams for both systems. As soon as the photographer attaches a lens to the camera body, the two MCUs communicate using the SPI subsystem. The lens MCU sends data about lens focal length, maximum aperture, and a few other settings to the main MCU.

The following happens when the photographer presses the shutter button. The main MCU calculates shutter speed and aperture settings based on light meter readings. The camera displays the shutter speed and aperture settings in the viewfinder and external LCD data displays. At the same time, the main MCU processes autofocusing data received from the Autofocus (AF) detection system. Again it sends commands to the lens MCU, which in turn drives the motors to move the lens optics. If the photographer presses the shutter release completely down, the main MCU performs sequencing control using the lens MCU when necessary. Generally, the sequence is to drive the mirror up, contract the aperture, open and close the shutter (film exposure), expand the aperture, drive the mirror down, and advance the film. Note that the shutter is always closed except during exposure.

The motors themselves employ new technology. It is beyond the scope of this book to describe them except to say that they are engineering marvels. The motors are miniature and can run on batteries. Yet they are fast and powerful enough for camera mechanics. The general communication sequence between the two MCUs is

Lens MCU: Send data to main MCU.

Main MCU: Command calculation and decision making; command output.

Lens MCU: Command evaluation and operational control.

The sequence repeats if necessary. The control categories are

Evaluative metering and auto exposure (AE)

Range sensing and auto focus (AF)

Shooting sequence control

Data management tasks

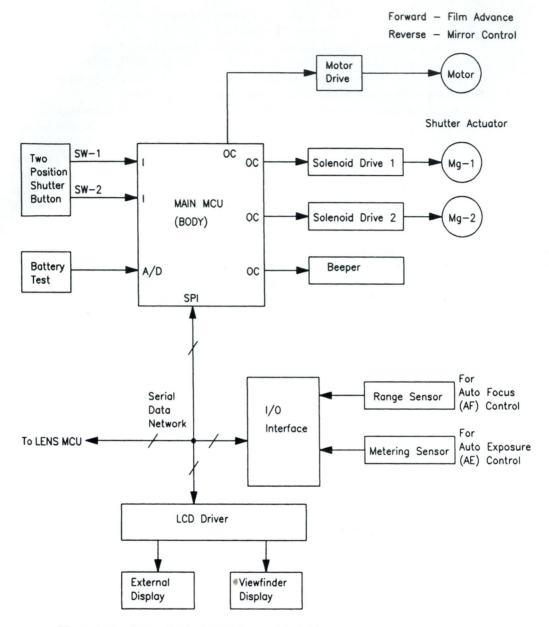

Figure 14.7 Camera Body Control System Block Diagram

In the following sections we will describe these operations with some simplifications. We will ignore flash operations.

Evaluative Metering and Auto Exposure (AE) Control. The scene to be photographed is known as the *frame*. This is the area seen in the viewfinder. The light intensity (luminosity) varies throughout a typical frame. Traditional light meters take an average of the lumi-

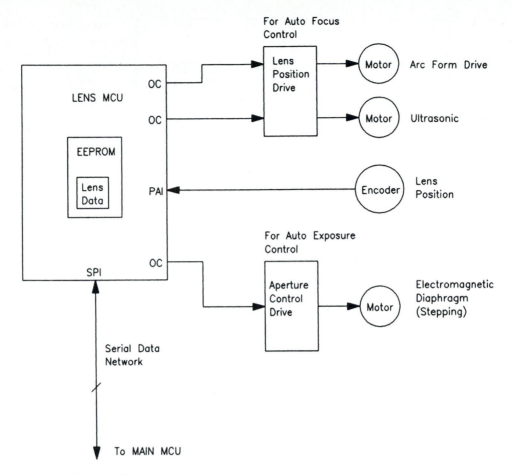

Figure 14.8 Lens Control System Block Diagram

nosity. The EOS camera uses a component with six sensors that measure luminance. Each measures light intensity for a section of the lens as shown in Figure 14.9. Based on this, the main MCU calculates exposure. Remember that an exposure setting is defined by the shutter opening time and aperture diameter (when contracted).

To understand why the frame is divided into six sections, consider the possible types of photographs. One may have only a background with more or less constant luminosity (main subject is entire scene). Another may have a few main subjects (such as people and animals) with a background (such as plants or room furniture).

The main MCU calculates the exposure setting using Boolean logic. To give you an idea of how the algorithm works, we present Listing 14.2.

```
*Listing 14.2
*Pseudocode to find correct exposure
*for a sample condition.

        if ((B - A == 0) && (C - B < 0))
                then set exposure based on A and B (primarily)
```

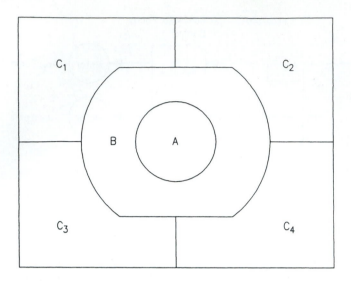

A: Luminance of center area
B: Luminance of intermediate ring
C: Luminance of four outer zones

Figure 14.9 Areas of Frame Used for Evaluative Metering

```
*This means that the subject is fairly large and relatively
*bright. The background is relatively dark.
```

If an average luminosity for the entire frame was used, the subject may be overexposed. Hence, it will appear too bright in the developed photograph.

Range Sensing and Auto Focus (AF). The range sensor is known as BASIS (Base-Stored Image Sensor).[3] It uses two secondary lenses (optics). Each lens focuses light on a solid-state device that accumulates charge when exposed to light. The camera is in focus when both devices have identical images. Note that the main MCU has already determined the subject to focus on after the autoexposure procedure.

The main MCU reads the sensor. If it detects that the images are not identical, it calculates how far to move the lens. It sends this data to the lens MCU. After receiving this data, the lens MCU drives the ultrasonic motor and stepping motor (arc form drive) to position the lens horizontally. It uses feedback from the encoder to do this. Listing 14.3 describes this procedure using pseudocode.

```
*Listing 14.3
*Pseudocode for autofocus (AF) routines

*Main MCU Routine
Main_AF
```

---

[3]The principle of operation is similar to that of a charge-coupled device (CCD). There is a possibility that BASIS will be used to pick up images for high-quality television.

```
        send AF_start command to lens MCU
        read range sensor
        calculate lens movement
        send lens movement data to lens MCU
        wait for lens MCU AF_verification
        send in-focus signal to viewfinder display
*Lens MCU Routine
Lens_AF
        wait for AF_start command from main MCU
        prepare for AF operation
        wait for lens movement data from main MCU
        read lens movement data
        calculate new position
        while lens position encoder != new position
                drive motors
        send AF_verification signal to main MCU
```

Shooting Sequence Control. In this section we describe only the one-shot mode. In servo mode, autofocus follows a subject until the photographer releases the shutter button. We also omit some other fine points. The camera uses a two-position shutter button (see Figure 14.7). If the photographer presses the button halfway (SW-1 on, SW-2 off), the camera performs only the autoexposure and focus functions. If the button is completely pressed (SW-1 and SW-2 on), the camera also completes the sequence to take a photograph.

Listing 14.4 shows the pseudocode for a simplified shooting sequence. Although two MCUs are involved, we show the code for both in one program to illustrate sequence continuity. We can think of the two MCUs as passing tokens to one another. One waits for the token to be passed by the other before it continues with the sequence.

```
*Listing 14.4
*Pseudocode for normal shooting sequence
*Most of the sequence controlled by the main MCU


        token = main
begin
        if battery check OK
                then continue
                else stop operation
        if SW-1 on
                then continue
                else exit
loop
        call AE and AF routines
        if SW-2 on
                then continue
                else if SW-1 on
                        then repeat loop
                        else exit
        if battery check OK
                then continue
                else stop operation
```

```
*arm shutter
        energize Mg-1, Mg-2
*drive mirror up
        run motor in reverse
        token = lens
*lens MCU contracts aperture diaphragm
        pulse aperture control drive
        token = main
*main MCU controls film exposure
*first open shutter
        deenergize Mg-1
        exposure time delay
*then close shutter
        deenergize Mg-2
        token = lens
* lens MCU expands aperture diaphragm
        pulse aperture control drive
        token = main
*bring mirror down
        run motor in reverse
*main MCU continues sequence
        time delay
        stop motor
        time delay
*advance film to next frame
        run motor forward until next frame
        repeat begin
```

Data Management Tasks. The camera's built-in microcontroller also provides many conveniences for the photographer. It stores data about each photograph (exposure) in EEPROM. The EEPROM in the 68HC11 performs many calculation-related data storage functions. One of its most important functions that is visible to the photographer is its ability to retain important data such as the frame count, exposure mode, and other camera settings even if the camera battery is removed.

The EEPROM in the main MCU has three main functions:

1. It acts as a buffer for AF data during processing, especially for predictive AF. (Servo Predictive AF is used in all EOS cameras except the 620, 650, 750, and 850.)

2. It acts as a nonvolatile storage medium for data directly related to the camera settings for the frame in use only.

3. It also stores service-related data such as the number of rolls of film taken and other internal camera settings related to AF, AE, and shutter speed accuracy, which may be calibrated by external devices. These external devices are attached directly to the 68HC11 for this purpose.

Some cameras (EOS 620, 630, and 650) have an optional accessory known as the Technical Back E. The camera body's main MCU transmits data to the Technical Back's MCU through a set of dedicated contacts in the back of the camera body. The data is then stored by the Technical Back.

Cameras that support a Technical Back E option record frame data as follows: When the film is rewound, an array of seven LED dots imprints the data on the film. This data includes date, time, exposure data, shutter speed, lens focal length, shooting mode, and frame counter. The camera also allows the photographer to enter comments using a special keyboard unit. The data can be displayed on the LCD display mounted on the back of the camera. The camera also has a communications port for connection to a personal computer.

■ 14.3 COMPUTER PERIPHERALS: DOT MATRIX PRINTER

Despite the attempt to have a paperless office, people still produce paper (hard-copy) documents using a computer and printer. A dot matrix printer is a relatively inexpensive device to produce hard copy. The application programs for things such as word processing, spreadsheet calculation, and drawing have software printer drivers to produce a variety of lettering styles (fonts and point size) and graphic symbols for many commonly used printers. An MCU uses most of its subsystems to control a dot matrix printer.

Another reason for choosing it as an application example is that hobbyists and students can build a printer control system as a project. There are many old printers available that current application programs do not support with software printer drivers. You may be able to buy one from a surplus store or auction. It is even possible to get one free. Once you have found an old printer, you can strip out its old control system but keep its power supply, sensors, actuators, and mechanical parts. Replace the control system with a modified EVBU board, similar board, or one of your own design. One warning, though: The project will require lots of work. We concern ourselves with printing text characters since this is the most common use for the dot matrix type printer.

Operation.   Figure 14.10 shows some operational features. The carriage motor moves the print head right or left. During this time, the print head prints the characters on paper. After printing a line, the paper feed motor moves the paper up. Then the print head moves horizontally to print the next line. The process continues until all lines are printed on a page. When a page is completed, the paper feed motor advances the paper to the next page.

Both motors are stepper motors (see Section 11.2.4). The paper feed motor positions the paper vertically and the carriage motor positions the print head horizontally.

The control panel contains switches for linefeed (LF), formfeed (FF), and on-line control. In printer terminology, they mean the following:

Linefeed (LF): Move paper up by one standard line height (4.2 mm).

Formfeed (FF): Move paper up by one standard page length.

On-line: Mode when printer communicates with computer.

If the on-line switch is active, the computer can send data to the printer using the Centronics parallel interface (see Section 9.8.1). With the on-line switch off, a user can manually control paper position using the linefeed and formfeed switches.

Print Mechanism.   Refer to Figure 14.10(b). The print head consists of solenoids (see Section 12.2.3). Each drives a stiff wire to impact the ink ribbon. This leaves a dot on the paper. A typical print head has nine solenoids arranged in a vertical column. A character is formed by firing different solenoids as the print head moves horizontally across the paper.

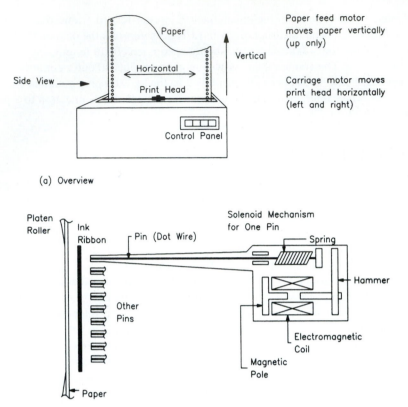

(a) Overview

(b) Print Head Solenoid Side View

Figure 14.10 Dot Matrix Printer

Figure 14.11 shows a pattern to produce the letter 'G' for a 9 × 11 matrix. For example, if the head is at column C1 and is moving left to right, the printer fires solenoids R2 to R6. Then the carriage motor advances the head to the C2 position and the printer fires solenoids R1 and R7. This process continues until the character is printed.

If ASCII characters are to be printed, the software has to convert each to a dot matrix pattern. If a graphical image is to be printed on the page, the entire image has to be converted into blocks of dot matrix patterns. Later, we'll see how the printer MCU uses a look-up table to find the dot matrix pattern (the dot matrix code) for a printable ASCII character.

General MCU Operations. Figure 14.12 shows one way that a 68HC11 could be used to control a dot matrix printer. In this case, a 68HC711E9 is used because of its internal 12K EPROM. The design could be modified to include more I/O by using the SPI subsystem and serial/parallel I/O chips. For example, DIP switches, control panel buttons, and stepper motor select outputs could be relocated to the external chips. This would allow the printer to use the SCI subsystem as a serial port in addition to the parallel interface port. For more sophisticated software, a 68HC711K4 can be used. It has a 24K EPROM with additional ports.

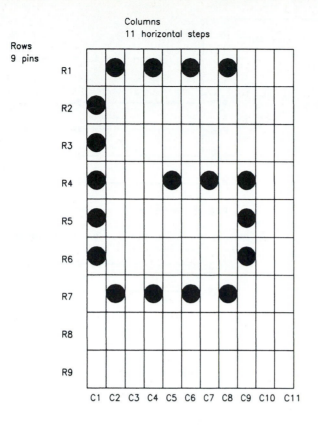

ASCII Character 'G' in 11 x 9 Dot Matrix Format

Figure 14.11 Dot Matrix Format

When the printer is powered up, it initializes the stack, registers, and outputs. It sends the print head to the home position, located on the left side. Periodically, it checks the panel buttons and performs the selected operations. If the printer is on-line, it waits for the computer to send data to it. When the computer user performs a print operation (print a file), the printer reads the data sent by the computer until its RAM buffer is full or until an end-of-file control character is received. If the RAM buffer is full, the printer asserts the BUSY line to tell the computer to suspend transmission. Then the printer prints the data stored in the buffer. When the buffer is empty, the printer deasserts BUSY to tell the computer to send more data until the buffer is full again or until an end-of-file character is received. Recall the Centronics interface handshake operation described in Section 9.8.1.

During the printing of ASCII characters stored in the RAM buffer, the printer software checks whether a character is printable or control. Control characters are the first 32 ASCII codes (see the Quick Reference [Appendix B]). Typically, a carriage return ($0D) is replaced by a space character ($20). A linefeed character ($0A) tells the printer to stop printing a line. It will then drive the paper feed motor to advance the paper and then proceed to print the next line.

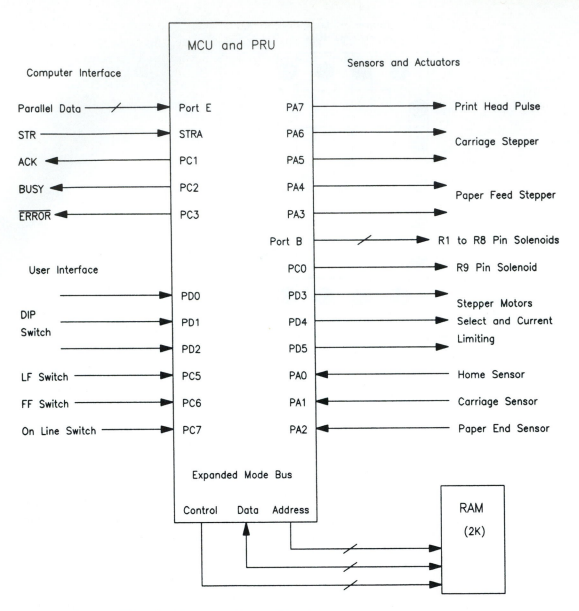

Figure 14.12 Block Diagram of Printer Control System

Many printers use an *escape sequence* for more software control. This is the escape character ($1B) followed by other characters. These sequences are not standard and differ depending on the printer model and manufacturer. For example, sending an escape character followed by 'G' tells the *IBM Proprinter II* to use double-strike printing for the characters that follow. An escape followed by 'H' will switch off double-strike printing.

If a printable character is fetched, the MCU looks up its dot matrix code. Then it steps the carriage motor and fires the print head solenoids as required. This puts the small dots on paper to make an image of the character. This process continues until the buffer is empty.

Then the printer gets more data from the computer. Listing 14.5 shows a simplified main program in pseudocode. It leaves out details such as determining print head direction and tracking when to perform a formfeed to advance to the next page.

```
*Listing 14.5
*Main printer program

    initialize registers, outputs, and variables
Again
    wait for strobe ;computer sends char(0)
    while buffer not full && char(i) ! = end-of-file
        fill RAM buffer ;get and store char(i)'s
    assert busy
    for i = 0 to last-in-buffer
        if char(i) == printable character
                then    ;solenoid routines
                    look up dot matrix code
                    print char(i)
            else
                    handle control char
    deassert busy
    repeat Again

Real Time Interrupt
    if on-line == 0
        then
                    stop printing
                    execute control panel routines
                    repeat Real Time Interrupt
        else
                    exit
```

Evidently, the MCU has many things to do. For a detailed description of the entire software, you can refer to reference 17 in Appendix C. It covers a design example using the Intel 8042 microcontroller. With some modifications, the same principles can be used for other microcontrollers.

Electromechanical Operations.   Now we look at how the other sensors and actuators are used. The stepper motors are of the four-phase type described in Section 11.2.4. The paper feed motor stepper drives two sprocket wheels using reduction gears. The rotation of the sprocket wheels move the paper up vertically. The carriage stepper motor moves the print head using reduction gears and a toothed drive belt.

The home sensor is an optical switch that switches to logic zero when the print head is in the leftmost position. The carriage sensor signal is sometimes known as the print timing signal. It is a slotted optical switch (see Figure 11.8). The encoder disk is connected to a shaft driven by the carriage motor. It provides a feedback signal that tells the MCU the carriage velocity and horizontal position (by counting pulses). The carriage sensor can also detect motor overload. The software can check if the disk turns too slowly. A very slow rotation indicates that the carriage motor is working too hard, hence in danger of overheating.

The paper end sensor is a photoreflector for detecting the point at which the printer runs out of paper. When the paper comes to the end, no reflected light is detected by the sensor's phototransistor. The software reacts to this signal by asserting $\overline{ERROR}$ to tell the computer that it cannot send more data.

Output PD3 switches rush or hold current for the paper feed motor. For a review of rush and hold current requirements of stepper motors, refer back to Section 11.2.4. Outputs PD4 and PD5 control a biasing circuit for the carriage motor. It enables one of two biasing voltages. This varies the distance the print head moves during each step signal.

Before triggering print wire solenoid pulses, it is necessary to accelerate the print head first. It takes a few pulses for the print head to reach constant speed. The MCU can detect a constant speed condition by checking if the stepper pulse outputs and carriage sensor signals are at the same frequency. As the print head moves across the paper, the pulses from the carriage sensor are used to time the printing of the vertical dots.

Nine outputs, port B, and one port C line (PC0) drive the print head solenoids. The print head pulse (PA7) triggers the print head solenoids to fire. The pulse is high for about 400 µs. Figure 14.13 shows a circuit that makes use of the TPIC2406 Intelligent Power Quad MOSFET Latch by Texas Instruments. This is the same MOSFET latch used to drive stepper motors (see Figure 11.6).

Normally, the pulse signal is low. This forces all latch outputs (DRAIN) high independent of the inputs. When the MCU has sent the dot wire data for a column, it asserts the pulse line for 400 µs. A logic high output will drive a corresponding output low to energize the corresponding solenoid coil.

Remember that the MCU fires the solenoid after the print head is moving at constant speed. When this is achieved, the solenoids are fired when a carriage sensor edge is detected. After starting the pulse, the MCU should send phase change signals to the carriage stepper. This will advance it to the next position.

Solenoid Software. Now we will target one detail of printer software. These are the solenoid routines referred to in Listing 14.5. Figure 14.14 shows how the look-up table is set up using the character 'G' as an example (also see Figure 14.11).

Listing 14.6 shows the sections of code to implement driving the solenoids to print a character in the left-to-right direction. An output compare interrupt (OC1) is used to trigger the 400-µs solenoid drive pulse. Subroutine RTHEAD would check the carriage sensor (PA1) for a high edge and then send the next step data for the carriage stepper motor. The code would be similar to that of Listing 11.5 with differences in I/O assignments. To print a character in the right-to-left direction, the software uses the dot matrix look-up table in the reverse order.

```
*Listing 14.6
*Excerpts of dot matrix printer application code applicable
*to firing print head solenoids in left-to-right print
*head travel direction.
*Note that registers remapped to Page0

DELAY4  EQU     800
*DOTMAT equals asc20h address
*get look-up address for ASCII character
*-----------------------------------
```

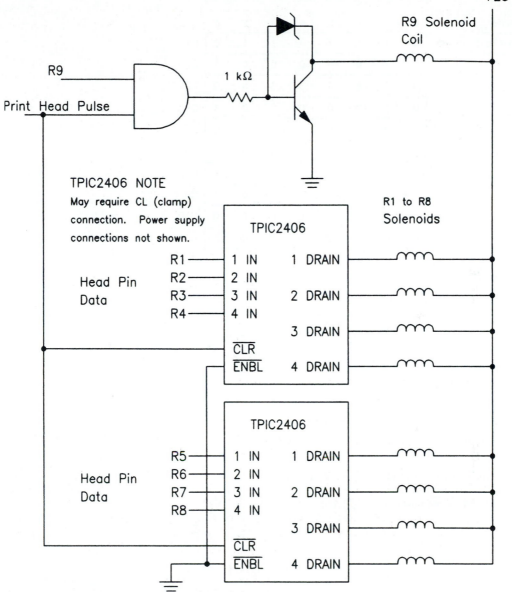

Figure 14.13 Solenoid Drive Circuits

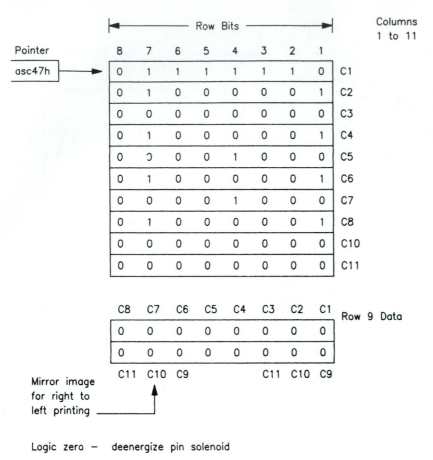

Logic zero – deenergize pin solenoid

Logic one – energize pin solenoid

Example shows character 'G', ASCII code **$47**.

Figure 14.14 Dot Matrix Look-Up Table

```
        ORG     0

R9      RMB     1           ;row 9 data work area

        ORG     $E000

* an excerpt from initialization part of code

        LDS     #$FF
        CLRA
        STAA    $103D       ;reference registers in page 0
        LDD     TCNT        ;prevent premature OC1
        STD     TOC1        ;set OC1 bits for OC1 to
        LDAA    #$80        ;drive OC1/PA7 low and assert
```

```
          STAA      OC1M              ;interrupt after
          CLRA                        ;successful compare
          STAA      OC1D
          BSET      TMSK1,$80
          BSET      DDRC,$01          ;PC0 output
          BSET      PACTL,$80         ;PA7 output
          CL1

*other code here

*excerpt of code to fetch a character from RAM buffer
          LDAB      0,X               ;get char from buffer

*section of code to handle control characters not shown
*note control chars in range of $00 to $1F

*at this point have a printable character
*now look up its dot matrix code
          SUBB      #$20              ;offset to start char $20 at 0
          LDAA      #13               ;multiply code by 13 since each char has
                                       13 bytes
          MUL
          ADDD      #DOTMAT           ;add it to base address
          XGDX                        ;load IX with address

          JSR       RPRNCHAR          ;print character

*more code follows

*some utility routine source listings
*-------------------------------------------

*Subroutine RPRNCHAR
*Get dot matrix code of character and print it
*Calling Registers
*     IX = address of first column
*No registers preserved for return
RPRNCHAR
          LDAB      #8                ;init column counter
          LDAA      11,X              ;store R9 info in RAM
          STAA      R9
RPRNCH1
*wait until next carriage sensor pulse detected
*and send next step signal to print head motor
          JSR       RTHEAD

          JSR       RNEWR9            ;set up R9 output

          LDAA      0,X               ;set up R1 to R8 output
```

```
        STAA    PORTB
        JSR     FIRE            ;drive solenoids
        INX                     ;point to next column
        DECB
        BNE     RPRNCH1         ;until first 8 done
        LDAA    4,X             ;get R9 info for C9, C10, C11
        STAA    R9
        JSR     RTHEAD          ;print C9
        JSR     RNEWR9
        LDAA    0,X
        STAA    PORTB
        JSR     FIRE
        JSR     RTHEAD          ;print C10
        JSR     RNEWR9
        LDAA    1,X
        STAA    PORTB
        JSR     FIRE
        JSR     RTHEAD          ;print C11
        JSR     RNEWR9
        LDAA    2,X
        STAA    PORTB
        JSR     FIRE
        RTS


*------------------------------------
*Subroutine FIRE
*Drives print solenoid pulse high
*OC1 interrupt will drive it low again after 400 us
*Only ACCB preserved
FIRE
        PSHB                    ;preserve ACCB
        BCLR    OC1M,$80        ;disable OC1
        BSET    PORTA,$80       ;drive pulse high
        LDD     TCNT            ;set up pulse width count
        ADDD    #DELAY4
        STD     TOC1
        BSET    OC1M,$80        ;enable OC1
        PULB                    ;restore ACCB
        RTS


*Subroutine RNEWR9
*Updates R9 output when head moves right
*Calling pseudo register
*     R9 = address of dot matrix code of R9
*No return registers
RNEWR9
        LDAA    R9              ;get current R9 info
        ROR     R9              ;point to next R9
        BCC     RNEW1           ;drive PC0 for current R9
        BSET    PORTC,$01
        RTS
```

```
RNEW1
      BCLR      PORTC,$01
      RTS

*Subroutine RTHEAD
*source code not shown
RTHEAD

*Service routine for OC1
RTOC1
*      ;automatically drives PA7 low
      LDAA      #$80            ;clear OC1F
      STAA      TFLG1
      RTI

*Excerpts of dot matrix look-up table

      ORG       $B600
DOTMAT
asc20h
*for character SP (space)
      FCB       00, 00, 00, 00, 00, 00, 00
      FCB       00, 00, 00, 00, 00, 00

*the dot matrix pattern for characters '!' up to 'F'
*are not shown but they would be placed here.

*now show our sample for character 'G'

asc47h
*for character 'G'
      FCB       $7E, $41, $00, $41, $08
      FCB       $41, $08, $41, $38, $00
      FCB       $00, $00, $00

*and similarly for 'H' and following printable characters
```

■ 14.4 SOCIETY AND MICROCONTROLLER TECHNOLOGY

Early in Chapter 1 we described the microcontroller as a Swiss army knife. The versatility of its I/O subsystems allows it to do many things. In addition to the three examples described in this chapter, many of the applications have been mentioned elsewhere in this book. The other applications not mentioned in this chapter include industrial uses such as robotics, programmable controllers, and chemical analyzers. Communications systems also use microcontrollers. A voice/data terminal for a telecommunications system uses an MCU for keyboard decoding and control of other chips to handle signaling. But like the general-purpose Swiss army knife, it is not the most suitable choice for some specialized applications. You would not use the knife to cut a block of wood if you had a large saw. Similarly,

Applications   603

specialized embedded controllers are more suitable to control the video graphic cards for computer monitors. Disk drive controllers are another example.

Unlike computer peripherals, many other things existed before they were re-designed to use MCUs. The automobile has existed since the late nineteenth century, and it worked fine. MCUs provide a combination of improved and less expensive control, but it can be overdone. Some cars used electronic voices to remind the driver of things such as using the seat belt. Since this annoyed customers, automobile manufacturers removed this feature.

There are other devices that would not exist without an MCU or some equivalent. A compact disc (CD) player uses an MCU to control movement of the pickup head that reads the disc data and to control the disk drive motor. The MCU decreases rotational speed as it moves the pickup outward from the center of the disk. It does this to maintain constant linear velocity of the data scanned by the pickup. Also, the pickup head reads the disk information to play music in the order selected by the consumer. It also controls the disk loading sequence and executes commands that the consumer has keyed in.

Like any other technology, microcontroller technology has led to some moral questions. Should one tamper with a control system? Some people who liked fast cars did not like the software that sacrificed performance for exhaust reduction. A few of them had enough programming and mechanical knowledge as well as equipment to modify EPROM and EEPROM data. They restored the original performance capabilities of the car. In doing so, they violated government exhaust reduction regulations. There is also the age-old question of "guns versus butter." The MCU can control the engine in a tank or a tractor.

How reliable is the software? If your favorite word processor has a glitch, it is an annoyance. If the antilock braking system in your car has a glitch, it can be fatal. Some people claim that commercially sold software typically has more than one mistake per 1000 lines of code. Most of the time the error is minor. Sometimes it is not. It may take years for the right combination of events to trigger the error. Think about it.

Due to the software reliability problem, MCU system developers are very careful. For example, an autopilot for an airliner may use three different types of MCU systems developed by three different teams. When in operation, the autopilot has three results to choose from. Normally, all results should be identical. If they are not, the majority is chosen (see reference 18 in Appendix C).

15 Eight-Bit Microcontroller Families

OBJECTIVES

After completing this chapter, you should be able to

- Describe basic characteristics of the 8-bit microcontroller industry.
- Describe characteristics of the Motorola 8-bit microcontroller families.
- Describe general characteristics and functionality of the Intel 8051 core.
- Choose a microcontroller for a given application.

In this chapter we will survey the variety of 8-bit microcontrollers available. Then we will look at how to choose an MCU.

The Core Approach. Most semiconductor manufacturers design their MCUs using a core approach. This means that different types of MCUs will use the same CPU *core* with some other common options. The different variations of MCUs have different memory types, memory sizes, I/O registers, and I/O subsystems. These variations are called *members* of an MCU *family*. Members of a family are software compatible, meaning that they will execute the same instructions. Section 15.1 looks at why and how the core approach is used by semiconductor manufacturers.

Families. So far, we have not considered variations of the 68HC11. In Section 15.2 we look at the members of the 68HC11 family and another Motorola 8-bit MCU family, the 68HC05. A brief description of the 68HC16 follows. It is a 16-bit extension of the 68HC11 family. Then we will look at the 68HC08 and 68HC12 families.

Section 15.3 covers some details of the Motorola HCS12 family. This is the next generation of the HC12 line, which in turn was the next generation of the HC11 line.

Section 15.4 looks briefly at other MCUs. Section 15.5 looks at how an *original equipment manufacturer* (OEM) can choose the MCU to control its product.

■ 15.1 MICROCONTROLLERS AND THE SEMICONDUCTOR INDUSTRY

A hierarchy of people are involved in the semiconductor industry. People such as ourselves are *consumers*. We wish to buy something useful, such as a coffeemaker, automobile, or weigh scale. The original equipment manufacturer (OEM) is the organization that designs and produces the product we wish to buy. In turn, the OEM is a *customer* of the semiconductor industry. The OEM chooses an MCU that best suits its needs. Company owners and managers wish to sell as many products as possible to consumers. To do this, they need to produce a product with the most features with the least cost. As a result, the competition between semiconductor manufacturers is fierce. An OEM may choose one MCU over the other because one is 25 cents cheaper (other factors being considered in the choice).

The semiconductor industry is facing challenges in silicon, manufacturing, and packaging technologies. Silicon geometries are shrinking. In other words, the doped regions (Section 6.1.3) are getting smaller. Integrated circuits include new functions, such as EEPROM and flash memories. New architectures are being developed. In manufacturing, IC quality and yield is being improved. *Yield* refers to the number of good circuits produced from a silicon slice. In packaging, there is a move toward higher densities, such as surface-mounted ICs. All of these improvements force semiconductor manufacturers to improve their products continually in order to compete.

A successful MCU manufacturer has to offer a core that is known by many potential customers or do a very good job of convincing them to use a new core. OEM designers have invested a lot of time to learn how to use an MCU and build a software library of functions (or subroutines) used to control their products. They have also invested money to purchase the required MCU-specific development tools, such as assemblers, compilers, simulators, and emulators. Therefore, it is difficult to convince them to change processor families without a good reason.

Semiconductor manufacturers use a modular design approach to expand a product line. This ensures upward software compatibility with future MCUs. Sometimes, another manufacturer may buy a license from the former to build its own variations based on a core. This is particularly true if the OEM industry as a whole has accepted the widespread use of certain cores, as is the case for Motorola's 68HC11, Intel's 8051, and Zilog's Z80, among others.

The MCU manufacturer also has to offer new family members each year. For example, a weigh scale manufacturer may have used a non-Motorola MCU in the past. The EEPROM feature of 68HC11 convinces the designers to switch to a 68HC11-based control system. EEPROM will be useful for storing custom calibration data. Even though the former MCU supplier may eventually offer a version with EEPROM or flash memory, the weigh scale manufacturer wants to build their new weigh scales as soon as possible. In this case the OEM has decided that a new feature was reason enough to warrant an investment in learning time and new development tools.

The moral is that semiconductor manufacturers have to offer new devices very often. To do this, the manufacturer has to devote time and money toward product development.

Deciding what to develop is risky. To make a profit on its investment, a manufacturer must sell large quantities of a new device. For example, it does not make sense to speculate that a new family member should have a complete 25-pin RS-232 serial subsystem. The manufacturer may design and market the chip, only to find out that nobody is interested in buying it.

Instead, the semiconductor manufacturers listen to their customers. Customers (OEMs) may ask for certain memory types and sizes, peripheral functions, and new peripherals. Let's say that a toy manufacturer wants an MCU with one PWM output and 8KB internal RAM. This does not mean that the manufacturer will proceed immediately to develop the product. However, if a motor control manufacturer asks for two PWM outputs with 4KB RAM and an automobile manufacturer asks for four PWM outputs, the situation changes. The semiconductor manufacturer may design a new MCU with four PWM outputs and 8KB RAM. The MCU will suit the needs of the three OEMs mentioned. The semiconductor manufacturer can now include the new MCU as a standard off-the-shelf item. Other OEMs may also decide to purchase quantities of the new MCU.

Manufacturers often *cut and paste* existing modules when designing new family members. Perhaps the semiconductor manufacturer had a version with 8KB RAM and another with two PWM outputs. The designers may use (cut) the 8KB RAM from the former and the PWM module twice from the latter. They then may add (paste) the modules to the core design. If a core approach is used to design new MCUs, the manufacturer can supply samples within six months after starting the design. Sometimes, the manufacturer must design a new module. Once designed, the new module is then available to designers to use for other new MCUs. For example, an OEM may request an LCD driver subsystem not currently available as a module. The semiconductor manufacturer has to design the module for the new MCU. This is a benefit because the LCD module is now available for other new members of the MCU family.

Reference 25 in Appendix C has good coverage of many microcontrollers including the 68HC11. For current information, we suggest that you contact the MCU manufacturers to check what they offer. Note that each member of an MCU family is identified by a part number. The manufacturer uses a system to assign part numbers. For example, Motorola part numbers that use 68HC711 instead of 68HC11 tell you that the MCU uses EPROM instead of ROM. Usually, we ask the sales representative to clarify the meaning of a part number; that's his or her job.

■ 15.2 MOTOROLA FAMILIES

Motorola offers several basic 8-bit families. The 68HC05 is an inexpensive core. The 68HC08 is an enhancement of the 68HC05. The 68HC11 is a higher-performance core. We will also look at the 68HC16 and 68HC12 families. Section 15.3 will overview the HCS12 family.

15.2.1 The 68HC11 Family

Our discussion of the 68HC11 has avoided mentioning differences between variations of the 68HC11. For specific information we have primarily used the 68HC11A8. Some other manufacturers (such as Mitsubishi and Toshiba) have supplied members of this family under a licensing agreement. The following will survey the primary features of some members. By the time you read this, there may be new family members. Again, please consult Motorola for current information. Note that we now include the "MC" prefix in the part numbers. Unless otherwise indicated, the members include other features listed for part number MC68HC11A8, although the "A" types are no longer produced.

MC68HC11A8:   8KB ROM, 256-byte RAM, 512-byte EEPROM, 64KB address range, SPI, SCI, 16-bit timer, three input captures, five output compares, eight A/D channels, real-time interrupt, watchdog timer, 38 I/O pins.

> *MC68HC11A0*:   No ROM, no EEPROM, 22 I/O pins.

> *MC68HC11A1*:   No ROM.

MC68HC811E2:   No ROM, 2KB EEPROM. Pin PA3 is bidirectional for use as an input capture or an output compare.

MC68HC11F1:   No ROM, 1KB RAM, enhanced timer and SCI subsystems, two or three input captures, four or five output compares, two new ports, F and G, nonmultiplexed address and data bus. It uses port B for the high address byte, port F for the low address byte, and port C for the data byte. Port A can also be a general-purpose I/O port. Port F is an output port or low address byte. Port G is an I/O port or a programmable chip select (CS) output.

MC68HC711K4:   24KB EPROM, 640-byte EEPROM, 768-byte RAM, enhanced SPI. A 68HC11F1 with a memory management unit (MMU) to address 1M external memory using port G for memory expansion. Port H is an I/O port or it can be used for chip selects and four PWM outputs.

MC68HC11E9:   12KB ROM, 512-byte RAM, enhanced timer, three or four input captures, four or five output compares.

> *MC68HC711E9*:   12KB EPROM instead of ROM.

> *MC68HC11E1*:   No ROM.

> *MC68HC11E20*:   20KB ROM, 768-byte RAM, HC11E9 pin compatible.

MC68HC11D3:   4KB ROM, 192-byte RAM, 40 pins, six timers, No A/D. This family member emphasizes a lower cost.

> *MC68HC711D3*:   4KB EPROM; also available with one-time programmable ROM (OTP).

> *MC68HC11ED0*:   512-byte RAM, HC11D3 pin compatible.

MC68HC11F1:   0KB ROM, 1KB RAM, 4-MHz nonmultiplexed expanded bus, chip selects.

MC68HC11M2:   32KB ROM, 1280-byte RAM, 0KB EEPROM, enhanced SCI, 2 SPI with chip select function.

MC68HC11N4:   24KB ROM, 768-byte RAM, 640-byte EEPROM, 2 PWM channels, 12 A/D, 2 D/A, math coprocessor, nonmultiplexed bus.

One-Time Programmable (OTP) ROM.   We will just say a few words about OTP ROM because it is also available with other MCUs. The OTP ROM offers some advantages over using custom ROM. It is available as a standard component because it is not necessary to specify the memory contents. There is no mask charge and minimum order quantity. More packaging options are available than for EPROM. It is also less expensive than EPROM. A disadvantage is that it cannot be erased and reused. For large quantities, MCUs using OTP ROM cost more than those using a custom mask ROM. To sell ROM-based MCUs cheaper, it is necessary to sell approximately 10,000 devices or more.

Low-Voltage.   Some family members can run on 3.0 V for low-power applications. Two of these are the MC68HC11D3 and MC68HC11K4.

Math Coprocessor.   Here we briefly describe the math coprocessor used in the N series. The module is called an arithmetic-logic unit (ALU). It is not the same thing as the ALU described in Section 1.3.3. The ALU consists of two 16-bit data registers (AREG and BREG), one 32-bit data register (CREG), one 8-bit control register (ALUC), and one 8-bit status register (ALUF). All registers are memory mapped. The ALU can perform signed 16-bit integer multiplication and division. It can also multiply 16-bit integers and accumulate a 32-bit product. To use the module, the software writes to the control register to specify the desired operation. Then it writes operands to the data registers. To start the operation, it writes to the lower data byte (BREG for multiply, AREG for integer division). Alternatively, it can start an operation by setting the trigger bit in the control register. This also happens to be the only way to start a fractional divide. The software must poll an arithmetic completion flag in the status register to determine if valid results are available in the data registers. Other flag bits indicate results such as zero, negative, overflow, zero remainder, and divide-by-zero error. The number of clock cycles per operation varies from 17 to 35. For more information, refer to the manual M68HC11 N Series Technical Data (Document MC68HC11N/D).

15.2.2 The 68HC05 Family

This is a popular family for low-cost applications. At least one member costs less than $1 per chip if ordered in bulk. In contrast, a 68HC11 may cost somewhere around $20 per chip, depending on the family member and quantities. When this book was written, Motorola offered more than 30 family members as standard parts. This was a result of a customer-specified integrated circuits (CSIC) program. Their promotional literature includes a chart of family members based primarily on "cat" names, such as Phonekat and Bearkat.

Figure 15.1 shows the core. Family members exhibit significant variation. Typically, the I/O subsystems use ports C and D. Some ports have fewer than eight lines. Some MCUs have fewer ports and use dedicated pins for some of the optional subsystem I/O. Perhaps it is more appropriate to call the entire selection a 68HC05 clan. Motorola divides the entire selection into families. The main features of the 68HC05 "families" are generalized as follows.

C Family.   These are generic 40-pin MCUs with RAM, ROM, parallel I/O, serial, and programmable timer subsystems.

B Family.   Members have 52 pins with EEPROM and many I/O subsystems.

M Family.   They contain four A/D channels and vacuum fluorescent display (VFD) drivers (see Section 9.5).

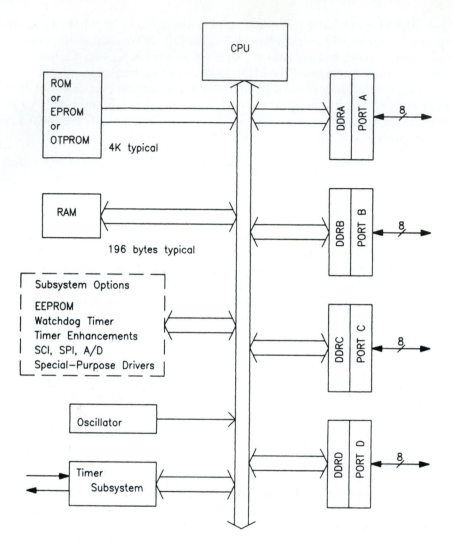

Figure 15.1   68HC05 Core Block Diagram

T Family.   Members contain on-screen display, A/D and D/A. Their primary applications are TVs and VCRs.

P/J Family.   This is a low-cost family with 20 or 28 pins. Typical ROM sizes are 1 or 2 kilobytes.

SC Family.   They have fewer pins and contain EEPROM. Their primary application is for smart cards.

L Family.   Members have a built-in LCD driver (see Section 9.3).

K Family.   Members feature a low pin count to make them cost competitive (under $1) with 4-bit microcontrollers.

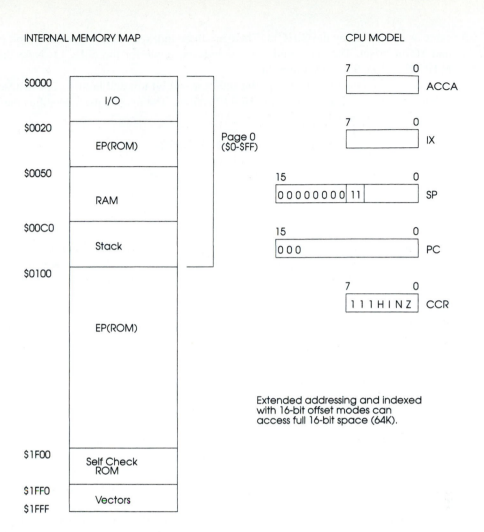

INTERNAL MEMORY MAP

$0000	I/O
$0020	EP(ROM)
$0050	RAM
$00C0	Stack
$0100	EP(ROM)
$1F00	Self Check ROM
$1FF0	Vectors
$1FFF	

Page 0 ($0-$FF)

CPU MODEL

ACCA (7 0)

IX (7 0)

SP (15 0) 0 0 0 0 0 0 0 0 11

PC (15 0) 0 0 0

CCR (7 0) 1 1 1 H I N Z

Extended addressing and indexed with 16-bit offset modes can access full 16-bit space (64K).

(Some versions extend to $3FFF)

Figure 15.2 68HC05 Programming Model

Other Families. There are other MCUs with features for telecommunications. Others have high current drivers useful for industrial control.

1.8-Volt Versions. This series is designated as 68HCL05; these versions are intended for battery-operated applications. Some members are 68HCL05C3, 68HCL05J1A, and 68HCL05P1.

Instruction Set and Addressing Modes. The 68HC05 is similar to the 68HC11 because both are descendants of the earlier 6800 microprocessor. Many instructions are similar. Like the 68HC11, the 68HC05 has instructions for data movement, data manipulation, and program control. Figure 15.2 shows the basic programming model. This MCU uses 10 different addressing modes. The inherent, immediate, direct, extended, and relative modes are the same as those used by the 68HC11. There are three indexed modes: no offset, 8-bit offset, and 16-bit offset. They are similar to the indexed mode for the 68HC11. Note that the 68HC05 uses an 8-bit index register.

The two remaining modes are bit set/clear and bit test and branch. The instructions using these modes differ from the 68HC11 versions. The opcode itself specifies the bit to be modified or tested. For example:

```
        BSET    0,$01   ;set bit 0 of address $01
*                       ;machine code is 10 01
        BCLR    0,$01   ;clear bit 0 of address $01
*                       ;machine code is 11 01
        BSET    1,$01   ;set bit 1 of address $01
*                       ;machine code is 12 01
```

The I/O registers are mapped to an address space accessible by direct addressing. This makes bit manipulation and checking quite efficient. Although many register addresses tend to be the same for all 68HC05 variations, you should consult the memory map for each type.

The 68HC05 has a small stack and does not have any instructions that directly manipulate the stack except for a reset stack (RSP) instruction. Other than this, only entering and returning from subroutines and interrupt service routines affect the stack.

15.2.3 The 68HC16 Family

Motorola introduced this family in 1991 as a *bridge* between the 8-bit 68HC11 family and the 32-bit 68300 family. It is hardware compatible with the 68300 MCU family and the 68000 microprocessor family. However, the 68HC16 family is software upward compatible with the 68HC11 family. This makes it useful for those who have invested in 68HC11 software and wish to expand their hardware capability. We have already examined its digital signal processing (DSP) feature (Section 13.5). In this section we will look at some of its other features.

The 68HC16 uses a 16-bit version of the 68HC11 CPU. It has three new multiply and two additional divide instructions, new addressing modes, and additional registers. These features make it easier for the MCU to use high-level languages and perform DSP functions. The 68HC16 uses a modular architecture. It uses some I/O subsystem modules taken from the 68300 family. These modules are more sophisticated than the 68HC11 versions. The queued serial module (QSM) provides two serial channels for asynchronous and synchronous communication. The system integration module (SIM) provides an external bus interface. The SIM includes programmable chip select logic, system failure protection, and a phase-locked-loop (PLL) oscillator subsystem. The PLL allows the MCU to vary its own clock speed. The maximum clock speed is 16.78 MHz. Other subsystems are the general-purpose timer (GPT), an enhanced HC11-type timer, RAM, and a 10-bit eight-channel A/D.

Figure 15.3 shows the programming model for the first member of the family. By using the address extension fields, the MCU can address 1 megabyte of data and 1 megabyte of program memory. This is done by using 4-bit extension fields for some registers. Fields ZK, YK, XK, and EK are part of the 16-bit address extension (K) register. Field SK is a separate 4-bit register.

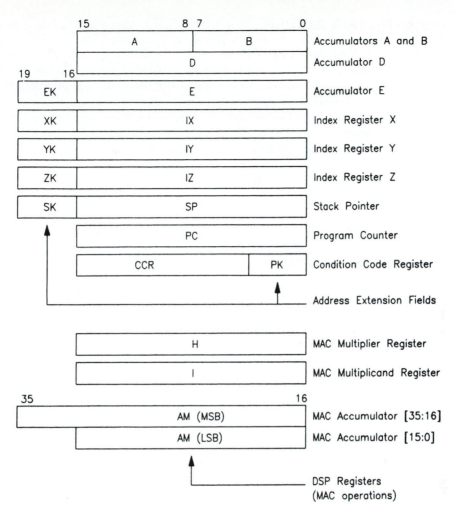

Figure 15.3 Programming Model of the 68HC16Z1

M68HC16Z1. This is the first member of the 68HC16 family.

M68HC16Y1. This is the second member of the 68HC16 family. It features a sophisti-cated time processing unit (TPU) and has 48KB ROM. The TPU allows the microcontroller to handle time-critical tasks without using the CPU software.

MC68HC916Y1 (Y1) and MC68HC916X1 (X1). Both feature 48KB of flash EEP-ROM. Designers can easily make modifications to the software in an established system. The 9Y1 is designed for automotive and motion control applications. The 9X1 is designed for communications, office automation, and consumer electronics products.

15.2.4 The 68HC08 Family

In 1993 Motorola introduced the 68HC08 family. The 68HC08 is upwardly object-code compatible with the 68HC05, meaning that it will run 68HC05 software. While the 68HC05 can be very cost effective, some customers want more power without switching to a new family, such as the 68HC11.

Subsystem options include modules for timers, direct memory access (DMA), serial communications, A/D and D/A converters, LCD and LED drivers, and RAM, ROM, and EPROM. Customers can request 68HC08 members using the same customer-specified integrated circuits (CSIC) approach that is used for the 68HC05 family. Motorola anticipates introducing modules for flash EEPROM, DSP, dual-tone multifrequency (DTMF) generator/receiver, and fuzzy logic. Motorola also supplies a simulator disk to help users evaluate the product. A tutorial (application note AN1218) helps a system designer utilize 68HC08 enhancements based on familiarity with the 68HC05. The following highlights some features taken primarily from Motorola's technical preview on the 68HC08 (document BR1101/D).

CPU Hardware.   The CPU uses a 16-bit index register (H:X), 16-bit stack pointer, and 16-bit program counter (see Figure 15.2). The 16-bit index register (H:X) is a combination of two 8-bit registers (H and X) that make up the high and low bytes, respectively. When the H index register is zero, the H:X register is functionally equivalent to the index register in the 68HC05. The 8-MHz bus provides a minimum instruction cycle time of 125 ns. An instruction prefetch mechanism reduces cycle time. Like the 68HC05, the 68HC08 includes low-power STOP and WAIT modes.

Software Features.   The CPU has 78 additional instructions, including enhanced stack manipulation. There are 16 addressing modes. Two of the modes use the stack pointer as an index register with 8- or 16-bit offsets. This will allow the implementation of efficient C cross compilers (see Section 13.6.6 on how C uses the stack).[1] There is an instruction to add a signed immediate value to the stack pointer (AIS). The following code shows how to set up space on the stack for local variables. Remember that subroutines using local variables are reentrant.

```
*Listing example to show how the 68HC08
*can create and use space for local variables.
SUBROUTINE
        AIS    #-10     ;Create 10 bytes of local space.

*                       ;Other subroutine instructions follow.
*                       ;Show example of stack pointer addressing
                        mode.
        LDA    18,SP    ;Load data from calling routine (old frame)
        STA    05,SP    ;and store it as a local variable.
```

---

[1]One industry trend is to design CPU hardware so that efficient software translators (such as C compilers) are possible.

```
  *                            ;More subroutine code
  \
    AIS      #10               ;Remove local space prior to return.
    RTS                        ;Return to calling routine.
```

The CPU includes a fast (five cycles) 8-bit multiply, which is similar to the MUL instruction for the 68HC11. It also has a 16-bit by 8-bit divide instruction (DIV). There are instructions that provide enhanced branching control. For example, the instruction compare and branch if equal (CBEQ) combines a compare (CMP) and branch (BEQ) for faster table look-up routines. Similarly, the decrement and branch if not equal instruction (DBNZ) provides a faster looping mechanism. It subtracts one from a loop counter (register or memory location) and branches if the result is not zero. There are memory-to-memory move instructions. New addressing modes include stack pointer, memory-to-memory, and indexed with post increment. The latter mode refers to incrementing the index register automatically so that it is pointing to the next location. This eliminates the necessity of using an increment index register (INX) instruction. The HC08 Central Processor Unit Reference Manual includes a section with 28 instruction set examples.

Direct Memory Access Module (DMA08).   This coprocessor module can control the address and data buses for byte transfers without using interrupts. It improves byte transfer efficiency by reducing required clock cycles by a factor of 12. The DMA08 can handle data from external peripherals while the CPU is in WAIT mode.

Timing Interface Module (TIM08).   This module is a 16-bit timer with a programmable prescaler and four channels. This module will eventually include customer specified timers. Each channel can be configured for input capture, output compare, or PWM. The counter can be free-running (like the 68HC11 TCNT register) or act as an up-counter.

Serial Communications Interface (SCI08).   This module supports full-duplex asynchronous communication (see Sections 8.6 and 10.1) similar to that in the SCI subsystem in the 68HC11. The module also features hardware parity and a DMA interrupt.

Serial Peripheral Interface (SPI08).   This module provides three-wire synchronous communications (see Sections 8.6, 10.5, and 10.6) similar to that in the SPI subsystem in the 68HC11. The module features a DMA interrupt option and a clock buffer ground pin to reduce electromagnetic interference (EMI) noise.

Clock Generation Module (CGM08).   This module uses an on-chip phase-locked loop (PLL). It uses the PLL to synthesize a 32-MHz clock signal required to generate 8-MHz clock pulses (note factor of 4). The module can operate using a 4-MHz external crystal.

System Control Module (SCM).   This module provides mode control selection and CPU configuration to specify reset, interrupt, low-power, low-voltage, and memory options. The intent is to provide flexibility for customer requirements.

15.2.5 The 68HC12 Family

In 1996 Motorola introduced a significant upgrade to the 68HC11 family. The 68HC12-style microcontrollers can use the same source code as the 68HC11. This microcontroller also uses the same programming model (as shown in Figure 2.6) and stacking operations. There are enhancements that we will summarize here. These include extra addressing modes and extra instructions. The extra instructions make it easier to produce efficient machine code when using high-level languages such as C. Other instructions directly support fuzzy logic as was already explained in Section 13.8.6.

Source Code Compatible. Source code for the 68HC11 can be assembled for the 68HC12. The object code is similar but not identical. There is source code compatibility even if some 68HC11 instructions do not exist for the 68HC12. The assembler for the 68HC12 recognizes the instructions and translates them into equivalent 68HC12 instructions. Some instructions are translated into smaller object code. In particular, Y-indexed instructions assemble into one less byte than the equivalent 68HC11 instructions.

16-Bit Architecture. The processor has a 16-bit data bus, unlike the 68HC11, which uses an 8-bit data bus. Some versions include the ability to handle an external 8-bit data bus. They use an on-chip integration module to manage the external interface. Recall the fetch-execute cycle mentioned in Section 2.4. The 68HC12 uses an *instruction queue.* It fetches three 16-bit words in sequence and begins execution. Thus, there is always a word available to the CPU without having to wait for extra fetches during the execution of some instructions. Stack access can be 2 bytes at a time. The top of the stack contains the "last-used" address instead of the "next-available" address as shown by the following.

Comparison of Sample Stacks After Execution of PULX

		68HC11 Stack	**68HC12 Stack**
SP before	00F0	available	data high byte
	00F1	data high byte	data low byte
SP after	00F2	data low byte	next item
	00F3	next item	next item

Enhanced Indexed Addressing. The 68HC12-type processor can use a higher number of different types of indexed addressing, which makes program loops and data manipulation operations more efficient (smaller and faster). In general, these enhancements make it possible to create more efficient C compilers.

The stack pointer can be used in all indexed operations. This is useful for stack frame operations for parameters and local variables in C function calls (see Section 13.6.6).

The program counter can be used in most indexed operations. Instructions can use accumulators A, B, and D as offsets. This is useful for table look-up operations. For example:

```
LDAA    B,X
```

is the same as

```
ABX
LDAA    0,X
```

Indexed addressing instructions can include automatic pre- or post-increment or decrement operations. Consider the following 68HC12 instruction: `STAA 1,-SP`. It initially decrements the stack pointer by one and then determines the effective address by adding an offset of one to the new stack pointer value. Hence, the instruction is equivalent to the `PSHA` instruction.

Instructions can use constant signed 5-, 8-, or 16-bit offsets. The 5-bit offset is useful because no extra object code bytes are required. In practice, only 5-bit offsets for a range of -16 to $+15$ are required. Indirect addressing is possible by using 16-bit offset indexed-indirect and accumulator D offset indexed-indirect. Indirect addressing refers to using the contents of an address as the address of the actual data to be retrieved or stored. Source code syntax uses square brackets to indicate indirect addressing. This is useful in situations in which a table may contain pointers to other tables. Consider the instruction

```
LDAA [2,X]
```

when index register X has a value of $C000 with the following data in memory.

The instruction adds an offset of 2 to calculate an *indirect* address $C002. It then uses the indirect address of $C002 (and $C003) to retrieve the final address of $C600. It then fetches the contents of $C600 to load the value of $1A into accumulator A.

Hexadecimal Address	Hexadecimal Contents
C000	C5
C001	00
C002	C6
C003	00
. . .	. . .
C600	1A

General Enhancements.   The microcontroller is faster due to a higher bus speed of 8 MHz, 16-bit-wide architecture, a reduced number of clock cycles for many instructions, a 20-bit ALU, and some specialized fast math instructions such as EMACS for signed 16-bit multiply and accumulate. There are additional instructions, such as

- Memory-to-memory move instructions
- Transfer instruction (TFR) that can move contents of any register to any other register
- Loop instructions that include counter updates and condition testing
- Long branches for branching to any address within a 64KB address space
- Minimum and maximum instructions for restricting data values within limits
- Fuzzy logic support (see Section 13.8.6)
- Table look-up with interpolation of values between values that are stored in the table
- Instructions to handle bank switching for subroutines in different blocks of memory

Bank switching refers to a method for expanding memory beyond the 64KB address space. With a 16-bit address bus, it is only possible to address 64KB possible locations. Bank switching is a technique used to assign blocks of available addresses to different physical sections of memory when the physical memory is larger than 64KB.

More Information. The 68HC12 CPU12 manual (Motorola document CPU12RM/AD) gives a detailed description of the CPU. The technical summary for family members (such as Motorola document MC68HC912B32TS/D) also gives information about hardware characteristics and memory-mapped registers.

■ 15.3 THE MOTOROLA HCS12 FAMILY

The HCS12 family is the next generation that follows the 68HC12 line. Hence, you can refer to Section 15.2.5 for many of its features. This section will outline more features. The prime enhancement over the 68HC12 line is the use of 0.25μ Flash memory. The symbol μ (pronounced mu) refers to the fabrication dimension of the smallest feature specified by the semiconductor mask (see Section 6.1.3). A micron is 10^{-6} meters, which is one-millionth of a meter (also one-thousandth of a millimeter). For comparison, note that the Intel Celeron processor uses 0.13μ features. Atmospheric particles are typically a few microns in diameter. Molecules such as water are only one-thousandth the size of a 0.25μ feature.

Possibly the family designation of HCS12 instead of 68HC12 is a marketing decision to emphasize the Flash memory use instead of ROM and of the degree of feature miniaturization. There is also more instruction set optimization to support C compilers.

The HCS12 family utilizes 32KB to 512KB Flash memory for the purpose of holding application code. The HCS12 core is upwardly compatible with the 68HC11 instruction set. However it is not machine code compatible. Most machine instructions are two or more bytes. It would be necessary to recompile or reassemble a 68HC11 program to run in a 68HC12 or HCS12 device. The basic CPU architecture is the same with the same programmer's model and interrupt stacking. Because it uses a CPU12 core, it features an instruction queue and has a Multiplexed External Bus Interface (MEBI), a Module Mapping Control (MMC), Interrupt control (INT), Breakpoints (BKP), and Background Debug Mode (BDM). Other features include Phase-Locked Loop (PLL), support for digital filtering, 10-bit analog-to-digital converters, enhanced capture timer, 8 Pulse-Width modulation (PWM) channels, and Inter-IC bus (IIC) that is compatible with the 12C standard. This section will provide an overview of some of these features.

Multiplexed External Bus Interface (MEBI). This subsystem provides access to some internal data that is inside the core to the outside using some ports. Essentially it governs the use of the port pins.

Module Mapping Control (MMC). This subsystem performs mapping and select operations for internal and external memory blocks involving bus-related data flow. It has registers to define the address space for internal components, and chip select signal generation for external devices. It also handles memory paging that is used to switch banks of memory into the same address space. The HCS12 family (as well as HC12) uses only a linear address space of 64KB as defined by its 16-bit address bus. However, it uses paging to swap areas of external memory into a page block within the 64KB space. This technique of extending memory capacity has advantages and disadvantages. The advantage is to maintain smaller instruction sizes. Otherwise instructions would have to include more address bits. The disadvantage of paging is extra overhead and software complexity for switching pages.

Interrupt Control (INT). This system provides more interrupts and masking options. It also features interrupt test registers which are useful for development and test purposes.

Breakpoints (BKP). This is useful for debugging software. Data and address values can be placed in setup registers. During program execution, a successful comparison will place the CPU in Background Debug Mode or initiate a software interrupt (SWI).

Background Debug Mode (BDM). This is a single-wire, background mode. Interfacing with the BDM is handled through a BKGD pin. Two other pins indicate when a low byte or a high byte of an instruction word is read into the instruction queue. There are a few registers that can be read to provide status information concerning CPU operation after a breakpoint occurs.

Inter-IC Bus (IIC). The system supports the 12C standard for two-wire serial interfacing between semiconductor devices. It is used where occasional communications is required over short distances between several devices. It is similar to SPI with one master device and several slaves, but it uses only two signals, Serial Data line (SDA) and a Serial Clock Line (SCL). Byte size communication occurs with a receiver acknowledging the transfer of each byte. It is also possible to specify device addresses; hence, it is not necessary to use additional address decoding hardware.

Normally, communication consists of four phases which are Start signal, slave address transmission, data transfer, and Stop signal. A Start signal from the Master indicates the beginning of a new data transfer of one or more bytes and brings all slaves out of their idle states. The first byte transferred is a 7-bit calling address followed by a read/write (R/W) bit. The slave with the matching address responds by sending back an acknowledgment bit. Data transfer proceeds byte-by-byte in the direction specified by the R/W bit. The receiving device acknowledges each transferred byte by pulling the SDA low at the ninth clock. Hence, each complete data byte transfer needs nine clock pulses. The master can terminate communications by asserting a Stop signal to free the bus. It does this by driving SDA from low to high while SCL is at a logical one.

■ 15.4 SURVEY OF OTHER FAMILIES

Table 15.1 is a partial list of manufacturers. For complete and current information about any MCU product, contact the closest sales representative of the manufacturer (if available).

■ 15.5 CHOOSING A MICROCONTROLLER

If you are part of a development team for an OEM product, you need to find a microcontroller. Possibly, you have been using one family in earlier applications. If so, you must decide whether a family member is suitable for the next application. Otherwise, you will have to find another one. Either way, you must still decide.

Costs. MCU manufacturers compete fiercely to sell their products. A lot is at stake. More than 10 times as many microcontrollers are sold as general-purpose microprocessors. This

TABLE 15.1   SOME MANUFACTURERS OF 8-BIT MICROCONTROLLERS

Manufacturer	Web Site
Advanced Micro Devices	http://www.amd.com/
Atmel	http://www.atmel.com/
California Micro Devices	http://www.calmicro.com/
Dallas Semiconductor Group	http://www.dalsemi.com/
Fujitsu Microelectronics, Inc.	http://www.fujitsu.com/
Harris Semiconductor	http://www.semi.harris.com/
Hitachi America, Inc.	http://www.hitachi.com/
Intel Corp.	http://www.intel.com/
Microchip Technology (for P/C micro)	http://www.microchip.com/
Mitsubishi Electric	http://www.mitsubishi.com/
Motorola, Inc.	http://www.mot.com/
National Semiconductor Corp.	http://www.national.com/
NEC Electronics, Inc.	http://www.nec.com/
Oki Semiconductors, Inc.	http://www.oki.com/
Philips Components	http://www.philips.com/
Rabbit Semiconductor	http://rabbitsemiconductor.com/
Rockwell International	http://www.rockwell.com/
SGS-Thompson Microelectronics, Inc.	http://www.st.com/
Siemens Components, Inc.	http://www.siemens.com/
Signetics Corp.	http://www.signetics.co.kr/
Texas Instruments	http://www.ti.com/
Toshiba America	http://www.toshiba.com/
Zilog, Inc.	http://www.zilog.com/

may be hard to believe when you walk into a store and see racks of computer magazines. One would think that more microprocessors are sold because of their use in personal computers. Perhaps microcontrollers do not get enough press coverage.

The following will give you an idea of the competition. If an OEM intends to develop a product to be sold in high volumes, a price difference of 10 cents per MCU is significant. With many MCUs selling for a few dollars a piece, you may wonder why it costs much more to buy the product produced by the OEM. Aside from taxes and profits, the consumer pays for packaging, other components, hardware design, and software development. The most significant cost is software. Approximately 50 to 75% of an MCU development project is to design the application software. It is what you cannot see that increases the price.

As a result, software support is a significant factor. For instance, what emulators, simulators, assemblers, and compilers are available? To reduce writing new software, OEM companies tend to use MCUs belonging to the same family whenever possible. If it is not possible, the OEM looks for the easiest upgrade path. For example, an OEM that has used 68HC11-type MCUs will find it attractive to use a 68HC16 type when more processing power is required. The OEM can use some of the existing software from previous products.

Analyzing the Application.   Before choosing an MCU, you will have to analyze the requirements of the application. Generally, you will have to look at the data processing, I/O, power consumption, and memory requirements.

Data Processing. For data processing, the processor must be able to perform *critical calculations.* These are calculations that must be completed within a time limit. If the application is engine control, the MCU must calculate when to fire the spark and send the signal before the piston reaches the firing position (Section 14.1). In this case you have to be sure that the CPU instruction set is fast and powerful enough.

Another consideration is the precision and range of data to be processed. This determines the bit size of the data. In an antenna positioning system, it may only be necessary to rotate the antenna in two-degree increments. Then, for full rotation, only 180 points are required. Hence, 8-bit data is sufficient to specify antenna position. Now, consider a robot whose arm must be positioned within a few seconds of arc.[2] Sixteen-bit data can represent increments of 20 seconds of arc. For very high resolutions (fractions of seconds of arc), it will be necessary to use floating-point.

One question is whether to use an MCU with 16- or 32-bit instructions. The alternative is to use subroutines for 16- or 32-bit math using an 8-bit MCU. If most of the calculations are low resolutions, it may be better to use subroutines to perform the higher-resolution calculations. The same may hold true if floating-point resolution is necessary. An assembly language library of floating-point subroutines or C language functions may do the job. This is cheaper than using floating-point hardware.

I/O. To determine I/O requirements, you will need to identify the quantity, amount, and type of signals. Also, you should consider the mechanical needs of the system to be controlled. One way to do this is to draw a block diagram of the overall system. For example, consider the diagram for a dot matrix printer (Figure 14.12). This type of diagram will tell you whether the MCU has the right type and amount of I/O pins for the system. If there is not enough, you will have to look at other options. The options are to add peripheral ICs or to use a different MCU.

As a case study for extra printer I/O, consider the following: Let's say that you want to add more operator keys and an LCD display to add a user programming feature. Some options are to add a parallel latch IC and LCD driver IC. This can be done by using the expanded bus or SPI subsystem. Another option is to add a 68HC05-type MCU with a built-in LCD driver. Then one can use the SPI for communication between the two processors. A third option is to use a different MCU, one that has more I/O. This leads to an interesting point. Some designs use a main processor to offload some tasks to other processors. A serially linked network of a 68HC11 with four or five 68HC05s may be cheaper than one 16-bit MCU.

Power. Some OEM products are powered by battery. The security system in an automobile is one example. It must function when the car is not being used. The MCU can run in low-power mode to minimize current drain from the car battery. When an intruder breaks into the car, he or she activates a sensor that wakes up the MCU to switch it to normal mode. Then the MCU executes the appropriate alarm sequence. To conserve battery power, most MCUs include low-power modes of some kind. Even some general-purpose microprocessors offer this feature. They have it because it is necessary to conserve power in portable (laptop or notebook) computers.

---

[2]One degree equals 60 minutes of arc and 1 minute of arc equals 60 seconds of arc.

Memory.   When considering an application's memory requirements, break it down into program and data memory. This categorizes it into nonvolatile (ROM, EPROM, OTP) and volatile (RAM) components. There is a hybrid category of customization. This would be nonvolatile or rarely modified data that is specific to a piece of equipment, such as serial number, calibration, and maintenance history. An EEPROM is a good candidate for this kind. Newer technologies, such as flash memory, may allow other economic options.

The volume of product expected to be sold determines what kind of program memory to select. In any case, an EPROM-based MCU is best suited for the product prototype. When the product is completely designed and produced, the production model may use a different kind of program memory. In order of small to large volumes, the choices are EPROM, OTP, and ROM. Careful economic analysis will decide which to use in any application. It is a speculative decision because the OEM has to forecast how many products will be sold.

To determine the size of memory, you can initially write a first-pass pseudocode for the application. This will give you an idea of the memory requirements. Still, this is a difficult estimate to make, and past experience may be the only guide. You will also need to estimate RAM space for variables, stack size, and scratchpad area for intermediate result of calculations. Some application software writers estimate RAM requirements by taking a percent of their estimated ROM (program memory) requirement. The language chosen will also be a factor. Assembly language may require 12 or more times as much ROM as RAM. A high-level language may require about 20 times as much ROM as RAM.

What Bit Size?   MCUs are available in 4-, 8-, 16-, and 32-bit types. When developing an MCU system, one factor in keeping the cost down is the bit size of the MCU. Those with a smaller bit size are cheaper per chip and require less expensive development tools and support chips. On the other hand, it is necessary for the MCU to be able to perform the necessary tasks. For this reason it may be necessary to use an MCU with a larger bit size.

To make the decision, it is necessary to analyze the application as described previously. Still we will risk the following generalizations. Note that there are exceptions. Use 4-bit MCUs when cost is the main consideration (for example, an electronic thermostat). Use 8-bit MCUs in applications that involve movement of byte data (for example, a dot matrix printer). Use 16-bit MCUs to improve *I/O throughput* (for example, a computer disk drive). We will say more about I/O throughput later. Other reasons are 16-bit math and better features for C programming. Use 32-bit MCUs when a very large addressing space is required (for example, a laser printer). In fact, some laser printers use two of them.

I/O Throughput.   This refers to how fast I/O requests from external devices can be handled. Another term is *I/O bandwidth*. When too much of the processor's time is spent handling I/O, we call the situation an *I/O bottleneck*. An I/O bottleneck will prevent an MCU from executing other tasks.

Most real-time control systems rely on interrupts to handle I/O. Most MCUs employ unique vectors for each interrupt source. Interrupt *latency time* is the time needed to handle the request. Its duration is from the time the request is made to the time service routine starts to execute. When many I/O subsystems are being used, multiple interrupt requests can occur. This may increase latency times beyond what is acceptable in the application. This is an I/O bottleneck.

Consider an automobile engine control system. The MCU may spend much of its time monitoring crankshaft revolution and generating spark timing pulses (Listing 14.1). Let's say that an SPI interrupt asserts itself to request interprocessor communication. The MCU may not be able to complete spark timing calculations and other tasks before the next timing signal is to be generated.

To minimize the possibility of an I/O bottleneck, the designer can choose a 16-bit MCU. Some 16-bit MCUs have reserved registers for use by the interrupt service routine. This eliminates the overhead in saving other register values onto the stack. It is also not necessary to restore the registers after completion of the service routine.

Board Design.   The choice of MCU will affect the design of the control system circuit board. If an MCU must operate in an expanded bus mode, the circuit board will require more components and wiring. The board will be more expensive and more susceptible to electrical noise and will consume more power. When considering the cost of system design alternatives, it is not enough simply to add the cost of individual components. When more components are added, wiring and manufacturing costs also increase. For a low-cost component, it costs more to put it on the circuit board than to buy the component itself.

One board design option is to test a design concept using a behavioral simulation program. Existing libraries include models for 68HC11, 8051, and others. Simulation speed is slow, so a designer may use it only to test parts of a program to evaluate design options. This will help in deciding whether one type of MCU is suitable. In our engine control example, the designer can simulate simultaneous interrupt requests during a critical time.

PC Compatible Controller.   Computers based on the 80386, 80486, or Pentium microprocessor are very popular. You may want to use a control system that can also run personal computer (PC) software. There are several third-party suppliers, such as National Instruments, who offer software to interact with popular data acquisition boards and instrumentation.

Intel introduced an embedded controller version of its 80386 microprocessor, the Intel 386EX. It is a fully static chip (can run at any speed up to its maximum) running on 3 V. The embedded controller has an intermodule bus with power management circuits. This controller is able to run PC software such as Windows and its applications. This means that it is relatively easy to build systems with graphical user interfaces. Visual programming environments such as Visual Basic can make this easier. Another advantage is the capability to reuse existing software. This is a big plus since software development is very expensive.

■ 15.6 SUMMARY

Necessity is the mother of invention. Perhaps it is simply convenience. As a society of users and consumers, we want equipment such as automobiles and appliances to work better. The original equipment manufacturers (OEMs) want to produce better versions so that they can sell more of them. The semiconductor manufacturer wants to sell the microcontrollers used to make the improvements. The end result is that somebody (the OEM) has to choose a microcontroller and know how to use it.

After considering technical factors, it is the nontechnical factors that may determine the microcontroller an OEM chooses. The following questions assume that we are the OEM: Does the semiconductor manufacturer have a good reputation? Are there enough trained assembly language programmers, or will we have to use C? Should we delay introduction of a new product while waiting for a new feature in our familiar family? Would it be quicker to invest in new development tools and training to use a different MCU family? Will this allow us to introduce the product to market faster? How much of an advantage will it be to introduce the product faster?

No doubt, the world will change. The question is: How? Technology develops. Sixteen-bit and 32-bit processors will become cheaper. Eight-bit processors will change as well. One noteworthy development is the Java programming language. There are developments in using Java for embedded control. This language promises the ability to write once and run anywhere. In fact, the original reason for developing Java was to add intelligence to consumer electronic devices. Another development is the use of the Internet for networking controllers.

Other developments are changes in processor architecture. It is beyond the scope of this book to include them. Check microcontroller-related publications for key terms, such as reduced instruction set computer (RISC), SPARC, parallel processing, and neural networks.

A Instruction Set Summary

■ DATA BOOK TABLE 10–1[1] MC68HC11A8
INSTRUCTIONS, ADDRESSING MODES,
AND EXECUTION TIMES

---

[1]*Data Book* means *Motorola Document MC68HC11A8/D, MC68HC11A8 HCMOS Single-Chip Microcontroller 68HC11 Data Book.*

Source Form(s)	Operation	Boolean Expression	Addressing Mode for Operand	Opcode	Operand(s)	Bytes	Cycle	Cycle by Cycle*	S	X	H	I	N	Z	V	C
ABA	Add Accumulators	A + B → A	INH	1B		1	2	2-1	-	-	↕	-	↕	↕	↕	↕
ABX	Add B to X	IX + 00:B → IX	INH	3A		1	3	2-2	-	-	-	-	-	-	-	-
ABY	Add B to Y	IY + 00:B → IY	INH	18 3A		2	4	2-4	-	-	-	-	-	-	-	-
ADCA (opr)	Add with Carry to A	A + M + C → A	A IMM	89	ii	2	2	3-1	-	-	↕	-	↕	↕	↕	↕
			A DIR	99	dd	2	3	4-1								
			A EXT	B9	hh ll	3	4	5-2								
			A IND,X	A9	ff	2	4	6-2								
			A IND,Y	18 A9	ff	3	5	7-2								
ADCB (opr)	Add with Carry to B	B + M + C → B	B IMM	C9	ii	2	2	3-1	-	-	↕	-	↕	↕	↕	↕
			B DIR	D9	dd	2	3	4-1								
			B EXT	F9	hh ll	3	4	5-2								
			B IND,X	E9	ff	2	4	6-2								
			B IND,Y	18 E9	ff	3	5	7-2								
ADDA (opr)	Add Memory to A	A + M → A	A IMM	8B	ii	2	2	3-1	-	-	↕	-	↕	↕	↕	↕
			A DIR	9B	dd	2	3	4-1								
			A EXT	BB	hh ll	3	4	5-2								
			A IND,X	AB	ff	2	4	6-2								
			A IND,Y	18 AB	ff	3	5	7-2								
ADDB (opr)	Add Memory to B	B + M → B	B IMM	CB	ii	2	2	3-1	-	-	↕	-	↕	↕	↕	↕
			B DIR	DB	dd	2	3	4-1								
			B EXT	FB	hh ll	3	4	5-2								
			B IND,X	EB	ff	2	4	6-2								
			B IND,Y	18 EB	ff	3	5	7-2								
ADDD (opr)	Add 16-Bit to D	D + M:M + 1 → D	IMM	C3	jj kk	3	4	3-3	-	-	-	-	↕	↕	↕	↕
			DIR	D3	dd	2	5	4-7								
			EXT	F3	hh ll	3	6	5-10								
			IND,X	E3	ff	2	6	6-10								
			IND,Y	18 E3	ff	3	7	7-8								
ANDA (opr)	AND A with Memory	A•M → A	A IMM	84	ii	2	2	3-1	-	-	-	-	↕	↕	0	-
			A DIR	94	dd	2	3	4-1								
			A EXT	B4	hh ll	3	4	5-2								
			A IND,X	A4	ff	2	4	6-2								
			A IND,Y	18 A4	ff	3	5	7-2								
ANDB (opr)	AND B with Memory	B•M → B	B IMM	C4	ii	2	2	3-1	-	-	-	-	↕	↕	0	-
			B DIR	D4	dd	2	3	4-1								
			B EXT	F4	hh ll	3	4	5-2								
			B IND,X	E4	ff	2	4	6-2								
			B IND,Y	18 E4	ff	3	5	7-2								
ASL (opr)	Arithmetic Shift Left	← □←□□□□□□□←0 C b7 b0	EXT	78	hh ll	3	6	5-8	-	-	-	-	↕	↕	↕	↕
			IND,X	68	ff	2	6	6-3								
			IND,Y	18 68	ff	3	7	7-3								
ASLA			A INH	48		1	2	2-1								
ASLB			B INH	58		1	2	2-1								
ASLD	Arithmetic Shift Left Double	□←□□ - - □□←0 C b15 b0	INH	05		1	3	2-2	-	-	-	-	↕	↕	↕	↕
ASR (opr)	Arithmetic Shift Right	→ □□→□□□□□□→□ b7 b0 C	EXT	77	hh ll	3	6	5-8	-	-	-	-	↕	↕	↕	↕
			IND,X	67	ff	2	6	6-3								
			IND,Y	18 67	ff	3	7	7-3								
ASRA			A INH	47		1	2	2-1								
ASRB			B INH	57		1	2	2-1								
BCC (rel)	Branch if Carry Clear	? C = 0	REL	24	rr	2	3	8-1	-	-	-	-	-	-	-	-
BCLR (opr) (msk)	Clear Bit(s)	M•(mm̄) → M	DIR	15	dd mm	3	6	4-10	-	-	-	-	↕	↕	0	-
			IND,X	1D	ff mm	3	7	6-13								
			IND,Y	18 1D	ff mm	4	8	7-10								
BCS (rel)	Branch if Carry Set	? C = 1	REL	25	rr	2	3	8-1	-	-	-	-	-	-	-	-
BEQ (rel)	Branch if = Zero	? Z = 1	REL	27	rr	2	3	8-1	-	-	-	-	-	-	-	-

*Cycle-by-cycle number provides a reference to Tables 10-2 through 10-8 which detail cycle-by-cycle operation.
   Example: Table 10-1 Cycle-by-Cycle column reference number 2-4 equals Table 10-2 line item 2-4.

Source Form(s)	Operation	Boolean Expression	Addressing Mode for Operand	Opcode	Operand(s)	Bytes	Cycle	Cycle by Cycle*	S	X	H	I	N	Z	V	C
BGE (rel)	Branch if ≥ Zero	? N ⊕ V = 0	REL	2C	rr	2	3	8-1	-	-	-	-	-	-	-	-
BGT (rel)	Branch if > Zero	? Z + (N ⊕ V) = 0	REL	2E	rr	2	3	8-1	-	-	-	-	-	-	-	-
BHI (rel)	Branch if Higher	? C + Z = 0	REL	22	rr	2	3	8-1	-	-	-	-	-	-	-	-
BHS (rel)	Branch if Higher or Same	? C = 0	REL	24	rr	2	3	8-1	-	-	-	-	-	-	-	-
BITA (opr)	Bit(s) Test A with Memory	A•M	A IMM	85	ii	2	2	3-1	-	-	-	-	↕	↕	0	-
			A DIR	95	dd	2	3	4-1								
			A EXT	B5	hh ll	3	4	5-2								
			A IND,X	A5	ff	2	4	6-2								
			A IND,Y	18 A5	ff	3	5	7-2								
BITB (opr)	Bit(s) Test B with Memory	B•M	B IMM	C5	ii	2	2	3-1	-	-	-	-	↕	↕	0	-
			B DIR	D5	dd	2	3	4-1								
			B EXT	F5	hh ll	3	4	5-2								
			B IND,X	E5	ff	2	4	6-2								
			B IND,Y	18 E5	ff	3	5	7-2								
BLE (rel)	Branch if ≤ Zero	? Z + (N ⊕ V) = 1	REL	2F	rr	2	3	8-1	-	-	-	-	-	-	-	-
BLO (rel)	Branch if Lower	? C = 1	REL	25	rr	2	3	8-1	-	-	-	-	-	-	-	-
BLS (rel)	Branch if Lower or Same	? C + Z = 1	REL	23	rr	2	3	8-1	-	-	-	-	-	-	-	-
BLT (rel)	Branch If < Zero	? N ⊕ V = 1	REL	2D	rr	2	3	8-1	-	-	-	-	-	-	-	-
BMI (rel)	Branch if Minus	? N = 1	REL	2B	rr	2	3	8-1	-	-	-	-	-	-	-	-
BNE (rel)	Branch if Not = Zero	? Z = 0	REL	26	rr	2	3	8-1	-	-	-	-	-	-	-	-
BPL (rel)	Branch if Plus	? N = 0	REL	2A	rr	2	3	8-1	-	-	-	-	-	-	-	-
BRA (rel)	Branch Always	? 1 = 1	REL	20	rr	2	3	8-1	-	-	-	-	-	-	-	-
BRCLR(opr) (msk) (rel)	Branch if Bit(s) Clear	? M• mm = 0	DIR	13	dd mm rr	4	6	4-11	-	-	-	-	-	-	-	-
			IND,X	1F	ff mm rr	4	7	6-14								
			IND,Y	18 1F	ff mm rr	5	8	7-11								
BRN (rel)	Branch Never	? 1 = 0	REL	21	rr	2	3	8-1	-	-	-	-	-	-	-	-
BRSET(opr) (msk) (rel)	Branch if Bit(s) Set	? (M̄)•mm = 0	DIR	12	dd mm rr	4	6	4-11	-	-	-	-	-	-	-	-
			IND,X	1E	ff mm rr	4	7	6-14								
			IND,Y	18 1E	ff mm rr	5	8	7-11								
BSET(opr) (msk)	Set Bit(s)	M + mm → M	DIR	14	dd mm	3	6	4-10	-	-	-	-	↕	↕	0	-
			IND,X	1C	ff mm	3	7	6-13								
			IND,Y	18 1C	ff mm	4	8	7-10								
BSR (rel)	Branch to Subroutine	See Special Ops	REL	8D	rr	2	6	8-2	-	-	-	-	-	-	-	-
BVC (rel)	Branch if Overflow Clear	? V = 0	REL	28	rr	2	3	8-1	-	-	-	-	-	-	-	-
BVS (rel)	Branch if Overflow Set	? V = 1	REL	29	rr	2	3	8-1	-	-	-	-	-	-	-	-
CBA	Compare A to B	A – B	INH	11		1	2	2-1	-	-	-	-	↕	↕	↕	↕
CLC	Clear Carry Bit	0 → C	INH	0C		1	2	2-1	-	-	-	-	-	-	-	0
CLI	Clear Interrupt Mask	0 → I	INH	0E		1	2	2-1	-	-	-	0	-	-	-	-
CLR (opr)	Clear Memory Byte	0 → M	EXT	7F	hh ll	3	6	5-8	-	-	-	-	0	1	0	0
			IND,X	6F	ff	2	6	6-3								
			IND,Y	18 6F	ff	3	7	7-3								
CLRA	Clear Accumulator A	0 → A	A INH	4F		1	2	2-1	-	-	-	-	0	1	0	0
CLRB	Clear Accumulator B	0 → B	B INH	5F		1	2	2-1	-	-	-	-	0	1	0	0
CLV	Clear Overflow Flag	0 → V	INH	0A		1	2	2-1	-	-	-	-	-	-	0	-
CMPA (opr)	Compare A to Memory	A – M	A IMM	81	ii	2	2	3-1	-	-	-	-	↕	↕	↕	↕
			A DIR	91	dd	2	3	4-1								
			A EXT	B1	hh ll	3	4	5-2								
			A IND,X	A1	ff	2	4	6-2								
			A IND,Y	18 A1	ff	3	5	7-2								

*Cycle-by-cycle number provides a reference to Tables 10-2 through 10-8 which detail cycle-by-cycle operation.
Example: Table 10-1 Cycle-by-Cycle column reference number 2-4 equals Table 10-2 line item 2-4.

TABLE A.1 (DATA BOOK TABLE 10–1) Continued

Source Form(s)	Operation	Boolean Expression	Addressing Mode for Operand	Opcode	Operand(s)	Bytes	Cycle	Cycle by Cycle*	S	X	H	I	N	Z	V	C
CMPB (opr)	Compare B to Memory	B – M	B IMM	C1	ii	2	2	3-1	-	-	-	-	↕	↕	↕	↕
			B DIR	D1	dd	2	3	4-1								
			B EXT	F1	hh ll	3	4	5-2								
			B IND,X	E1	ff	2	4	6-2								
			B IND,Y	18 E1	ff	3	5	7-2								
COM (opr)	1's Complement Memory Byte	$FF – M → M	EXT	73	hh ll	3	6	5-8	-	-	-	-	↕	↕	0	1
			IND,X	63	ff	2	6	6-3								
			IND,Y	18 63	ff	3	7	7-3								
COMA	1's Complement A	$FF – A → A	A INH	43		1	2	2-1	-	-	-	-	↕	↕	0	1
COMB	1's Complement B	$FF – B → B	B INH	53		1	2	2-1	-	-	-	-	↕	↕	0	1
CPD (opr)	Compare D to Memory 16-Bit	D – M:M + 1	IMM	1A 83	jj kk	4	5	3-5	-	-	-	-	↕	↕	↕	↕
			DIR	1A 93	dd	3	6	4-9								
			EXT	1A B3	hh ll	4	7	5-11								
			IND,X	1A A3	ff	3	7	6-11								
			IND,Y	CD A3	ff	3	7	7-8								
CPX (opr)	Compare X to Memory 16-Bit	IX – M:M + 1	IMM	8C	jj kk	3	4	3-3	-	-	-	-	↕	↕	↕	↕
			DIR	9C	dd	2	5	4-7								
			EXT	BC	hh ll	3	6	5-10								
			IND,X	AC	ff	2	6	6-10								
			IND,Y	CD AC	ff	3	7	7-8								
CPY (opr)	Compare Y to Memory 16-Bit	IY – M:M + 1	IMM	18 8C	jj kk	4	5	3-5	-	-	-	-	↕	↕	↕	↕
			DIR	18 9C	dd	3	6	4-9								
			EXT	18 BC	hh ll	4	7	5-11								
			IND,X	1A AC	ff	3	7	6-11								
			IND,Y	18 AC	ff	3	7	7-8								
DAA	Decimal Adjust A	Adjust Sum to BCD	INH	19		1	2	2-1	-	-	-	-	↕	↕	↕	↕
DEC (opr)	Decrement Memory Byte	M – 1 → M	EXT	7A	hh ll	3	6	5-8	-	-	-	-	↕	↕	↕	-
			IND,X	6A	ff	2	6	6-3								
			IND,Y	18 6A	ff	3	7	7-3								
DECA	Decrement Accumulator A	A – 1 → A	A INH	4A		1	2	2-1	-	-	-	-	↕	↕	↕	-
DECB	Decrement Accumulator B	B – 1 → B	B INH	5A		1	2	2-1	-	-	-	-	↕	↕	↕	-
DES	Decrement Stack Pointer	SP – 1 → SP	INH	34		1	3	2-3	-	-	-	-	-	-	-	-
DEX	Decrement Index Register X	IX – 1 → IX	INH	09		1	3	2-2	-	-	-	-	-	↕	-	-
DEY	Decrement Index Register Y	IY – 1 → IY	INH	18 09		2	4	2-4	-	-	-	-	-	↕	-	-
EORA (opr)	Exclusive OR A with Memory	A ⊕ M → A	A IMM	88	ii	2	2	3-1	-	-	-	-	↕	↕	0	-
			A DIR	98	dd	2	3	4-1								
			A EXT	B8	hh ll	3	4	5-2								
			A IND,X	A8	ff	2	4	6-2								
			A IND,Y	18 A8	ff	3	5	7-2								
EORB (opr)	Exclusive OR B with Memory	B ⊕ M → B	B IMM	C8	ii	2	2	3-1	-	-	-	-	↕	↕	0	-
			B DIR	D8	dd	2	3	4-1								
			B EXT	F8	hh ll	3	4	5-2								
			B IND,X	E8	ff	2	4	6-2								
			B IND,Y	18 E8	ff	3	5	7-2								
FDIV	Fractional Divide 16 by 16	D/IX → IX; r → D	INH	03		1	41	2-17	-	-	-	-	-	↕	↕	↕
IDIV	Integer Divide 16 by 16	D/IX → IX; r → D	INH	02		1	41	2-17	-	-	-	-	-	↕	0	↕
INC (opr)	Increment Memory Byte	M + 1 → M	EXT	7C	hh ll	3	6	5-8	-	-	-	-	↕	↕	↕	-
			IND,X	6C	ff	2	6	6-3								
			IND,Y	18 6C	ff	3	7	7-3								
INCA	Increment Accumulator A	A + 1 → A	A INH	4C		1	2	2-1	-	-	-	-	↕	↕	↕	-
INCB	Increment Accumulator B	B + 1 → B	B INH	5C		1	2	2-1	-	-	-	-	↕	↕	↕	-
INS	Increment Stack Pointer	SP + 1 → SP	INH	31		1	3	2-3	-	-	-	-	-	-	-	-

*Cycle-by-cycle number provides a reference to Tables 10-2 through 10-8 which detail cycle-by-cycle operation.
  Example: Table 10-1 Cycle-by-Cycle column reference number 2-4 equals Table 10-2 line item 2-4.

628   Appendix A

TABLE A.1 (*DATA BOOK* TABLE 10–1) *Continued*

Source Form(s)	Operation	Boolean Expression	Addressing Mode for Operand	Opcode	Operand(s)	Bytes	Cycle	Cycle by Cycle*	S	X	H	I	N	Z	V	C
INX	Increment Index Register X	$IX + 1 \rightarrow IX$	INH	08		1	3	2-2	-	-	-	-	-	↕	-	-
INY	Increment Index Register Y	$IY + 1 \rightarrow IY$	INH	18 08		2	4	2-4	-	-	-	-	-	↕	-	-
JMP (opr)	Jump	See Special Ops	EXT	7E	hh ll	3	3	5-1	-	-	-	-	-	-	-	-
			IND,X	6E	ff	2	3	6-1								
			IND,Y	18 6E	ff	3	4	7-1								
JSR (opr)	Jump to Subroutine	See Special Ops	DIR	9D	dd	2	5	4-8	-	-	-	-	-	-	-	-
			EXT	BD	hh ll	3	6	5-12								
			IND,X	AD	ff	2	6	6-12								
			IND,Y	18 AD	ff	3	7	7-9								
LDAA (opr)	Load Accumulator A	$M \rightarrow A$	A IMM	86	ii	2	2	3-1	-	-	-	-	↕	↕	0	-
			A DIR	96	dd	2	3	4-1								
			A EXT	B6	hh ll	3	4	5-2								
			A IND,X	A6	ff	2	4	6-2								
			A IND,Y	18 A6	ff	3	5	7-2								
LDAB (opr)	Load Accumulator B	$M \rightarrow B$	B IMM	C6	ii	2	2	3-1	-	-	-	-	↕	↕	0	-
			B DIR	D6	dd	2	3	4-1								
			B EXT	F6	hh ll	3	4	5-2								
			B IND,X	E6	ff	2	4	6-2								
			B IND,Y	18 E6	ff	3	5	7-2								
LDD (opr)	Load Double Accumulator D	$M \rightarrow A, M + 1 \rightarrow B$	IMM	CC	jj kk	3	3	3-2	-	-	-	-	↕	↕	0	-
			DIR	DC	dd	2	4	4-3								
			EXT	FC	hh ll	3	5	5-4								
			IND,X	EC	ff	2	5	6-6								
			IND,Y	18 EC	ff	3	6	7-6								
LDS (opr)	Load Stack Pointer	$M:M + 1 \rightarrow SP$	IMM	8E	jj kk	3	3	3-2	-	-	-	-	↕	↕	0	-
			DIR	9E	dd	2	4	4-3								
			EXT	BE	hh ll	3	5	5-4								
			IND,X	AE	ff	2	5	6-6								
			IND,Y	18 AE	ff	3	6	7-6								
LDX (opr)	Load Index Register X	$M:M + 1 \rightarrow IX$	IMM	CE	jj kk	3	3	3-2	-	-	-	-	↕	↕	0	-
			DIR	DE	dd	2	4	4-3								
			EXT	FE	hh ll	3	5	5-4								
			IND,X	EE	ff	2	5	6-6								
			IND,Y	CD EE	ff	3	6	7-6								
LDY (opr)	Load Index Register Y	$M:M + 1 \rightarrow IY$	IMM	18 CE	jj kk	4	4	3-4	-	-	-	-	↕	↕	0	-
			DIR	18 DE	dd	3	5	4-5								
			EXT	18 FE	hh ll	4	6	5-6								
			IND,X	1A EE	ff	3	6	6-7								
			IND,Y	18 EE	ff	3	6	7-6								
LSL (opr)	Logical Shift Left	[diagram: ← C b7 ... b0 ← 0]	EXT	78	hh ll	3	6	5-8	-	-	-	-	↕	↕	↕	↕
			IND,X	68	ff	2	6	6-3								
			IND,Y	18 68	ff	3	7	7-3								
LSLA			A INH	48		1	2	2-1								
LSLB			B INH	58		1	2	2-1								
LSLD	Logical Shift Left Double	[diagram: C b15 ... b0 ← 0]	INH	05		1	3	2-2	-	-	-	-	↕	↕	↕	↕
LSR (opr)	Logical Shift Right	[diagram: 0 → b7 ... b0 C]	EXT	74	hh ll	3	6	5-8	-	-	-	-	0	↕	↕	↕
			IND,X	64	ff	2	6	6-3								
			IND,Y	18 64	ff	3	7	7-3								
LSRA			A INH	44		1	2	2-1								
LSRB			B INH	54		1	2	2-1								
LSRD	Logical Shift Right Double	[diagram: 0 → b15 ... b0 C]	INH	04		1	3	2-2	-	-	-	-	0	↕	↕	↕
MUL	Multiply 8 by 8	$A \times B \rightarrow D$	INH	3D		1	10	2-13	-	-	-	-	-	-	-	↕

*Cycle-by-cycle number provides a reference to Tables 10-2 through 10-8 which detail cycle-by-cycle operation.
  Example: Table 10-1 Cycle-by-Cycle column reference number 2-4 equals Table 10-2 line item 2-4.

Source Form(s)	Operation	Boolean Expression	Addressing Mode for Operand	Opcode	Operand(s)	Bytes	Cycle	Cycle by Cycle*	S	X	H	I	N	Z	V	C
NEG (opr)	2's Complement Memory Byte	0 − M → M	EXT	70	hh ll	3	6	5-8	-	-	-	-	↕	↕	↕	↕
			IND,X	60	ff	2	6	6-3								
			IND,Y	18 60	ff	3	7	7-3								
NEGA	2's Complement A	0 − A → A	A INH	40		1	2	2-1	-	-	-	-	↕	↕	↕	↕
NEGB	2's Complement B	0 − B → B	B INH	50		1	2	2-1	-	-	-	-	↕	↕	↕	↕
NOP	No Operation	No Operation	INH	01		1	2	2-1	-	-	-	-	-	-	-	-
ORAA (opr)	OR Accumulator A (Inclusive)	A + M → A	A IMM	8A	ii	2	2	3-1	-	-	-	-	↕	↕	0	-
			A DIR	9A	dd	2	3	4-1								
			A EXT	BA	hh ll	3	4	5-2								
			A IND,X	AA	ff	2	4	6-2								
			A IND,Y	18 AA	ff	3	5	7-2								
ORAB (opr)	OR Accumulator B (Inclusive)	B + M → B	B IMM	CA	ii	2	2	3-1	-	-	-	-	↕	↕	0	-
			B DIR	DA	dd	2	3	4-1								
			B EXT	FA	hh ll	3	4	5-2								
			B IND,X	EA	ff	2	4	6-2								
			B IND,Y	18 EA	ff	3	5	7-2								
PSHA	Push A onto Stack	A → Stk, SP=SP−1	A INH	36		1	3	2-6	-	-	-	-	-	-	-	-
PSHB	Push B onto Stack	B → Stk, SP=SP−1	B INH	37		1	3	2-6	-	-	-	-	-	-	-	-
PSHX	Push X onto Stack (Lo First)	IX → Stk, SP=SP−2	INH	3C		1	4	2-7	-	-	-	-	-	-	-	-
PSHY	Push Y onto Stack (Lo First)	IY → Stk, SP=SP−2	INH	18 3C		2	5	2-8	-	-	-	-	-	-	-	-
PULA	Pull A from Stack	SP=SP+1, A ← Stk	A INH	32		1	4	2-9	-	-	-	-	-	-	-	-
PULB	Pull B from Stack	SP=SP+1, B ← Stk	B INH	33		1	4	2-9	-	-	-	-	-	-	-	-
PULX	Pull X from Stack (Hi First)	SP=SP+2, IX ← Stk	INH	38		1	5	2-10	-	-	-	-	-	-	-	-
PULY	Pull Y from Stack (Hi First)	SP=SP+2, IY ← Stk	INH	18 38		2	6	2-11	-	-	-	-	-	-	-	-
ROL (opr)	Rotate Left	C b7 ← b0 C	EXT	79	hh ll	3	6	5-8	-	-	-	-	↕	↕	↕	↕
			IND,X	69	ff	2	6	6-3								
			IND,Y	18 69	ff	3	7	7-3								
ROLA			A INH	49		1	2	2-1								
ROLB			B INH	59		1	2	2-1								
ROR (opr)	Rotate Right	C b7 → b0 C	EXT	76	hh ll	3	6	5-8	-	-	-	-	↕	↕	↕	↕
			IND,X	66	ff	2	6	6-3								
			IND,Y	18 66	ff	3	7	7-3								
RORA			A INH	46		1	2	2-1								
RORB			B INH	56		1	2	2-1								
RTI	Return from Interrupt	See Special Ops	INH	3B		1	12	2-14	↕	↓	↕	↕	↕	↕	↕	↕
RTS	Return from Subroutine	See Special Ops	INH	39		1	5	2-12	-	-	-	-	-	-	-	-
SBA	Subtract B from A	A − B → A	INH	10		1	2	2-1	-	-	-	-	↕	↕	↕	↕
SBCA (opr)	Subtract with Carry from A	A − M − C → A	A IMM	82	ii	2	2	3-1	-	-	-	-	↕	↕	↕	↕
			A DIR	92	dd	2	3	4-1								
			A EXT	B2	hh ll	3	4	5-2								
			A IND,X	A2	ff	2	4	6-2								
			A IND,Y	18 A2	ff	3	5	7-2								
SBCB (opr)	Subtract with Carry from B	B − M − C → B	B IMM	C2	ii	2	2	3-1	-	-	-	-	↕	↕	↕	↕
			B DIR	D2	dd	2	3	4-1								
			B EXT	F2	hh ll	3	4	5-2								
			B IND,X	E2	ff	2	4	6-2								
			B IND,Y	18 E2	ff	3	5	7-2								
SEC	Set Carry	1 → C	INH	0D		1	2	2-1	-	-	-	-	-	-	-	1
SEI	Set Interrupt Mask	1 → I	INH	0F		1	2	2-1	-	-	-	1	-	-	-	-
SEV	Set Overflow Flag	1 → V	INH	0B		1	2	2-1	-	-	-	-	-	-	1	-

* Cycle-by-cycle number provides a reference to Tables 10-2 through 10-8 which detail cycle-by-cycle operation.
　　Example: Table 10-1 Cycle-by-Cycle column reference number 2-4 equals Table 10-2 line item 2-4.

Source Form(s)	Operation	Boolean Expression	Addressing Mode for Operand	Opcode	Operand(s)	Bytes	Cycle	Cycle by Cycle*	S	X	H	I	N	Z	V	C
STAA (opr)	Store Accumulator A	A → M	A DIR	97	dd	2	3	4-2	-	-	-	-	↕	↕	0	-
			A EXT	B7	hh ll	3	4	5-3								
			A IND,X	A7	ff	2	4	6-5								
			A IND,Y	18 A7	ff	3	5	7-5								
STAB (opr)	Store Accumulator B	B → M	B DIR	D7	dd	2	3	4-2	-	-	-	-	↕	↕	0	-
			B EXT	F7	hh ll	3	4	5-3								
			B IND,X	E7	ff	2	4	6-5								
			B IND,Y	18 E7	ff	3	5	7-5								
STD (opr)	Store Accumulator D	A → M, B → M + 1	DIR	DD	dd	2	4	4-4	-	-	-	-	↕	↕	0	-
			EXT	FD	hh ll	3	5	5-5								
			IND,X	ED	ff	2	5	6-6								
			IND,Y	18 ED	ff	3	6	7-7								
STOP	Stop Internal Clocks		INH	CF		1	2	2-1	-	-	-	-	-	-	-	-
STS (opr)	Store Stack Pointer	SP → M:M + 1	DIR	9F	dd	2	4	4-4	-	-	-	-	↕	↕	0	-
			EXT	BF	hh ll	3	5	5-5								
			IND,X	AF	ff	2	5	6-8								
			IND,Y	18 AF	ff	3	6	7-7								
STX (opr)	Store Index Register X	IX → M:M + 1	DIR	DF	dd	2	4	4-4	-	-	-	-	↕	↕	0	-
			EXT	FF	hh ll	3	5	5-5								
			IND,X	EF	ff	2	5	6-8								
			IND,Y	CD EF	ff	3	6	7-7								
STY (opr)	Store Index Register Y	IY → M:M + 1	DIR	18 DF	dd	3	5	4-6	-	-	-	-	↕	↕	0	-
			EXT	18 FF	hh ll	4	6	5-7								
			IND,X	1A EF	ff	3	6	6-9								
			IND,Y	18 EF	ff	3	6	7-7								
SUBA (opr)	Subtract Memory from A	A − M → A	A IMM	80	ii	2	2	3-1	-	-	-	-	↕	↕	↕	↕
			A DIR	90	dd	2	3	4-1								
			A EXT	B0	hh ll	3	4	5-2								
			A IND,X	A0	ff	2	4	6-2								
			A IND,Y	18 A0	ff	3	5	7-2								
SUBB (opr)	Subtract Memory from B	B − M → B	B IMM	C0	ii	2	2	3-1	-	-	-	-	↕	↕	↕	↕
			B DIR	D0	dd	2	3	4-1								
			B EXT	F0	hh ll	3	4	5-2								
			B IND,X	E0	ff	2	4	6-2								
			B IND,Y	18 E0	ff	3	5	7-2								
SUBD (opr)	Subtract Memory from D	D − M:M + 1 → D	IMM	83	jj kk	3	4	3-3	-	-	-	-	↕	↕	↕	↕
			DIR	93	dd	2	5	4-7								
			EXT	B3	hh ll	3	6	5-10								
			IND,X	A3	ff	2	6	6-10								
			IND,Y	18 A3	ff	3	7	7-8								
SWI	Software Interrupt	See Special Ops	INH	3F		1	14	2-15	-	-	-	1	-	-	-	-
TAB	Transfer A to B	A → B	INH	16		1	2	2-1	-	-	-	-	↕	↕	0	-
TAP	Transfer A to CC Register	A → CCR	INH	06		1	2	2-1	↕	↕	↕	↕	↕	↕	↕	↕
TBA	Transfer B to A	B → A	INH	17		1	2	2-1	-	-	-	-	↕	↕	0	-
TEST	TEST (Only in Test Modes)	Address Bus Counts	INH	00		1	**	2-20	-	-	-	-	-	-	-	-
TPA	Transfer CC Register to A	CCR → A	INH	07		1	2	2-1	-	-	-	-	-	-	-	-
TST (opr)	Test for Zero or Minus	M − 0	EXT	7D	hh ll	3	6	5-9	-	-	-	-	↕	↕	0	0
			IND,X	6D	ff	2	6	6-4								
			IND,Y	18 6D	ff	3	7	7-4								
TSTA		A − 0	A INH	4D		1	2	2-1	-	-	-	-	↕	↕	0	0
TSTB		B − 0	B INH	5D		1	2	2-1	-	-	-	-	↕	↕	0	0
TSX	Transfer Stack Pointer to X	SP + 1 → IX	INH	30		1	3	2-3	-	-	-	-	-	-	-	-
TSY	Transfer Stack Pointer to Y	SP + 1 → IY	INH	18 30		2	4	2-5	-	-	-	-	-	-	-	-

*Cycle-by-cycle number provides a reference to Tables 10-2 through 10-8 which detail cycle-by-cycle operation.
  Example: Table 10-1 Cycle-by-Cycle column reference number 2-4 equals Table 10-2 line item 2-4.

TABLE A.1 *(DATA BOOK* TABLE 10–1) *Continued*

Source Form(s)	Operation	Boolean Expression	Addressing Mode for Operand	Machine Coding (Hexadecimal)		Bytes	Cycle	Cycle by Cycle*	Condition Codes							
				Opcode	Operand(s)				S	X	H	I	N	Z	V	C
TXS	Transfer X to Stack Pointer	IX – 1 → SP	INH	35		1	3	2-2	-	-	-	-	-	-	-	-
TYS	Transfer Y to Stack Pointer	IY – 1 → SP	INH	18 35		2	4	2-4	-	-	-	-	-	-	-	-
WAI	Wait for Interrupt	Stack Regs & WAIT	INH	3E		1	***	2-16	-	-	-	-	-	-	-	-
XGDX	Exchange D with X	IX → D, D → IX	INH	8F		1	3	2-2	-	-	-	-	-	-	-	-
XGDY	Exchange D with Y	IY → D, D → IY	INH	18 8F		2	4	2-4	-	-	-	-	-	-	-	-

* Cycle-by-cycle number provides a reference to Tables 10-2 through 10-8 which detail cycle-by-cycle operation.

     Example: Table 10-1 Cycle-by-Cycle column reference number 2-4 equals Table 10-2 line item 2-4.

** Infinity or Until Reset Occurs

*** 12 Cycles are used beginning with the opcode fetch. A wait state is entered which remains in effect for an integer number of MPU E-clock cycles (n) until an interrupt is recognized. Finally, two additional cycles are used to fetch the appropriate interrupt vector (14 + n total).

dd = 8-Bit Direct Address ($0000 – $00FF) (High Byte Assumed to be $00)
ff   = 8-Bit Positive Offset $00 (0) to $FF (255) (Is Added to Index)
hh = High Order Byte of 16-Bit Extended Address
ii   = One Byte of Immediate Data
jj   = High Order Byte of 16-Bit Immediate Data
kk = Low Order Byte of 16-Bit Immediate Data
ll   = Low Order Byte of 16-Bit Extended Address
mm = 8-Bit Bit Mask (Set Bits to be Affected)
rr   = Signed Relative Offset $80 (– 128) to $7F (+ 127)
       (Offset Relative to the Address Following the Machine Code Offset Byte)

(Copyright of Motorola, Used by Permission.)

B Quick Reference

■ **MC68HC11 BLOCK DIAGRAM**

     See Figure B.1.

■ **DATA BOOK AND PROGRAMMING REFERENCE GUIDE TABLES AND FIGURES**[1]

     Table B.1: Register and Control Bit Assignments
     Table B.2: Interrupt Vector Assignments
     Figure B.2: Interrupt Stacking Order
     Figure B.3: Stacking Operations
     Table B.3: 68HC11 Signals (Pins)
     Figure B.4: Pin Assignments for 52-Pin PLCC and CLCC
     Hexadecimal and Decimal Conversion
     ASCII Chart
     Cross-References to Other Data

---

[1]*Data Book* means *Motorola Document MC68HC11E/D Rev2, M68HC11E Family Technical Data.*
Programming Reference Guide means *Motorola Document MC68HC11ERG/AD M68HC11*
E Series Programming Reference Guide.

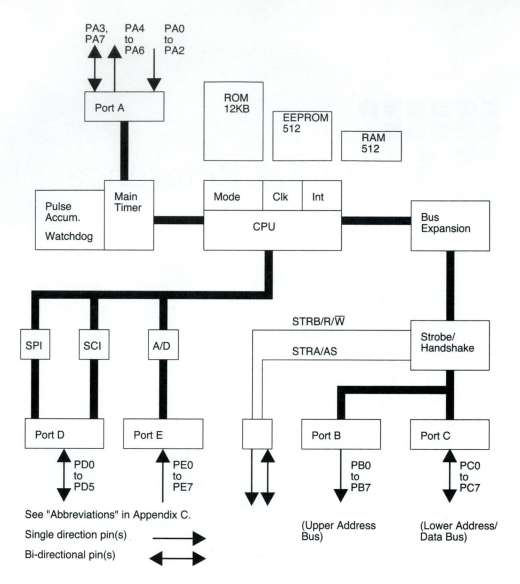

Figure B.1 MC68HC11E9 Block Diagram

TABLE B.1A

M68HC11 E Series Registers (1 of 2)

The 128-byte register block can be remapped to any 4K boundary.

Addr	Bit 7	6	5	4	3	2	1	Bit 0	Name
$1000	PA7	PA6	PA5	PA4	PA3	PA2	PA1	PA0	PORTA
$1001									Reserved
$1002	STAF	STAI	CWOM	HNDS	OIN	PLS	EGA	INVB	PIOC
$1003	PC7	PC6	PC5	PC4	PC3	PC2	PC1	PC0	PORTC
$1004	PB7	PB6	PB5	PB4	PB3	PB2	PB1	PB0	PORTB
$1005	PCL7	PCL6	PCL5	PCL4	PCL3	PCL2	PCL1	PCL0	PORTCL
$1006									Reserved
$1007	DDC7	DDC6	DDC5	DDC4	DDC3	DDC2	DDC1	DDC0	DDRC
$1008	—	—	PD5	PD4	PD3	PD2	PD1	PD0	PORTD
$1009	—	—	DDD5	DDD4	DDD3	DDD2	DDD1	DDD0	DDRD
$100A	PE7	PE6	PE5	PE4	PE3	PE2	PE1	PE0	PORTE
$100B	FOC1	FOC2	FOC3	FOC4	FOC5	—	—	—	CFORC
$100C	OC1M7	OC1M6	OC1M5	OC1M4	OC1M3	—	—	—	OC1M
$100D	OC1D7	OC1D6	OC1D5	OC1D4	OC1D3	—	—	—	OC1D
$100E	Bit 15	14	13	12	11	10	9	Bit 8	TCNT (Hi)
$100F	Bit 7	6	5	4	3	2	1	Bit 0	TCNT (Lo)
$1010	Bit 15	14	13	12	11	10	9	Bit 8	TIC1 (Hi)
$1011	Bit 7	6	5	4	3	2	1	Bit 0	TIC1 (Lo)
$1012	Bit 15	14	13	12	11	10	9	Bit 8	TIC2 (Hi)
$1013	Bit 7	6	5	4	3	2	1	Bit 0	TIC2 (Lo)
$1014	Bit 15	14	13	12	11	10	9	Bit 8	TIC3 (Hi)
$1015	Bit 7	6	5	4	3	2	1	Bit 0	TIC3 (Lo)
$1016	Bit 15	14	13	12	11	10	9	Bit 8	TOC1 (Hi)
$1017	Bit 7	6	5	4	3	2	1	Bit 0	TOC1 (Lo)
$1018	Bit 15	14	13	12	11	10	9	Bit 8	TOC2 (Hi)
$1019	Bit 7	6	5	4	3	2	1	Bit 0	TOC2 (Lo)
$101A	Bit 15	14	13	12	11	10	9	Bit 8	TOC3 (Hi)
$101B	Bit 7	6	5	4	3	2	1	Bit 0	TOC3 (Lo)
$101C	Bit 15	14	13	12	11	10	9	Bit 8	TOC4 (Hi)
$101D	Bit 7	6	5	4	3	2	1	Bit 0	TOC4 (Lo)
$101E	Bit 15	14	13	12	11	10	9	Bit 8	TI4/O5 (Hi)
$101F	Bit 7	6	5	4	3	2	1	Bit 0	TI4/O5 (Lo)
$1020	OM2	OL2	OM3	OL3	OM4	OL4	OM5	OL5	TCTL1
$1021	EDG4B	EDG4A	EDG1B	EDG1A	EDG2B	EDG2A	EDG3B	EDG3A	TCTL2

(Copyright of Motorola, Used by Permission.)

TABLE B.1B

	Bit 7	6	5	4	3	2	1	Bit 0	
$1022	OC1I	OC2I	OC3I	OC4I	I4/O5I	IC1I	IC2I	IC3I	TMSK1
$1023	OC1F	OC2F	OC3F	OC4F	I4/O5F	IC1F	IC2F	IC3F	TFLG1
$1024	TOI	RTII	PAOVI	PAII	—	—	PR1	PR0	TMSK2
$1025	TOF	RTIF	PAOVF	PAIF	—	—	—	—	TFLG2
$1026	DDRA7	PAEN	PAMOD	PEDGE	DDRA3	I4/O5	RTR1	RTR0	PACTL
$1027	Bit 7	6	5	4	3	2	1	Bit 0	PACNT
$1028	SPIE	SPE	DWOM	MSTR	CPOL	CPHA	SPR1	SPR0	SPCR
$1029	SPIF	WCOL	—	MODF	—	—	—	—	SPSR
$102A	Bit 7	6	5	4	3	2	1	Bit 0	SPDR
$102B	TCLR	SCP2[1]	SCP1	SCP0	RCKB	SCR2	SCR1	SCR0	BAUD
$102C	R8	T8	—	M	WAKE	—	—	—	SCCR1
$102D	TIE	TCIE	RIE	ILIE	TE	RE	RWU	SBK	SCCR2
$102E	TDRE	TC	RDRF	IDLE	OR	NF	FE	—	SCSR
$102F	R7/T7	R6/T6	R5/T5	R4/T4	R3/T3	R2/T2	R1/T1	R0/T0	SCDR
$1030	CCF	—	SCAN	MULT	CD	CC	CB	CA	ADCTL
$1031	Bit 7	6	5	4	3	2	1	Bit 0	ADR1
$1032	Bit 7	6	5	4	3	2	1	Bit 0	ADR2
$1033	Bit 7	6	5	4	3	2	1	Bit 0	ADR3
$1034	Bit 7	6	5	4	3	2	1	Bit 0	ADR4
$1035	—	—	—	PTCON	BPRT3	BPRT2	BPRT1	BPRT0	BPROT
$1036	MBE	—	ELAT	EXCOL	EXROW	T1	T0	PGM	EPROG[2]
$1037									Reserved
$1038									Reserved
$1039	ADPU	CSEL	IRQE	DLY	CME	—	CR1	CR0	OPTION
$103A	Bit 7	6	5	4	3	2	1	Bit 0	COPRST
$103B	ODD	EVEN	ELAT[3]	BYTE	ROW	ERASE	EELAT	EPGM	PPROG
$103C	RBOOT	SMOD	MDA	IRVNE	PSEL3	PSEL2	PSEL1	PSEL0	HPRIO
$103D	RAM3	RAM2	RAM1	RAM0	REG3	REG2	REG1	REG0	INIT
$103E	TILOP	—	OCCR	CBYP	DISR	FCM	FCOP	TCON	TEST1
$103F	EE3[4]	EE2[4]	EE1[4]	EE0[4]	NOSEC	NOCOP	ROMON[5]	EEON	CONFIG

1. MC68HC(7)11E20 only
2. MC68HC711E20 only
3. MC68HC711E9 and MC68S711E9 only
4. MC68HC811E2 only
5. Not applicable to MC68HC811E2. For devices with disabled ROM array (MC68HC11E0, MC68HC11E1, MC68L11E0, or MC68L11E1) this bit must never be set to one.

(Copyright of Motorola, Used by Permission.)

TABLE B.2 INTERRUPT VECTOR ASSIGNMENTS

Vector Address	Interrupt Source	CCR Mask Bit	Local Mask
FFC0, C1 – FFD4, D5	Reserved	—	—
FFD6, D7	SCI serial system • SCI receive data register full • SCI receiver overrun • SCI transmit data register empty • SCI transmit complete • SCI idle line detect	I	RIE RIE TIE TCIE ILIE
FFD8, D9	SPI serial transfer complete	I	SPIE
FFDA, DB	Pulse accumulator input edge	I	PAII
FFDC, DD	Pulse accumulator overflow	I	PAOVI
FFDE, DF	Timer overflow	I	TOI
FFE0, E1	Timer input capture 4/output compare 5	I	I4/O5I
FFE2, E3	Timer output compare 4	I	OC4I
FFE4, E5	Timer output compare 3	I	OC3I
FFE6, E7	Timer output compare 2	I	OC2I
FFE8, E9	Timer output compare 1	I	OC1I
FFEA, EB	Timer input capture 3	I	IC3I
FFEC, ED	Timer input capture 2	I	IC2I
FFEE, EF	Timer input capture 1	I	IC1I
FFF0, F1	Real-time interrupt	I	RTII
FFF2, F3	$\overline{\text{IRQ}}$ (external pin)	I	None
FFF4, F5	$\overline{\text{XIRQ}}$ pin	X	None
FFF6, F7	Software interrupt	None	None
FFF8, F9	Illegal opcode trap	None	None
FFFA, FB	COP failure	None	NOCOP
FFFC, FD	Clock monitor fail	None	CME
FFFE, FF	$\overline{\text{RESET}}$	None	None

(Copyright of Motorola, Used by Permission.)

Memory Location	CPU Registers
SP	PCL
SP–1	PCH
SP–2	IYL
SP–3	IYH
SP–4	IXL
SP–5	IXH
SP–6	ACCA
SP–7	ACCB
SP–8	CCR

Figure B.2 Interrupt Stacking Order (Copyright of Motorola, Used by Permission.)

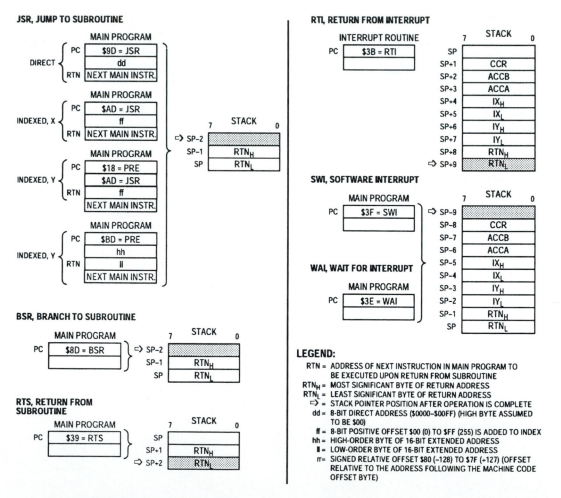

Figure B.3 Stacking Operations (Copyright of Motorola, Used by Permission.)

TABLE B.3   68HC11 SIGNALS (PINS)[a]

Signal (pin)	Function	Signal (pin)	Function
ADx	Data pin x	PDx	Port D pin x
ANx	Analog channel x	PEx	Port E pin x
Ax	Address pin x	R/W	Read/Write
EXTAL	Crystal	RXD	Receive
ICx	Input capture pin x	SCK	Serial clock
IRQ	Interrupt request	SS	Slave select
LIR	Load instruction register	STRA	Strobe A
MISO	Master in, slave out	STRB	Strobe B
MODA	Mode A pin	TXD	Transmit
MODB	Mode B pin	V_{DD}	Positive supply voltage
MOSI	Master out, slave in A/D	V_{RH}	High reference voltage for
OCx	Output compare pin x	V_{RL}	Low reference voltage for A/D
PAI	Pulse accumulator input	V_{SS}	Negative reference (ground)
PAx	Port A pin x	XIRQ	X-interrupt request
PBx	Port B pin x	XTAL	Crystal
PCx	Port C pin x		

[a]See also Pin Assignments.

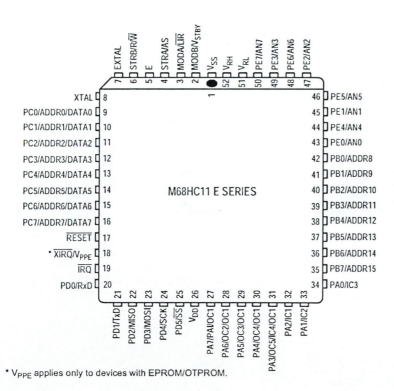

Figure B.4   Pin Assignments for 52-Pin PLCC and CLCC (Copyright of Motorola, Used by Permission.)

Hexadecimal and Decimal Conversion

How to Use:

Conversion to Decimal: Find the decimal weights for corresponding hexadecimal characters beginning with the least significant character. The sum of the decimal weights is the decimal value of the hexadecimal number.

Conversion to Hexadecimal: Find the highest decimal value in the table which is lower than or equal to the decimal number to be converted. The corresponding hexadecimal character is the most significant. Subtract the decimal value found from the decimal number to be converted. With the difference, repeat the process to find subsequent hexadecimal characters.

15 Byte 8		7 Byte 0	
15 Char 12	11 Char 8	7 Char 4	3 Char 0
Hex	Dec	Hex	Dec	Hex	Dec	Hex	Dec
0	0	0	0	0	0	0	0
1	4,096	1	256	1	16	1	1
2	8,192	2	512	2	32	2	2
3	12,288	3	768	3	48	3	3
4	16,384	4	1,024	4	64	4	4
5	20,480	5	1,280	5	80	5	5
6	24,576	6	1,536	6	96	6	6
7	28,672	7	1,792	7	112	7	7
8	32,768	8	2,048	8	128	8	8
9	36,864	9	2,304	9	144	9	9
A	40,960	A	2,560	A	160	A	10
B	45,056	B	2,816	B	176	B	11
C	49,152	C	3,072	C	192	C	12
D	53,248	D	3,328	D	208	D	13
E	57,344	E	3,584	E	224	E	14
F	61,440	F	3,840	F	240	F	15

ASCII Chart

ASCII CHARACTER SET (7-Bit Code)								
M S Digit / **L S Digit**	**0**	**1**	**2**	**3**	**4**	**5**	**6**	**7**
0	NUL	DLE	SP	0	@	P		p
1	SOH	DC1	!	1	A	Q	a	q
2	STX	DC2	"	2	B	R	b	r
3	ETX	DC3	#	3	C	S	c	s
4	EOT	DC4	$	4	D	T	d	t
5	ENQ	NAK	%	5	E	U	e	u
6	ACK	SYN	&	6	F	V	f	v
7	BEL	ETB	'	7	G	W	g	w
8	BS	CAN	(	8	H	X	h	x
9	HT	EM	)	9	I	Y	i	y
A	LF	SUB	*	:	J	Z	j	z
B	VT	ESC	+	;	K	[	k	{
C	FF	FS	'	<	L	\	l	|
D	CR	GS	-	=	M	]	m	}
E	SO	RS	.	>	N	^	n	~
F	SI	US	/	?	O	_	o	DEL

■ CROSS-REFERENCES TO OTHER DATA

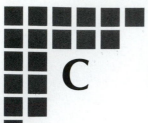

C Further Information

■ REFERENCES

1. Cannon, D. L., and G. Luecke. 1982. *Understanding electronic control of energy systems.* Fort Worth, Tex.: Radio Shack.

2. *Encyclopaedia Britannica,* Vol. 4, 15th ed.

3. Moralee, D. 1981. Microprocessor architectures: Ten years of development. *Electronics and Power* 27: 214–221.

4. Adamson, T. 1990. *Structured C for technology.* Columbus, Oh.: Merrill.

5. Lipovsky, G. J. 1988. *Single- and multiple-chip microcomputer interfacing.* Austin, Tex.: Prentice Hall, Englewood Cliffs, N.J./Motorola, Inc.

6. Glass, Brett L. 1990. The SCSI bus, Parts 1 and 2. *Byte* 15: 267–274, 291–298.

7. Bartree, Thomas C. 1987. *Data communications, networks and systems.* Indianapolis: Howard W. Sams.

8. Currie, W. Scott. 1989. *LANs explained.* Chichester, West Sussex, England: Ellis Horwood.

9. Floyd, Thomas L. 1990. *Digital fundamentals.* 3d ed. Columbus, Oh.: Merrill.

10. Tompkins, Willis J., and John G. Webster, eds. 1988. *Interfacing sensors to the IBM PC.* Englewood Cliffs, N.J.: Prentice Hall.

11. Loveday, G. C., and B. D. Brighouse. 1987. *Microprocessors in engineering systems.* Marshfield, Mass.: Pitman Publishing.

12. Staugaard, Andrew C., Jr. 1988. *8-bit microprocessor interfacing and applications.* Lexington, Mass.: D. C. Heath.

13. Schultz, Warren. 1988. *Understanding SENSEFETs.* Motorola Application Note, AN1001/D. Austin, Tex.: Motorola, Inc.

14. Dorf, Richard C. 1986. *Modem control systems.* 4th ed. Reading, Mass.: Addison-Wesley.

15. Waite, Mitchell, and Stephen Prata. 1990. *The Waite Group's new C primer plus.* Indianapolis: Howard W. Sams.

16. Auslander, David M., and Cheng H. Tham. 1990. *Real-time software for control: Programming examples in C.* Englewood Cliffs, N. J.: Prentice Hall.

17. Scott, Christopher. 1989. *Complex control with the UPI-42H.* Intel Application Note AP-161, from *Intel Microprocessor and Peripheral Handbook,* Vol. II, *Peripheral.* Santa Clara, Cal.: Intel.

18. Cook, Rick. 1991. Embedded systems in control. *Byte* 16: 153–160.

19. Sohie, Guy R. L. 1989. Implementation of Fast Fourier Transforms on Motorola's DSP56000/DSP56001 and DSP96002 Digital Signal Processors. Motorola Application Literature APR4/D. Austin, Tex.: Motorola, Inc.

20. Papamichalis, Panos, and John So. 1986. Implementation of fast Fourier transform algorithms with the TMS32020. *Digital Signal Processing Applications with the TMS320 Family,* Vol. I. Dallas, Tex.: Texas Instruments, Inc.

21. *8-bit embedded controller handbook,* order number 270645-001. 1989. Santa Clara, Cal.: Intel.

22. Cox, Earl. 1994. *The fuzzy systems handbook: A practioners's guide to building, using, and maintaining fuzzy systems.* New York: Academic Press.

23. Kosko, Bart. 1992. *Neural networks and fuzzy systems: A dynamical approach to machine intelligence.* Englewood Cliffs, N. J.: Prentice Hall.

24. McNeill, Daniel, and Paul Freiburger. 1993. *Fuzzy logic.* New York: Simon & Schuster.

25. Hintz, Kenneth, and Daniel Tabak. 1992. *Microcontrollers, architecture, implementation, and programming.* New York: McGraw-Hill.

26. Krol, Ed. 1993. *The whole Internet user's guide & catalog.* Sebastopol, Cal.: O'Reilly & Associates.

27. Jones, Joseph L., and Anita M. Flynn. 1993. *Mobile robots.* Wellesley, Mass.: A. K. Peters Ltd.

■ SOME TRADE MAGAZINES

Circuit Cellar INK: The Computer Applications Journal
Circuit Cellar Incorporated
4 Park Street, Suite 20
Vernon, CT 06066-3233
(800) 269-6301
subscribe@circellar.com
http://www.circellar.com

EDN Electronics Design News
Computer Center
P.O. Box 5563
Denver, CO 80217-0844
(303) 470-4445
http://www.ednprodmag.com/

Embedded Systems Programming
(847) 647-8602 or (888) 847-6177 (U.S. only)
http://www.embedded.com/mag.shtml

■ MANUALS

Motorola, *HC11 E Series,* order number MC68HC11E/D (referred to as the *Data Book*).

Motorola, *M68HC11 Reference Manual,* order number M68HC11RM/AD (referred to as the *Reference Manual*).

Motorola, *M68HC11E Series Programming Reference Guide,* order number MC68HC11A8RG/AD (referred to as the *Programming Reference Guide*).

Motorola, ASEMBLER.DOC, a text file included with the EVB/EVBU program disk. It documents the AS11 cross assembler.

■ SOURCES

Motorola

Students and instructors may use University Support for educational discounts:
http://mot-sps.com/support/university/

Other Suppliers of Support Software

Avocet Systems, Inc.
120 Union Street
Rockport, ME 04856
Tel.: (800) 448-8500
     (207) 236-9055
http://www.avocetsystems.com/avocet/

Coactive Aesthetics
4000 Bridgeway
Sausalito, CA 94965-1445
Tel.: (415) 289-1722
Fax: (415) 289-1320
coactive@coactive.com
http://www.coactive.com/

ImageCraft
P.O. Box 64226
Sunnyvale, CA 94086-4226
Tel.: (408) 749-0702
Fax: (408) 739-6994
icc11@imagecraft.com, info@imagecraft.com
http://www.imagecraft.com/

Evaluation Boards

Axiom Manufacturing
717 Lingco Drive, Suite #209
Richardson, TX 75081
Tel: (972) 994-9676
Fax: (972) 994-9170
sales@axman.com
http://www.axman.com/

HVW Technologies Inc.
Suite 473, 300-8120 Beddington Blvd., N.W.
Calgary, Alberta T3K 2A8, CANADA
Tel: (403) 730-8603
Fax: (403) 730-8903
Info@HVWTech.com
http://www.hvwtech.com/

Technological Arts
26 Scollard St.
Toronto, Ontario M5R 1E9 CANADA
Tel: (416) 963-8996
Fax: (416) 963-9179
Sales@technologicalarts.com
http://www.technologicalarts.com/

Wytec Company
Suite 238, 360-23 W. Shick Road
Bloomingdale, IL 60108-1414
Tel: (630) 894-1440
Fax: (509) 461-4330
Sales@wytec.com
http://www.wytec.com

Other Resources

http://fleming0.flemingc.on.ca/~pspasov/MCU/mcu.htm

# D Conventions

C CONVENTIONS

We will use C language syntax occasionally as comments for assembly language source code.

Arithmetic Operators

Operator	Action
−	Subtraction, unary, minus
+	Addition
×	Multiplication (note that C actually uses *; we use ×)
/	Division
%	Modulo division
− −	Decrement
+ +	Increment

Pointer Operator (*)

The * operator takes the value of the variable it precedes and uses that value as the *address* of the information in memory. It means "at address" or "contents of address." For example:

$$*y = 100$$

means: Place the value of 100 at address y. In the case of the 68HC11, it could mean put the value of 100 at the address referenced by the y index register. If y = 20, then 100 would be put in address 20.

Relational and Bitwise Operators

Relational operator	Meaning
>	Greater than
>=	Greater than or equal
<	Less than
<=	Less than or equal
==	Equal
!=	Not equal

Bitwise operator	Meaning
&	AND
|	OR
^	XOR
~	One's complement
>>	Right shift
<<	Left shift

■ LOGIC LEVELS FOR DIGITAL SIGNALS

Active High

If a signal name appears without a line over it, it means that it causes the named action when the signal is at a logic high. It is known as *active high*. For example, CS1 is a chip select signal that selects the chip when it is high.

Active Low

If a signal name appears with a line above it, it means that it causes the named action when the signal is at a logic low. For example, $\overline{CS2}$ is a chip select signal that selects the chip when it is low.

Another Example

$R/\overline{W}$ causes the read (R) action when it is high and the write (W) action when it is low.

■ NUMBERING SYSTEMS

A prefix or a suffix will indicate a numbering system in most cases. Generally, Motorola assembly languages use prefix notation. The following shows different ways of representing the number *eighteen*.

System	Prefix	Suffix	Prefix example	Suffix example
Hexadecimal	$	H	$12	12H
	0x		0x12	
Decimal			18	18
Octal	@	Q	@22	22Q
Binary	%	B	%00010010	00010010B

Unless specified otherwise, a number without a prefix or suffix is decimal. A number using the H suffix must begin with a zero if its most significant digit (MSD) is a letter (e.g., 0C5H). Exceptions are object, machine, or hex code listings in program documentation. Otherwise, comments will indicate the numbering system used. In C language hex numbers are represented by the 0x (zero-ex) prefix.

ASCII

An ASCII character is shown within single quotes. For example; 'a' becomes $61. An ASCII (or character) string is a sequence of ASCII characters. It is shown within double quotes. For example, "String" becomes the sequence $53, $74, $72, $69, $6E, $67.

■ MEMORY

Addressing

The lowest address is considered to be the top of memory. The highest address is the bottom of memory. Each unique address references a byte. If data is more than 1 byte long, it is addressed using the most significant byte address. Multibyte data is stored from top to bottom in the order of the most significant byte (lower address) to the least significant byte (higher address). This is Motorola's convention, which is also known as "little-endian." For example, the 16-bit word, $7C45, at address $2030 is stored as follows.

The most significant byte, $7C, is stored in address $2030.
The least significant byte, $45, is stored in address $2031.

68HC11 Memory Map

This book uses the default addresses for memory-mapped registers ($1000 to $103F) and internal RAM ($00 to $FF). Note that you can change the default addresses to start at any 4KB boundary ($0000, $1000, $2000, . . . , $F000). To do this, change the INIT register.

Address Contents

Parentheses enclosing a number indicates the contents of an address. For example, "the contents of address $1B is $5C" is written as ($1B) = $5C. See also Section 2.5.9.

■ CROSS ASSEMBLY

After Chapter 4, most assembly language listings are based on using the Motorola free AS11 cross assembler. Sometimes we refer to the Avocet AVMAC11 cross assembler. Normally, we do not show all the details for the sake of brevity. Most listings are part of a larger program that has the following template. It shows the necessary overall structure, stack initialization, register symbol definition, and interrupt vectors. You may have to modify these depending on the system you are using.

```
*TEMPLATE LISTING
*The book's cross assembly convention
*------------------------------------

*Opening comments

*Register and Vector definition files if necessary, see Appendix E

*Insert HC11REG.ASM here ;memory mapped registers
*Insert HC11VEC.ASM here ;vector jump table

*Can insert by pasting or concatenation
*------------------------------------

*PAGE0 segment
      ORG    0

*Area for page 0 if used
*------------------------------------

*Demo segment
      ORG    $100   ;for typical EVBU system
*use different address if applicable

*Necessary initialization
*This is the first section of code to be executed.

*Set up the stack if your development system
*does not already do this for you. The following
*shows how to emulate the EVB/EVBU user stack.

      LDS    #$47   ;this is the user stack

*Set up any time-protected bits if applicable.
*(must be done within 64 E cycles after a RESET)
*Put this code in EEPROM.
*If you are using the EVB board,
*reset the board with jumper header J4
*connected to pins 2 and 3.
```

```
*For the EVBU board, reset board with jumper header J2
*connected to pins 1 and 2
*To return to the BUFFALO
*monitor, use the instruction: JMP $E00A

*This is a reminder to show how to set up the
*vector jump table using IRQ as an example.
*Also see Section 8.4.2, Vector Jump Table,
*and see Figure 3.8. Assumes file HC11VEC.ASM inserted.

        LDAA    #$7E            ;load JMP opcode
        STAA    JIRQ            ;and store
        LDX     #RIRQ           ;load service routine address
        STX     JIRQ+1          ;and store

*Continue to set up vector jump table for other interrupts
*if applicable

        LDX     #REGBAS         ;point to memory-mapped registers

*Other initialization code follows if applicable

*Area for more Demo code

*-------------------------------------
*Utility segment
*This can follow the Demo segment.
*The book examples often use a starting address of $180

*Area for Utility code

*Demonstrate beginning code for interrupt (IRQ) service
*routine. The first instruction follows label RIRQ.

RIRQ
*RIRQ is the label to start IRQ service routine
*rest of service routine code follows it.

*                               ;up until the return instruction

        RTI             ;return from RIRQ

*Area for more Utility code

*-------------------------------------
```

```
*Table segment

*Area for look-up and data tables when applicable

*--------------------------------------
                END           ;optional pseudo-op to end assembly
*--------------------------------------
```

Note that most chapter listings assume that symbol definitions for registers are defined (see HC11REG.ASM in Appendix E). Interrupt vectors are defined by using a jump table as specified by header file HC11VEC.H (see Appendix E). This is the same method as that used by the EVB or EVBU boards.

If necessary, you may convert the listings to other formats.

DATA MANIPULATION

Set	To make a bit a logic 1
Clear	To make a bit a logic 0
Set up	To specify the value of all the bits in a binary word: in other words, to create a specific binary number

MEASUREMENT QUANTITIES

See Tables D.1 to D.3.

TABLE D.1 SOME UNITS OF MEASUREMENT

Unit	Quantity	Abbreviation
Ampere	Current	A
Degree Celsius	Temperature	°C
Celsius degrees	Temperature change	C°
Farad	Capacitance	F
Hertz$^{(S^{-1})}$	Frequency	Hz
Ohm	Resistance	Ω
Second	Time	s
Volt	Voltage	V
Watt	Power	W

TABLE D.2 SOME PREFIXES AND ABBREVIATIONS FOR DECIMAL MULTIPLIERS[a]

Multiplier	Prefix	Abbreviation
10^{12}	Tera	T
10^{9}	Giga	G
10^{6}	Mega	M
10^{3}	Kilo	k
10^{-3}	Milli	m
10^{-6}	Micro	μ
10^{-9}	Nano	n
10^{-12}	Pico	p

[a]Examples of use:

$$1 \text{ million ohms} = 1 \text{ megohm} = 1 M\Omega$$

$$\frac{1 \text{ ampere}}{1000} = 1 \text{ milliampere} = 1 \text{ mA}$$

$$\frac{1 \text{ second}}{10^6} = 1 \text{ microsecond} = 1 \ \mu s$$

TABLE D.3 COMMON MEMORY UNITS

Unit	Abbreviation
Byte	
Kilobyte (2^{10} bytes $\cong 10^3$ bytes)	KB
Megabyte (2^{20} bytes $\cong 10^6$ bytes)	MB
Gigabyte (2^{30} bytes $\cong 10^9$ bytes)	GB

 **E** # Header and Library
File Source Listings

■ HC11REG.H

```
*FILE HC11REG.H - library file for memory-mapped registers
*to be used by an application source code. Use it
*by concatenating it to beginning of the application source code
*or by pasting it in beginning of the application source code.
*For register information, refer to Table B.1 (Appendix B)
*These are MC68HC11A8 and E9 part registers.
*Modify listing as required for other 68HC11 parts.


*Registers will be addressed with Ind,X mode

REGBAS     EQU     $1000        ;Starting address for
*                               ;register block

PORTA      EQU     $00
PIOC       EQU     $02
PORTC      EQU     $03
PORTB      EQU     $04
PORTCL     EQU     $05
DDRC       EQU     $07
PORTD      EQU     $08
DDRD       EQU     $09
PORTE      EQU     $0A
CFORC      EQU     $0B
OC1M       EQU     $0C
OC1D       EQU     $0D
```

```
TCNT      EQU     $0E
TIC1      EQU     $10
TIC2      EQU     $12
TIC3      EQU     $14
TOC1      EQU     $16
TOC2      EQU     $18
TOC3      EQU     $1A
TOC4      EQU     $1C
TOC5      EQU     $1E
TCTL1     EQU     $20
TCTL2     EQU     $21
TMSK1     EQU     $22
TFLG1     EQU     $23
TMSK2     EQU     $24
TFLG2     EQU     $25
PACTL     EQU     $26
PACNT     EQU     $27
SPCR      EQU     $28
SPSR      EQU     $29
SPDR      EQU     $2A
BAUD      EQU     $2B
SCCR1     EQU     $2C
SCCR2     EQU     $2D
SCSR      EQU     $2E
SCDR      EQU     $2F
ADCTL     EQU     $30
ADR1      EQU     $31
ADR2      EQU     $32
ADR3      EQU     $33
ADR4      EQU     $34
BPROT     EQU     $35        ;EEPROM Block Protect for E9 part, not A8
OPTION    EQU     $39
COPRST    EQU     $3A
PPROG     EQU     $3B
HPRIO     EQU     $3C
INIT      EQU     $3D
TEST1     EQU     $3E
CONFIG    EQU     $3F

*END OF FILE HC11REG.ASM
```

■ HC11VEC.H

```
*FILE HC11VEC.H - library file vectors
*to be used by an application source code. Use it
*by concatenating it to beginning of application source code
*or by pasting it into beginning of application source code.
```

```
*These vectors are identical to those
*for EVB and EVBU board programs.

*VECTOR JUMP TABLE SEGMENT
*-----------------------------------
JUMP       EQU     $00C4
           ORG     JUMP

JSCI       RMB     3
JSPI       RMB     3
JPAIE      RMB     3
JPAO       RMB     3
JTOF       RMB     3
JTOC5      RMB     3
JTOC4      RMB     3
JTOC3      RMB     3
JTOC2      RMB     3
JTOC1      RMB     3
JTIC3      RMB     3
JTIC2      RMB     3
JTIC1      RMB     3
JRTII      RMB     3
JIRQ       RMB     3
JXIRQ      RMB     3
JSWI       RMB     3
JILLOP     RMB     3
JCOP       RMB     3
JCMF       RMB     3

*VECTORS segment
*-----------------------------------
VECTORS    EQU     $FFD6
           ORG     VECTORS

VSCI       FDB     JSCI
VSPI       FDB     JSPI
VPAIE      FDB     JPAIE
VPAO       FDB     JPAO
VTOF       FDB     JTOF
VTOC5      FDB     JTOC5
VTOC4      FDB     JTOC4
VTOC3      FDB     JTOC3
VTOC2      FDB     JTOC2
VTOC1      FDB     JTOC1
VTIC3      FDB     JTIC3
VTIC2      FDB     JTIC2
VTIC1      FDB     JTIC1
VRTII      FDB     JRTII
VIRQ       FDB     JIRQ
```

```
VXIRQ     FDB     JXIRQ
VSWI      FDB     JSWI
VILLOP    FDB     JILLOP
VCOP      FDB     JCOP
VCMF      FDB     JCMF
VRES      FDB     $E000        ;external or power-up RESET
*                              ;address for BUFFALO monitor

*END OF FILE HC11VEC.ASM
```

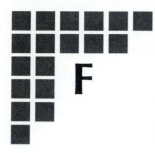

F Digital Logic and Binary Codes

This section will provide a very quick overview of the subject of digital logic and binary codes. It is essential background for understanding some topics. If you wish to read a more comprehensive treatment of the subject, consult any of the many books available, such as *Digital Fundamentals* by Thomas L. Floyd (reference 9 in Appendix C).

■ A DIGITAL LOGIC PRIMER

As currently practiced, digital logic uses entities (variables) with only two conditions or choices (states).[1] A variable is one or the other. The two states for a variable may be any of the following:

true	false
on	off
high	low
up	down
one	zero

and so on. Hopefully, you get the idea. The variables are usually called *bits*. "Bit" is the short form for "binary digit." *Binary* refers to the fact that a bit can have only one of two possible values.

---

[1]One type of digital logic is multivalued logic in which variables can have more than two states. At the time of this writing, it is primarily an area of academic research. Another type is fuzzy logic. It uses degrees of membership instead of a strict true or false membership. Examples of fuzzy variable conditions are very hot, hot, warm, cool, cold, and very cold. Section 13.8 covers fuzzy logic using free software available for the 68HC11.

The logic *gate* is the fundamental building block for circuits that manipulate these variables. These circuits are known as digital circuits. A gate provides an output that depends on the condition of the inputs. There may be one, two, or more inputs, but a gate has only one output. A gate is for a digital circuit what a cell is for a living organism. Even the most complex microprocessor is a vast system of interconnecting gates.

A schematic symbol represents each type of gate. A *truth table* defines the function of each type of gate. No, a truth table is not something to prevent you from lying. It is a mathematical definition expressed in the form of rows and columns. It shows the gate output for all possible combinations of inputs.

In Figure F.1, note that gate schematics with a bubble perform an inverting (not) function. For example, not 0 is 1 and not 1 is 0. Generally, gates can have more than the two inputs shown in Figure F.1. Electronically, a logic 0 is nominally 0 V. A logic 1 is +5 V. This

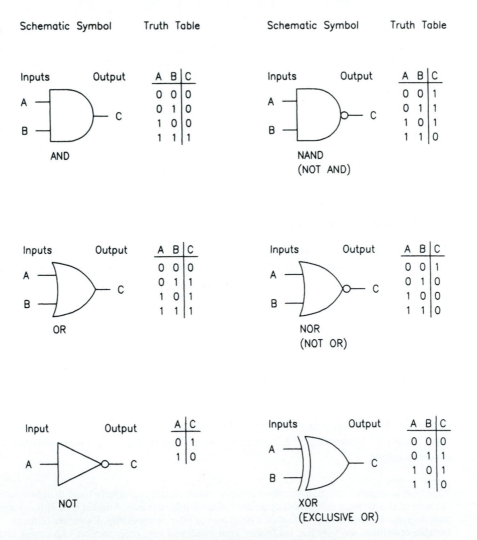

Figure F.1 Basic Logic Gates: Schematic Symbol and Truth Table

TABLE F.1 BOOLEAN EXPRESSIONS

Function	Equation
AND	$C = A \cdot B$
OR	$C = A + B$
NOT	$C = \overline{A}$
NAND	$C = \overline{A \cdot B}$
NOR	$C = \overline{A + B}$
XOR	$C = A \oplus B$

is a generalization. Some systems use different values or types of signals to represent the two binary values. The logic gates perform binary operations known as Boolean functions or equations. Boolean logic is a branch of mathematics. In Section 1.2 we mentioned that George Boole developed this branch of mathematics. It uses symbolic equations as shown in Table F.1.

To illustrate what this means, consider the following logic AND situation. Depending on your car, if you sit in it AND do not buckle up your seat belt, a warning alarm turns on. The system uses sensors to detect both conditions. We can represent this as a Boolean equation:

$$ALARM = SEAT \cdot \overline{NOBUCKLE} \tag{F.1}$$

The alarm goes on if the seat sensor is on AND the buckle sensor is off.

Another control example is a pedestrian crosswalk on a busy street. Each side has a button. Pressing either button causes a red traffic light to turn on. A Boolean equation for this system is

$$LIGHT = BUTTON\_1 + BUTTON\_2 \tag{F.2}$$

This means "turn on the light if BUTTON_1 OR BUTTON_2 is on." Actually, the situation is more complicated. The light should remain on for a short time period after a button is released.

PRACTICE

A system has the following sensor conditions:

$$SENS1 = 0, SENS2 = 1, SENS3 = 0, SENS4 = 0, SENS5 = 1$$

What is the result of the following Boolean equations?

a. $(\overline{SENS1} + SENS2) \cdot SENS3$
b. $(\overline{SENS4 \cdot SENS5}) + SENS1$
c. $(\overline{SENS1 + SENS2}) \cdot SENS5$

Solution

a. $(0 + 1) \cdot 0 = 1 \cdot 0 = 0$

b. $\overline{(0.1)} + 0 = (1 \cdot 1) + 0 = 1 + 0 = 1$

c. $\overline{(0 + 1)} \cdot 1 = \overline{1} \cdot 1 = 0 \cdot 1 = 0$

Recall that we had to specify what a logic level represents. In equations (F.1) and (F.2), both physical outputs turn on if the output bit is high (active high). However, it does not have to behave this way. We can use an active-low output. For example, to turn on the alarm, clear bit $\overline{\text{ALARM}}$ to 0 instead of setting it to 1. Similarly for inputs, variable $\overline{\text{NOBUCKLE}}$ is active low because a low indicates that the driver has not buckled up.

Note that some systems may use negative logic. This means that a high voltage represents a logic 0 and a low voltage represents a logic 1. Chapter 10 covers the RS-232 data communications standard. It uses negative logic to represent binary data.

PRACTICE

A chemical process requires that the vat solution remain above a minimum temperature. The vat is large; hence, it uses three low-temperature switches located at different sections. If a temperature goes below the minimum, the corresponding switch changes to the active logic state. An *active-high* low-temperature switch will change to a *high* logic state when the temperature goes low. An *active-low* switch will go *low* when the temperature goes low. The alarm is an *active-high* output, meaning that it goes *high* to indicate an alarm condition. Write the Boolean equation to turn on an alarm:

a. If any temperature is low, the switches are active low.
b. If any temperature is low, the switches are active high.
c. If most of the temperature switches detect a low temperature, switches are active high.
d. If all temperature switches detect a low temperature, switches are active low.

Solution

a. $\text{ALARM} = \overline{\text{TEMP1}} + \overline{\text{TEMP2}} + \overline{\text{TEMP3}}$

Alternatively, $\overline{\text{ALARM}} = \text{TEMP1} \cdot \text{TEMP2} \cdot \text{TEMP3}$, meaning that the alarm is off only if all sensors are high.

b. $\text{ALARM} = \text{TEMP1} + \text{TEMP2} + \text{TEMP3}$
c. $\text{ALARM} = \text{TEMP1} \cdot \text{TEMP2} + \text{TEMP1} \cdot \text{TEMP3} + \text{TEMP2} \cdot \text{TEMP3}$
d. $\text{ALARM} = \overline{\text{TEMP1}} \cdot \overline{\text{TEMP2}} \cdot \overline{\text{TEMP3}}$

Alternatively, $\overline{\text{ALARM}} = \text{TEMP1} + \text{TEMP2} + \text{TEMP3}$, meaning that the alarm is off if any sensor is high.

That is enough said about Boolean equations for now. Next we look at how binary information is represented.

■ NUMBER SYSTEMS

We all know the base 10 number system (although you may not realize it). These are numbers such as -2.5, 16.9, -529, and so on. Computers and related equipment such as microcontrollers internally represent information using bits. This is because they are digital machines. Sometimes the information that the bits represent is numbers. In this case the numbers are base 2 numbers. They are also known as binary numbers. For simplicity, we consider only integers at this time.

Recall that in decimal numbers, the position of each digit represents a *weight,* which is a multiplication by 10 for every left move to the next digit. For example, decimal 2536:

$$2536 = 6 \times 10^0 + 3 \times 10^1 + 5 \times 10^2 + 2 \times 10^3$$
$$= 6 \times 1 + 3 \times 10 + 5 \times 10 \times 10 + 2 \times 10 \times 10 \times 10$$
$$= 6 + 30 + 500 + 2000$$

Similarly, in binary numbers, each leftward bit position has a weight that represents multiplication by 2. For example, binary 1010:

$$1010 = 0 \times 2^0 + 1 \times 2^1 + 0 \times 2^2 + 1 \times 2^3$$
$$= 0 \times 1 + 1 \times 2 + 0 \times 2 \times 2 + 1 \times 2 \times 2 \times 2$$
$$= 0 + 2 + 0 + 8 = \text{decimal 10 (ten)}$$

Hence, binary 1010 represents decimal 10. Note that binary numbers only use the digits (bits) 1 and 0. Table F.2 shows binary weights used in bytes. We will cover fractional values later.

Since using binary numbers is cumbersome, people often use another number system to represent binary numbers. This is the hexadecimal system or base 16 system. It is commonly known as hex. This system uses 16 possible digits. The first letters of the alphabet are used to represent digits greater than nine (9). For example, hex 2A5:

$$2A5 = 5 \times 16^0 + A \times 16^1 + 2 \times 16^2$$
$$= 5 \times 1 + 10 \times 16 + 2 \times 16 \times 16$$
$$= 5 + 160 + 512 = 677$$

TABLE F.2 BIT WEIGHTS IN UNSIGNED BYTES

Bit position	7	6	5	4	3	2	1	0
Integer								
Bit weight	2^7	2^6	2^5	2^4	2^3	2^2	2^1	2^0
Value	128	64	32	16	8	4	2	1
Fraction								
Bit weight	2^{-1}	2^{-2}	2^{-3}	2^{-4}	2^{-5}	2^{-6}	2^{-7}	2^{-8}
Value[a]	0.5000	0.2500	0.1250	0.0625	0.0313	0.0156	0.0078	0.0039

[a]Fraction values are rounded to four places.

Tables F.3 and F.4 show the relationships between the three number systems. For the binary numbers, note that each group of 4 bits, starting from the right, represents a hex digit.

Conventions

In practice, binary numbers are 8-, 16-, or 32-bit numbers. A *byte* is defined as an 8-bit binary number (or code). A *nibble* is a 4-bit binary number. At this point, refer to Figure 1.7. For now ignore the words *address* and *data*. The top shows a 16-bit binary number (2 bytes). The bit at the farthest right is the least significant bit (LSB). The bit at the farthest left is the most significant bit (MSB). It represents a multiplication by the highest power of 2. As explained in Section 1.4.1, the percentage sign (%) and dollar sign ($) precede a number to indicate whether it is a binary or hex number, respectively. Careful comparison of the binary number with Table F.3 should reveal that %0010101001110011 is indeed $2A73. The long sequence of 1s and 0s also illustrates why people use hex numbers to represent binary numbers. Remember: Each group of 4 bits (nibble) from the right represents a hex digit. The reader may wish to review Section 1.4.1 quickly at this time. Note that Appendix D also lists numbering system conventions.

Identifying Bits

Sometimes, a bit in a byte (8 bits) or 16-bit word (2 bytes) represents a physical quantity such as the SEAT and BUCKLE inputs and ALARM output of equation (F.1). In Figure 1.7, note that bits are also addressed by their position relative to the LSB (rightmost bit). The LSB is bit 0. The next one is 1, the next is 2, and so on until the MSB. The numbering starts from 0 because each bit position represents a weight with an increasing power of 2. Hence, from the right to left, we have 2^0, 2^1, 2^2, and so on (see Table F.2). For example, the 16-bit word in Figure 1.7 has bit $0 = 1$, bit $1 = 1$, bit $2 = 0$, and so on.

TABLE F.3 DECIMAL, HEXADECIMAL, AND BINARY CONVERSION FOR NIBBLES

Decimal	Hex	Binary
0	0	0000
1	1	0001
2	2	0010
3	3	0011
4	4	0100
5	5	0101
6	6	0110
7	7	0111
8	8	1000
9	9	1001
10	A	1010
11	B	1011
12	C	1100
13	D	1101
14	E	1110
15	F	1111

TABLE F.4   DECIMAL, HEXADECIMAL, AND BINARY
CONVERSION FOR BYTES

Decimal	Hex	Binary
0	00	0000 0000
1	01	0000 0001
2	02	0000 0010
3	03	0000 0011
4	04	0000 0100
5	05	0000 0101
6	06	0000 0110
7	07	0000 0111
8	08	0000 1000
9	09	0000 1001
10	0A	0000 1010
11	0B	0000 1011
12	0C	0000 1100
13	0D	0000 1101
14	0E	0000 1110
15	0F	0000 1111
16	10	0001 0000
17	11	0001 0001
⋮		
127	7F	0111 1111
128	80	1000 0000
129	81	1000 0001
⋮		
254	FE	1111 1110
255	FF	1111 1111

■ NUMBER CONVERSIONS

It may not be necessary to perform conversions when working with microcontrollers. In Chapter 4 we explain that many assemblers will convert binary and decimal numbers as well as ASCII codes to hex. If it is necessary to perform conversions, you can use one of several methods. One method is to use a calculator that can perform number conversions. Books about digital logic (e.g., reference 9 in Appendix C) describe other methods. In the remainder of this section, we will explain how to use Table F.3 and the Quick Reference (Appendix B) as an alternative method.

As we have seen earlier, it is simple to convert between binary and hex by using 4-bit groups (nibbles) from right to left. For assistance to convert between hex and binary, you can use Table F.3. For assistance to convert between hex and decimal, you can use "Hexadecimal and Decimal Conversion" in the Quick Reference (Appendix B).

For example, convert %1100101001 to hex. Using Table F.3, work from right to left (in order of least significant bit to most significant bit):

$$\%1001 = \$9; \%0010 = \$2; \%0011 = \$3$$

Note that we added leading zeros to complete the leftmost nibble. The hex number is $329.

For another example, convert $1D5C to binary and decimal. Using Table F.3, work from right to left:

$$\$C = \%1100; \$5 = \%0101; \$D = \%1101; \$1 = \%0001$$

Remember, there are 4 bits per hex digit. The binary result is then

$$\%0001110101011100 = \%1110101011100$$

Using the "Hexadecimal and Decimal Conversion" in Appendix B, work from right to left:

$$\$C \rightarrow 12; \$5 \rightarrow 80; \$D \rightarrow 3328; \$1 \rightarrow 4096$$

The arrow is shorthand for "represents." Now add the decimal values.

$$12 + 80 + 3328 + 4096 = 7516$$

Converting from decimal to hex is more difficult. Using the "Hexadecimal and Decimal Conversion" in Appendix B, find the next-lowest decimal number in the chart. This represents the most significant hex digit. Then subtract the two. Find the next lowest number relative to the difference. This represents the next significant hex digit. Repeat the process until you have found the least significant digit. For example, convert 15,253 to hex.

Using the Appendix B chart, we have

$$
\begin{array}{ll}
15{,}253 - 12{,}288 = 2965; & 12{,}288 \rightarrow \$3 \\
2965 - 2816 = 149; & 2816 \rightarrow \$B \\
149 - 144 = 5; & 144 \rightarrow \$9 \\
5; & 5 \rightarrow \$5
\end{array}
$$

The hex result is $3B95.

PRACTICE

a. Convert the following to decimal and binary: $A250; $5C; $123.
b. Convert the following to hex and decimal: %110100101001; %101000000110101; %0110110.
c. Convert the following to hex and binary: 4578; 512; 17,981.

Solution

a. 41,552; 92; 291
   % 1010001001010000; %1011100; %100100011
b. $D29; $5035; $36
   3369; 20,533; 54
c. $11E2; $200; $463D
   %1000111100010; %1000000000; %100011000111101

■ BINARY RANGES

With a fixed number of bits, there is a fixed quantity of possible binary integers. Consider flipping coins. With one coin there are two possibilities, heads or tails. With one bit, you can have a 0 or a 1. With two coins, you can have four possibilities: heads/heads, heads/tails, tails/heads, or tails/tails. With three coins there are eight possibilities, with four coins there are 16 possibilities, and so on. A binary number is like a sequence of coins in which each coin represents a bit. We can use n to represent the bit size. This is the number (quantity) of bits. Then the bit position of the most significant bit (MSB) is $n - 1$. The quantity of possible binary numbers becomes

$$Possible\_Quantity\_of\_Binary\_Numbers = 2^n \tag{F.3}$$

For example, with eight bits ($n = 8$), there are $2^8 = 256$ possible combinations. To find the largest possible hex integer, use the following equation.

$$Largest\_Unsigned\_Integer = 2^n - 1 \tag{F.4}$$

In Chapter 2 (Section 2.7) we explain the use of signed integers in microcontroller arithmetic. Table F.5 shows the range of some common unsigned integer bit sizes.

Sometimes, we need to use positive and negative numbers. In that case we can split the unsigned range evenly down the middle using the MSB as the sign bit. Half of the numbers have an MSB of 0, for non-negative numbers. The other half has an MSB of 1, making them negative. For example, an 8-bit unsigned range of 0 to 255 becomes -128 to $+127$. The number zero is considered positive because its MSB is 0. Table F.6 shows the range of signed integers. The hex representation may seem confusing because *seemingly* larger hex numbers represent negative numbers. It turns out that binary arithmetic for signed numbers is easily implemented using the system just described. In Section 2.7 (Figure 2.17 in particular) we explain why this is so.

TABLE F.5   RANGE OF UNSIGNED INTEGERS

Integer size in bits (n)	Range	
	Hex	Decimal
4	0 to $F	0 to 15
8	0 to $FF	0 to 255
12	0 to $FFF	0 to 4095
16	0 to $FFFF	0 to 65,535
20	0 to $FFFFF	0 to 1,048,575
24	0 to $FFFFFF	0 to 16,777,215
32	0 to $FFFFFFFF	0 to 4,294,967,295

TABLE F.6 RANGE OF SIGNED INTEGERS[a]

Integer size in bits (n)	Range (negative, positive)	
	Hex	Decimal
4	$8 to $F, 0 to $7	-8 to -1, 0 to $+7$
8	$80 to $FF, 0 to $7F	-128 to -1, 0 to $+127$
12	$800 to $FFF, 0 to $7FF	-2048 to -1, 0 to $+2047$
16	$8000 to $7FFF, 0 to $7FFF	$-32,768$ to -1, 0 to $+32,767$

[a]Note that the MSB $= 0$ for positive numbers and MSB $= 1$ for negative numbers. See also Figure 2.17.

Fractions and Normalizing

Fractions.   Now consider binary fractions. Refer back to Table F.2. Each bit position from right to left has a weight that represents a negative power of 2. For example, the byte hex 74 (binary 01110100):

$$01110100 = 0 \times 2^{-1} + 1 \times 2^{-2} + 1 \times 2^{-3} + 1 \times 2^{-4} + 0 \times 2^{-5} + 1 \times 2^{-6}$$
$$+ 0 \times 2^{-7} + 0 \times 2^{-8}$$

$$= 0 \times \frac{1}{2} + 1 \times \frac{1}{4} + 1 \times \frac{1}{8} + 1 \times \frac{1}{16} + 0 \times \frac{1}{32} + 1 \times \frac{1}{64} + 0 \times \frac{1}{128} + 0 \times \frac{1}{256}$$

$$= 0 + 0.25 + 0.125 + 0.0625 + 0 + 0.015625 + 0 + 0 = \text{decimal } 0.453125$$

Scaling.   In many control situations it is useful to work with fractions. You may prefer to think of fractions as percentages of full scale. We use percentages because there are limits to what we can measure. There are also limits to the number of bits used in binary data. Consider a speedometer. It may have a range (visible scale) of 0 to 200 kilometers per hour. A speed of 100 kilometers per hour will be read as half the speed range. The hex number $80 (representing 50% or $0.5 \times$ full scale) can represent this speed. Consider also a fuel tank with a total storage capacity of 30 liters. Twelve liters of fuel is the same as saying the tank is 40% full ($0.40 \times 30 = 12$). Hence, the hex number $66 ($104/256 \cong 0.4$) can represent the amount of fuel remaining in the tank. Since measurements have limits, signals are often expressed as a percentage of a measurement range. Figure F.2 illustrates some measurement scales and their relationship with hex numbers.

Normalizing.   It is often convenient to express full scale as the number 1 and zero scale as the number 0. Any number between the limits is a fraction. This use of fractions between 0 and 1 is known as *normalizing*. Therefore, there is no need to consider whether we are working with micrometers with precision tooling or kilometers when navigating across the country. Normalization is useful in arithmetic operations such as multiplication and square root.

To normalize an analog number, divide it by the *span*. The span is the difference between the maximum and minimum values of the range. In general, a scale may start at a nonzero offset, such as the 4 to 20 mA scale commonly used in process control. Use equation (F.5) to normalize a number as a fraction.

$$X_{\text{fraction}} = \frac{X_{\text{analog}} - \text{offset}}{\text{span}} \tag{F.5}$$

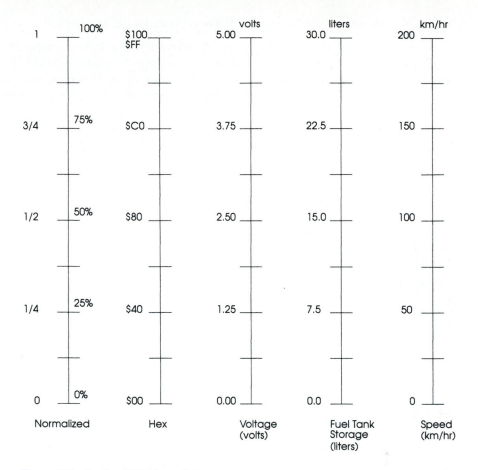

Figure F.2 Scaling With Examples

where span = maximum analog value − minimum analog value

offset = minimum analog value

n = number of bits

X = number prior to normalization

X_{fraction} = normalized number as a decimal fraction

As illustrated by Figure F.2, we restrict our discussion to zero offsets in this section. Use equation (F.6) to express the normalized number in the form of a hex or binary integer.

$$X_{\text{normalized}} = \frac{(2^n - 1)(X_{\text{analog}} - \text{offset})}{\text{span}} = (2^n - 1)X_{\text{fraction}} \qquad \text{(F.6)}$$

where span = maximum analog value − minimum analog value

offset = minimum analog value

n = number of bits

X = number prior to normalization

$X_\text{normalized}$ = normalized number expressed in hex, binary (or decimal integer)

Note that you could also express a normalized number as a decimal integer. For 8-bit numbers the normalized equivalent in decimal would be anywhere between 0 and 255. Note that numbers normalized as fractions differ from binary fractions determined using bit weights (see Table F.2) by one bit. For a sufficiently large bit size (n), we can make them identical as follows

$$X_\text{normalized} = (2^n - 1)X_\text{fraction} \cong 2^n X_\text{fraction}$$ (F.7)

or

$$X_\text{fraction} = \frac{X_\text{normalized}}{2^n}$$ (F.8)

PRACTICE

a. Calculate the decimal equivalents of the following normalized 8-bit hex numbers: $A7, $94, $26.
b. Calculate the decimal equivalents of the following normalized 16-bit hex numbers: $A700, $0940, $0026.

Solution

Use equation (F.8) to determine the fraction.

a. 167/256 = 0.65234, 148/256 = 0.5781, 38/256 = 0.1484
b. 42,752/65,536 = 0.65234, 2368/65,536 = 0.03613, 38/65,536 = 0.00058

Multiplication. Consider multiplication first. The product (result) of multiplying two numbers greater than 1 is larger than either of the numbers. Note that overflow is possible with binary integers. Multiplication of two 8-bit integers requires a 16-bit integer for the product. For example, $32 \times 53 = 1696$. Using hex, this is $20 \times $35 = $06A0. An alternative is to use fractions. The same operation using fractions is

$$\$20 \times \$35 = \frac{32}{256} \times \frac{53}{256} = \frac{1696}{65,536}$$
$$= 0.125 \times 0.207$$
$$= 0.02588$$

We can represent this result as an 8-bit number.

$$\frac{1696}{65,536} = \frac{6.625}{256} \cong \frac{07}{256}$$

Thus, $20 \times \$35 = \$07$ when using fractions (or normalized numbers).

Rounding Error. With normalized numbers, we can multiply any pair of 8-bit numbers and obtain an 8-bit result. However, this comes at a cost of losing accuracy. Consider the following.

$$\$FA \times \$04 = \frac{250}{256} \times \frac{4}{256} = \frac{1000}{65,536}$$
$$= 0.977 \times 0.0156$$
$$= 0.01526$$

Again, we can represent this result as an 8-bit number.

$$\frac{100}{65,536} = \frac{3.90625}{256} \cong \frac{04}{256}$$

Thus, $\$FA \times \$04 = \$04$ with 8-bit normalization! This is known as *rounding* or *round-off error*. It is the result of using a fixed number of bits to represent numerical quantities. The weight of the least significant bit limits the accuracy. This situation worsens when three or more 8-bit numbers are multiplied. More bits will improve the situation at the expense of using more memory. The use of floating-point numbers is necessary when large ranges or high accuracy is required. Section 2.7.7 describes floating-point numbers.

PRACTICE

a. Multiply the following hex numbers using 8-bit normalization.

$$\$26 \times \$7B \times \$C0 \times \$E8$$

b. Multiply the same numbers without normalization and then divide the product by 2^{32}. Compare the result with the answer in (a).

Solution

a. $\frac{38}{256} \times \frac{123}{256} \times \frac{192}{256} \times \frac{232}{256} \cong \frac{18}{256} \times \frac{192}{256} \times \frac{232}{256} \cong \frac{13}{256} \times \frac{232}{256} \cong \frac{12}{256} \cong 0.0469$

b. $\frac{38 \times 123 \times 192 \times 232}{2^{32}} = \frac{2.08198656 \times 10^8}{4.294967296 \times 10^9} \cong 0.0485^*$

---

*When the fractional part is exactly 0.5, one can round it to the nearest even number. For example, 0.5 rounds to 0, and 1.5 rounds to 2. Note that the software must do this since unmodified microcontroller operations will simply truncate (strip off lesser significant bits).

The difference between the two is $0.0485 - 0.0469 = 0.0016$. This is less than the bit weight of the LSB of an 8-bit fraction (see Table F.2). Hence, there is no difference if the latter product is converted to 8 bits. Both products are approximately 12/256.

Square Root. Now consider square root. In Section 2.5.10 we describe an application for using square root. In this case, normalization increases accuracy. First consider 8-bit integers from 0 to 255. The range of their square roots is 0 to 16. Thus, there are fewer possible square root values! For example, the square root of numbers from 21 to 30 would be rounded to 5. Now consider normalization. The square root of any number between 0 and 1 lies between 0 and 1.

Let's try some examples. Find the square root of $80. First, normalize $80.

$$\$80 \rightarrow \frac{128}{256} = 0.5$$

Then find the square root.

$$\sqrt{0.5} = 0.707 \times \frac{256}{256} \cong \frac{181}{256} \rightarrow \$B5$$

Hence, the square root of the normalized number $80 is $B5. Find the square root of $DD. Normalize $DC.

$$\$DC \rightarrow \frac{220}{256} = 0.859$$

Find the square root.

$$\sqrt{0.859} = 0.927 \times \frac{256}{256} \cong \frac{237}{256} \rightarrow \$ED$$

The square root of the normalized number $DC is $ED. Figure 2.15 (Section 2.5.10) uses this example to illustrate a look-up table application.

PRACTICE

   a. Using a "normalized number" approach, find the square root of the following 8-bit hex numbers: $A7, $94, $26.

   b. Similarly, find the square root of the following 16-bit hex numbers: $A700, $0940, $0026.

These are the same numbers as those used in the Practice questions for the subsection "Normalizing."

Solution

a.

$$\sqrt{\frac{167}{256}} \cong \frac{207}{256} \to \$CF, \quad \sqrt{\frac{148}{256}} \cong \frac{195}{256} \to \$C3, \quad \sqrt{\frac{38}{256}} \cong \frac{99}{256} \to \$63$$

b.

$$\sqrt{\frac{42,752}{65,536}} \cong \frac{52,932}{65,536} \to \$CEC4, \quad \sqrt{\frac{2368}{65,536}} \cong \frac{12,457}{65,536} \to \$30A9,$$

$$\sqrt{\frac{38}{65,536}} \cong \frac{1578}{65,536} \to \$062A$$

■ OTHER CODES

So far, we have said that a sequence of bits represents an integer number. We will see also that a collection of bits can represent things as well. Hence, consider the collection to be a code. In Section 1.1.2 we said that a binary code can represent keys on a keyboard, the ASCII code. The ASCII code can also represent the output characters that a printer puts on paper. In Chapter 2 we show how signed integers (negative and positive) are represented. Also, Chapter 2 shows that fractional numbers with decimal points can be represented (floating point) using a binary code. The instructions telling a computer what to do is another binary code.[2] Ultimately, a computer program is a long sequence of binary codes called the *machine code*. If that isn't enough, there are all sorts of different codes used to encode data. Different application programs and peripherals in various computer systems may use different codes.

---

[2]Users of DOS with PCs may note that some files store information using the ASCII code. This is called *text file format*. Some executable files (.COM, .EXE) are stored using machine code, which is sometimes called *binary file format*.

G Basic Waveforms

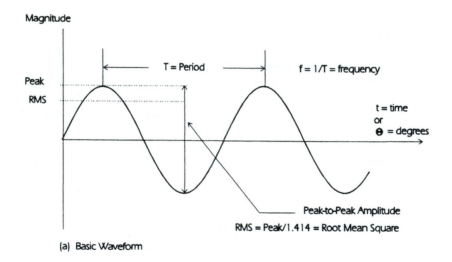

(a) Basic Waveform

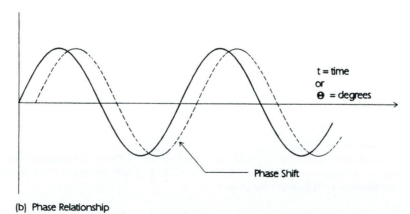

(b) Phase Relationship

Figure G.1 Sine Wave

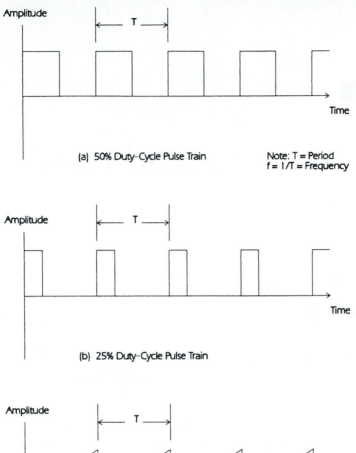

(a) 50% Duty-Cycle Pulse Train

Note: T = Period
f = 1/T = Frequency

(b) 25% Duty-Cycle Pulse Train

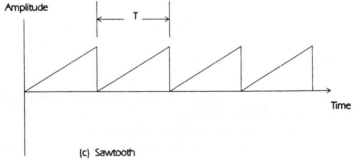

(c) Sawtooth

Figure G.2   Some Other Waveforms

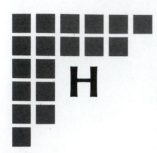
H Internet and CD-ROM Resources

■ INTERNET

Following are some universal resource locators (URLs) that refer to Web document locations. It is possible that some URLs may change after the publication date.

> Author's Microcontroller Resources Page. Check this page primarily for updates.
> http://fleming0.flemingc.on.ca/~pspasov/mcu/mcu.htm
>
> Motorola Semiconductor's Products Link
>
> The main link is http://e-www.motorola.com/index.html
>
> The navigation path to find 68HC11 information should be
> Motorola: Semiconductors: Microcontrollers: Products: 8-Bit (M68HC05, M68HC08, M68HC11): M68HC11 Family
>
> The navigation path to find 68HC12 information should be
> Motorola: Semiconductors: Microcontrollers: Products: 16-Bit (HCS12, M68HC12, DSP56800, M68HC16): M68HC12 Family
>
> The navigation path to find HCS12 information should be
> Motorola: Semiconductors: Microcontrollers: Products: 16-Bit (HCS12, M68HC12, DSP56800, M68HC16): HCS12 Family

Cross Assembler—AS11

www.programmersheaven.com
Various Web sites such as the above contain archive copies of the AS11 cross assembler. These sites contain files, such as AS11.EXE, ASLINK.ECXE, and ASSEMBLER.DOC.

C Cross Compiler—ICC11

www.imagecraft.com

You can download a fully functional 30-day demo of ICC11 V6, which has been supplied by Richard Man of ImageCraft. After downloading, run the setup program to install the program.

Library.   The files are in the Lib folder (after installation). These are object files, which can be linked by the GNU linker. ICC11 will link them as required when producing executable code (s record files).

Compiler Options.   To set options, choose Compiler from the Options menu. The Text section can define the address range for code. The Data section can define the address range for data. When compiling to executable, the compiler produces object code from the C source code and links it with a startup file and specified library files. The default startup file is `crt11.o` and `libc11.a` is a library file included by default. It is possible to specify other library files. Also, you can create your own startup file if it is in the correct object format.

Internet Samples.   Some samples submitted by ImageCraft customers:

HC11 hc11i2c.zip code implementing 12C driver by Grant Beattie.

ucos11.zip uCOS for HC11, written for V4.5, needs to be modified for V5 or above.

libhb.zip MIT Handyboard library, by Chuck McManis and Richard Man.

libmb.zip MIT Miniboard library, by Chuck McManis and Richard Man.

libfl.zip Peter Dunster's F1 library, by Peter Dunster and Richard Man.

usb11.zip James Lye's USB sample.

Tools for the 68HC12 and HCS12

A good starting point is www.programmersheaven.com
One link with free tools for both is at
http://www.geocities.com/englere_geo/

Eagle Layout Editor

www.cadsoft.de

This site contains the install program for a demo version. It can edit schematics and board layouts. Hence, it can be used to view schematic files (.sch). The program is supplied by CadSoft Computer GmbH. The site also includes a layout sample for the 68HC12.

CAMCAD Shareware

www.rsi-inc.com

This site contains a viewer for various CAD formats including those for PCB designs. Hence, it can view HPGL files for PCB design (.hpl). The program is supplied by Router Solutions, Inc.

FuzzyTech

www.fuzzytech.com.
This site contains the demo version of FuzzyTech, a program for developing fuzzy logic software for target processors such as the 68HC11. Typically, the demo has restrictions such as 2 inputs, 1 output, 1 rule block, and others.

Motorola Knowledge-Base Generator (KBG) for Fuzzy Logic

e.g., www.programmersheaven.com
Various Web sites such as the above contain archive copies of the KBG. These sites contain the Motorola FUzzy Design GEnerator (FUDGE) and associated files, such as FTOOL.EXE, FUDGE.EXE, and FUZZ11B3.ASM.

Wytec Evaluation Board

www.wytec.com or www.evbplus.com/
This site contains information about obtaining the Wytec EVB replacement development board, enhanced development boards such as for the 68HC12 and HCS12.

■ CD-ROM RESOURCES

Motorola PDF Files—Folder Motorola

This folder contains assorted documents and application notes in PDF format.

Simulator—Folder THRSIM11

The disk includes a special demo version of the THRSim11 simulator, which has been supplied by Harry Broeders. This demo is a special version made for this book and differs from the free demo which can be downloaded from the Internet. You can purchase the full version at http://www.hc11.demon.nl/thrsim11/thrsim11.htm.

Installation of Demo Version.   Run the setup program in folder THRSIM11DEMO.

Familiarization.   Run THRSim11 DEMO. The Help system does provide instructions; however, it assumes some prior familiarization with 68HC11 assembly language programming. If you are just beginning this may not be the case. Instead you can try one of the programs in the CODE folder of the disk.

Open the file CODE\L2-1.ASM. You should see a window with the code of Listing 2.1. Modify the source as follows.

```
*Listing 2.1
*This is a demo program to introduce the format of an
*assembly language program for the 68HC11. It reads the
*input of the coolant temperature sensor of Figure 1.1 and
*subtracts an offset (a fixed number). It then stores
```

```
*this result for further processing. Another part of the
*program (not shown) will multiply this number to obtain the
*temperature in degrees Celsius. Then it sends the result to the
*dashboard display for the driver.

        ORG     $FF00               ;specify start address
        LDAA    COOLANT_TEMP        ;get coolant temperature
        SUBA    #CT_OFFSET          ;subtract offset
        STAA    STORE_TEMP          ;store it for further processing
        BRA     *

*Following not in book code since it involves assembler directives
*that won't be covered until Chapter 4.

COOLANT_TEMP    EQU         $1031
CT_OFFSET EQU   $20
STORE_TEMPEQU   $1004

*This sample is source code only.
*Corresponding machine code is not shown.
```

Save the file as L2-1DEMO.ASM. The demo program has a restricted memory map set up for

ROM	$FF00 to $FFFF
I/O Registers	$1000 to $103F
LCD Device	$1040, $1041
RAM	$00 to $FF

Modify Address References. This means that you will have to modify most of the program examples in the book to use the addresses that correspond to the demo memory map. Hence, the code will start at address $FF00. For STORE-TEMP we will use address $1004 instead of $D004. *Some* of the I/O registers can be used as substitutes for using extended addressing mode. To practice extended addressing with the demo you will have to use the I/O registers since some of these registers are writeable; hence, they behave like RAM. Suitable substitute addresses for both reading and writing data are $1016 to $101D. Some substitute addresses such as $1031 are okay if used only for reading data, which is the case for the program.

BRA *. You should also add this instruction. This is an instruction that we will use as the last instruction in the demonstration program. Its purpose is to act as the end of the program.

Assemble the Program. To be able to execute the code, you need to assemble it. To do this, select Assemble from the File menu (or click the Assemble button, which is fourth from the left in the toolbar). Now a new window will appear showing the listing, which is the source code shown with the associated machine code to the left. The first line of code is highlighted because the Program Counter is pointing to the beginning of the program. The simulator highlights the instruction that is about to be executed—not the instruction that just has been executed. At this point you can minimize the source window (not the listing window) if you want.

View Registers and Memory.   Before running the program you will want to set up windows to display the contents of registers and memory. To view the registers, select Registers from the View menu and then select the submenu item CPU Registers. The CPU Registers window will appear showing the contents of the registers. The registers correspond to those shown in the programming model of Figure 2.6. These registers are:

A	Accumulator A
B	Accumulator B
D	Accumulator D
X	Index Register X
Y	Index Register Y
SP	Stack Pointer
PC	Program Counter
CC	Condition Code

Except for CC, the contents are shown in hexadecimal. The contents of CC are shown in binary. You can also see the same information displayed in a status bar at the bottom of the window.

To view memory, select Memory from the View menu and then select the submenu item Memory List. A Set Memory dialog window pops up. Enter a start address of 1000 (numbers interpreted as hex) and click OK. View a second memory location by repeating these steps except that you enter a start address of 102d. Note that the memory windows also show symbolic names such as STORE_TEMP and COOLANT_TEMP, which correspond to addresses $1004 and $1031, respectively.

Reposition the Windows.   One feature of the simulator is that it is necessary to resize and position the windows in order to view them. Do so now. If you wish, use Figure H.1 as a guide. Note that Figure H.1 shows the simulator after the program has run, which is why the last program line is highlighted.

Modify Registers and Memory.   You can enter data into the registers and memory as required. For example, double-click A in the CPU Registers window, enter aa, and press Enter. Accumulator A should now have the contents aa. Double-click address $1031 in the Memory List window and enter 47. Address $1031 should now have the data $47.

Single Step.   One of the ways to execute the program is to execute the program one line at a time so that you can examine the result of executing each instruction. This is known as single stepping. Before you begin, check that the first instruction in the listing is highlighted. The program counter (PC) should have the contents $ff00.

To single step, select Step from the Execute menu or press the F7 key. The second instruction should be highlighted. The program counter should contain $ff03 meaning that it is pointing to the second instruction, which is at address $ff03. Note that accumulator A now has the data $47. The result of executing LDAA COOLANT_TEMP is to load accumulator A with the contents of address $1031.

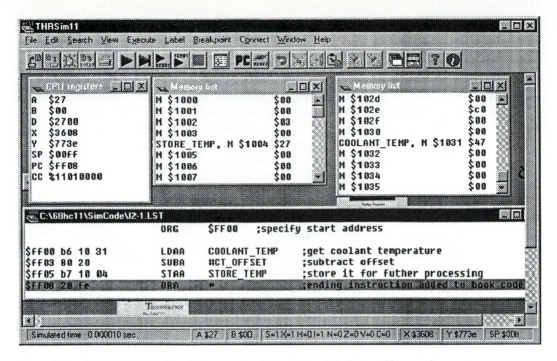

Figure H.1 THRSim11 Window. This example shows Listing 2.1. Some address assignment changes are due to the demo version requirements.

Note: If you want to repeat execution of this instruction you will have to manually change the contents of the program counter back to $ff00. Also manually change accumulator A to anything other than $47. This is so you can observe the effect of the executing instruction.

Single step the next instruction (press F7). The third instruction is highlighted. The result of executing the second instruction is that accumulator A now contains $27. The SUBA instruction subtracted $20 from the previous contents of accumulator A and put the result back into accumulator A. In other words, $47 − $20 = $27. The program counter now holds the value $ff05 so it is pointing at the third instruction.

Single step the next instruction. The fourth and final instruction is highlighted. The result of executing the third instruction is that address $1004 now contains $27. The STAA instruction stored the value in accumulator A into memory with an address of $1004. The program counter now holds the value $ff08 so it is pointing at the fourth instruction.

Single step again. The program counter does not change and the BRA * remains highlighted. This instruction causes repeated execution of itself.

Repeating Execution. To repeat execution, you can manually change the program counter back to the start address of the program (address $ff00). Alternatively, select the source window and assemble the program again. Modify accumulator A and address $1031 to try different values. We caution you not to modify other memory locations and registers at this stage—not until you become more familiar with the 68HC11 so you can interpret the results.

Exit.   To exit the simulator, select Exit from the File menu.

Experimenting With Other Programs.   If you are using the full version of the simulator, you will not need to modify the program examples in the book. With the demo version, modification is usually necessary in order to work within the memory map of the simulator. The simulator can help you understand some of the programming examples. As you become more familiar with microcontroller technology, you can explore other features of the simulator.

Here is a list of the first few program modifications necessary for using the demo version of the simulator.

Listing 2.4
Start code at $FF00, i.e., ORG $FF00

Listing 2.5
Start code at $FF00, i.e., ORG $FF00

Listing 2.6
Start code at $FF00, i.e., ORG $FF00
Use data address $1016 instead of $6D00

I/O Example.   This example illustrates the use of some I/O ports and assumes that you have read up to Section 2.5.9 inclusive.

Run the simulator and select New from the File menu. Enter the following program and save it as S2-5-9.ASM.

```
ORG      $FF00
LDAA     #$A5        ;load data to output
STAA     $1004       ;send it to Port B
LDAA     $1003       ;read data from Port C
BRA      *
```

Assemble the program. View the CPU registers and view the memory list with a start address of $1000.

To view port B, select Pins from the View menu and then select the submenu item PB Pins. To view port C, select Pins from the View menu and then select the submenu item PC Pins. Position and resize the windows as required.

You can modify pin values. Click on a port C pin. Double-click it. This will toggle the value from 0 to 1 (or 1 to 0). Note that the corresponding hex data in address $1003 changes. Change some other port C pins to have a variety of 1s and 0s. Single step the first two instructions. Note that the STAA instruction drives the port B pins to correspond to the data sent to the port. Single step the next two instructions. Note that accumulator A is loaded with data that corresponds to the signals at the port C pins.

Other Program Modifications.   Here is a partial list of more program modifications necessary for using the demo version of the simulator.

Listing 2.8
Start code at $FF00, i.e., ORG $FF00
Use an IX value of $00 instead of $2000
Use an IY value of $80 instead of $B020

Listing 2.9
Start code at $FF00, i.e., ORG $FF00
Eliminate LDAB instruction. Instead manually load accumulator B.
Use RAM from $00 to $FF as the substitute EPROM space from $B600 to $B6FF.
Also LDX instruction should load #$0 instead of #$B600.

Listing 2.9a, 2.10, 2.11, 2.12
Start code at $FF00, i.e., ORG $FF00

Listing 2.13
Start code at $FF00, i.e., ORG $FF00
Use different source and destination blocks such as $00 and $50 instead of $C000 and $C500.

Others
When required, start code at $FF00, i.e., ORG $FF00
Other changes to match the demo version memory map may be necessary.

Glossary

ac   Alternating current. In household terms, this is electricity whose voltage changes polarity periodically. In electronics, it often refers to the transistor or chip signal component that changes. The shape of the waveform is a sine wave (Appendix G). *See also* dc.

Active Level (or Signal)   See "Logic Levels for Digital Signals" in Appendix D.

Address   Binary number that indicates the location of information in memory.

Addressing Mode   Specifies where an instruction (opcode) finds data. It is part of the operand.

Algorithm   Description of a step-by-step procedure for solving a computation problem.

Analog   A signal that can vary continuously from one level to another without making discrete (discontinuous) steps. A number representing an analog signal.

ANSI C   Standard that defines the language and a standard library for C. It has more features than K&R C.

Architecture   Fundamental level of design of a computer. It describes the properties of a CPU that relate directly to the way it executes programs—as seen by assembly language programmer. This includes word size, register organization, instruction set, and addressing modes.

Array   A set of sequentially indexed elements having the same data type.

ASCII   American Standard Code for Information Interchange. A recognized standard 7-bit binary code to represent printable characters and control codes used for data communications. It is the first 128 characters of the ANSI character set (see Appendix B).

Assemble   To convert assembly language source code into object code.

Assembler   Program that converts assembly language source code into object code.

Assembly Language   Words and rules for giving instructions to a CPU in terms of data transfers or operations with registers and memory addresses. It is the mnemonic form of the machine language for a CPU. Each type of CPU uses its own assembly language. *See also* Architecture.

682

Assert   To drive a signal to its active level. See "Logic Levels for Digital Signals" in Appendix D.

Asynchronous   Not clocked. Communication between devices without a common clock for a time reference.

Baud   Frequency of the smallest signaling unit in serial data communications.

BCD   Binary-coded decimal. Starting from the right (LSB), each group of 4 bits (nibble) codes a decimal digit.

Binary   Base 2 number system that uses only digits 1 and 0. To obtain the value of a binary number, move from right to left to multiply each bit by an increasing power of 2. Then add the result. *See also* Bit.

Bit   Binary digit. It is either a 1 or 0.

Bit Size   The number (quantity) of bits used in a binary numbering system. Typically, it is a power of 2.

Block Diagram   This shows the functional parts of a computer. It is not a physical diagram. Its purpose is similar to that of a schematic for describing circuits.

Boolean Logic   Rules that evaluate binary inputs to a binary output of 1 (True) or 0 (False) using operations such as AND, OR, and NOT. Software instructions can implement Boolean logic. Digital circuits or other switching circuits implement it in hardware.

BUFFALO   Monitor program resident in and used by the EVB or EVBU board.

Buffer   Device that isolates or amplifies a signal from another device so that it can be read by yet another device.

Bus   Typically, it is the extension of a microprocessor's pins plus some other signals. It is the link between different components in a computer system. Many recent "standard" buses are defined precisely and are much more than the extension of a microprocessor's pins.

Byte   Eight-bit binary number. It represents a two-digit hex number. Often represents an ASCII code (MSB = 0).

C   (1) High-level programming language common in microcontroller applications. (2) Carry bit (see Figure 2.6).

Call   The action of passing control from a higher-level program module to a lower-level program module.

Calling Register   Register used by a subroutine to get its input data.

Character   The fundamental unit of asynchronous data communication. A 7- or 8-digit binary number, normally coded in ASCII. It is also the C variable type for 8-bit integers.

Clear   To change the value of a bit or all bits to logic 0.

Clock Circuit   Circuit that produces a fixed-frequency (usually, square) signal that triggers each new CPU operation.

CMOS   Complementary metal-oxide semiconductor. It refers to one type of transistor used to build computer logic circuits. Their main characteristic is low power consumption compared to previous methods of fabricating circuits.

Compiler   Program that converts a high-level language source code into object code.

Complement   The inverse (logic NOT) of all the bits in a binary number. The complement of a 1 bit is 0. The complement of a 0 bit is 1.

Computer   Automatic electronic machine that performs calculations. A computer is capable of accepting data, performing operations according to instructions (program), and providing results of the operations. Microprocessors and microcontrollers are specific types of computers.

Conditional Assembly   Using instructions to tell the assembler how to assemble the source code based on the value of some variables.

Configuration   Process of initializing I/O subsystem registers.

Contention   Two or more devices trying to drive an output line at the same time.

CPU   Central processing unit. It is the part of the computer, microprocessor, or microcontroller that controls the transfer of data and execution of instructions.

Cross Assembly   To assemble the source code for a target system (CPU) using a computer with a different processor (CPU) than the target system.

CSMA/CD   Carrier Sense Multiple Access with Collision Detection. This is a networking technique to determine which devices are trying to access the network and how to handle simultaneous access attempts. *See also* Ethernet.

Data   Information. In the context of this book, it is usually a byte that represents a value such as a number, ASCII code, or condition of an I/O device. It is not an address or an instruction.

Data Type   Refers to the way a program variable stores its value in memory. In C, for example, the *character* data type uses 1 byte and the *integer* data type uses 2 bytes.

dc   Direct current. In household terms, this is electricity whose voltage stays constant. In electronics, it often refers to the transistor or chip signal component that stays constant. *See also* ac.

Debug   Getting rid of bugs. It is the process of finding what is wrong with a system.

Declaration   A group of program statements in a high-level language that defines the name and type of variables.

Decrement   Subtract one from a number.

Demultiplex   To receive one signal from a communication channel having several signals.

Dereference   In the C language, it is the asterisk (*) operator that is used to declare a pointer variable. This operator takes the value of the variable it precedes and uses that value as the address of the information in memory. It is also known as an indirection operator. For example, if y is 5, then *y means at address 5.

Development System   Collection of software and hardware required to create MCU programs.

Digital   A signal that can vary from one level to another only by making discrete steps. A number representing a digital signal.

Diode   A two-terminal electronic circuit element that passes current in only one direction under normal operating conditions.

Direct Memory Access (DMA)   A technique that allows data transfer to and from memory without direct control by the central processing unit.

Duty Cycle   The percentage of time that a signal is at the high level in an alternating waveform.

EEPROM   Electrically erasable read-only memory. Similar to EPROM except that one can erase individual bytes using a sequence of program instructions.

Embedded Controller   This is sometimes synonymous with *microcontroller*. Typically, it refers to any single-chip device that includes a microprocessor, memory, and I/O interfaces. They include specialized devices designed to control specific applications. These include computer communications, graphic displays, memory management, and other systems that require the processor to respond within tight time constraints. A microcontroller is considered to be a type of embedded controller.

Enable   Allow an action to occur when a signal or event occurs. If an interrupt is enabled, the corresponding interrupt signal or event will cause the service routine to execute.

EPROM   Erasable read-only memory. A semiconductor memory device that stores data in a nonvolatile form. It can only be modified by exposing the entire device to ultraviolet light for a period of time. Then new data can be stored by following a special programming procedure.

Ethernet   Refers to similar standards for the physical and data link layer for local area networks (LAN) that uses CSMA/CD. Many Ethernet LANs use twisted pair cable at speeds of 10 Mbps, 100 Mbps, or higher.

EVB   Evaluation board (in general). This book uses the term to reference specifically the M68HC11EVB supplied by Motorola.

EVBU   Universal evaluation board for the 68HC11 supplied by Motorola. It is a scaled-down version of the EVB.

Fetch/Execute Cycle   Process that a CPU uses to execute every instruction.

File   Collection of information kept in a computer disk.

Flash (E)EPROM   A semiconductor memory device that stores data in a nonvolatile fashion. It is possible to modify blocks of data (not individual bytes) using a special programming procedure. Typically less expensive than EEPROM.

Frame   *See* Stack Frame.

Frequency   The number of times per second that a signal changes.

Function   A subroutine (program module) that returns a value using the name of the function.

Fuzzification   The process of converting analog (continuous) values into a grade of membership in fuzzy logic.

Fuzzy Logic   Unlike Boolean logic, it uses degrees of membership instead of strict (crisp) true or false values. Examples of fuzzy variable conditions are: very hot, hot, warm, tepid, cool, cold, and very cold.

Fuzzy Logic Inference   The process of making decisions using fuzzy logic.

Global Variable   Variable in a computer program that can be read or modified by any part of the program (including subroutines and interrupt service routines). They are stored in an absolute location in RAM.

Grade of Membership   In fuzzy logic, this indicates the percentage of belonging to a membership function or fuzzy state. For example, someone might consider 20 degrees Celsius to be 70% warm and 30% hot.

Hardware   Physical parts and I/O connections of a computer system. It includes the components and their wiring and mounting.

HCMOS   High-speed complementary metal-oxide semiconductor. It refers to one type of transistor used to build computer logic circuits. Its main characteristic is low power consumption compared to previous methods of fabricating circuits. Unlike CMOS, it can switch logic states at a faster speed.

Header File   Source code file that is inserted into another using the "Include" preprocessor directive.

Hex   Short form for hexadecimal (it is not a witch's curse).

Hex File   File with ASCII data in a standard format that specifies the machine code and its address. Often, a development system converts it into an executable form (binary file) and then loads it into a target system (MCU) for execution.

Hexadecimal   Base 16 numbering system. It is used as shorthand for writing binary numbers because it is relatively easy to convert them back and forth.

IC   Integrated circuit. A semiconductor chip usually containing complex circuits.

IEEE   Institute of Electronic and Electrical Engineers, an international organization. One of its functions is to develop standards in the areas of electrical/electronic and computer technology.

IEEE-488   Standard for general purpose instrumentation bus (GPIB).

IEEE-754   Standard for binary floating-point arithmetic.

Impedance   Voltage drop across an element (or circuit) divided by the current flowing through it. The impedance value includes magnitude and phase relationship, which is why it is used for ac signals. Effectively, it is a measure of how much a circuit element restricts current flow through it. Note that resistance is a special case of impedance when the phase shift between voltage and current is zero, which is the case for dc signals.

Include   Preprocessor directive for the assembler (or compiler). It tells the assembler to insert a specified file into the file it is presently assembling.

Increment   Add one to a number.

Indirection   This is another term used for the Dereference operator (*).

Inhibit   Prevent an action from occurring, particularly an interrupt. If an interrupt is inhibited, the corresponding interrupt signal or event will not cause the service routine to execute.

Interface   Device or system used to match the output on one device or system to the input of another so that they can exchange information.

Interfacing   Process of using an interface.

Internet   A standard way for one computer to communicate with another computer anywhere in the world. Technically, it is a collection of packet switching networks connected by routers using the Transmission Control Protocol/Internet Protocol (TCP/IP) Internet Protocol Suite.

Interrupt   Forcing the processor to suspend execution of current instructions and immediately execute a special set of instructions (the service routine). After the special instructions are executed, the processor continues from where it left off.

I/O   Input/output. It refers to data connections with respect to the microprocessor (or another component).

I/O Port   Collection of pins (usually eight for a byte of data) that carry I/O signals.

K&R C   Original version of C developed by Kernighan and Ritchie, which has also become accepted as a C standard.

Library   Collection (file) of subroutines that are available for use by any program.

Link   To combine one or more object files to create a hex file.

Local Variable   Variable in a computer program that can only be used by the specific subroutine during a specific call (its lifetime). Usually, it is stored in the stack during execution of the subroutine. When the subroutine finishes execution, the address where the local variable was stored may be overwritten by another part of the program.

Machine Code   *See* Machine Language.

Machine Language   Only language that a CPU can understand. It is a binary code of instructions for the CPU to execute.

Macro   Section of source code, written once, that has a unique label. When the assembler or compiler sees the label in other parts of the source code, it replaces it with the section of code.

Masked   To prevent an interrupt request from being acknowledged by the CPU. *See also* Inhibit.

MCU   (1) Microcontroller unit; this book uses MCU as an abbreviation for microcontroller; *see also* Microcontroller. (2) Microcomputer unit.

Membership Function   A numerical definition of fuzzy logic states. *See also* Grade of Membership.

Microcontroller (MCU)   Single chip used to control other devices. It contains a CPU, memory, and I/O circuits. It is a general-purpose device suitable for a variety of applications. Typically, an MCU must respond to inputs within a short period of time to produce appropriate outputs. An MCU is a type of embedded controller.

Microprocessor (μP)   Single chip that contains the CPU or most of the computer. Microcontrollers are special types of microprocessors.

Modem   Modulator-Demodulator Device used by computers to communicate over telephone lines.

Modulation   To change a signal into a form suitable for transmission over a long distance.

Monitor   Program that is part of the development system. It allows a programmer to examine and modify memory and register contents of a processor. It also allows the programmer to start and stop execution of processor instructions. Usually, it includes a simple (nonsymbolic) assembler and disassembler.

Multiplex   To send more messages simultaneously on a single channel.

Multitasking   The process of executing several distinct programs using a time-sharing technique such that it appears that all are executing simultaneously.

Network   Communications system for computers and peripherals.

Nibble   Four-bit binary number. Often represents a hex or BCD digit.

Noise   Change imposed upon an electrical signal from the environment without changing the inputs to the device supplying the signal.

Normalize   Scaling a range of numbers to a scale of 0 to 1.

Nyquist Frequency   Twice the highest-frequency component of a time-varying signal. The signal itself must have a finite maximum frequency component (bandlimited). *See also* Sampling Theorem.

Object Code   Sequence of binary numbers and other information that represents machine code. It is not in an executable form since the start address may be unknown. Object code files are linked to produce executable (machine) code. Do not confuse object code with object-oriented programming (OOP), which is something entirely different.

Ohm's Law   The voltage drop across a component is equal to the component's resistance multiplied by the current flowing through it. The common mathematical form is $V = IR$.

Opcode   First part of a machine language instruction. It tells the CPU what to do, including addressing mode. In source code it is a mnemonic that does not include the addressing mode.

Operand   Second part of a machine language instruction that specifies where to find the data. It may be 1 or 2 bytes, representing data or the address of the data. In source code it specifies the addressing mode.

Page   A block of 256 bytes.

Parallel   Transferring several bits at a time using a line for each bit. Typical parallel transfers are in multiples of eight.

Parameter   In the C language, it is the variable name used for passing an input value to a function.

PCbug11   Monitor program resident in computer RAM used to control the EVBU board.

Pointer   A storage location in memory whose contents represent the address of another value.

Prebyte   First byte of a 2-byte opcode in machine code.

Preprocessor   Part of the assembler or compiler that executes special instructions to set things up before converting the source code into object code.

Procedure   The term used for a subroutine used in higher-level languages. In some languages it differs from a function since it returns values as parameters.

Processor   Reference to any type of microcontroller or microprocessor.

Program   Collection of instructions telling the CPU what to do.

Protocol   Set of rules regulating the exchange of data between two devices or programs.

Pseudocode   A shorthand method for writing program code without worrying about programming language rules (syntax). It is more like a human language than actual code.

Pseudo-op   Instructions in the source code telling the assembler what to do, not the CPU.

RAM   Random access memory. A type of semiconductor memory where the CPU can store temporary information. The stored data is lost when power is removed (volatile).

RC   Refers to a resistor and capacitor series connection in an electric circuit. The value of resistance multiplied by capacitance is equal to the time constant.

Reentrant Subroutine   A subroutine that may be called (entered) again before the same subroutine has completed execution from a previous call. To execute correctly, a reentrant subroutine must use local variables.

Register   Place where data can be modified for a specific purpose. Often, it is a specific place in hardware.

Reset   Forcing the MCU (or CPU) into an initial state. When referring to a bit, it means the same as clear.

Resistance   *See* Impedance.

Return Register   Register used by a subroutine to place its output data.

RMS   Root mean square. Effective average value of a periodic signal that ignores its change in polarity (or direction). Mathematically, it is defined as the square root of the average of the square of the signal values.

ROM   Read-only memory. A type of semiconductor memory that the CPU can only read. The stored data is permanent and will not be lost if power is removed (nonvolatile).

Row   A block of 16 bytes whose start address has a least significant hex digit of zero; for example, $B630.

Rule Evaluation   In fuzzy logic, it is the process of computing fuzzy outputs given fuzzy inputs. Rules typically are IF-THEN statements with fuzzy AND operators. Some also use fuzzy OR operators.

Run Away   This refers to incorrect operation of the fetch/execute cycles in a central processing unit (CPU). Typically, the cycle is out of step by fetching a byte intended to be an operand and interpreting it to be an operand. At this point, the software could be doing almost anything, such as overwriting critical data. In the 68HC11, the Computer Operating Properly (COP) reset and the Power-On Reset (POR) are used to force the CPU to restart the sequence of fetch/execute cycles at a known point.

Sampling Theorem   To obtain complete information about an analog signal using digital samples at regular time intervals, it is necessary to sample it at a rate equal to or greater than the Nyquist frequency. *See also* Nyquist Frequency.

Segment   A block of source code.

Serial   Transferring one bit at a time using a single line (or medium).

Set   To change the value of a bit or all bits to logic 1.

Signed Numbers   Binary number whose MSB determines whether it is negative or positive.

Software   The nonhardware components of the computer. It is the programs (set of instructions) that tell the computer what to do. It also refers to associated documentation that explains the programs.

Source Code   A computer or microcontroller program written by a person using text information to indicate instructions. It is not executable. A translator (assembler or compiler) translates it to produce machine code. Source code is typically stored on a disk as a text (ASCII) file.

Stack   Block of RAM used to store temporary data that can be accessed using a pointer.

Stack Frame   The part of the stack used to store local variables during the execution of a subroutine.

State   In relation to a process or a processor, it defines a condition in which a specific set of inputs will generate a specific set of outputs. In relation to digital logic, it refers to the condition of a bit; whether it is logic 1 or logic 0.

Static Variable   Variable in a computer program that is stored in a reserved area of RAM instead of using the stack. It is not possible to use the area for other variables.

Subroutine   Section of code, written once, that can be executed repeatedly from other parts of a program.

Synchronous   Clocked. Communication between devices using a common clock signal for a time reference.

Syntax   Grammar rules of a computer language. Instead of saying "turn on the light," the program statement might be something like: light = on.

Time Constant   Property of a system that represents how an output changes in response to a sudden (step) input change. It is the time required for an output to increase to 63% of its final value or to decrease to 37% of its initial value. The time it takes for an output to reach its final value is approximately five times the time constant of the system.

Token   In software design, it is a symbol representing a name or entity. In some computer networks, it is a group of bits used to signal network access by a station.

Topology   Shape of a network.

Transceiver   Device that transmits and receives signals.

Transistor   A three-terminal electronic circuit element with nonlinear characteristics. For a bipolar junction transistor (BJT), the voltage between the base and emitter terminal determines the amount of current flowing between the collector and emitter terminals. For a field-effect transistor (FET), the voltage between the gate and source terminals determine the amount of current flowing between the drain and source terminals. *See also* CMOS and HCMOS (they are special types of FETs).

Two's Complement   A coding method to represent negative binary numbers. One way to find the two's complement of a number is to subtract it from 2^n, where n is the bit size.

Upwardly Compatible   This refers to a newer computer that can also run programs written for an older computer. However, the newer computer usually includes new instructions; hence, the older computer may not run programs written for the newer computer. We say that the new computer is upwardly compatible from the old. For example, the 68HC11 is upwardly compatible from the 6801.

Value   In the context of a C language function, it is the numeric (or string) quantity resulting from executing that function. Typically this returned value will be used in another part of the program. For example, the value of square(2), the square of 2, is 4.

Variable   In higher-level languages, it is a place-holder for program values that can be modified during program execution. In other words, it is a location in RAM.

Vector   The start address of a reset or interrupt service routine. Usually it is a 2-byte number that is stored in ROM or EPROM. Since vectors are binary numbers stored in memory, the vectors are part of the software design. Also known as interrupt or reset vector.

Vector Address   The address where the vector is stored. The 68HC11 vectors are a fixed part of the chip's hardware design.

Volatile   Refers to memory that loses data when power is removed from the memory device. Nonvolatile memory retains data when power is removed.

Word   Largest number of bits that a CPU can process inside its registers. For the 68HC11, it is a 16-bit binary number. Note that the 68HC11 is an 8-bit microcontroller because it usually handles 8-bit data.

■ ABBREVIATIONS

A	Ampere
ac	Alternating current
ACCA	Accumulator A
ACCB	Accumulator B
ACCD	Accumulator D
ACIA	Asynchronous communication interface adapter
ACK	Acknowledge
A/D	Analog-to-digital converter
ADC	Analog-to-digital converter
ALU	Arithmetic and logic unit
ANSI	American National Standards Institute
AS	Address strobe
ASCII	American Standard Code for Information Interchange
BASIS	Base-stored image sensor
BCD	Binary-coded decimal
BDC	Bottom dead center
BIOS	Basic input-output system
BP	Backplane
BPS	Bits per second
BS	Backspace
BW	Bandwidth
C	Carry/borrow bit in CCR, capacitor, C language, Celsius
CCD	Charge-coupled device
CCR	Condition code register
CD	Compact disc
CMOS	Complementary metal-oxide semiconductor
CMRR	Common-mode rejection ratio
COM	Communications port
COP	Computer operating properly
CPU	Central processing unit
CRC	Cyclic redundancy check
CRT	Cathode ray tube

CS	Chip select
CSIC	Customer-specified integrated circuit
CTS	Clear to send
D/A	Digital-to-analog converter
DAC	Data accepted, digital-to-analog converter
DAV	Data available
dB	Decibel
dc	Direct current
DCD	Data carrier detect
DCE	Data communications equipment
DIP	Dual-in-line package
DMA	Direct memory access
DOS	Disk operating system
DPSK	Differential phase shift keying
DPTR	Data pointer
DSP	Digital signal processor
DTE	Data terminal equipment
DTMF	Dual-tone multifrequency
DTR	Data terminal ready
E	System clock signal
ECM	Engine control module
EEPROM	Electrically erasable read-only memory
EIA	Electronic Industries Association
EMI	Electromagnetic interference
EOS	Electro-optical system
EPROM	Erasable read-only memory
EVB	Evaluation board
EVB2	*See* EVBII
EVBII	Modernized evaluation board
EVBU	Universal evaluation board
FET	Field-effect transistor
FF	Formfeed
FFT	Fast Fourier transform
FG	Frame ground
FIDE	Fuzzy inference development environment
FIFO	First in, first out
FIR	Finite impulse response
FP	Frontplane in LCD device, frame pointer in C language
FSK	Frequency shift keying
G	Gate enable; ground
GPIB	General-purpose instrumentation bus
GPT	General-purpose timer
H	Half-carry bit in CCR
HCMOS	High-speed CMOS
HDLC	High-level data link control
I	I-interrupt mask bit in CCR
IC	Integrated circuit; input capture

IEEE	Institute of Electronic and Electrical Engineers
IIR	Infinite impulse response
I/O	Input/output
IRQ	Interrupt request
ISDN	Integrated services digital network
IX	Index register X
IY	Index register Y
K	Kilo; kilohm
Kb	Kilobit
KB	Kilobyte
KBG	Knowledge-base generator
LAN	Local area network
LF	Linefeed
LIFO	Last in, first out
LIR	Load instruction register
LON	Local operating network
LSB	Least significant bit
LSD	Least significant digit
LVDT	Linear variable displacement transformer
MAC	Media Access Control
MAC	Multiply and accumulate
MCU	Microcontroller unit; microcomputer unit
MMU	Memory management unit
MOSFET	Metal-oxide semiconductor field-effect transistor
MSB	Most significant bit
MSD	Most significant digit
μP	Microprocessor
N	Negative bit in CCR
NAK	Negative (or not) acknowledge
NPN	Negative-positive-negative bipolar transistor
NRZ	Non return to zero
OC	Output compare
OEM	Original equipment manufacturer
OPT	One-time programmable
OSI	Open systems interconnection
PC	Program counter; personal computer
PCM	Pulse-code modulation
PIA	Peripheral interface adapter
PLC	Programmable logic controller
PLCC	Plastic-leaded chip carrier
PLL	Phase-locked loop
PO	Paper out
POR	Power-on reset
PPI	Programmable peripheral interface
PRU	Port replacement unit
PSK	Phase shift keying

PSW	Program status word
PTM	Programmable timer module
pv	Process variable
PWM	Pulse-width modulation
QAM	Quadrature amplitude modulation
QSM	Queued serial module
R	Read; resistor
RAM	Random access memory
RC	Resistor-capacitor
REQ	Request
RFD	Ready for data
RI	Ring indicator
RISC	Reduced instruction set computer
RMS	Root mean square
ROM	Read-only memory
RTE	Real-time executive
RTI	Return from interrupt
RTII	Real-time interrupt
RTS	Ready to send
R/W	Read/write
RxD	Receive
S	Select; stop bit in CCR
SCI	Serial communications interface
SCR	Silicon-controlled rectifier
SCSI	Small computer systems interface
SDLC	Synchronous data link control
SFR	Special function registers
SG	Signal ground
SIM	System integration module
SIU	Serial interface unit
SLR	Single-lens reflex
SOIC	Small outline integrated circuit
sp	Set point
SP	Stack pointer
SPI	Serial peripheral interface
SSDA	Synchronous serial data adapter
STB	Strobe
T	Period, time constant
TCP/IP	Transmission Control Protocol/Internet Protocol
TDC	Top dead center
TTL	Transistor transistor logic
TV	Television
TxD	Transmit
U	Unit
UART	Universal asynchronous receiver-transmitter
USART	Universal synchronous and asynchronous receiver-transmitter

UV	Ultraviolet
V	Overflow bit in CCR, volt
V$_{CC}$	Positive supply voltage
VCR	Videocassette recorder
VFD	Vacuum fluorescent display
W	Write
X	X-interrupt mask bit in CCR
XIRQ	X-interrupt request
Z	Zero bit in CCR; impedance
Z$_0$	Characteristic impedance

Index